Theory of Arched Structures

Igor A. Karnovsky

Theory of Arched Structures

Strength, Stability, Vibration

Springer

Igor A. Karnovsky
Northview Pl 811
V3J 3R4 Coquitlam
Canada
ovlebed@gmail.com

ISBN 978-1-4614-0468-2 e-ISBN 978-1-4614-0469-9
DOI 10.1007/978-1-4614-0469-9
Springer New York Dordrecht Heidelberg London

Library of Congress Control Number: 2011937582

Printed on acid-free paper

Springer is part of Springer Science+Business Media (www.springer.com)

In memory of Prof. Anatoly B. Morgaevsky

Preface

In modern engineering, as a basis of construction, arches have a diverse range of applications. Today the theory of arches has reached a level that is suitable for most engineering applications. Many methods pertaining to arch analysis can be found in scientific literature. However, most of this material is published in highly specialized journals, obscure manuals, and inaccessible books. This is not surprising, as the intensive development of arch theory, particularly stability and vibration have mostly occurred in the 1940s to the 1960s. Therefore, most engineers lack the opportunity to utilize these developments in their practice.

The author has committed to the goal of presenting a book which encompasses essential and tested methods on fundamental methods of arch analysis and equally important problems.

> *The objective of the Book is to provide to readers with detailed procedures for analysis of the strength, stability, and vibration of various types of arched structures, using exact analytical methods of classical Structural Analysis.*

In 2004, professor L.A. Godoy published the article "*Arches: A Neglected Topic in Structural Analysis Courses*." This in-depth investigation highlights a deep rift between the modern level of development of arch theory and the level of presentation of this theory in existing material on structural analysis.

In 2009, the author of this book, with co-author O. Lebed published the textbook "*Advanced Methods of Structural Analysis*" (Springer), in which arch theory is presented in a much greater depth and volume than in existing textbooks. However, the issue of producing a single book which covers both general and specialized problems of arches remained unsolved. The book presented here sheds light on issues of strength, stability, and vibrations, as well as special problems of arches and arched structures.

In this book special attention is directed toward the discussion of fundamental properties of structures. An engineer who is armed with fundamental knowledge and means of computation is essentially set to succeed in modern day engineering. Solutions of problems of strength, stability, and vibrations of arches in most cases are broken down to basic formulas which can be easily applied to engineering practice.

This book is based on the author's experience as a teacher and consultant in structural mechanics. It is intended for senior undergraduate students in structural engineering and for postgraduate students who are concerned with different problems of arches structures. The book will be a useful reference for engineers in the structural industry.

Vancouver, Canada Igor A. Karnovsky

Distribution of Material in the Book

This book contains an introduction, four parts (nine chapters), and an appendix.

The first part "Strength" contains three chapters. Chapter 1 is devoted to fundamental methods of determining displacement of elastic structures in general accompanied by examples specifically for arches.

Chapter 2 covers the analysis of three-hinged arches, while analysis of redundant arches is considered in Chap. 3; in these chapters a special attention is dedicated to the analysis of arched structures using influence lines.

Second part "Stability" contains two chapters. Chapter 4 provides analytical methods of the stability of arches. These methods are based on the integration of differential equations.

Chapter 5 presents Smirnov's matrix method and approximate method. Approximate method is based on the approximation of the arch by straight members with subsequent application of the precise displacement method in canonical form.

The third part, "Vibration" contains two chapters. Chapter 6 deals with computation of eigenvalues and eigenfunctions for arches. For analysis of the circular uniform arch, Lamb's differential equation is used; for analysis of parabolic uniform arch the Rabinovich's model is applied. The frequency of vibration for arches with different ratio "rise/span" of an arch are presented on the basis of this model.

Chapter 7 presents forced vibrations of arches.

The fourth part of the book, "Special Topics" holds the goal of presenting introductory information regarding problems which until now have only been discussed in specialized literature. Chapter 8 contains the static nonlinear problems. They are plastic analysis of the arches and arched structures with one-sided constraints. Chapter 9 is devoted to dynamical stability of arches, and dynamics of arched structures subjected to moving inertial load.

Finally, the appendix contains the fundamental tabulated data essential for engineering practice involving arches.

Sections 2.1, 2.2, 2.4, and 2.6 were written by Olga Lebed.

Acknowledgments

First of all I must thank professor L.A. Godoy (University of Puerto Rico at Mayaguez) and Steven Elliot (Senior Editor, Springer) who stimulated the creation of this book.

I wish to express deep gratitude to Olga Lebed (Condor Rebar Consultants, Vancouver, Canada) for the very constructive discussions of the book, organizational assistance, and writing four sections of the book.

A number of friends and colleagues have helped me directly or indirectly; my sincere gratitude to professors Valeriy A. Baranenko (Chemical Technology University, Ukraine), Jurij G. Kreimer (Civil Engineering University, Ukraine), Vladimir M. Ovsjanko (Polytechnical University, Belorussia), P. Eng Tat'jana Volina (Ukraine), as well as to members of the Russian National Library (Moscow) for helping with the pursuit of difficult-to-access literature and documents.

I wish to express deep gratitude to Evgeniy Lebed (University of British Columbia, Canada) for productive discussions and helpful criticism as well as assisting with many numerical calculations and validation of results.

Many thanks to Tamara Moldon (Vancouver, Canada) for editing assistance.

I would like to thank all the staff of Springer who contributed to this project.

Finally, I would like to thank my relatives, many friends, and colleagues, who have supported me through all stages of research and development of this book.

The author appreciates comments and suggestions to improve the current edition. All constructive criticism will be accepted with gratitude.

Contents

Part III Vibration Analysis

Introduction

Arches and arched structures have a wide range of uses in bridges, arched dams and in industrial, commercial, and recreational buildings. They represent the primary structural components of important and expensive structures, many of which are unique. Current trends in architecture heavily rely on arched building components due to their strengths and architectural appeal.

Complex structural analysis of arches is related to the analysis of the arches strength, stability, and vibration. This type of multidimensional analysis aims at ensuring the proper functionality of an arch as one of the fundamental structural elements.

Terminology

We start our consideration from terminology for a bridge arch (Fig. 1a). The arch is supported by abutments. The heels and crown are the lowest and highest points of the arch, respectively; supports may be rolled, pinned, or fixed. Horizontal distance between two heels is span l, a vertical distance between heels line and crown is rise f. Extrados is the top outer surface of the arch. Intrados is the lower inner surface of the arch. A body of the arch itself may be solid or with webbed members.

As a bridge trusses, the bridge arches are connected using arch bracing. All structural members over the arch are called overarched construction. Deck and arch are connected by vertical members called posts. If the roadway is located below an arch, then vertical members are called hangers. If movement of vehicles is at the intermediate level, then a loaded deck is partially connected with arch by poles and partially by hangers. The posts are compressed, while the hangers are extended.

For structural analysis, a real structure has to be presented in the idealized and simplified form using the axial line of the structural components. For this, a so-

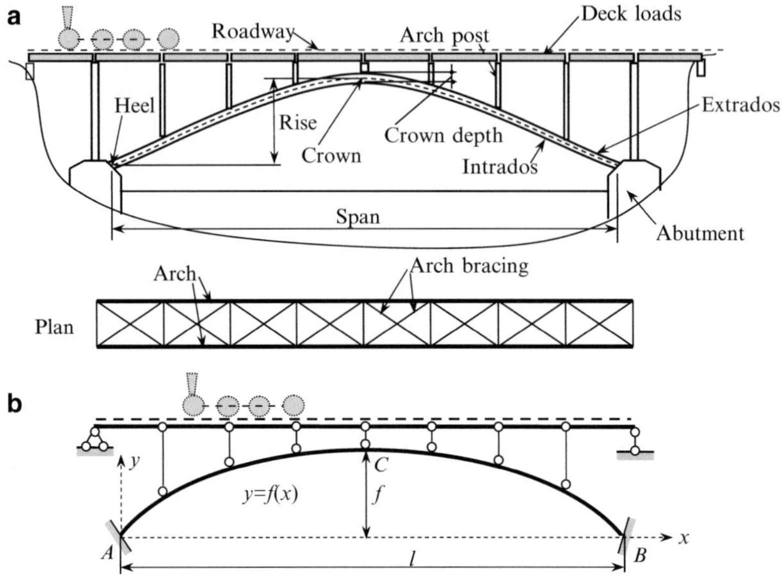

Fig. 1 (**a**, **b**) Components of an arch bridge and design diagrams for a deck-arch bridge

called design diagram of the real structure is used. Design diagram is a critically important concept of structural analysis. Design diagram of a real structure reflects the most important and primary features of the structure such as types of members, types of supports, types of joints, while some features of secondary importance (shapes of cross-sections of members, existence of local reinforcements or holes, size of supports and joints, etc.) are ignored.

Few general rules of representing a real structure by its design diagram are:

- A structure is presented as a set of simple structural members
- Real supports are replaced by their idealized supports
- Any connection between members of a structure are replaced by idealized joints
- Cross-section of any member is characterized by its area or/and moment of inertia

It is obvious that a real structure may be represented using different design diagrams.

An arch with overarched members and its design diagram is shown in Fig. 1b. Design diagram also contains information about the shape of the neutral line of the arch. Usually this shape is given by the expression $y = f(x)$.

Note that posts or hangers are connected to the arch itself by means of hinges.

In bridge construction the arches are subdivided into deck-bridge arch (Fig. 1), through-bridge arch, and arch with deck at some intermediate level (Fig. 2). Also, double-deck bridges exist with the lower deck designed for a railway, and the upper deck is utilized for a roadway.

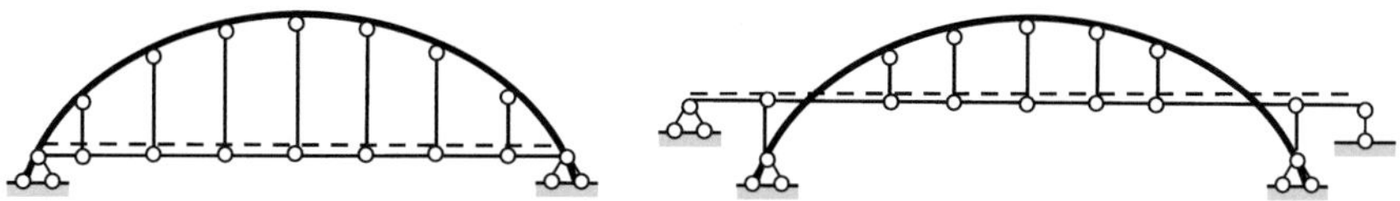

Fig. 2 Design diagrams of the through-bridge arch and arch with deck at intermediate level

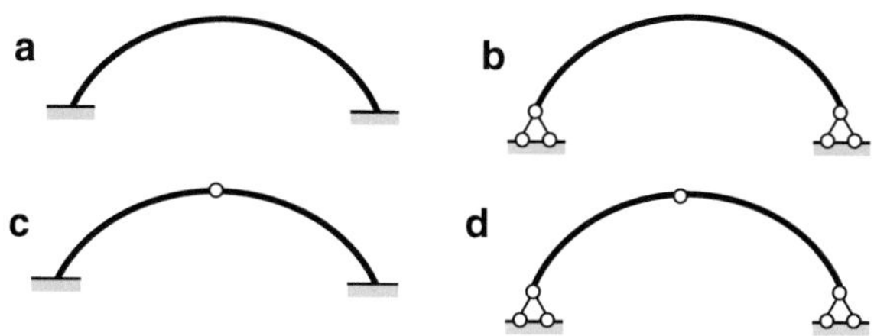

Fig. 3 Design diagrams of arches: (**a**) hingeless arch; (**b**) two-hinged arch; (**c**) one-hinged arch; (**d**) three-hinged arch

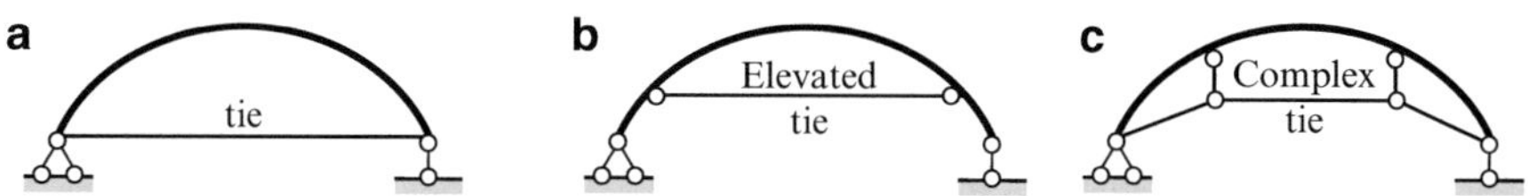

Fig. 4 Arches with tie

Based on their design, arches are divided into hingeless (arch with fixed ends), one-hinged, two-hinged, and three-hinged ones (Fig. 3a–d). All arches presented in Fig. 3, except for the three-hinged arch (d), are statically indeterminate (redundant) ones.

A tie is an additional member which allows us to reinforce an arch. A single tie may be installed on the level of the supports (Fig. 4a), or elevated (b). The tie may also be complex (c). Prestressed tie allows us to control the internal forces in the arch itself.

The aches may be constructed with supports at different elevations. In this case they are called askew arches.

Peculiarities of Arch Behavior

Since posts have hinges at the ends (Fig. 2), then only axial force arises in them. If the posts with fixed ends are thin elements with small flexural stiffness, then they cannot perceive and transmit the bending moments. In both cases, the loads from deck are transferred through posts (hangers) on the arch as concentrated forces.

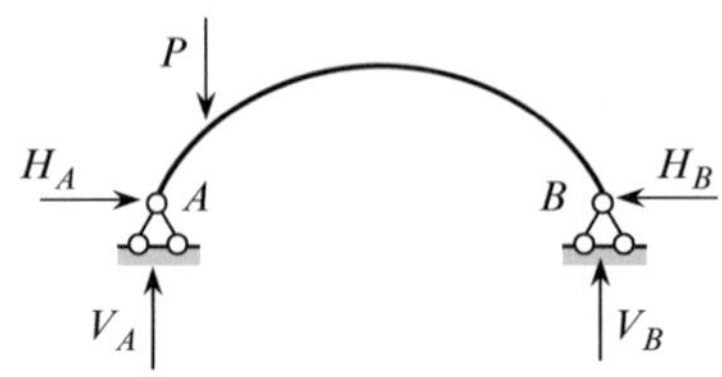

Fig. 5 Reactions of the arch

The fundamental feature of an arched structure is that horizontal reactions appear even if the structure is subjected to vertical load only. These horizontal reactions $H_A = H_B = H$ are called a *thrust* (Fig. 5). If structure has a curvilinear axis but thrust does not exist then this structure cannot be treated as an arch. The presence of thrust leads to a fundamental difference in behavior between arches and beam – the bending moments in arches are smaller than in beams of the same span and loads. Advantages of arches over beams increase as the length of a span increases.

Presence of thrust demands reinforcement of the part of a structure which is subjected to horizontal force.

However, the thrust may be absorbed by a tie; with this, supports of the arch are only subjected to vertical forces.

In addition to the bending moments and shear forces that arise in beams, axial compressive forces are also present in arches. These forces may cause a loss of stability of the arch.

There are advantages and disadvantages of each type of arches. Different design diagrams of the arches may be compared, taking into account different criteria. These include differences in their deformability, internal forces, critical loads, frequencies of vibration, sensitivity of arches to settling of supports, temperature changes, fabrication errors, etc.

Three-hinged arches have less rigidity than two-hinged and hingeless arches. Breaks in elastic curve over a hinge leads to additional forces in the cases where a moving load is present. In the cases when a structure is built on weak soil, three-hinged arches are preferred over hingeless arches since additional stresses caused by the settling of supports do not arise in these structures [Bro99], [Sch80].

Figure 6 shows characteristic distribution of the maximum bending moments in different arches in the presence of a moving load; each arch (diagrams a–d) has a unique bending moment (diagram e) [Kis60]. It is evident that a one-hinged arch (curve c) is the least efficient in regards to bending moment at its supports. In hingeless arches (curve d), the distribution of bending moment is most favorable because of its smoothness.

In the three-hinged arch (a), internal forces arise as a result of external load only. The rest of the arches (b–d) are sensitive to the displacements of supports, changes in temperature, and errors of fabrication. For masonry or concrete arches, material shrinkage should be taken into account, since this property of material leads to additional stresses.

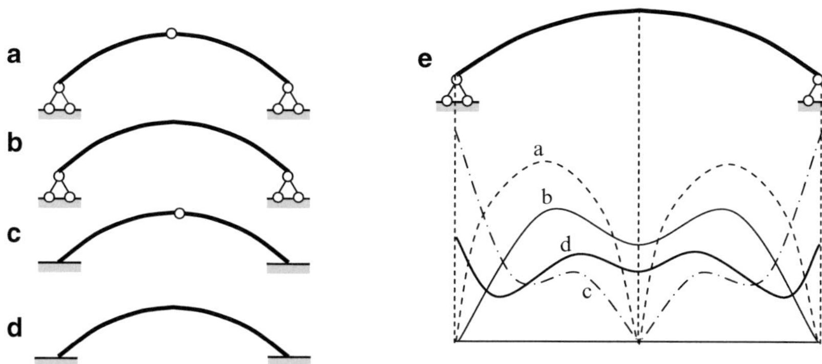

Fig. 6 (**a–d**) Types of arches; (**e**) approximate distribution of maximum bending moments across the span of different types of arches. In Fig. 6e design diagram as two-hinged arch is shown arbitrary

Initial Data for Structural Analysis

A comprehensive structural analysis includes the strength, stability, and vibration analysis. Strength analysis (static analysis) deals with the determination of internal forces and deflections of the arch due to action of static loads only. Stability analysis deals with the determination of loads which leads new forms of equilibrium (the loss of stability) of the arch. Vibration analysis considers determination of frequencies of free vibration of arch, as well as determination of internal forces and displacements of the arch subjected to specific external disturbing loads.

For analysis of arches, the following data have to be clearly outlined and specified: type of arch (hingeless, two-hinged, etc.); its shape (circle, parabolic, etc.); its dimensions (span and rise); location of supports (same or different elevation); presence of the tie, its type (single or complex), and its location. In the case of an arched bridge, it is necessary to show location of a loaded deck (Figs. 1–2), location of the hangers (or/and posts), and ways of theirs connections with arch itself and with loaded deck.

Computation of internal forces for two-hinged and hingeless arches requires knowing the law of change of cross-sectional area $A(x)$ and corresponding moment of inertia $I(x)$, along the axis of the arch. For a tie it is necessary to present the ratio $EI_{\text{arch}}/EA_{\text{tie}}$. For computation of deflections for all types of arches it is necessary to know $A(x)$ and $EI(x)$.

Assumptions

Some of the common assumptions made in this book include the following:

1. Material of the arch obeys Hooke's law (physically linear statement)
2. Deflections of the arches are small compared with the span of the arch (geometrically linear statement). The cases of nonlinear statement are specifically mentioned.
3. All constraints, which are introduced into the arched structure are two-sided, i.e., each constraint prevents displacements in two directions. The case of one-sided constraints is specifically mentioned.
4. In the case of elastic supports the relationship between deflection of constraint and corresponding reaction is linear.
5. The load is applied in the longitudinal plane of symmetry of the arch. The case of out-of-plane loading is specifically mentioned.

Besides the above assumptions, supplementary assumptions are introduced in corresponding parts of the book.

Some remarks related to structural analysis of the arches:

1. Since arches are represented by curvilinear rods, then their analysis, strictly speaking, should be performed using the theory of the curvilinear rods. However, curvature of the arches used in the construction is small ($R/h>10$), therefore, the curvature of the arch may be neglected and deflections of the arch are assumed to be calculated as for straight rods [Kis60].
2. The superposition principle is valid under assumptions 1–4. In the case of one-sided constraints the superposition principle requires special treatment.

Shape of the Arches

As it is shown below, distribution of internal forces in arches depends on the shape of the central line of an arch. According to their shapes, arches are divided into the circular arch, parabolic arch, etc. Equation of the central line and some necessary formulae for circular and parabolic arches are presented below. For both cases, origin of coordinate axis is located at point A as shown in Fig. 7.

Circular arch. Ordinate y of any point of the central line of the circular arch is calculated by the formula

$$y=\sqrt{R^2-\left(\frac{l}{2}-x\right)^2}-R+f;\quad R=\frac{f}{2}+\frac{l^2}{8f}, \tag{1}$$

where x is the abscissa of the same point of the central line of the arch; R is the radius of curvature of the arch; f and l are the rise and span of the arch.

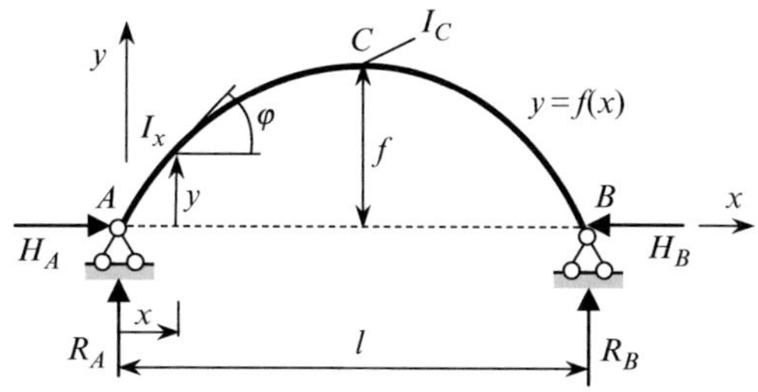

Fig. 7 Design diagram of two-hinged arch

The angle φ between the tangent to the center line of the arch at point (x, y) and horizontal axis is determined as follows:

$$\sin\varphi = (l - 2x)\frac{1}{2R}; \quad \cos\varphi = (y + R - f)\frac{1}{R}. \tag{2}$$

Parabolic arch. Ordinate y of any point of the central line of the parabolic arch is

$$y = 4fx(l - x)\frac{1}{l^2}. \tag{3}$$

Trigonometric functions of the angle between the tangent to the center line of the arch at point (x, y) and a horizontal axis are as follows:

$$\tan\varphi = \frac{dy}{dx} = \frac{4f}{l^2}(l - 2x); \quad \cos\varphi = \frac{1}{\sqrt{1+\tan^2\varphi}};$$
$$\sin\varphi = \cos\varphi \times \tan\varphi. \tag{3a}$$

For the left half-arch the functions $\sin\varphi > 0$, $\cos\varphi > 0$, and for the right half-arch the functions $\sin\varphi < 0$ and $\cos\varphi > 0$.

Length S of half-axis of symmetrical arch and length of the axis of the arch S_k from the origin (point A) to an arbitrary point k with coordinates $x_k = \xi_k l$, $y_k = \eta_k l$ arc

$$S = \frac{l}{4}\left[\sec\varphi_0 + \frac{1}{4m}\ln(4m + \sec\varphi_0)\right],$$

$$S_k = S - \frac{l}{16m}\left(\frac{\tan\varphi_k}{\cos\varphi_k} + \ln\frac{1+\sin\varphi_k}{\cos\varphi_k}\right). \tag{4}$$

where φ_0 is a slope at the support A; parameter $m = f/l$.

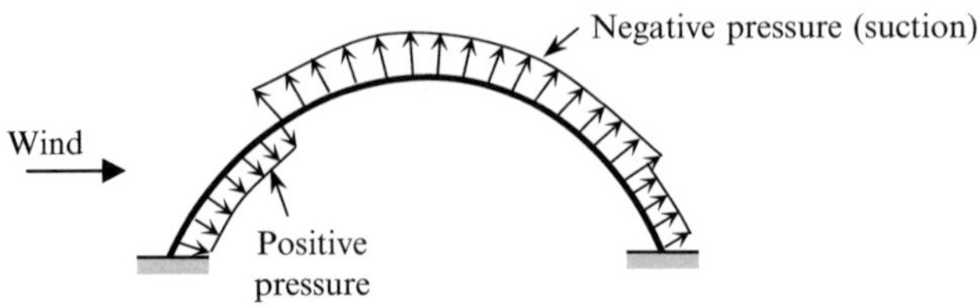

Fig. 8 Pressure of the wind on the surface of the arch

Catenary arch. Ordinate y of any point of the central line of the catenary arch as a function of load may be calculated by the formula which is presented in Sect. 2.3.2.

More expressions $y(x)$ for different arch shapes are presented in Tables A.1–A.5 [Kar01].

Strictly speaking, the concept of arch shape includes not only equation of central line as shown above, but also the law of flexural rigidity along the axis of the arch [Kis60]. The flexural rigidity $EI(x)$ may be constant or variable along the axis of the arch depending on expected distribution of internal forces, requirements of a constructive nature and asthetic considerations. Usually the variable rigidity of the arch $EI(x)$ expresses in terms of rigidity of the arch at crown, EI_C, where E is a modulus of elasticity, I_C is a moment of inertia of a cross section at the crown C of an arch. This will be considered in more details in Sect. 3.1.

Loads

Arches, as main structural components, are subject to a variety of loads depending on the purpose of the arch and conditions of its operation.

For arches in public and industrial buildings the main loads are deadweight, live-load, and snow. These loads act in the longitudinal plane of symmetry of the arch and lead to in-plane bending. A significant load for arched structures is a wind pressure. The wind leads to the positive and negative loads onto the arch. A simplified scheme of the wind pressure is shown in Fig. 8.

In the case of a tall arch, the in-plane wind loads leads to significant internal forces in the arch. If a tall arch has a small own weight, then the formation of the negative reactions is possible; this dangerous phenomenon leads to the separation of the arch from abutment.

Pressure of the wind, which is directed perpendicular to the plane of the arch, leads to out-of-plane bending of the arch. These loads are absorbed by bracing between arches.

A dangerous phenomenon is observed in the case of an arched cover with open sides. Wind pressure, which is parallel to an open aperture, flows around them and creates a vacuum inside. As a result, the positive pressure onto the arch increases and suction decreases.

For arched bridges the main loads, which lead to the in-plane bending of the arch, are the following: deadweight, vertical loads from vehicles, and horizontal load caused by their longitudinal deceleration. Also, in the case of a bridge with curvature in the horizontal plane, one should take into account horizontal loads, which are caused by moving vehicles in a curvilinear trajectory.

The settlement of supports may induce in-plane and out-of-plane bending. Out-of-plane bending also arises by horizontal out-of-plane wind pressure, and seismic loads. Asymmetric location of the load with respect to the longitudinal plane of symmetry also leads to out-of-plane bending of the arch.

Some types of loads have a distinctly dynamic nature. Among them are seismic loads, wind gusts, moving inertial loads and their deceleration, impacts of wheels on the joints of rails on railway bridges. In the case of road bridges one should take into account the roughness of their surface.

If the shell is reinforced with ribs and is immersed into a liquid, then the pressure on the shell is transmitted on ribs and each rib can be considered as an arch due to a uniformly distributed radial load.

Determination of loads on the arch and the consideration of all possible combinations of loads is an important part of engineering analysis

Part I
Strength Analysis

Chapter 1
Deflections of Elastic Structures

This chapter describes some effective methods for computing different types of deflections of deformable structures. The structure may be subjected to different actions, such as variety of external loads, change of temperature, settlements of supports, and errors of fabrication. Advantages and disadvantages of each method and field of their effective application are discussed. Much attention is given to a graph multiplication method which is a most effective method for bending structures. Fundamental properties of deformable structures are described by reciprocal theorems.

1.1 General

Any load which acts on the structure leads to its deformation. It means that a structure changes its shape, the points of the structure displace, and relative position of separate points of a structure changes. There are other reasons of the deformation of structures. Among them is a settlement of supports, change of temperature, etc. Large displacements could lead to disruption of a structure functioning properly and even its collapse. Therefore, an existing Building Codes establish limit deflections for different engineering structures. Ability to compute deflections is necessary for the estimation of rigidity of a structure, for comparison of theoretical and actual deflections of a structure, as well as theoretical and allowable deflections. Besides that, computation of deflections is an important part of analysis of any statically indeterminate structure. Computation of deflections is also an integral part of a dynamical analysis of the structures. Thus, the computation of deflections of deformable structures caused by different reasons is a very important problem of structural analysis.

Outstanding scientists devoted their investigations to the problem of calculation of displacements [Tim53]. Among them are Bernoulli, Euler, Clapeyron, Castigliano, Maxwell, Mohr, and others. They proposed a number of in-depth and ingenious ideas for the solution of this problem. At present, methods for

I.A. Karnovsky, *Theory of Arched Structures: Strength, Stability, Vibration*,
DOI 10.1007/978-1-4614-0469-9_1,

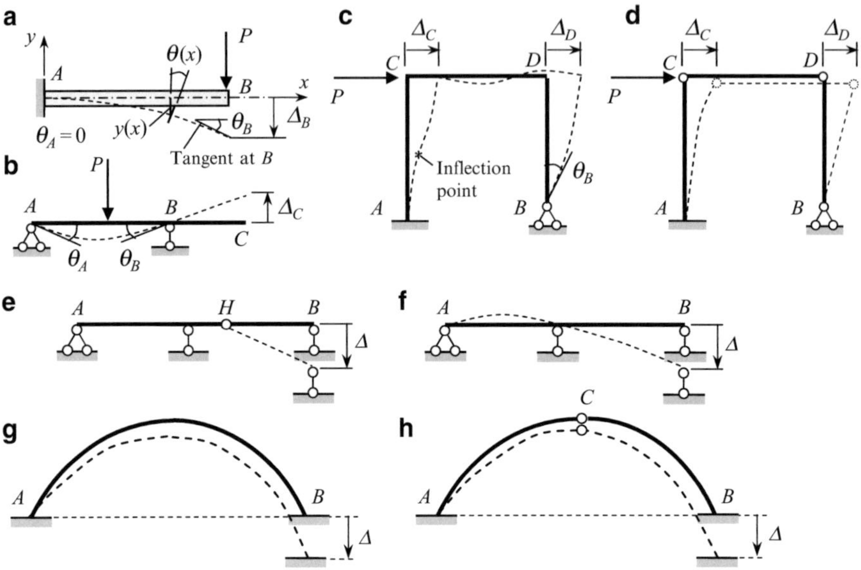

Fig. 1.1 (**a–d**) Deflected shapes of some structures. (**e–h**) Deflected shapes of beams and arches caused by the settlement of support *B*

computation of the displacements are developed with sufficient completeness and commonness for engineering purposes and are brought to elegant simplicity and perfection.

The deformed shape of a bend structure is defined by transversal displacements $y(x)$ of every points of a structural member. The slope of the deflection curve is given by $\theta(x) = \mathrm{d}y/\mathrm{d}x = y'(x)$. Deflected shapes of some structures are presented in Fig. 1.1. In all cases, elastic curves (EC) reflect the deformable shape of the neutral line of a member; the EC are shown by dotted lines in exaggerated scale.

A cantilever beam with load *P* at the free end is presented in Fig. 1.1a. All points of the neutral line have some vertical displacements $y(x)$. Equation $y = y(x)$ is the EC equation of a beam. Each section of a beam has not only a transversal displacement, but also an angular displacement $\theta(x)$ as well. Maximum vertical displacement Δ_B occurs at *B*; maximum slope θ_B also happens at the same point. At the fixed support *A*, both linear and angular displacements Δ_A and θ_A are zero.

The simply supported beam with overhang is subjected to vertical load *P* as shown in Fig. 1.1b. The vertical displacements at supports *A* and *B* are zero. The angles of rotation θ_A and θ_B are maximum, but have different directions. Since overhang *BC* does not have external loads, the elastic curve along the overhang presents the straight line, i.e., the slope of the elastic curve θ within this portion is constant. The angles of rotation of sections, which are located infinitely close to the left and right of support *B* are equal.

Figure 1.1c shows the frame due to action of horizontal force *P*. At fixed support *A* the linear and angular displacements are zero, while at pinned support *B* the angle of rotation $\theta_B \neq 0$. The joints C and D have the horizontal displacements Δ_C and Δ_D; under special assumptions these displacements are equal. Joints C and D have angular displacements θ_C and θ_D (they are not labeled on the sketch). The linear and angular displacements of joints C and D lead to deformation of the vertical members as shown on the sketch. Since support A is fixed, then the left member AC has an inflection point.

Figure 1.1d shows the frame with hinged ends of the cross-bar CD; the frame is subjected to horizontal force P. In this case, the cross-bar and column BD has a *displacement* but does not have *deflection* and members move as absolutely rigid one – the motion of the member CD is a translation, while the member BD rotates around point B. Thus, it is a possible displacement of the member without the relative displacements of its separate points. So a displacement is not always accompanied by deflections, however, deflections are impossible without displacement of its points.

Figure 1.1e, f shows the shapes of the beams caused by settlement of support. A new form of statically determinate beam (Fig. 1.1e) is characterized by displacement of portion H–B as absolutely rigid body, i.e., without *deflection* of the beam. In case 1.1f, a new form of the beam occurs with the deflection of the beam.

Figure 1.1g, h shows the deflected shapes of the arches caused by settlement of support. Elastic curve in the case of the hingeless arch is a monotonic function, while in case of a one-hinged arch (Fig. 1.1h) this property of elastic curve of deformable axis of the arch is disrupted at hinge C.

There are two principle analytical approaches to computation of displacements. The first of them is based on the integration of the differential equation $EI(\mathrm{d}^2y/\mathrm{d}x^2) = -M(x)$ of the elastic curve of a beam. Modification of this method leads to the initial parameters method. The second approach is based on the fundamental energetic principles. The following precise analytical methods represent the second group: Castigliano theorem, dummy load method (Maxwell–Mohr integral), Graph multiplication method (Vereshchagin rule), and elastic load method.

All methods from both groups are exact and based on the following assumptions:

1. Structures are physically linear (material of a structures obey Hook's law).
2. Structures are geometrically linear (displacements of a structures are much less than their overall dimensions).

1.2 Initial Parameters Method

Initial parameters method presents a modification of double integration method in case when a beam has several portions and as a result, expressions for bending moments are different for each portion. Initial parameter method allows us to obtain an equation of the elastic curve of a beam with any type of supports (rigid or elastic) and, most important, for any number of portions of a beam.

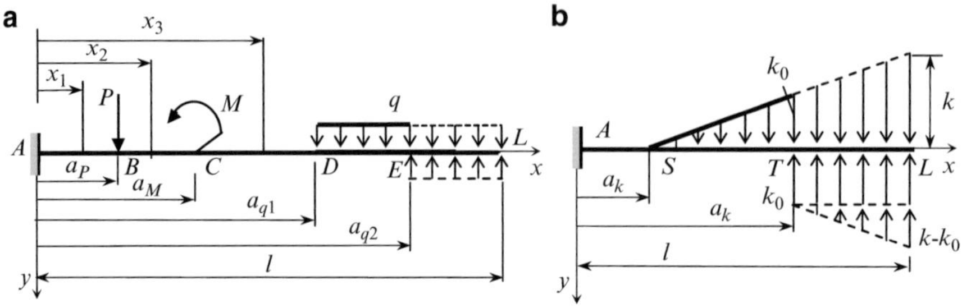

Fig. 1.2 Initial parameters method notation

Fundamental difference between the initial parameter and the double integration method, as it is shown below, lies in the following facts:

1. Initial parameters method does not require setting up the expressions for bending moments for different portions of a beam, formulating corresponding differential equations and their integration. Instead, the method uses a once-derived expression for displacement. This expression allows us to calculate slope, bending moments, and shear along the beam and is called the universal equation of elastic curve of a beam.
2. Universal equation of the elastic curve of a beam contains only two unknown parameters for *any* number of portions.

A general case of a beam under different types of loads and the corresponding notational convention is presented in Fig. 1.2a. The origin is placed at the extreme left end point of a beam, the x-axis is directed along the beam, and y-axis is directed downward. Support A is shown as fixed, however, it can be any type of support or even free end. Load q is distributed along the portion DE. Coordinates of points of application of concentrated force P, couple M, and initial point of distributed load q are denoted as a with corresponding subscript P, M, and q. This beam has five portions (AB, BC, CD, DE, and EL), which leads to the ten constants of integrating using the double integration method.

The initial parameter method requires the following rules to be entertained:

1. Abscises x for all portions should be reckoned from the origin; in this case, the bending moment expression for each next portion contains all components related to the previous portion.
2. Uniformly distributed load may start from any point but it must continue to the very right point of the beam. If a distributed load q is interrupted (point E, Fig. 1.2a), then this load should be continued till the very right point and action of the added load must be compensated by the same, but oppositely directed load. The same rule remains for load which is distributed by triangle law. If load is located within the portion S–T (Fig. 1.2b), it should be continued till the very right point L of the beam and action of the added load must be compensated by the same but oppositely directed loads (uniformly distributed load with intensity

Table 1.1 Bending moments in unified form for different type of loading

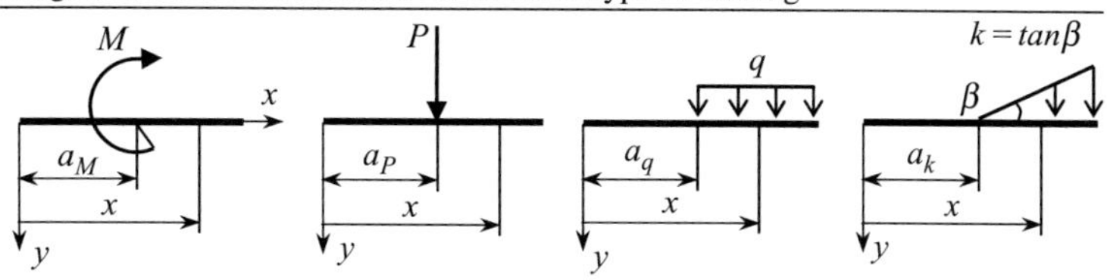

$M(x)$	$\pm M(x-a_M)^0$	$\pm P(x-a_P)^1$	$\pm \frac{q(x-a_q)^2}{2}$	$\pm \frac{k(x-a_k)^3}{2\times 3}$
$\int M(x)\mathrm{d}x$	$\pm M(x-a_M)$	$\pm \frac{P(x-a_P)^2}{2}$	$\pm \frac{q(x-a_q)^3}{2\times 3}$	$\pm \frac{k(x-a_k)^4}{2\times 3\times 4}$
$\int \mathrm{d}x \int M(x)\mathrm{d}x$	$\pm \frac{M(x-a_M)^2}{2}$	$\pm \frac{P(x-a_P)^3}{2\times 3}$	$\pm \frac{q(x-a_q)^4}{2\times 3\times 4}$	$\pm \frac{k(x-a_k)^5}{2\times 3\times 4\times 5}$

k_0 and load distributed by triangle law with maximum intensity k–k_0 at point L). Both of these compensated loads start at point T and do not interrupt until the extremely right point L.

3. All components of a bending moment within each portion should be presented in unified form using the factor (x–a) in specified power, as shown in Table 1.1. For example, the bending moment for the second and third portions (Fig. 1.2a) caused by the active loads only are

$$M(x_2) = -P(x_2 - a_P),$$
$$M(x_3) = -P(x_3 - a_P) - M(x_3 - a_M)^0.$$

4. Integration of differential equation should be performed *without opening the parenthesis*.

All of these conditions are called Cauchy–Clebsch conditions [Tim53].

Initial parameters method is based on the equation $EIy'' = -M(x)$. Integrating it twice leads to the following expressions for slope and linear displacement

$$EI\theta = -\int M(x)\mathrm{d}x + C_1,$$
$$EIy = -\int \mathrm{d}x \int M(x)\mathrm{d}x + C_1x + D_1. \tag{1.1}$$

The transversal displacement and slope at $x = 0$ are $y = y_0$, $\theta = \theta_0$. These displacements are called the initial parameters. Equations (1.1) allow getting the constants in terms of initial parameters

$$D_1 = EIy_0 \text{ and } C_1 = EI\theta_0.$$

Table 1.2 Initial parameters method. Parameter n for the specific loads

Type of load	F	n
Couple	M	2
Concentrated force	P	3
Uniformly distributed load	q	4
The load distributed by triangle law	k	5

Finally (1.1) may be rewritten as

$$EI\theta = EI\theta_0 - \int M(x)\mathrm{d}x,$$

$$EIy = EIy_0 + EI\theta_0 x - \int \mathrm{d}x \int M(x)\mathrm{d}x. \tag{1.2}$$

These equations are called the initial parameter equations for uniform beam. For practical purposes, the integrals from (1.2) should be calculated for special types of loads using the above rules 1–4. These integrals are presented in Table 1.1.

Combining (1.2) and data in Table 1.1 allows us to write the general expressions for the linear displacements $y(x)$ and slope $\theta(x)$ for a uniform beam:

$$EIy(x) = EIy_0 + EI\theta_0 x - \sum \pm \frac{F(x - a_F)^n}{n!}, \tag{1.3}$$

$$EI\theta(x) = EI\theta_0 - \sum \pm \frac{F(x - a_\mathrm{F})^{n-1}}{(n-1)!}, \tag{1.4}$$

where EI is a flexural rigidity of the beam; F is any load (concentrated, couple, or a distributed one); y_0 and θ_0 are transversal displacement and slope at $x = 0$; a_F is the distance from the origin of the beam to the point of application of a concentrated force, couple, or to the starting point of the distributed load and n is the parameter, which depends on the type of the load. Types of load F and corresponding parameter n are presented in Table 1.2.

Equation (1.3) is called the universal equation of elastic curve of a beam. This equation gives an easiest way of deriving the equation of elastic curve of uniform beam and calculating displacement at any specified point. This method is applicable for a beam with arbitrary boundary conditions, subjected to any types of loads.

Notes

1. The negative sign before the symbol Σ corresponds to the y-axis directed downward.

2. Summation is related only to loads, which are located to the left of the section x. It means that we have to take into account only those terms, for which the difference $(x - a)$ is positive.
3. Reactions of supports and moment of a clamped support must be taken into account as well, like any active load.
4. Consideration of all loads including reactions must start at the very left end and move to the right.
5. Sign of the load factor $\pm F(x - a_F)^n/n!$ coincides with the sign of bending moment due to the load, which is located at the left side of the section x.
6. Initial parameters y_0 and θ_0 may be given or be unknown, depending on boundary conditions.
7. Unknown parameters (displacements or forces) are to be determined from the boundary conditions and conditions at specified points, such as the intermediate support and/or intermediate hinge.

For positive bending moments at x due to couple M, force P, and uniformly distributed load q, the expanded equations for displacement and slope are

$$EIy(x) = EIy_0 + EI\theta_0 x - \frac{M(x - a_M)^2}{2!} - \frac{P(x - a_P)^3}{3!} - \frac{q(x - a_q)^4}{4!}, \tag{1.5}$$

$$EI\theta(x) = EI\theta_0 - M(x - a_M) - \frac{P(x - a_P)^2}{2} - \frac{q(x - a_q)^3}{6}. \tag{1.6}$$

Expressions for bending moment and shear force may be obtained by formula

$$M(x) = -EIy''(x), \quad Q(x) = -EIy'''(x). \tag{1.7}$$

Advantages of the Initial Parameters Method

1. Initial parameters method allows to obtain the *expression* for elastic curve of the beam. The method is very effective in case of large number of portions of a beam.
2. Initial parameters method do not require to form the expressions for bending moment at different portions of a beam and integration of differential equation of elastic curve of a beam; a procedure of integration was once used for deriving the universal equation of a beam and then only simple algebraic procedures are applied according to (1.3).
3. The number of unknown initial parameters is always equals two and does not depend on the number of portions of a beam.
4. Initial parameters method may be effectively applied for beams with elastic supports and beams subjected to displacement of supports. Also, this method may be applied for statically indeterminate beams. Detailed examples of application of initial parameters method are considered in [Kar10].

1.3 Maxwell–Mohr Integral

The Maxwell–Mohr procedure presents a universal method for computation of displacement at any point of any deformable structure. Also, the Maxwell–Mohr procedure allows calculating mutual linear and angular displacements. Different sources, which may cause displacements of a structure, are considered. They are different types of loads, and change of temperature.

1.3.1 Deflection Due to External Loads

A general expression for displacements of any deformable structure may be written as

$$\Delta_{kp} = \sum \int_0^s \frac{M_p \bar{M}_k}{EI} \mathrm{d}s + \sum \int_0^s \frac{N_p \bar{N}_k}{EA} \mathrm{d}s + \sum \int_0^s \mu \frac{Q_p \bar{Q}_k}{GA} \mathrm{d}s. \tag{1.8}$$

Summation is related to all elements of a structure. Fundamental expression (1.8) is known as Maxwell–Mohr integral. The following notations are used: Δ_{kp} is displacement of a structure in the kth direction in P condition, i.e., displacement in direction of unit load (first index k) due to the given load (second index p); M_p, N_p, and Q_p are the internal forces (bending moment, axial and shear force, respectively) in P condition; and $\overline{M}_k$, $\overline{N}_k$, $\overline{Q}_k$ are the internal forces due to the unit load, which acts in the kth direction and corresponds to the required displacement. A and I are the area and moment inertia of a cross-section; E and G are Young's and shear modulus of elasticity; η is nondimensional parameter depends on the shape of the cross section. For rectangular cross section $\eta = 1.2$, for circular cross section $\eta = 10/9$. The unit load (force, couple, etc.) also termed the dummy load.

Proof. For bending systems, the Castigliano's theorem for computation of linear and angular displacements at point k may be presented as follows [Cra00]:

$$y_k = \int \frac{M(x)}{EI} \frac{\partial M(x)}{\partial P_k} \mathrm{d}x, \quad \theta_k = \int \frac{M(x)}{EI} \frac{\partial M(x)}{\partial M_k} \mathrm{d}x,$$

where $M(x)$ is bending moment at section x; P_k and M_k are force and couple at section k.

Both formulas may be simplified. For this purpose, let us consider the simply supported beam subjected to force P and couple M (Fig. 1.3).

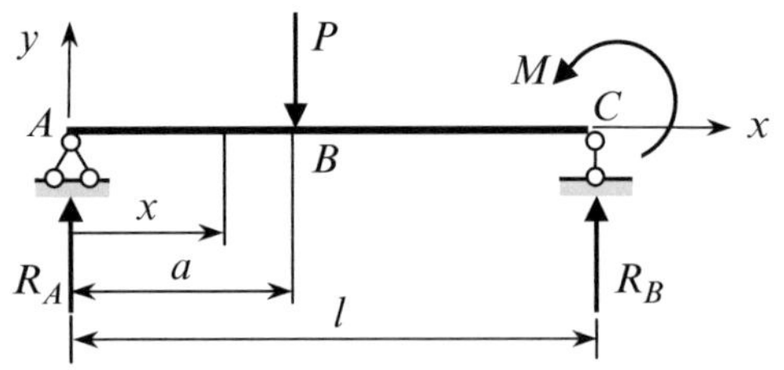

Fig. 1.3 Simply supported beam loaded by P and M

Reaction $R_A = P(l - a/l) + M(1/l)$ and the bending moment for the left and right portions of the beam are

$$M(x) = R_A x = P\frac{(l-a)}{l}x + M\frac{x}{l}, \quad (x \le a)$$
$$M(x) = R_A x - P(x-a) = P\frac{a}{l}(l-x) + M\frac{x}{l}, \quad (x \ge a).$$

Both expressions present the *linear* functions of the loads P and M. In general case, suppose a structure is subjected to the set of concentrated loads P_1, P_2,. . ., couples M_1, M_2,. . ., and distributed loads q_1, q_2,. . .. This condition of structure is called as P condition (also known as the actual or loaded condition). In case of P condition, a bending moment at any section x is a *linear function* of these loads

$$M(x) = a_1P_1 + a_2P_2 + \cdots + b_1M_1 + b_2M_2 + \cdots + c_1q_1 + c_2q_2 + \cdots$$

where coefficients a_i, b_i, and c_i depend on geometrical parameters of the structure, position of loads, and location of the section x.

If it is required to find displacement at the point of application of P_1, then, as an intermediate step of Castigliano's theorem we need to calculate the partial derivative of bending moment $M(x)$ with respect to force P_1. This derivative is $\partial M(x)/\partial P_1 = a_1$. According to expression for $M(x)$, this parameter a_1 may be considered as the bending moment at section x caused by *unit dimensionless force* ($P_1 = 1$). State of the structure due to action of unit dimensionless load (unit force or unit couple) is called *unit state*. Thus, calculation of partial derivative in Castigliano's formula may be changed by calculation of a bending moment caused by unit dimensionless load

$$y_k = \int \frac{M(x)}{EI}\frac{\partial M(x)}{\partial P}\,\mathrm{d}x = \int \frac{M(x)\overline{M}_k}{EI}\,\mathrm{d}x,$$

where $\overline{M}_k$ is bending moment in the unit state. Keep in mind that $\overline{M}_k$ is always a *linear function* and represents the bending moment due to a unit load, which corresponds to the required displacement.

In a similar way, terms, which take into account the influence of normal and shear forces, may be transformed. Thus, displacements caused by any combination of loads may be expressed in terms of internal stresses developed by given loads and *unit* load, which corresponds to required displacement. That is the reason why this approach is termed the dummy load method.

For different types of structures, relative contribution of first, second, and third terms of (1.8) in the total displacement Δ_{kp} is different. For practical calculation, depending on type and shape of a structure, the following terms from (1.8) should be taken into account:

(a) For trusses – only second term.
(b) For beams, arches and frames with ratio of height of cross section to span 0.2 or less – only first term.

(c) For beams with ratio of height of cross section to span more than 0.2 – the first and third terms.
(d) For gently sloping arches – the first and second terms.
(e) For arches with ratio of radius of curvature to height of cross section 5 or more – all terms.

In case of trusses, the displacement should be calculated by formula

$$\Delta_{kp} = \sum \int_0^l \frac{N_p \overline{N}_k}{EA} \mathrm{d}s. \tag{1.8a}$$

Since all elements are straight ones, and axial stiffness are constant along all length of each elements, then this formula may be presented as

$$\Delta_{kp} = \sum \frac{N_p \overline{N}_k}{EA} l. \tag{1.8b}$$

Strictly speaking, integral equation (1.8) is applicable only for structure which contains the straight members. Effect of curvature will be discussed further. □

Procedure for Computation of Deflections Using Maxwell–Mohr Integral

1. Express internal forces in P condition for an arbitrary cross section in terms of its position x.
2. Construct the unit condition. For this we should apply unit load (dummy load), which corresponds to the required displacement:
   (a) For linear displacement, a corresponding dummy load represents the unit force, which is applied at the point where displacement is to be determined and acts in the same direction.
   (b) For angular displacement, a corresponding dummy load is the unit couple, which is applied at the point where angle of rotation is to be determined.
   (c) For mutual linear displacement of two sections, a corresponding dummy load represents two unit forces, which are applied at the points where displacement is to be determined and act in the opposite directions.
   (d) For mutual angular displacement of two sections, a corresponding dummy load represents two unit couples, which are applied at given sections and act in the opposite directions.
3. Express the internal forces in unit condition for an arbitrary cross section in terms of its position x.
4. Calculate Maxwell–Mohr integral.

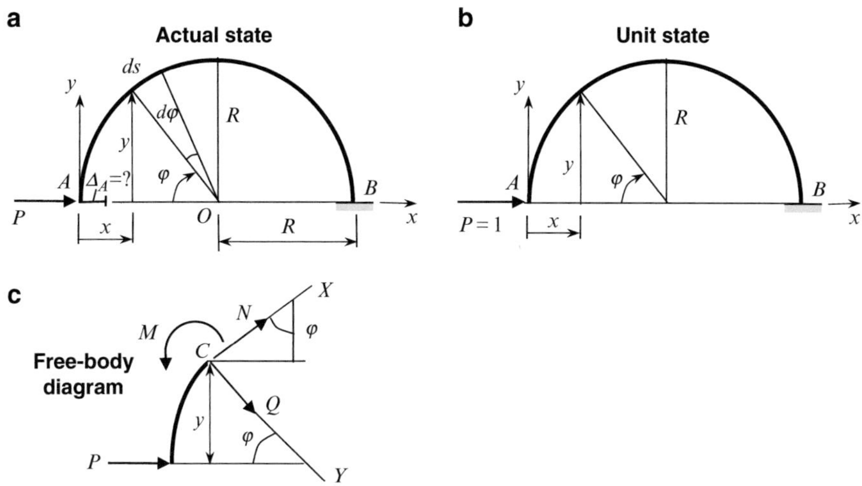

Fig. 1.4 Design diagram of the curvilinear bar, unit state and free-body diagram

Table 1.3 Internal forces of the circular bar for actual and unit state

State	Bending moment	Axial force	Shear force
Actual	$M_p = -Py$	$N_p = -P \sin \varphi$	$Q_p = -P \cos \varphi$
Unit	$\bar{M} = -1 \times y$	$\bar{N} = -1 \times \sin \varphi$	$\bar{Q} = -1 \times \cos \varphi$

Positive sign of displacement means that the real displacement coincides with the direction of the unit load, or work performed by unit load along the actual direction is positive.

Now we demonstrate the application of general formula (1.8) for calculation of displacement of curvilinear bar.

Example 1.1. A circular uniform bar with central angle 180° is clamped at point B and carrying horizontal force P at point A as shown in Fig. 1.4a. Calculate the horizontal displacement Δ_A of point A.

Solution. For the given problem, the unit state labeled by index k presents the force $P = 1$ applied at point A in the same direction as force P for the actual state (Fig. 1.4b). Free-body diagram is presented in Fig. 1.4c.

For computation of bending moment M at point C, it is convenient to use the x and y axis with origin at A, while for computation of shear Q and axial force N at same point C it is convenient to use the X–C–Y axis (Fig. 1.4c). Equilibrium conditions $\sum M_C = 0, \quad \sum X = 0, \quad \sum Y = 0$ lead to the following expressions of internal forces for actual and unit states (Table 1.3).

This data lead to the following expression for required displacement:

$$\Delta_A = \frac{1}{EI} \int_0^{\pi R} Py^2 \mathrm{d}s + \frac{1}{EA} \int_0^{\pi R} P \sin^2 \varphi \, \mathrm{d}s + \frac{\mu}{GA} \int_0^{\pi R} P \cos^2 \varphi \, \mathrm{d}s. \qquad \text{(a)}$$

Now let us represent y and ds in terms of polar coordinates as follows: $ds = R\,d\varphi$, $y = R\sin\varphi$ (Fig. 1.4a). Changing limits of integration ($\pi R \to \pi$), becomes

$$\Delta_A = \frac{PR^3}{EI}\int_0^{\pi}\sin^2\varphi\,d\varphi + \frac{PR}{EA}\int_0^{\pi}\sin^2\varphi\,d\varphi + \frac{\mu PR}{GA}\int_0^{\pi}\cos^2\varphi\,d\varphi. \quad \text{(a)}$$

Since $\int_0^{\pi}\sin^2\varphi\,d\varphi = \int_0^{\pi}\cos^2\varphi\,d\varphi = \pi/2$, then the final result for required displacement is

$$\Delta_A = \frac{PR^3}{EI}\frac{\pi}{2} + \frac{PR}{EA}\frac{\pi}{2} + \frac{\mu PR}{GA}\frac{\pi}{2} = \frac{PR^3}{EI}\frac{\pi}{2}\left(1 + \frac{I}{A}\frac{1}{R^2} + \frac{\mu EI}{GA}\frac{1}{R^2}\right). \quad \text{(b)}$$

All terms in the parenthesis take into account the bending moment, axial forces, and shear, respectively.

Discussion

Let us compare the displacements due to bending moments, axial forces, and shearing forces. For this purpose, let us replace A, I, and μ by their values, which correspond to the rectangular cross section of the bar as follows: $A = bh$, $I = (bh^3/12)$. Shear modulus of elasticity is $G = E/2(1+v)$. If $h = 2b$, the Poisson's coefficient $v = 0.25$ and coefficient $\mu = 1.2$, then

$$\Delta_A = \frac{PR^3}{EI}\frac{\pi}{2}\left(1 + \frac{1}{12}\frac{h^2}{R^2} + \frac{1}{4}\frac{h^2}{R^2}\right).$$

Even if the ratio $(h/R) = 0.1$, then

$$\Delta_A = \frac{PR^3}{EI}\frac{\pi}{2}(1 + 0.000833 + 0.0025).$$

Therefore, in our case, the displacements due to axial and shearing forces constitute about 0.08 and 0.25% of the displacement due to bending moment.

Let us assume that shear and axial forces for the bar in Fig. 1.4a are neglected; it is easy to show that in this case the vertical and angular displacements for point A are as follows:

$$\Delta_A^{\text{vert}} = \frac{2PR^3}{EI}(\uparrow), \quad \theta_A = \frac{2PR^2}{EI}\,(\text{clockwise}).$$

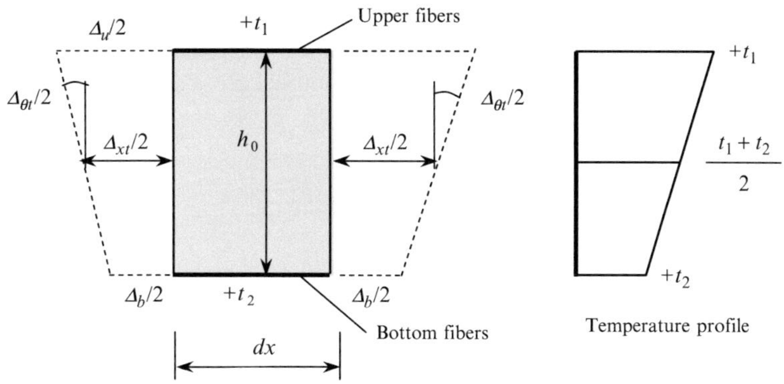

Fig. 1.5 Distribution of temperature and displacements within the height of cross section

1.3.2 Deflections Due to Change of Temperature

It is very often in engineering practice that the members of a structure undergo the thermal effects. In case of statically determinate structures, the change of temperature leads to displacements of points of a structure *without* an appearance of temperature internal forces, while in case of statically indeterminate structures the change of temperature causes an appearance of temperature internal forces. Often these internal forces may approach significant values. Analysis of any statically indeterminate structure subjected to change of temperature is based on the calculation of displacement of statically determinate structure. So, calculation of displacements due to change of temperature is a very important problem for analysis of both statically determinate and indeterminate structures.

The first two terms of Maxwell–Mohr's integral equation (1.8) may be rewritten as follows:

$$\Delta_{kp} = \sum \int_0^l \bar{M}_k \Delta_{0p} + \sum \int_0^l \bar{N}_k \Delta_{xp}, \tag{1.9}$$

where $\Delta_{\theta p} = M_p\,dx/EI$ is the mutual angular displacement of both sections faced apart at distance dx due to the given load and $\Delta_{xp} = N_p\,dx/EA$ is the mutual axial displacement of both sections faced apart at a distance dx due to the given load.

These terms may be easily computed for the case of temperature change. Let us consider elementary part of a structure with length dx. The height of the cross section of the member is h_0. The upper and bottom fibers of the member are subjected to temperature increase t_1 and t_2, respectively, above some reference temperature. Corresponding distribution of temperature (temperature profile) is presented in Fig. 1.5. If the change of temperature for bottom and uppers fibers are equal ($t_1 = t_2$), then this case presents the uniform change of temperature; if $t_1 \neq t_2$ then this case is referred as nonuniform change of temperature [Rab60].

The expansion of the upper and bottom fibers equals to $\Delta_u = \alpha t_1 \,\mathrm{d}x$ and $\Delta_b = \alpha t_2 \,\mathrm{d}x$, respectively; these expressions contain coefficient of thermal expansion α of member material. In the case of symmetrical cross section, the expansion of the fiber at the mid-height equals to

$$\Delta_{xt} = \alpha \frac{t_1 + t_2}{2} \mathrm{d}x. \tag{1.9a}$$

The mutual angle of rotation of two plane sections, which are located apart from each other on distance $\mathrm{d}x$

$$\Delta_{\theta t} = \alpha \frac{|t_1 - t_2|}{h_0} \mathrm{d}x. \tag{1.9b}$$

Now we can substitute (1.9a) and (1.9b) into (1.9). Finally, the displacement in kth direction due to change of temperature may be presented in the following form:

$$\Delta_{kt} = \sum \int_0^s \alpha \frac{t_1 + t_2}{2} \bar{N}_k \,\mathrm{d}s + \sum \int_0^s \alpha \frac{|t_1 - t_2|}{h_0} \bar{M}_k \,\mathrm{d}s, \tag{1.10}$$

where $\bar{M}, \bar{N}$ are bending moment and axial force due to the unit generalized force in kth direction; this force should be corresponding to required temperature displacements.

A difference $t_1 - t_2$ is a temperature gradient; a half-sum $(t_1 + t_2)/2$ is a temperature at the centroid of the symmetric cross section (the axis of symmetry coincides with neutral axis). If the cross section is nonsymmetrical about its neutral axis, then the term $(t_1 + t_2)/2$ must be replaced by $t_2 + [(t_1 - t_2)/2]y$, where y is the distance of the lower fiber to the neutral axis.

The term $(t_1 + t_2)/2$ means that a bar is subjected to uniform thermal effect; in this case, all fibers are expanded by the same values. The term $|t_1 - t_2|/h_0$ means that a bar is subjected to nonuniform thermal effect; in this case a bar is subjected to bending in such way that the fibers on the neutral line have no thermal elongation. So, the first and second terms in (1.10) present displacements in kth direction due to uniform and nonuniform change of temperature, respectively. Integrals $\int \bar{M}_k \,\mathrm{d}s$ and $\int \bar{N}_k \,\mathrm{d}s$ present the areas of bending moment and axial force diagram in unit condition, which corresponds to required displacement.

The presentation of Maxwell–Mohr integral in form (1.10) allows us to calculate any displacement (linear, angular, mutual linear, mutual angular) caused by uniform or nonuniform change of temperature. This formula does not take into account the influence of shear. The procedure of summation in formula (1.10) must be carried over all members of the system. The signs at all terms in this formula will be obtained as follows: if the displacements of the element induced by both the change of temperature and by the unit load occur at the same direction, then the corresponding term of the equation will be positive.

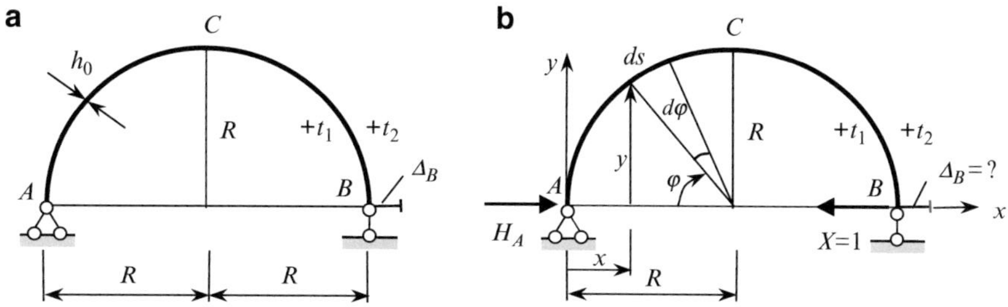

Fig. 1.6 Curvilinear bar. Design diagram and unit state

Procedure for analysis is as follows:

1. Construct the unit state. For this, we should apply unit generalized force X, which corresponds to the required displacement.
2. Construct the bending moment and axial force diagrams in the unit state.
3. For each member of a structure to compute the term $\int \overline{N}_k \, dx$, which is the area of axial force diagram in the unit state.
4. For each member of a structure to compute the term $\int \overline{M}_k \, dx$, which is the area of bending moment diagram in the unit state.
5. Apply formula (1.10).

Example 1.2. Determine the horizontal displacement of point B of the uniform semicircular bar in Fig. 1.6a, if the indoor and outdoor temperature rises by t_1°C and t_2°C, respectively. The height of cross-section bar is h_0.

Solution. A temperature effect related to curvilinear bar, therefore the general expression for temperature displacement, should be presented in the form (1.10), i.e., in terms of curvilinear coordinate s instead of x as for straight member. Unit load $X = 1$ (Fig. 1.6b) corresponds to required horizontal displacement at B. In the unit state reaction $H_A = 1$ and internal forces due to unit load X are as follows:

$$\bar{N}_k = -1 \times \sin\varphi; \quad \bar{M}_k = -1 \times y = -1 \times R \sin\varphi.$$

Thus (1.10) for displacement at B becomes

$$\Delta_B = \alpha \frac{t_1 + t_2}{2} \int_0^{\pi R} (-\sin\varphi) \mathrm{d}s + \alpha \frac{|t_1 - t_2|}{h_0} \int_0^{\pi R} (-R \sin\varphi) \mathrm{d}s.$$

Integration is performed along a curvilinear road of length πR. In the polar coordinates $\mathrm{d}\varphi = \mathrm{d}s/R$, $\sin\varphi = y/R$ the limits of integration become $0 - \pi$:

$$\Delta_B = \alpha \frac{t_1 + t_2}{2} \int_0^{\pi} (-\sin\varphi) R \, \mathrm{d}\varphi + \alpha \frac{|t_1 - t_2|}{h_0} \int_0^{\pi} (-R \sin\varphi) R \, \mathrm{d}\varphi.$$

Thus, for required displacement we get the following expression

$$\Delta_B = \alpha \frac{t_1 + t_2}{2}(-2R) + \alpha \frac{|t_1 - t_2|}{h_0}\left(-2R^2\right) = -\alpha R(t_1 + t_2) - 2\alpha R^2 \frac{|t_1 - t_2|}{h_0}. \quad \text{(a)}$$

Negative sign in (a) means that unit force X produces negative work on the real displacement, i.e., the displacement of the point B due to temperature changes is directed from left to right.

For uniform change of temperature (i.e., when gradients for indoor and outdoor temperatures are the same), i.e., $t_1 = t_2$, a difference $t_1 - t_2 = 0$ and only first term of (1.10) or (b) should be taken into account. In this case, the horizontal displacement equals to $\Delta_B = -2\alpha Rt$.

Summary

1. Maxwell–Mohr integral presents the fundamental and power method for the calculation of arbitrary displacements of any elastic structure. Displacements may be the result of any types of loads and change of temperature.
2. In order to calculate any displacement, it is necessary to consider two states of a structure, i.e., the given and unit states. Unit state presents the same structure, but loaded by unit generalized force corresponding to the required displacement.
3. The terms of (1.8), which should be taken into account depend on the type of structure (discussed in Sect. 1.3.1).
4. For both states, given and unit, it is necessary to set up expressions for corresponding internal forces and calculate the required displacement by the Maxwell–Mohr integral.

1.4 Graph Multiplication Method

Graph multiplication method presents most effective way for computation of any displacement (linear, angular, mutual, etc.) of bending structures, particularly for framed structures. The advantage of this method is that the integration procedure according to Maxwell–Mohr integral is replaced by elementary algebraic procedure on two bending moment diagrams in the actual and unit states. This procedure was developed by Vereshchagin (1925) and is often referred as the Vereshchagin rule [Rab60].

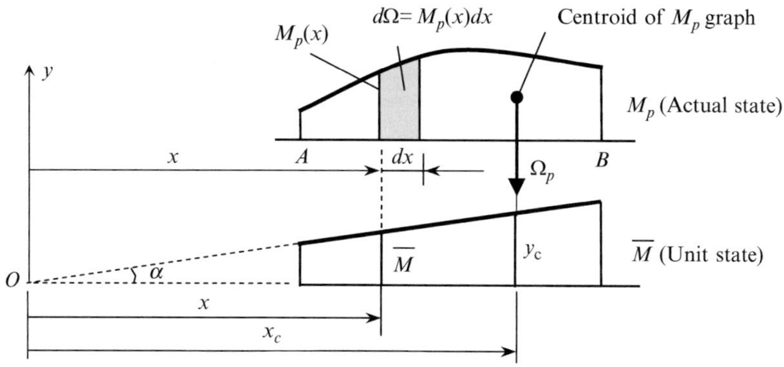

Fig. 1.7 Graph multiplication method. Bending moment diagrams M_p and $\overline{M}$ in actual and unit states

1.4.1 Vereshchagin Rule

Let us consider some portion *AB* which is a part of a bending structure; the flexural stiffness, *EI*, within of this portion is constant. The bending moment diagrams for this portion in actual and unit state are M_P and $\overline{M}$. Both diagrams for portion *AB* are presented in Fig. 1.7. In general case, a bending moment diagram M_p in the actual state is bounded by curve, but for special cases may be bounded by straight line (if a structure is subjected to concentrated forces and/or couples). However, it is obvious that in the *unit state* the bending moment diagram $\overline{M}$ is always bounded by the *straight line*. Just this property of unit bending moment diagram allows us to present the Maxwell–Mohr integral for bending systems in the simple form.

Ordinate of the bending moment in actual state at section x is $M_p(x)$. Elementary area of a bending moment diagram in actual state is $\mathrm{d}\Omega = M_p(x)\,\mathrm{d}x$. Since $M = x\tan\alpha$, then integral in Maxwell–Mohr formula may be presented as (coefficient $1/EI$ by convention is omitted)

$$\int M_p\overline{M}\,\mathrm{d}x = \int M_p(x\tan\alpha)\mathrm{d}x = \tan\alpha\int x\,\mathrm{d}\Omega.$$

Integral $\int x\,\mathrm{d}\Omega$ represents the static moment of the area of the bending moment diagram in actual state with respect to axis Oy. It is well known that a static moment may be expressed in terms of total area Ω and coordinate of its centroid x_c by formula $\int x\,\mathrm{d}\Omega = \Omega_p x_c$. It is obvious that $x_c\tan\alpha = y_c$. Therefore, the Maxwell–Mohr integral may be presented as follows:

$$\frac{1}{EI}\int M_p\overline{M}\,\mathrm{d}x = \frac{\Omega_p y_c}{EI}. \tag{1.11}$$

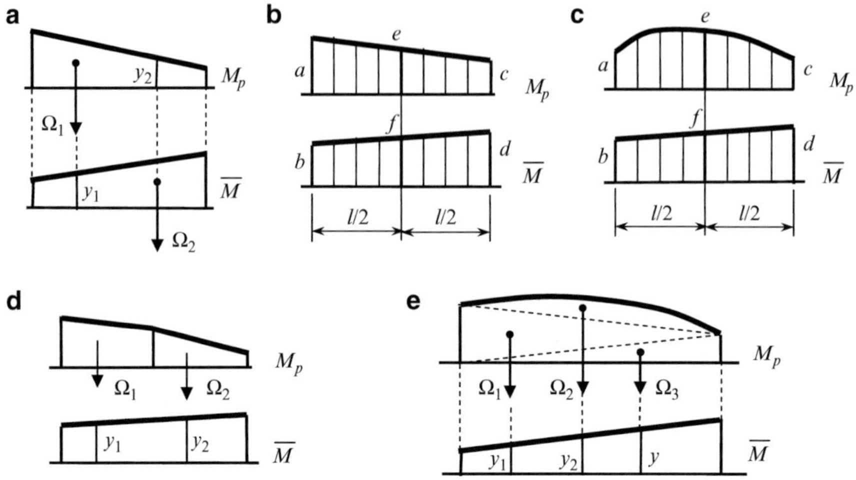

Fig. 1.8 Multiplication of two bending moment diagrams

The procedure of integration $\int M_p\overline{M}\,dx = \Omega_p y_c$ is called the "multiplication" of two graphs.

> The result of multiplication of two graphs, at least one of which is bounded by a straight line (bending moment diagram in unit state), equals to area Ω of the bending moment diagram M_p in actual state multiplied by the ordinate y_C from the unit bending moment diagram $\overline{M}$, which is located under the centroid of the M_p diagram.

It should be remembered, that *the ordinate y_C must be taken from the diagram bounded by a straight line*. The graph multiplication procedure (1.11) may be presented by conventional symbol ($\times$) as

$$EI\Delta_{kp} = \int M_p\overline{M}_k\,dx = M_p \times \overline{M}_k. \tag{1.12}$$

It is obvious that the same procedure may be applicable to calculation of similar integrals, which appear in Maxwell–Mohr integral, i.e., $\int N_p\overline{N}\,dx$ and $\int Q_p\overline{Q}\,dx$.

1.4.2 Trapezoid and Simpson Rules

If the structure in the actual state is subjected to concentrated forces and/or couples, then both the bending moment diagrams in actual and unit states are bounded by the straight lines (Fig. 1.8a). In this case, the multiplication procedure of two diagrams is commutative. It means that the area Ω could be calculated on any of the two diagrams and corresponding ordinate y_C will be measured from the second one, i.e., $\Omega_1 y_1 = \Omega_2 y_2$. This expression may be rewritten in terms of specific ordinates, as presented in Fig. 1.8b.

In this case, the displacement as a result of the multiplication of two graphs may be calculated using the two following rules:

1. Trapezoid rule allows calculating the required displacement in terms *of extreme* ordinates

$$\Delta = \frac{l}{6EI}(2ab + 2cd + ad + bc), \tag{1.13}$$

   where the crosswise end ordinates has unity coefficients. This formula is precise.
2. Simpson's rule allows calculating the required displacement in terms of *extreme and middle* ordinates

$$\Delta = \frac{l}{6EI}(ab + 4ef + cd). \tag{1.14}$$

Equation (1.14) may also be used for the calculation of displacements, if the bending moment diagram in the actual condition is bounded by a *curve* line. If the bending moment diagram M_p is bounded by quadratic parabola (Fig. 1.8c), then the result of multiplication of two bending moment diagrams by formula (1.14) is exact; this case occurs if a structure is carrying uniformly distributed load. If the bending moment diagram M_p is bounded by cubic parabola, then the procedure (1.14) leads to the approximate result.

If a graph M_p is bounded by a broken line, then both graphs have to be divided by several portions as shown in Fig. 1.8d. In this case, the result of multiplication of both graphs is

$$\int M_p\overline{M}\,\mathrm{d}x = \Omega_1 y_1 + \Omega_2 y_2.$$

Sometimes it is convenient to subdivide the curved bending moment diagram by a number of "good" shapes, for example, in Fig. 1.8e. In this case

$$\int M_p\overline{M}\,\mathrm{d}x = \Omega_1 y_1 + \Omega_2 y_2 + \Omega_3 y_3.$$

1.4.3 Signs Rule

According to (1.12) the displacement will be positive, when the area of the diagram M_p and the ordinate y_C of the diagram $\overline{M}$ have the same sign. If ordinates in (1.13) or (1.14) of bending moment diagram for actual and unit states are placed on the *different sides* of the basic line, then result of their multiplication is negative.

The positive result indicates that displacement occurs in the direction of applied unit load.

Procedure for computation of deflections by graph multiplication method is as follows:

1. Draw the bending moment diagram M_p for the actual state of the structure.
2. Create a unit state of a structure. For this apply a unit load at the point where the deflection is to be evaluated. For computation of linear displacement we need to apply unit force $P = 1$, for angular displacement to apply unit couple $M = 1$, etc.
3. Draw the bending moment diagram $\overline{M}$ for the unit state of the structure. Since the unit load (force, couple) is dimensionless, then the ordinates of unit bending moment diagram $\overline{M}$ in case of force $F = 1$ and couple $M = 1$ are units of length (m) and dimensionless, respectively.
4. Apply the graph multiplication procedure using the most appropriate form: Vereshchagin rule (1.11), trapezoid rule (1.13), or Simpson's formula (1.14).

Useful tables for multiplication of two bending moment diagrams are presented in [Kar10].

Example 1.3. A cantilever beam *AB*, length l, carrying a uniformly distributed load q (Fig. 1.9). Flexural rigidity EI is constant. Compute (a) the angle of rotation θ_A; (b) the vertical displacement Δ_A at the free end.

Solution. Analysis of the structure starts from construction of bending moment diagram M_p due to given external load. This diagram is bounded by quadratic parabola and maximum ordinate equals $ql^2/2$.

(a) *Angle of rotation at point A*. The unit state presents the same beam subjected to unit couple $M = 1$ at the point where it is required to find angular displacement; direction of this couple is arbitrary (Fig. 1.9a). It is convenient that both unit and actual state and their bending moment diagrams locate one under another.

The next step is "multiplication" of two bending moment diagrams. The area of square parabola is $\Omega = (1/3) \times (ql^2/2) \times l$. Centroid of this diagram is located on the distance $l/4$ from fixed support. Corresponding ordinate y_C from diagram $\overline{M}$ of unit state is 1. Multiplication procedure is presented in Table 1.4.

This table also contains computation of required displacement using the Simpson rule. Ordinates a and b are taken from the bending moment diagrams for actual and unit states, respectively, at the left end of a beam (point *A*); ordinates e and f are taken at the middle of the beam *AB*, and ordinates c and d at the right end (point *B*).

(b) *Vertical displacement at point A*. The bending moment diagram M_p for actual state is shown in Fig. 1.9b; this diagram for problems (a) and (b) is same. The unit state presents the same structure with concentrated force $P = 1$, which acts at point *A*; direction of the unit force is chosen in arbitrary way. The unit state with corresponding bending moment diagram $\overline{M}$ is presented in Fig. 1.9b.

Computation of displacements using Vereshchagin rule in general form and by Simpson rule are presented in Table 1.5.

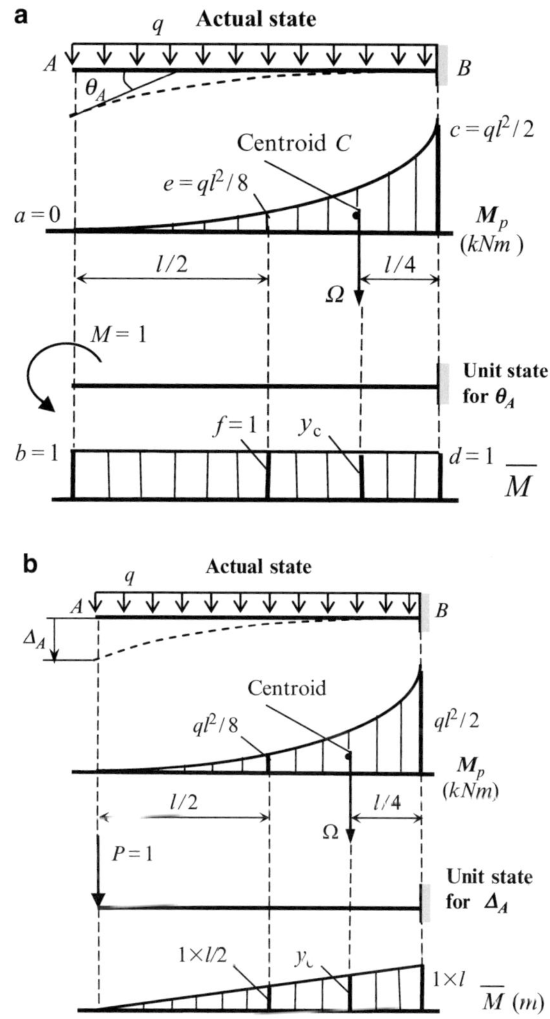

Fig. 1.9 (**a**) Actual state, unit state for θ_A and corresponding bending moment diagrams. (**b**) Actual state, unit state for Δ_A and corresponding bending moment diagrams

Table 1.4 Graph multiplication procedures

	General formula (1.11)	Simpson rule (1.14)
Displacement	$\Delta = \frac{1}{EI}\Omega y_C$	$\Delta = \frac{l}{6EI}(ab + 4ef + cd)$
Angular $\theta_A = \frac{M_P \times \overline{M}}{EI}$	$\theta_A = \frac{1}{EI} \times \underbrace{\frac{1}{3}\frac{ql^2}{2}l}_{\Omega} \times \underbrace{1}_{y_C} = \frac{ql^3}{6EI}$	$\theta_A = \frac{l}{6EI}\left(\underbrace{0 \times 1}_{ab} + \underbrace{4\frac{ql^2}{8} \times 1}_{4ef} + \underbrace{\frac{ql^2}{2} \times 1}_{cd}\right) = \frac{ql^3}{6EI}$

Table 1.5 Graph multiplication procedures

	General formula (1.11)	Simpson rule (1.14)
Displacement	$\Delta = \frac{1}{EI}\Omega y_C$	$\Delta = \frac{l}{6EI}(ab + 4ef + cd)$
Linear		
$\Delta_A = \frac{M_P \times \overline{M}}{EI}$	$\Delta_A = \frac{1}{EI} \times \underbrace{\frac{1}{3}\frac{ql^2}{2} l}_{\Omega} \times \underbrace{\frac{3}{4} \times 1 \times l}_{y_C} = \frac{ql^4}{8EI}$	$\Delta_A = \frac{l}{6EI}\left(\underbrace{0 \times 0}_{ab} + \underbrace{4\frac{ql^2}{8} \times 1 \times \frac{l}{2}}_{4ef} + \underbrace{\frac{ql^2}{2} \times 1 \times l}_{cd}\right) = \frac{ql^4}{8EI}$

1.5 Maxwell–Mohr Formula for Curvilinear Rods

Strictly speaking, the three-term Maxwell–Mohr formula is only valid for straight rods. The formula reduces to only an approximation for curvilinear rods. We show a modification of this formula that can be applied to curvilinear rods.

Consider an infinitely small element of a curvilinear rod with curvature ρ and length ds. The corresponding central angle is denoted by dφ. The cross-sectional area of the element is A. A force N is applied at the ends of the element, as shown in Fig. 1.10.

The total shrinkage (elongation) of this element is equal to $N\,\mathrm{d}s/EA$. The axial strain is equal to N/EA. If we neglect the change in radius of curvature, then we can assume that the angle of rotation of the two ends of the elements will be $(N/EA)\mathrm{d}\varphi = (N/EA)(\mathrm{d}s/\rho)$. When $N = 1$ the angle of rotation of ends sections becomes $\mathrm{d}s/EA\rho$ (unit displacement). By the law of reciprocal unit displacements (for more details see Sect. 1.8, and [Dar89], [Kar10]) it follows that if two unit couples $M = 1$ act on the cross-sectional faces of the element ds, then the same deformation occurs in the axial direction of the element, i.e., $\delta_{ik} = \delta_{ki}$. In the case of an arbitrary couple M, the axial deformation becomes $M\,\mathrm{d}s/EA\rho$. This elongation (shrinkage) is due to the fact that the sections rotate about a neutral axis, which does not pass through their center of gravity. Therefore, bending moments induce axial deformations, while axial forces cause bending in curvilinear rods. So, bending moments lead to additional work being done by axial forces, while axial forces lead to additional work which is performed by bending moments. Therefore, the refined formula for displacements takes the form [Rab54a]

$$\Delta_{iP} = \sum \int \overline{M}_i \left(\frac{M_p}{EI} + \frac{N_p}{EA\rho}\right) \mathrm{d}s + \sum \int \overline{N}_i \left(\frac{N_p}{EA} + \frac{M_p}{EA\rho}\right) \mathrm{d}s + \sum \int \mu \frac{\overline{Q}_i Q_P}{GA} \mathrm{d}s. \tag{1.15}$$

Here, as before, the horizontal bar indicates that the corresponding force is related to a unit state.

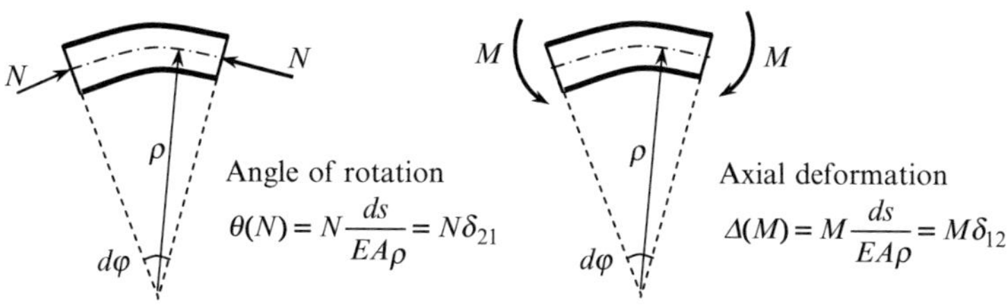

Fig. 1.10 Angular strain due by axial forces and axial strain due by couples

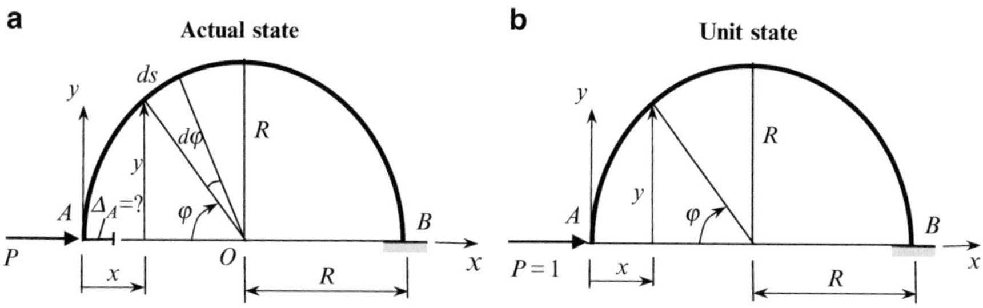

Fig. 1.11 (**a**) Design diagram of the curvilinear bar, (**b**) unit state

Example 1.4. A horizontal force P is applied to a semicircular uniform rod of R (Fig. 1.11a). Determine the horizontal displacement of the point at which the force is applied.

Solution. We limit ourselves to only the first term of the general formula (1.15). The actual and unit states are shown in Fig. 1.11. Corresponding internal forces are

$$M_p = -Py, \quad N_p = -P \sin\varphi, \quad \overline{M}_1 = -1 \times y.$$

The first term in the general formula is broken up into two integrals:

$$\Delta_{ip} = \int \overline{M}_i \left(\frac{M_p}{EI} + \frac{N_P}{EA\rho} \right) \mathrm{d}s = \int \overline{M}_i \frac{M_p}{EI} \mathrm{d}s + \int \overline{M}_i \frac{N_p}{EAR} \mathrm{d}s. \qquad \text{(a)}$$

In polar coordinates $\mathrm{d}s = R\,\mathrm{d}\varphi$, $y = R \sin\varphi$ so the formula takes the form

$$\Delta_{ip} = \int_0^{\pi} \overline{M}_i \frac{M_p}{EI} R\,\mathrm{d}\varphi + \int_0^{\pi} \overline{M}_i \frac{N_p}{EAR} R\,\mathrm{d}\varphi$$
$$= \int_0^{\pi} (-1y) \frac{(-Py)}{EI} R\,\mathrm{d}\varphi + \int_0^{\pi} (-1y) \frac{(-P \sin\varphi)}{EAR} R\,\mathrm{d}\varphi.$$

Substitute $y = R\sin\varphi$ and evaluate the integrals. As a result we get

$$\Delta_{ip} = \frac{PR^3}{EI} \times \frac{\pi}{2} + \frac{PR}{EA} \times \frac{\pi}{2} = \frac{PR^3}{EI} \times \frac{\pi}{2}\left(1 + \frac{r^2}{R^2}\right), \tag{b}$$

where $r = \left(\sqrt{I/A}\right)$ is the radius of inertia of the cross-section of the curvilinear rod. For given ratio r/A we can evaluate effect of second term in the parenthesis and compare with result in Example 1.1.

1.6 Elastic Loads Method

Elastic load method allows *simultaneous* computation of displacements for a *set* of points of a structure. This method is based on conjugate beam method. This method allows us to consider beams of variable cross sections. For trusses this method leads to precise results. For arches this method is approximate, however, very effective.

1.6.1 Computation of Elastic Load

Let us consider a part of any actual structure; the vertical displacements y for specified points, labeled as $n-2, n-1, n, n+1, n+2$, etc., are to be determined. Corresponding displacement diagram is presented in Fig. 1.12a.

Now let us consider a *fictitious* structure subjected to any loads W applied at point where we are required to find displacements (Fig. 1.12b); these fictitious loads of the fictitious structure are not yet known. Displacement of a real structure and moment of fictitious structure are related as $y = M_f/EI$. Therefore, the problem is to find such loads W so that the bending moment diagram of fictitious structure would be proportional to the vertical displacement diagram of the actual structure. Loads W are called elastic loads, i.e., they are such fictitious loads so that bending moment diagram of fictitious structure coincides (with accuracy to constant multiplier $1/EI$) with displacements diagram of the real structure [Kar10].

Corresponding fictitious bending moment diagram of a structure is shown in Fig. 1.12b; index "f – fictitious" for bending moments M and shear Q is omitted. Shear forces within the portions λ_n and λ_{n+1} are

$$Q_n = \frac{M_n - M_{n-1}}{\lambda_n}; \quad Q_{n+1} = \frac{M_{n+1} - M_n}{\lambda_{n+1}}.$$

Equilibrium equation for a free-body diagram of an infinitesimal element of a structure in vicinity of a point n (Fig. 1.12c) leads to the following expression for applied load W_n:

$$W_n = Q_n - Q_{n+1}.$$

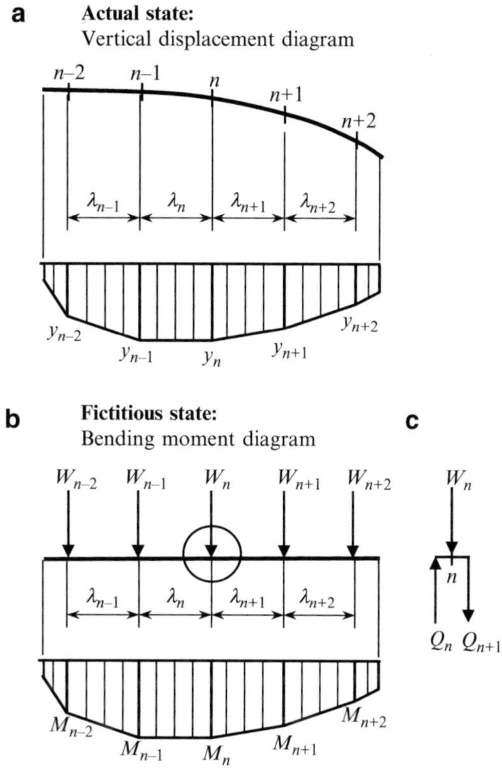

Fig. 1.12 (**a**, **b**) Resemblance of vertical displacement and bending moment; (**c**) free-body diagram in vicinity of point n

In terms of bending moments, this expression may be rewritten as follows:

$$W_n = \frac{M_n - M_{n-1}}{\lambda_n} - \frac{M_{n+1} - M_n}{\lambda_{n+1}}$$
$$= -M_{n-1}\frac{1}{\lambda_n} + M_n\left(\frac{1}{\lambda_n} + \frac{1}{\lambda_{n+1}}\right) - M_{n+1}\frac{1}{\lambda_{n+1}}. \quad (1.16)$$

This formula allows us to calculate fictitious load W_n at point n if bending moments at points $n-1$, n, and $n+1$ are known. If this formula is used for calculating all forces W_i and after which the resulting forces will be used for construction of bending moment diagram, then this diagram will be same as the diagram in Fig.1.12b.

Thus, the elastic load at point n becomes

$$W_n = -y_{n-1}\frac{1}{\lambda_n} + y_n\left(\frac{1}{\lambda_n} + \frac{1}{\lambda_{n+1}}\right) - y_{n+1}\frac{1}{\lambda_{n+1}}. \quad (1.17)$$

Our fundamental goal is calculating the displacement using elastic loads; on the other hand, for calculation of the elastic loads according to (1.17), it is necessary to

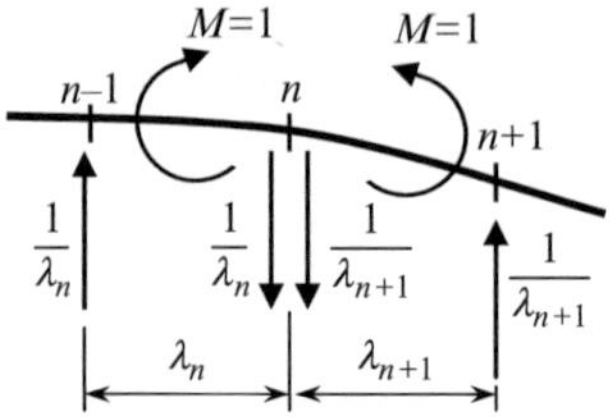

Fig. 1.13 Unit state

know displacements. In order to get away from this "closed circle" we need to modify formula (1.17).

For this purpose, we need to understand the physical meaning of this expression. Let us calculate the mutual angle of two portions at the point n. The unit state of the given structure contains two unit couples $M = 1$ with opposite directions as shown in Fig. 1.13. The couple $M = 1$, which acts at the left portion, may be replaced by two forces of same magnitude $1/\lambda_n$, but having opposite signs; similarly, the right couple $M = 1$ may be replaced by two forces $1/\lambda_{n+1}$.

Now we can see that the right-hand side of (1.17) represents the work done by forces within the real displacements. Indeed, the first term $(-y_{n-1}(1/\lambda_n))$ represents the work performed by force $1/\lambda_n$ of the unit state on the real displacement y_{n-1}; negative sign means that this force and real displacement have the opposite directions. The second term of (1.17) represents the work, which is produced by two forces $1/\lambda_n$ and $1/\lambda_{n+1}$ of the unit state on the real displacement y_n. Similarly, the last term (1.17) represents the work done by force $1/\lambda_{n+1}$ of the unit state within the real displacement y_{n+1}.

Since forces $1/\lambda_n$ and $1/\lambda_{n+1}$ are the result of two unit couples $M = 1$, then expression (1.17) may be considered as a work done by these unit couples on the mutual angle of both portions at point n. On the other hand, the work done by unit couples $M = 1$ on the mutual angle may be expressed in terms of *internal* forces using the Maxwell–Mohr integral:

$$
\begin{aligned}
&-y_{n-1}\frac{1}{\lambda_n}+y_n\left(\frac{1}{\lambda_n}+\frac{1}{\lambda_{n+1}}\right)-y_{n+1}\frac{1}{\lambda_{n+1}}\\
&=M\left[\sum\int_0^l \overline{M}\frac{M_P\,\mathrm{d}x}{EI}+\sum\int_0^l \overline{N}\frac{N_P\,\mathrm{d}x}{EA}+\sum\eta\int_0^l \overline{Q}\frac{Q_P\,\mathrm{d}x}{GA}\right],
\end{aligned}
$$

where expression in the parenthesis in right-hand side represents the mutual angle.

Since $M = 1$, then finally the right-hand side of (1.17) may be rewritten as follows:

$$
\begin{aligned}
-y_{n-1}\frac{1}{\lambda_n}+y_n\left(\frac{1}{\lambda_n}+\frac{1}{\lambda_{n+1}}\right)-y_{n+1}\frac{1}{\lambda_{n+1}}&=\sum\int_0^l \overline{M}\frac{M_P\,\mathrm{d}x}{EI}+\sum\int_0^l \overline{N}\frac{N_P\,\mathrm{d}x}{EA}\\
&\quad+\sum\eta\int_0^l \overline{Q}\frac{Q_P\,\mathrm{d}x}{\mathrm{GA}}.
\end{aligned}
$$

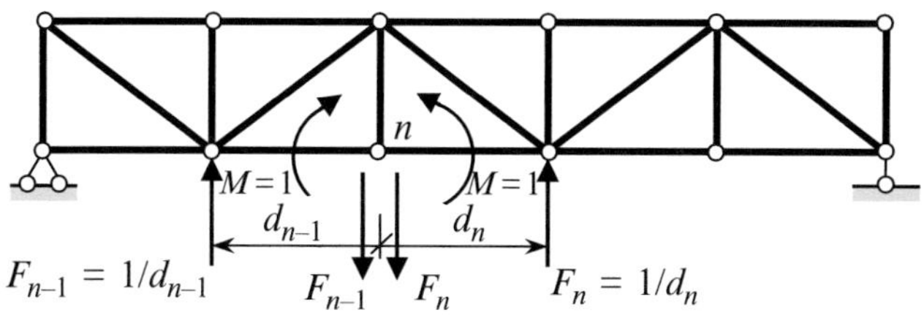

Fig. 1.14 Unit state for calculation of elastic load at joint n

As a result, the general formula for elastic loads in terms of internal forces is

$$W_n = \sum \int_0^l \overline{M} \frac{M_P \, \mathrm{d}x}{EI} + \sum \int_0^l \overline{N} \frac{N_P \, \mathrm{d}x}{EA} + \sum \eta \int_0^l \overline{Q} \frac{Q_P \, \mathrm{d}x}{GA}. \tag{1.18}$$

This formula resembles the Maxwell–Mohr integral. The fundamental difference comes from the fact that the unit state is constructed differently. In other words, the right-hand side of (1.18) is similar to that of (1.8), while the left-hand side of (1.18) represents an elastic load and the left-hand side of (1.8) is a displacement.

For beams, only the first term of (1.18) should be taken into account; the accuracy of elastic curve depends on the number of points (number of elastic loads). For arches it is necessary to take into account the first and third terms of (1.18). The elastic curve will be approximate. Construction of displacement of joint points of a truss requires only the second term of (1.18). In this case, expression for elastic loads becomes

$$W_n = \sum \frac{\overline{N}_n N_P l}{EA}. \tag{1.19}$$

Application of elastic loads method is the most interesting and effective to a truss. The procedure for computing displacements is outlined below.

1. Calculate the axial forces N_P in all elements of the truss caused by given loads.
2. Calculate the elastic load at a joint n. For this
   (a) Show a fictitious truss. If a real truss is simply supported then the fictitious truss is also simply supported.
   (b) Apply two unit couples $M = 1$ to members, which are adjacent to the joint n. Present each couple using forces $F_{n-1} = (1/d_{n-1})$ for span d_{n-1} and $F_{n-1} = (1/d_n)$ for span d_n, as shown in Fig. 1.14.
   (c) Calculate the axial forces $\overline{N}_n$ in all elements of the truss caused by forces in Fig. 1.14.
   (d) Calculate the elastic load W_n at the joint n by (1.19).

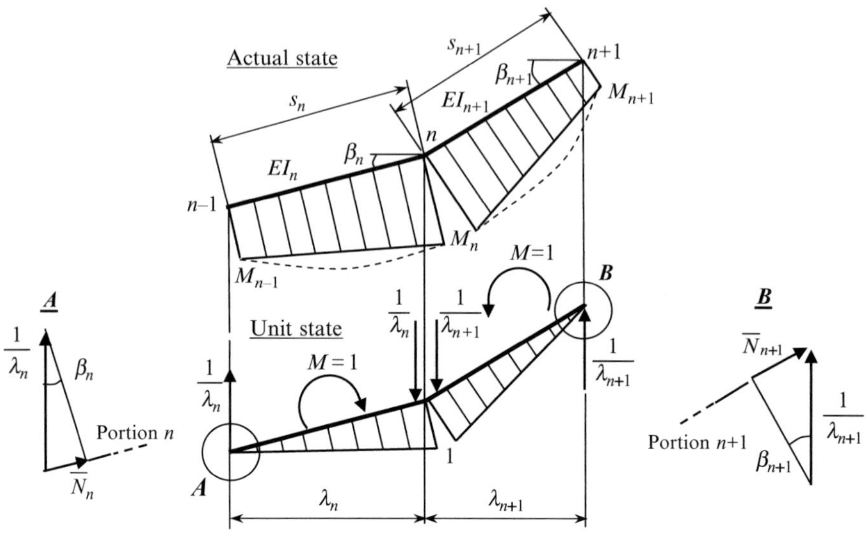

Fig. 1.15 Deriving of the expanded formula for elastic load

3. Calculate the elastic loads W for remaining joints of the truss chord, as explained in procedure 2.
4. Show the fictitious beam subjected to all elastic loads W. If the elastic load is positive, then it should be directed downward, i.e., in the same direction as the adjacent forces of neighboring couples.
5. Construct the bending moment diagram for fictitious beam on the tensile fibers. This diagram presents displacements of all joints of the entire real truss.

Detailed examples and advantages of this method are discussed in [Kar10].

1.6.2 *Expanded Form for Elastic Loads*

Finally, we present the important expanded expression of elastic loads for beams and rigid frames. This expression will take into account not only bending moments, but also axial forces. We consider a case when inclined members with length of s_n and s_{n+1} are subjected to bending moments and axial forces (Fig. 1.15). We assume that the axial forces N_n and N_{n+1} within portions n and $n + 1$ are constant. Axial forces in the unit state are $\overline{N}_n = -\sin\beta_n/\lambda_n$ and $\overline{N}_{n+1} = \sin\beta_{n+1}/\lambda_{n+1}$ for the left and right portions, respectively. Calculation of these forces is shown in Fig. 1.15, joints A and B.

Within the limits of the left portion n, the calculation of the first term of (1.18) may be performed using multiplication of two bending moment diagrams (trapezoid rule 1.13) in actual and unit states. For left portion we have

$$\sum \int_0^l \overline{M} \frac{M_P \, \mathrm{d}x}{EA} = \frac{s_n}{6EI_n}(2M_{n-1} \times 0 + 2M_n \times 1 + M_{n-1} \times 1 + M_n \times 0)$$
$$= \frac{s_n}{6EI_n}(M_{n-1} + 2M_n).$$

The second term of (1.18) may be presented as

$$\sum \int_0^l \overline{N} \frac{N_P \, \mathrm{d}x}{EA} = -\frac{1}{EA_n} \times \frac{\sin \beta_n}{\lambda_n} N_n s_n = -\frac{1}{EA_n} \times \frac{\sin \beta_n}{\lambda_n} N_n \frac{\lambda_n}{\cos \beta_n}$$
$$= -\frac{N_n}{EA_n} \times \tan \beta_n.$$

Within the limits of the right portion $n + 1$, the calculation of the first and second terms of (1.18) may be performed by similar manner.

Finally, the elastic loads should be calculated by the following expanded formula:

$$W_n = \frac{s_n}{6EI_n}(M_{n-1} + 2M_n) + \frac{s_{n+1}}{6EI_{n+1}}(2M_n + M_{n+1}) - \frac{N_n}{EA_n} \tan \beta_n$$
$$+ \frac{N_{n+1}}{EA_{n+1}} \tan \beta_{n+1}. \tag{1.20}$$

We can see that during the calculation of elastic load W_n, the sum of the integrals is affecting only two adjacent elements at a node n [Sni66]. This formula is known as the expanded formula for elastic loads and may be effectively applied for the calculation of displacements of arches. Thus, for the calculation of elastic load at point n we need to know the bending moments in actual state, at three consecutive points ($n - 1, n, n + 1$) and axial forces within the adjacent portions.

Example 1.5. Design diagram of a cantilever beam is presented in Fig. 1.16a. Compute the deflections of the beam at the points 0, 1, 2.

Solution. Subdivide the beam into two equal parts (0–1 and 1–2). The bending moment diagram for actual beam is shown in Fig. 1.16b. Fictitious beam and elastic loads W_0 and W_1 are shown in Fig. 1.16c. For calculation of W_0 we need to know bending moments at three consecutive points; dotted line shows additional portion of the beam with ends points -1 and 0; the length and stiffness of this portion are λ_0 and $EI_0 = \infty$, respectively. The elastic loads are

$$W_0 = \frac{\lambda_0}{6EI_0}(M_{-1} + 2M_0) + \frac{\lambda_1}{6EI_1}(2M_0 + M_1)$$
$$= \frac{\lambda_0}{\infty}(M_{-1} + 2M_0) + \frac{l}{12EI}\left(2Pl + \frac{Pl}{2}\right) = \frac{5Pl^2}{24EI},$$

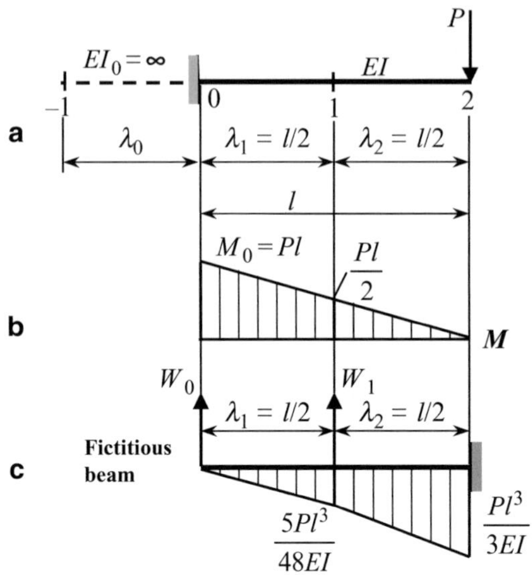

Fig. 1.16 (**a**) Design diagram; (**b**) bending moment diagram of the real beam; (**c**) fictitious beam, bending moment diagram, and displacement of the real beam

$$\begin{aligned} W_1 &= \frac{\lambda_1}{6EI_1}(M_0 + 2M_1) + \frac{\lambda_2}{6EI_2}(2M_1 + M_2) \\ &= \frac{l}{12EI}\left(Pl + 2\frac{Pl}{2}\right) + \frac{l}{12EI}\left(2\frac{Pl}{2} + 0\right) = \frac{Pl^2}{4EI}. \end{aligned}$$

Now these elastic loads should be applied to the fictitious beam. Since the bending moment diagram is traced on the tensile fibers and ordinates of M diagram are located above the axis, the elastic loads should be directed upward. Corresponding bending moment diagram of fictitious beam is presented in Fig. 1.16c. At the same time, this diagram presents the elastic displacements of the real beam. Displacements at the points 0, 1, and 2 are exact. This result may be obtained using graph multiplication method; in this case two unit states should be constructed.

Example 1.6. Design diagram of a nonuniform beam is presented in Fig. 1.17a. Determine the displacements at the free end point and at the point where force P is applied.

Solution. Subdivide the beam into three parts length λ_1, λ_2, and λ_3. The points with required displacements are labeled as 0, 1 and 2. The bending moment diagram M_p for actual beam is shown in Fig. 1.17b. Fictitious (conjugate) beam with elastic loads W_1 and W_2 are shown in Fig. 1.17c. Load W_0 at free end of the entire beam is not shown.

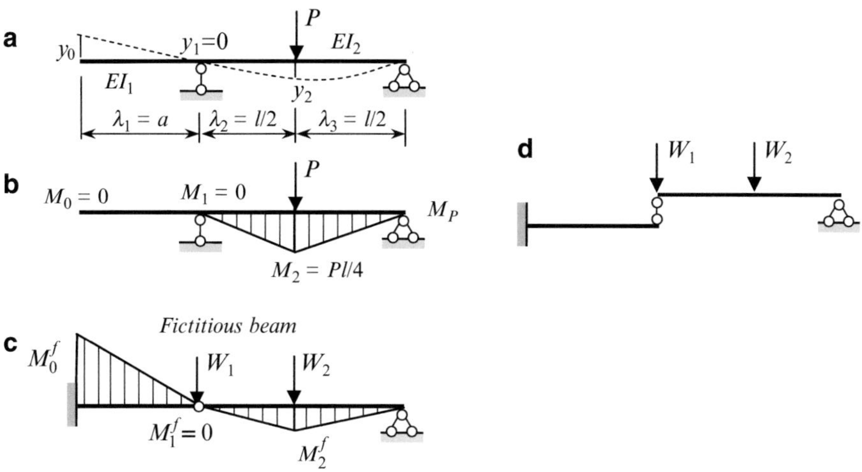

Fig. 1.17 Computation of displacement by elastic loads method

Elastic loads according to (1.20) are

$$W_1 = \frac{\lambda_1}{6EI_1}(M_0 + 2M_1) + \frac{\lambda_2}{6EI_2}(2M_1 + M_2)$$
$$= \frac{a}{6EI_1}(0 + 2 \times 0) + \frac{l}{2 \times 6EI_2}\left(2 \times 0 + \frac{Pl}{4}\right) = \frac{Pl^2}{48EI_2},$$

$$W_2 = \frac{\lambda_2}{6EI_2}(M_1 + 2M_2) + \frac{\lambda_3}{6EI_2}(2M_2 + M_3)$$
$$= \frac{l}{2 \times 6EI_2}\left(0 + 2 \times \frac{Pl}{4}\right) + \frac{l}{2 \times 6EI_2}\left(2 \times \frac{Pl}{4} + 0\right) = \frac{Pl^2}{12EI_2}.$$

The fictitious structure is a Gerber–Semikolenov beam [Kar10]; corresponding interaction scheme is shown in Fig. 1.17d. Bending moments of the fictitious beam are

$$M_2^f = \frac{W_2 l}{4} = \frac{Pl^2}{12EI_2}\frac{l}{4} = \frac{Pl^3}{48EI_2},$$

$$M_0^f = -\left(W_1 + \frac{W_2}{2}\right)a = -\frac{Pl^2 a}{16EI_2}.$$

Vertical displacements of the initial beam are

$$y_0 = M_0^f = -\frac{Pl^2 a}{16EI_2}; \quad y_1 = M_1^f = 0; \quad y_2 = M_2^f = \frac{Pl^3}{48EI_2}.$$

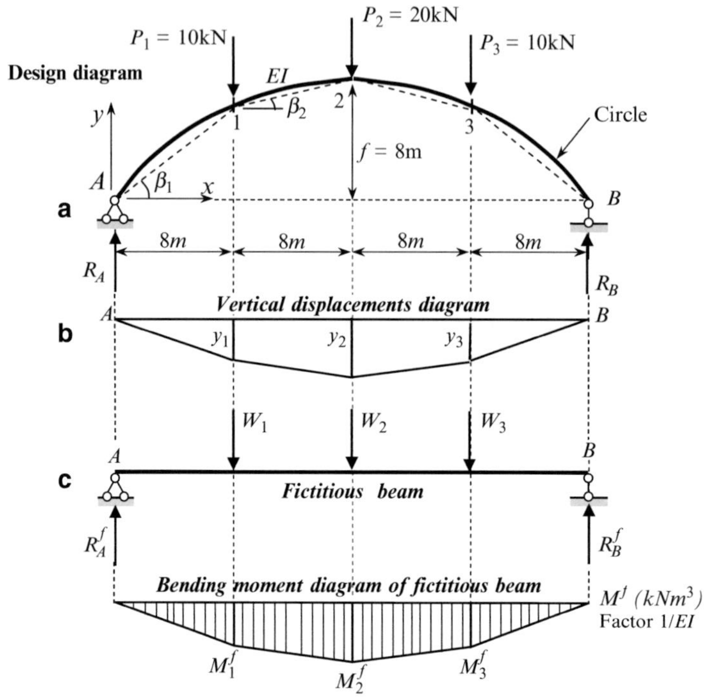

Fig. 1.18 Curvilinear simply supported bar. (**a**) Design diagram; (**b**) vertical displacements; (**c**) fictitious beam and corresponding bending moment diagram

Example 1.7. Curvilinear simply supported uniform circular rod AB is subjected to loads P_i (Fig. 1.18a). Determine the vertical displacements at points 1, 2, 3.

Solution. Required displacements are shown in Fig. 1.18b. Let us replace the curvilinear bar by a set of straight members of the same flexural rigidity; they are shown by dotted line in Fig. 1.18a. For curvilinear rod with parameters $l = 32$ m and $f = 8$ m, according to (1) the coordinates for point 1 are $x = 8$ m, $y = 6.33$ m; the lengths of straight elements are $s_{A-1} = 10.2$ m and $s_{1-2} = 8.33$ m; $\tan \beta_1 = 0.7912$, $\tan \beta_2 = 0.2087$.

Bending moments in loaded state are $M_1 = 160$ kNm, $M_2 = 240$ kNm, $M_3 = M_1$.

Fictitious beam is shown in Fig. 1.18c. According to (1.20), elastic loads consist of two parts

$$W_n = W_n(M) + W_n(N),$$

where

$$W_n(M) = \frac{s_n}{6EI_n}(M_{n-1} + 2M_n) + \frac{s_{n+1}}{6EI_{n+1}}(2M_n + M_{n+1})$$

$$W_n(N) = -\frac{N_n}{EA_n}\tan\beta_n + \frac{N_{n+1}}{EA_{n+1}}\tan\beta_{n+1}.$$

It is easy to check that the axial forces for entire structure are $N_{A-1} = -32.267\,\text{kN}$ and $N_{1-2} = -49.07\,\text{kN}$.

The components of a first elastic load are

$$W_1(M) = \frac{s_{A-1}}{6EI}(M_A + 2M_1) + \frac{s_{1-2}}{6EI}(2M_1 + M_2)$$
$$= \frac{10.2}{6EI}(0 + 2 \times 160) + \frac{8.1724}{6EI}(2 \times 160 + 240) = \frac{1,307\,\text{kN}\,\text{m}^2}{EI},$$

$$W_1(N) = -\frac{N_{A-1}}{EA}\tan\beta_1 + \frac{N_{1-2}}{EA}\tan\beta_2 = -\frac{32.267}{EA}0.7912 - \frac{49.07}{EA}0.2087$$
$$= -\frac{35.77\,\text{kN}}{EA}.$$

The first elastic load becomes

$$W_1 = \frac{1,307}{EI} - \frac{35.77}{EA} = \frac{1,307}{EI}\left(1 - 0.0274\frac{I}{A}\right).$$

Similarly, for second elastic load we get

$$W_2(M) = \frac{s_{1-2}}{6EI}(M_1 + 2M_2) + \frac{s_{2-3}}{6EI}(2M_2 + M_3)$$
$$= \frac{6.33}{6EI}(160 + 2 \cdot 240) + \frac{8.1724}{6EI}(2 \times 240 + 160) = \frac{1,547}{EI},$$

$$W_2(N) = -\frac{N_{1-2}}{EA}\tan\beta_2 - \frac{N_{2-3}}{EA}\tan\beta_3 = -\frac{49.07}{EA}0.2087 \times 2 = -\frac{20.48}{EA},$$

$$W_2 = \frac{1,547}{EI}\left(1 - 0.0132\frac{I}{A}\right).$$

It is clear that $W_3 = W_1$. If axial forces are neglected, then the second terms in the formulas for elastic loads should be omitted.

Reactions of fictitious beam, corresponding bending moments and required displacements are

$$R_A^f = R_B^f = W_1 + \frac{W_2}{2},$$

$$M_1^f = R_A^f \times 8, \quad M_2^f = R_A^f \times 16 - W_1 \times 8,$$

$$y_1 = M_1^f, \quad y_2 = M_2^f, \quad y_3 = y_1.$$

1.7 Differential Relationships for Curvilinear Rods

This section contains two groups of differential relations for curvilinear rods. The first group establishes relationships between internal forces while the second group establishes the relation between displacements and strains.

1.7.1 Relationships Between Internal Forces

The following internal forces arise in planar arches: bending moment M, shear force Q, and axial force N. We establish a relationship among them. An infinitely small element i–j, with length ds, central angle dφ, and radius of curvature ρ is shown in Fig. 1.19. This element is subjected to normal, tangential and moment loads distributed among the entire length ds. We denote their intensities by q, p, and m, respectively. The following internal forces act on the ends of the element: M, Q, and N at section i, and the same forces with their elementary gain dM, dQ, and dN at section j. Positive directions of internal forces are shown in Fig. 1.19. The x and y axis are directed along the tangent and normal to the element at section i. The point O denotes the intersection point of two tangents from sections i and j. The angle between these two tangents is equal to the central angle dφ. We construct the equilibrium equations for this element [Rzh82].

1. Projection of all forces onto the x axis is

$$\sum X = N - (N + \mathrm{d}N)\cos \mathrm{d}\varphi + (Q + \mathrm{d}Q)\sin \mathrm{d}\varphi - p\,\mathrm{d}s = 0.$$

Since $\sin \mathrm{d}\varphi \cong \mathrm{d}\varphi$ and $\cos \mathrm{d}\varphi \cong 1$, we get $(\mathrm{d}N/\mathrm{d}s) = Q(\mathrm{d}\varphi/\mathrm{d}s) - p$. And since $(\mathrm{d}\varphi/\mathrm{d}s) = (1/\rho)$, then

$$\frac{\mathrm{d}N}{\mathrm{d}s} - \frac{Q}{\rho} + p = 0. \tag{1.21}$$

Derivative of the axial force along the axis is directly proportional to the shear force. The coefficient of proportionality is the curvature of the rod.

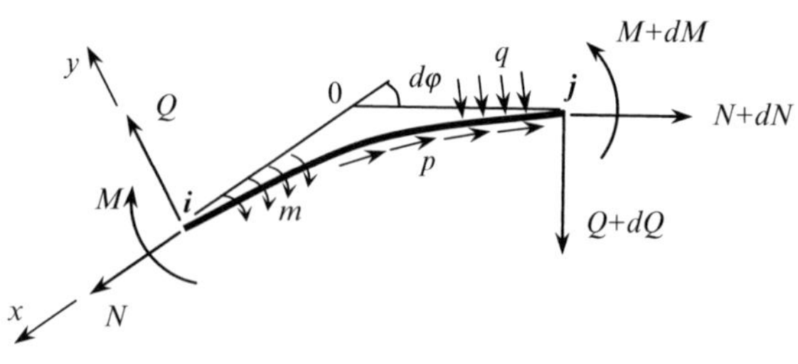

Fig. 1.19 Free-body diagram of the curvilinear member i–j

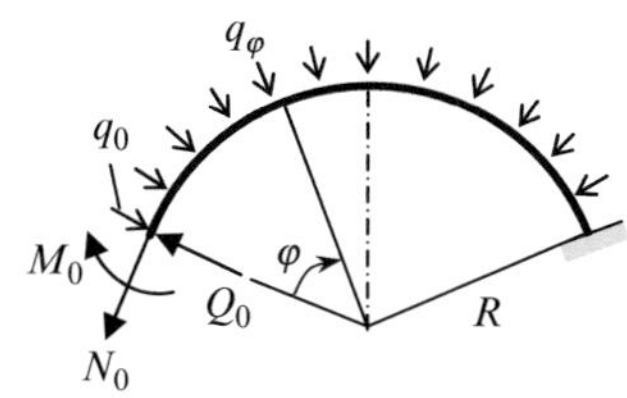

Fig. 1.20 Design diagram of circular rod

2. Projection of all forces onto the y axis is

$$\sum Y = Q - (Q + \mathrm{d}Q)\cos \mathrm{d}\varphi - (N + \mathrm{d}N)\sin \mathrm{d}\varphi - q\,\mathrm{d}s \times \cos\frac{\mathrm{d}\varphi}{2} = 0.$$

Neglecting the infinitely small second order term we get

$$\frac{\mathrm{d}Q}{\mathrm{d}s} + \frac{N}{\rho} + q = 0. \tag{1.22}$$

3. The sum of moments about point 0 is

$$\sum M_0 = -M + (M + \mathrm{d}M) - Q\frac{\mathrm{d}s}{2} - (Q + \mathrm{d}Q)\frac{\mathrm{d}s}{2} - q\,\mathrm{d}s \times \xi + m\,\mathrm{d}s = 0.$$

Here, ξ is the distance from the resultant load q on the element $\mathrm{d}s$ to the point O. It is clear that this is also an infinitely small quantity. After simplification we get

$$\frac{\mathrm{d}M}{\mathrm{d}s} - Q + m = 0. \tag{1.23}$$

Integrate (1.21) (1.23) and limit ourselves to a circular bar with only a uniform radial load q_0. We also assume that loads N, Q, and M act on the ends of the bar, as shown in Fig. 1.20.

In polar coordinates, with the chosen direction of initial forces (1.21)–(1.23) take the form

$$\frac{\mathrm{d}N}{\mathrm{d}\varphi} = Q, \tag{1.21a}$$

$$\frac{\mathrm{d}Q}{\mathrm{d}\varphi} = -N - qR, \tag{1.22a}$$

$$\frac{\mathrm{d}M}{\mathrm{d}\varphi} = QR. \tag{1.23a}$$

Equations (1.21a)–(1.23a) are known as Kirchhoff's equations [Kir76], [Lov20]. Differentiate (1.22a) by φ and take into account (1.21a). The equation becomes

$$\frac{d^2Q}{d\varphi^2} + Q = 0. \tag{1.23b}$$

Boundary conditions:

1. When $\varphi = 0$, shear $Q = Q_0$.
2. When $\varphi = 0$, $(dQ/d\varphi) = -N_0 - q_0R$.

Integrating (1.23b) leads to an expression for the shear force expressed in terms of the initial parameters N_0 and Q_0.

$$Q = Q_0 \cos\varphi - (N_0 + q_0R)\sin\varphi. \tag{1.24}$$

Substitute this expression into (1.21a) and (1.23a). After integrating we get

$$N = N_0 \cos\varphi + Q_0 \sin\varphi - q_0R(1 - \cos\varphi), \tag{1.25}$$

$$M = M_0 + Q_0 R \sin\varphi - \left(N_0R + q_0R^2\right)(1 - \cos\varphi). \tag{1.26}$$

Equations (1.24)–(1.26) allow us to determine the distribution of internal loads in a circular rod of radius R subjected to a uniform hydrostatic radial load q_0 in terms of the initial parameters N_0, Q_0, and M_0.

Another way of calculating the internal forces consists of bringing (1.21)–(1.23) to one resolving equation for a curvilinear rod. From (1.22) we express N as $N = -\rho(dQ/ds) - \rho q$ and substitute this expression into (1.21)

$$-\frac{d}{ds}\left(\rho\frac{dQ}{ds}\right) - \frac{d}{ds}(\rho q) - \frac{Q}{\rho} + p = 0.$$

Substitute the expression for Q from (1.23) into the above equation. As a result we get

$$\frac{d}{ds}\left(\rho\frac{d^2M}{ds^2}\right) + \frac{1}{\rho}\frac{dM}{ds} - p + \frac{d(\rho q)}{ds} + \frac{d}{ds}\left(\rho\frac{dm}{ds}\right) - \frac{m}{\rho} = 0. \tag{1.27}$$

In the case of a rod with a circular shape ($\rho = R = \text{const}$), (1.27) takes the form

$$R\frac{d^3M}{ds^3} + \frac{1}{R}\frac{dM}{ds} - p + R\frac{dq}{ds} + R\frac{d^2m}{ds^2} - \frac{m}{R} = 0. \tag{1.28}$$

Noticing that $ds = R\,d\varphi$, (1.28) can be written in terms of polar coordinates

$$\frac{d^3M}{d\varphi^3} + \frac{dM}{d\varphi} = R^2\left(p - \frac{dq}{d\varphi}\right) + R\left(m - \frac{d^2m}{d\varphi^2}\right). \tag{1.29}$$

Table 1.6 Differential relationships for curvilinear bar

	Internal forces			
Displacement	N	Q	M	Loads
u	d/ds	$-1/\rho$	0	p
υ	$1/\rho$	d/ds	0	q
Ψ	0	-1	d/ds	m
Strain	ε	γ	χ	

Integrating this equation allows us to find an expression for the bending moment as a function of the angle. Following this, we can find an expression for the shear force using (1.23) and then find the axial force using (1.21).

Later we use (1.29) for static analysis of symmetric circular two-hinged arches with constant cross section and modified equation (1.27) for stability analysis of redundant arches.

1.7.2 Relationships Between Displacements and Strains

Equations (1.21)–(1.23) may be presented in tabulated form (Table 1.6) using the differential operators.

The top row contains forces N, Q, and M. The last column contains loads p, q, and m, in accordance to Fig. 1.19. Three subsequent rows represent the three differential equations (1.21)–(1.23). For example, the first row may be rewritten in the form $(\mathrm{d}N/\mathrm{d}s) - (Q/\rho) + 0 \times M + p = 0$, which corresponds to the differential equation (1.21).

The left most column contains the displacements which correspond to the type of load. Tangential displacement u corresponds to longitudinal force p, normal displacement υ corresponds to transversal load q, and the angle of rotation Ψ corresponds to the moment load m.

The bottom row contains the strains of member. Longitudinal strain ε corresponds to the longitudinal force N, shear strain γ corresponds to shear force Q, and additional warp of the axis of the rod χ to its initial curvature corresponds to the bending moment M.

Through this table we can easily understand the relationships between displacements and strains [Rzh82]. To achieve this, we apply a matrix transpose and obtain

$$-\frac{\mathrm{d}u}{\mathrm{d}s} + \frac{\upsilon}{\rho} + 0 \times \psi + \varepsilon = 0,$$

$$-\frac{u}{\rho} - \frac{\mathrm{d}\upsilon}{\mathrm{d}s} - 1 \times \psi + \gamma = 0,$$

$$0 \times u + 0 \times \upsilon - \frac{\mathrm{d}\psi}{\mathrm{d}s} + \chi = 0.$$

From these expressions we get the relations

$$\begin{aligned}
\varepsilon &= \frac{\mathrm{d}u}{\mathrm{d}s} - \frac{\upsilon}{\rho},\\
\gamma &= \frac{u}{\rho} + \frac{\mathrm{d}\upsilon}{\mathrm{d}s} + \psi,\\
\chi &= \frac{\mathrm{d}\psi}{\mathrm{d}s}.
\end{aligned} \tag{1.30}$$

Special Cases

1. Shear strain γ is neglected. In this case (1.30) become

$$\varepsilon = \frac{\mathrm{d}u}{\mathrm{d}s} - \frac{\upsilon}{\rho}; \quad \psi = -\frac{u}{\rho} - \frac{\mathrm{d}\upsilon}{\mathrm{d}s}; \quad \chi = \frac{\mathrm{d}\psi}{\mathrm{d}s} = -\frac{\mathrm{d}}{\mathrm{d}s}\left(\frac{u}{\rho}\right) - \frac{\mathrm{d}^2\upsilon}{\mathrm{d}s^2}. \tag{1.31}$$

2. Shear strain γ and axial strain ε are neglected. Equations (1.31) lead to

$$\upsilon = \rho\frac{\mathrm{d}u}{\mathrm{d}s}; \quad \psi = -\frac{u}{\rho} - \frac{\mathrm{d}}{\mathrm{d}s}\left(\rho\frac{\mathrm{d}u}{\mathrm{d}s}\right); \quad \chi = \frac{\mathrm{d}\psi}{\mathrm{d}s} = -\frac{\mathrm{d}}{\mathrm{d}s}\left(\frac{u}{\rho}\right) - \frac{\mathrm{d}^2}{\mathrm{d}s^2}\left(\rho\frac{\mathrm{d}u}{\mathrm{d}s}\right). \tag{1.32}$$

3. For a straight rod ($\rho = \infty, \quad s = x$). Equations (1.30) become

$$\varepsilon = \frac{\mathrm{d}u}{\mathrm{d}x}; \quad \gamma = \frac{\mathrm{d}\upsilon}{\mathrm{d}x} + \psi; \quad \chi = \frac{\mathrm{d}\psi}{\mathrm{d}x}.$$

   This leads to

$$\varepsilon = \frac{\mathrm{d}u}{\mathrm{d}x}; \quad \psi = \gamma - \frac{\mathrm{d}\upsilon}{\mathrm{d}x}; \quad \chi = \frac{\mathrm{d}\psi}{\mathrm{d}x} = \frac{\mathrm{d}\gamma}{\mathrm{d}x} - \frac{\mathrm{d}^2\upsilon}{\mathrm{d}x^2}. \tag{1.33}$$

4. For a straight elastic rod ($\rho = \infty, \quad s = x; \quad \chi = M/EI; \quad \gamma = \eta(Q/GA)$) the second and third equations of (1.33) lead to

$$\chi = \frac{M}{EI} = \frac{\mathrm{d}\psi}{\mathrm{d}x} = \frac{\mathrm{d}}{\mathrm{d}x}\left(\gamma - \frac{\mathrm{d}\upsilon}{\mathrm{d}x}\right) = \frac{\eta}{GA}\frac{\mathrm{d}Q}{\mathrm{d}x} - \frac{\mathrm{d}^2\upsilon}{\mathrm{d}x^2}$$

   or

$$\frac{\mathrm{d}^2\upsilon}{\mathrm{d}x^2} = -\frac{M}{EI} + \frac{\eta}{GA}\frac{\mathrm{d}Q}{\mathrm{d}x}.$$

5. For a circular bar ($\rho = R = \text{const}, \quad \mathrm{d}s = R\,\mathrm{d}\varphi$) (1.30)–(1.32) lead to the following results:

$$\varepsilon = \frac{1}{R}\left(\frac{\mathrm{d}u}{\mathrm{d}\varphi} - \upsilon\right); \quad \gamma = \frac{1}{R}\left(u + \frac{\mathrm{d}\upsilon}{\mathrm{d}\varphi}\right) + \psi; \quad \chi = \frac{1}{R}\frac{\mathrm{d}\psi}{\mathrm{d}\varphi}, \tag{1.30a}$$

$$\varepsilon = \frac{1}{R}\left(\frac{\mathrm{d}u}{\mathrm{d}\varphi} - \upsilon\right); \quad \psi = -\frac{1}{R}\left(u + \frac{\mathrm{d}\upsilon}{\mathrm{d}\varphi}\right); \quad \chi = \frac{\mathrm{d}\psi}{\mathrm{d}s} = -\frac{1}{R^2}\left(\frac{\mathrm{d}u}{\mathrm{d}\varphi} + \frac{\mathrm{d}^2\upsilon}{\mathrm{d}\varphi^2}\right), \tag{1.31a}$$

$$\upsilon = \frac{\mathrm{d}u}{\mathrm{d}\varphi}; \quad \psi = -\frac{1}{R}\left(u + \frac{\mathrm{d}^2u}{\mathrm{d}\varphi^2}\right); \quad \chi = -\frac{1}{R^2}\left(\frac{\mathrm{d}u}{\mathrm{d}\varphi} + \frac{\mathrm{d}^3u}{\mathrm{d}\varphi^3}\right) = -\frac{1}{R^2}\left(\upsilon + \frac{\mathrm{d}^2\upsilon}{\mathrm{d}\varphi^2}\right). \tag{1.32a}$$

1.7.3 Lamb's Equation

Let us consider a circular rod of a constant cross-section, subjected to uniform radial load q; in axial direction the rod is nondeformable. In this case, the equilibrium equations can be combined into one sixths order ordinary differential equations with constant coefficients, relative to tangential displacements u

$$\frac{\mathrm{d}^6u}{\mathrm{d}\varphi^6} + 2\frac{\mathrm{d}^4u}{\mathrm{d}\varphi^4} + \frac{\mathrm{d}^2u}{\mathrm{d}\varphi^2} + \frac{qR^3}{EI}\left(\frac{\mathrm{d}^4u}{\mathrm{d}\varphi^4} + \frac{\mathrm{d}^2u}{\mathrm{d}\varphi^2}\right) = 0,$$

where angle φ determines the position of the point on the nondeformable axis of the rod. This equation is known as Lamb's equation [Lam1888], [Rek73], [Rzh55].

Let the central angle be 2α. If the angle φ is measured from vertical line, then $-\alpha \le \varphi \le \alpha$. The positive displacement u is directed along the tangent of the circle in the direction in which φ increases.

Boundary conditions: For the fixed end, the tangential and radial displacements, and slope are $u = 0, \quad \partial u/\partial\varphi = 0, \quad \partial^2u/\partial\varphi^2 = 0$, respectively. For hinged end $u = 0, \quad \partial u/\partial\varphi = 0, \quad \partial^3u/\partial\varphi^3 = (R^2/EI)M_s$. The last condition takes into account external moment M_s at the support.

Lamb's equation will be used for static and dynamic stability analysis of a circular uniform arches.

Another form of the solution is possible; one can integrate equations ((1.21)–(1.23)) and take the constant of integration to represent the initial parameters [Bir68]. It is clear that the corresponding solution will represent a generalization of the initial parameter method (Sect. 1.2) for circular uniform rod. The books [Bir68, vol. 1], [Uma72-73], [Roa75] contain numerous tabulated data for computation of internal forces and displacements of circular uniform arches for cases of in-plane and out-of-plane loading.

Note the equation of plane curvilinear rod in a general case (nonuniform cross section, variable radius of curvature) may be found in [Rzh55]. This differential equation of the sixth order with respect to tangential displacement u takes into account the tangential and radial distributed loads.

1.8 Reciprocal Theorems

Reciprocal theorems reflect fundamental properties of any linear statistical determinate or indeterminate elastic systems. These theorems find extensive application in the analysis of redundant structures [Kar10]. Primary investigations were performed by Betti (1872), Maxwell (1864), Lord Rayleigh (1873–1875), Castigliano (1872), and Helmholtz (1886) [Tim53], [Tod60].

1.8.1 Theorem of Reciprocal Works (Betti Theorem)

Let us consider elastic structure subjected to loads P_1 and P_2 separately; let us call it as first and second states (Fig. 1.21). Set of displacements Δ_{mn} for each state are shown below. The first index m indicates the direction of the displacement and the second index n denotes the load, which causes this displacement.

Thus, Δ_{11} and Δ_{12} are displacements in the direction of load P_1 due to load P_1 and P_2, respectively, Δ_{21} and Δ_{22} are displacements in the direction of load P_2 due to load P_1 and P_2, respectively.

Let us calculate the strain energy of the system by considering consequent applications of loads P_1 and P_2, i.e., state 1 is *additionally* subjected to load P_2. Total work done by both of these loads consists of three parts:

1. Work done by the force P_1 on the displacement Δ_{11}. Since load P_1 is applied *statically* (from zero to P_1 according to triangle law), then $W_1 = P_1\Delta_{11}/2$.
2. Work done by the force P_2 on the displacement Δ_{22}. Since load P_2 is applied statically, then $W_2 = P_2\Delta_{22}/2$.
3. Work done by the force P_1 on the displacement Δ_{12}; this displacement is caused by load P_2. The load P_1 approached its maximum value P_1 before application of P_2. Corresponding P_1–Δ_1 diagram is shown in Fig. 1.21, so $W_3 = P_1\Delta_{12}$.

Since potential energy U equals to the total work, then

$$U = \frac{1}{2}P_1\Delta_{11} + \frac{1}{2}P_2\Delta_{22} + P_1\Delta_{12}. \tag{1.34}$$

On the other hand, considering of application of load P_2 first and then P_1, i.e., if state 2 is additionally subjected to load P_1, then potential energy U equals

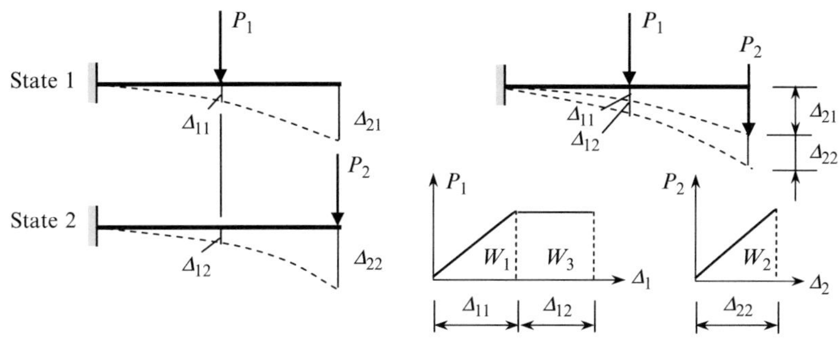

Fig. 1.21 Two state of the elastic structure. Computation of work done by the load P_1 and additional load P_2

$$U = \frac{1}{2}P_2\Delta_{22} + \frac{1}{2}P_1\Delta_{11} + P_2\Delta_{21}. \tag{1.35}$$

Since strain energy does not depend on the order of loading, then the following fundamental relationship is obtained

$$P_1\Delta_{12} = P_2\Delta_{21} \quad \text{or} \quad W_{12} = W_{21}, \tag{1.36}$$

The theorem of reciprocal works (1.36) said that *in any elastic system the work performed by load of state 1 along displacement caused by load of state 2 equals to work performed by load of state 2 along displacement caused by load of state 1.*

1.8.2 Theorem of Reciprocal Displacements (Maxwell Theorem)

Let us consider two states of elastic structure subjected to *unit loads* $P_1 = 1$ and $P_2 = 1$. Displacements caused by unit loads is called the *unit displacements* and denoted by letter δ_{mn}. The first index m indicates the direction of the displacement and the second index n denotes the unit load, which causes this displacement.

Thus, δ_{11} and δ_{12} are displacements in the direction of load P_1 due to load $P_1 = 1$ and $P_2 = 1$, respectively; δ_{21} and δ_{22} are displacements in the direction of load P_2 due to load $P_1 = 1$ and $P_2 = 1$, respectively.

In case of unit loads, the theorem of reciprocal works $P_2\Delta_{21} = P_1\Delta_{12}$ leads to the following fundamental relationship $\delta_{12} = \delta_{21}$. In general,

$$\delta_{nm} = \delta_{mn}. \tag{1.37}$$

This equation shows that *in any elastic system, displacement along nth load caused by unit mth load equals to displacement along mth load caused by unit nth load*. The term "displacement" refers to linear or angular displacements, and the term "load" means force or moment.

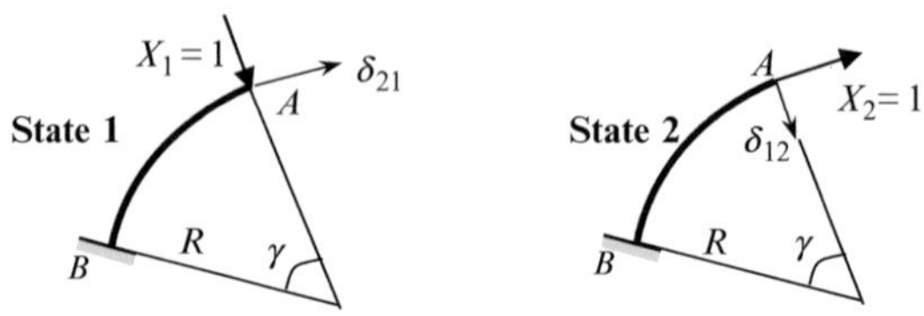

Fig. 1.22 Theorem of reciprocal unit displacements

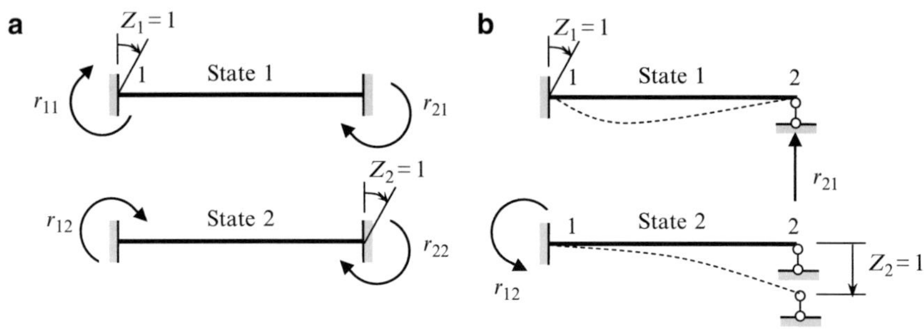

Fig. 1.23 Theorem of reciprocal unit reactions: (**a**) clamped–clamped beam; (**b**) clamped–pinned beam

This theorem is demonstrated by the following example. Fixed-free circular bar is subjected to unit *radial* load X_1 in the first state and unit *axial* load X_2 in the second state (Fig. 1.22). According to Table A.7, we get $\delta_{12} = \delta_{21} = (R^3/2EI)(1 - \cos\gamma)^2$. This table contains expressions for other unit displacements.

Theorem of reciprocal displacements will be widely used for analysis of redundant arches by the Force method.

1.8.3 Theorem of Reciprocal Reactions (Rayleigh First Theorem)

Let us consider two states of elastic structure subjected to *unit displacements* of supports. They are $Z_1 = 1$ and $Z_2 = 1$ (Fig. 1.23a). Reactions caused by unit displacements are called the *unit reactions* and denoted by r_{mn}. The first index m indicates constrain where unit reaction arises and the second index n denotes constrain, which is subjected to unit displacement.

Thus r_{11} and r_{12} are reactions in the constrain 1 due to displacement $Z_1 = 1$and $Z_2 = 1$, respectively; r_{21} and r_{22} are reactions in the constrain 2 due to displacement $Z_1 = 1$ and $Z_2 = 1$, respectively.

The theorem of reciprocal works $r_{11} \times 0 + r_{21} \times 1 = r_{12} \times 1 + r_{22} \times 0$ leads to the following relationships $r_{21} = r_{12}$. In general

$$r_{nm} = r_{mn}. \tag{1.38}$$

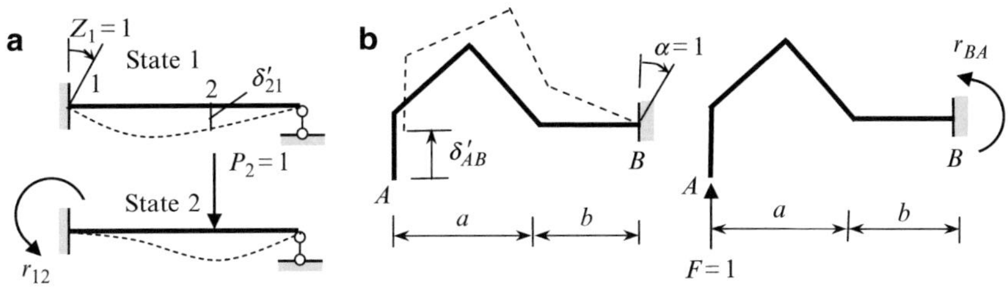

Fig. 1.24 Theorem of reciprocal of unit displacements and reactions

The theorem of reciprocal reactions said that *in any elastic system reaction* r_{nm}, *which arises in nth constrain due to unit displacement of constrain mth, equals reaction* r_{mn}, *which arises in mth constrain due to unit displacement of constrain nth.*

This is demonstrated by the following example (Fig. 1.23b). Unit displacements of the clamped–pinned beam are as follows: $Z_1 = 1$ is a unit angle of rotation of the clamped support and $Z_2 = 1$ is a vertical linear displacement of the pinned support. Unit reactions are as follows: r_{21} is a vertical reaction in constrain 2 caused by unit angular displacement of support 1 and r_{12} is a moment in constrain 1 caused by unit vertical linear displacement of support 2. It is known [Kar10] that unit reactions are $r_{21} = r_{12} = 3EI/l^2$.

1.8.4 Theorem of Reciprocal Displacements and Reactions (Rayleigh Second Theorem)

Let us consider two states of elastic structure subjected to *unit displacement* $Z_1 = 1$ and *unit load* $P_2 = 1$ (Fig. 1.24a). Displacement δ'_{21} occurs in direction of load P_2 due to unit *displacement* Z_1. Reaction r_{12} arises in constrain 1 due to unit load P_2.

The theorem of reciprocal work in extended form should be presented as follows:

$$-r_{12} \times 1 = 1 \times \delta'_{21}$$

so we get that $-r_{12} = \delta'_{21}$. In general,

$$r_{jk} = -\delta'_{kj}. \tag{1.39}$$

The theorem of reciprocal unit displacements and reactions said that *reaction in jth constrain due to unit load of kth direction and displacement in kth direction due to unit displacement of jth constrain are equal in magnitude but opposite in sign.*

This theorem is illustrated in Fig. 1.24b. In order to find a vertical displacement at the point A due to unit rotation of the support B, apply unit force $F = 1$ along required displacement. Moment at fixed support due to force $F = 1$ is $r_{BA} = -F\ (a + b)$. Since $F = 1$, therefore the vertical displacement is $\delta'_{AB} = a + b$.

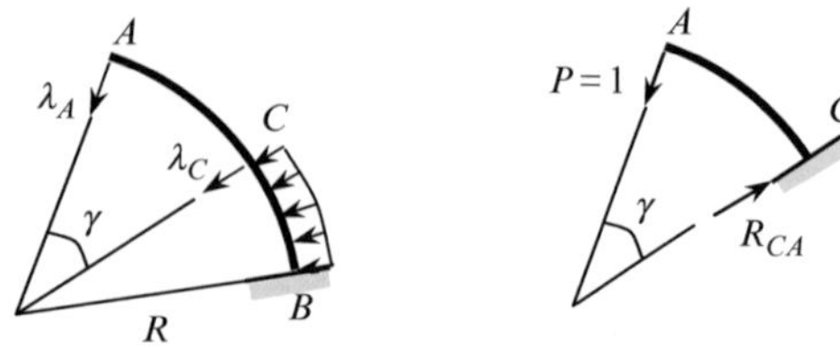

Fig. 1.25 Design diagram of the fixed-free circular rod

1.8.5 Transfer Matrix

Let us consider fixed-free circular rod of radius R loaded by a radial distributed load within the portion CB (Fig. 1.25). The radial, tangential, and angular displacements λ_C, ξ_C, ψ_C of the section C are known. The problem is to determine the displacements of the free end A, assuming that portion AC displaces as absolutely rigid body.

Apply the unit force $P = 1$ in direction of λ_A. Reaction R_{CA}, which corresponds to displacement λ_C is $R_{CA} = -P \cos\gamma$. The negative sign shows that direction of R_{CA} is opposite to the displacement λ_C. According to theorem of reciprocal displacements and reactions, we get $\lambda_A = \lambda_C \cos\gamma$. Similarly, the vector displacement at the section A in terms of transfer matrix and displacement vector of the section C may be represented as follows [Uma72]:

$$\begin{bmatrix} \lambda \\ \xi \\ \psi \end{bmatrix}_A = \begin{bmatrix} \cos\gamma & -\sin\gamma & R\sin\gamma \\ \sin\gamma & \cos\gamma & R(1-\cos\gamma) \\ 0 & 0 & 1 \end{bmatrix} \cdot \begin{bmatrix} \lambda \\ \xi \\ \psi \end{bmatrix}_C. \tag{1.40}$$

1.9 Boussinesq's Equation

Boussinesq's differential equation (1883) describes the behavior of a circular rod with constant cross-sectional dimensions.

1.9.1 Two Forms of Boussinesq's Equation

Let us consider a circular rod of radius R. For deriving of differential equation we consider two cases.

Case 1. Axial force is neglected. Bending moment $M = \chi EI$; taking into account equation (1.32a) we get the differential equation with respect to radial displacement

$$\frac{d^2 \upsilon}{d\varphi^2} + \upsilon = -\frac{MR^2}{EI}. \tag{1.41}$$

Let the clamped-free rod is subjected to uniform radial load q (Fig. 1.20). Expression (1.26) for bending moment $M = M_0 + Q_0 R \sin\varphi - (N_0 R + q_0 R^2) \times (1 - \cos\varphi)$ is substituted into (1.41) and integrated with respect to φ. We get expression for radial displacement in terms of initial parameters

$$\upsilon = \upsilon_0 \cos\varphi + \theta_0 R \sin\varphi - u_0 \sin\varphi - M_0 c(1 - \cos\varphi) - Q_0 cR \\ \times \frac{\sin\varphi - \varphi\cos\varphi}{2} - (q_0 R - N_0) cR\beta, \tag{1.42}$$

$$c = \frac{R^2}{EI}, \quad \beta = 1 - \cos\varphi - \frac{\varphi \sin\varphi}{2}.$$

We can apply this expression for hingeless arch subjected to uniformly distributed radial load q_0. Since the initial parameters are $\upsilon_0 = \theta_0 = u_0 = 0$ and $M_0 = Q_0 = 0, \quad N_0 = -q_0 R$, then for radial displacement we get

$$\upsilon = -(q_0 R - N_0) cR\beta = -2q_0 \frac{R^4}{EI}\beta.$$

If a central angle is 180° then for $\varphi = 90°$ (crown) parameter $\beta = 0.2146$ and $\upsilon = 0.4292(q_0 R^4/EI)$.

Case 2. Axial force is taken into account. According to (1.31a), the curvature χ of the rod, in addition to the initial curvature is

$$\chi = \frac{d\psi}{ds} = -\frac{1}{R^2}\left(\frac{du}{d\varphi} + \frac{d^2\upsilon}{d\varphi^2}\right)$$

so the expression for bending moment becomes

$$M = \chi EI = -\frac{EI}{R^2}\left(\frac{du}{d\varphi} + \frac{d^2\upsilon}{d\varphi^2}\right). \tag{a}$$

The axial deformation is $\varepsilon = (1/R)((du/d\varphi) - \upsilon)$, so

$$N = \varepsilon EA = \frac{EA}{R}\left(\frac{du}{d\varphi} - \upsilon\right). \tag{b}$$

From (a) and (b), we extract the $du/d\varphi$ term and equate the right-hand sides. As a result we arrive at the Boussinesq equation [Sni66]

$$\frac{d^2\upsilon}{d\varphi^2} + \upsilon = -\frac{MR^2}{EI} - \frac{NR}{EA}. \tag{1.43}$$

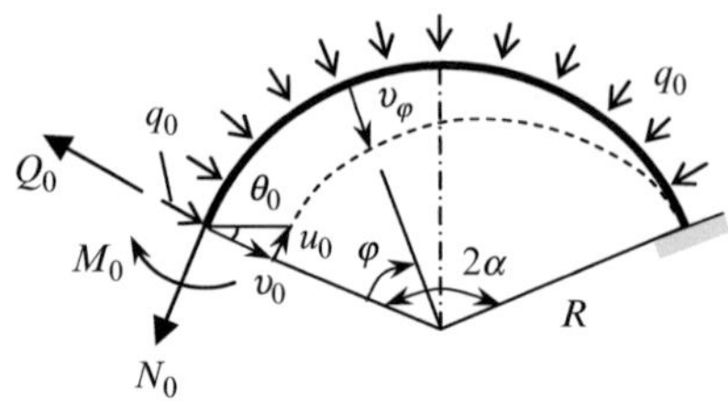

Fig. 1.26 Design diagram of uniform circular rod

Table 1.7 Parameter f in terms of λ and α

	α			
λ	$\pi/6$	$\pi/4$	$\pi/3$	$\pi/2$
0.001	0.9984	0.9995	0.9994	1.0
0.005	0.9997	0.9975	0.9969	1.0
0.01	0.9968	0.9950	0.9938	1.0
0.05	0.9834	0.9742	0.9680	1.0

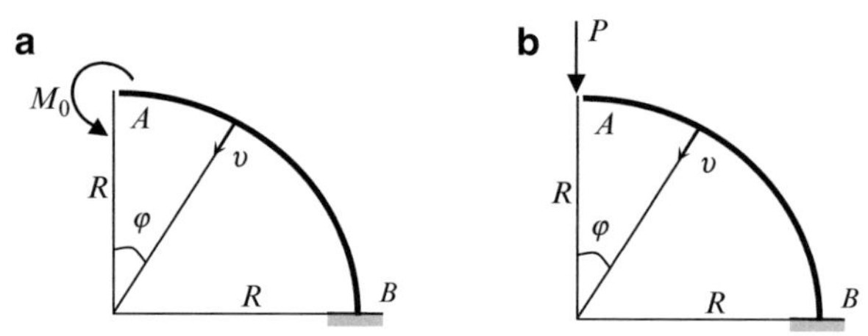

Fig. 1.27 Design diagram of a circular uniform rod

If we substitute the expressions for bending moment (1.26) and axial force (1.25) into (1.42) and integrate, we obtain an expression for radial displacement υ. Using relationship $\upsilon = \mathrm{d}u/\mathrm{d}\varphi$, we can find the tangential displacement u and then we can find the angle of rotation ψ and additional warp of the axis of the rod χ to its initial curvature. For this we use the second and third equation of (1.32a). We employ an important result for design diagrams in Fig. 1.26. Assume the central angle is 2α and the cross-sectional area is A.

The axial force works out to be $N_0 = -q_0Rf$, where $f = (n - \lambda(\alpha + 1))/(n - \lambda)$ and $\lambda = (r^2/R^2)$, $r^2 = I/A$, $n = (\alpha - 1)\tan\alpha - 1 + (\alpha + 2\tan\alpha)\cot\alpha$. Table 1.7 presents the parameter f in terms of λ and α.

In a large range of values of λ and α, the parameter f is very close to unity, so we can take Boussinesq's equation to be in the form of (1.41), with a high degree of precision.

1.9.2 Displacements of a Circular Rod

We use Boussinesq's equation to determine radial displacements of a uniform circular rod with a central angle of 90º (Fig. 1.27a, b).

Loading in Fig. 1.27a

Bending moment $M = -M_0$, and Bousssinesq equation (1.41) takes the form

$$\frac{d^2 \upsilon}{d\varphi^2} + \upsilon = \frac{M_0 R^2}{EI}. \tag{1.44}$$

This equation has the solution $\upsilon = \upsilon_1 + \upsilon^*$. Since the right-hand side of the equation in constant, we look for a particular solution in the form $\upsilon^* = k$, where k is a some unknown constant. To determine the unknown constant, substitute this expression back into the differential equation. As a result we get $k = (M_0 R^2)/EI$, and the full solution is $\upsilon = A \cos \varphi + B \sin \varphi + k$, where A and B are constants of integration.

Boundary conditions:

1. When $\varphi = \pi/2$ the radial displacement is $\upsilon = 0$. This condition leads to $B = -k = -(M_0 R^2)/EI$.
2. When $\varphi = \pi/2$ (fixed support) we have $d\upsilon/d\varphi = 0$, so $-A \sin \varphi + B \cos \varphi = 0$. This condition leads to $A = 0$.

The final expression for the radial displacement is

$$\upsilon = \frac{M_0 R^2}{EI}(1 - \sin \varphi).$$

On the free end the vertical displacement is $\Delta_{\text{vert}} = \dfrac{M_0 R^2}{EI}(\downarrow)$.

Loading in Fig. 1.27b

In this case, the bending moment $M = -PR \sin \varphi$ and Boussinesq' equation (1.41) is

$$\frac{d^2 \upsilon}{d\varphi^2} + \upsilon = -\frac{M_\varphi R^2}{EI} = \frac{PR^3}{EI} \sin \varphi. \tag{1.45}$$

Its solution has the form $\upsilon = \upsilon_1 + \upsilon^*$. Note that the coefficient of φ in the right-hand side and the second term in the left-hand side are both equal to unity ($1 \times \upsilon$ and $\sin(1 \times \varphi)$). So we look for a particular solution in the form of $\upsilon^* = k\varphi \cos \varphi$. Substitute this expression into the differential equation to determine the unknown coefficient. As a result we obtain $k = -(PR^3/2EI)$, and the full solution is $\upsilon = A \cos \varphi + B \sin \varphi - (PR^3/2EI)\varphi \cos \varphi$.

Boundary conditions:

1. When $\varphi = \pi/2$ the radial displacement is $\upsilon = 0$. This condition leads to $B = 0$.
2. When $\varphi = \pi/2$

$$\frac{d\upsilon}{d\varphi} = -A \sin \varphi - k \cos \varphi + k\varphi \sin \varphi = 0$$

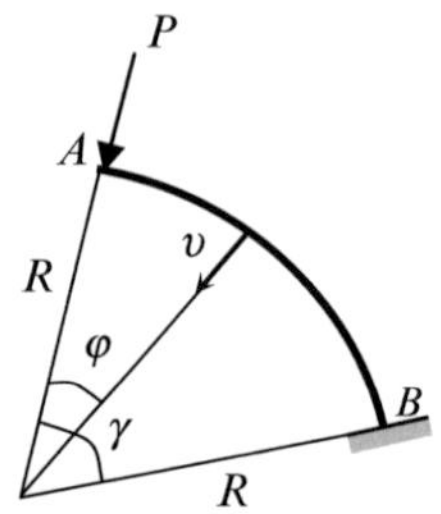

Fig. 1.28 Uniform circular rod carrying the radial force P

and this leads to $A = P\pi R^3/4EI$. The final expression of the radial displacement is given by

$$\upsilon = \frac{P\pi R^3}{4EI}\left(1 - 2\frac{\varphi}{\pi}\right)\cos\varphi.$$

On the free end the vertical displacement is $\Delta_{\text{vert}} = \dfrac{P\pi R^3}{4EI}(\downarrow)$.

Now let us consider a more general case. A radial load P is applied to the free end of the uniform clamped-free rod of radius R and central angle γ (Fig. 1.28).

Bending moment is $M = -PR\sin\varphi$. Boussinesq' equation (1.41) becomes

$$\frac{d^2\upsilon}{d\varphi^2} + \upsilon = -\frac{M_\varphi R^2}{EI} = \frac{PR^3}{EI}\sin\varphi.$$

Its solution gain has the form $\upsilon = \upsilon_1 + \upsilon^*$. As before, we look for a particular solution in the form $\upsilon^* = k\varphi\cos\varphi$. To determine the unknown constant we plug this expression back into the differential equation. As a result we get $k = -(PR^3/2EI)$, and the full solution becomes

$$\upsilon = A\cos\varphi + B\sin\varphi - \frac{PR^3}{2EI}\varphi\cos\varphi. \tag{a}$$

Boundary conditions:

1. When $\varphi = \gamma$, the radial displacement of support B is $\upsilon = 0$, so

$$\upsilon = A\cos\gamma + B\sin\gamma - \frac{PR^3}{2EI}\gamma\cos\gamma = 0. \tag{b}$$

2. When $\varphi = \gamma$ the slope $d\upsilon/d\varphi = 0$. The slope at any point is

$$\frac{d\upsilon}{d\varphi} = -A\sin\varphi + B\cos\varphi - \frac{PR^3}{2EI}(\cos\varphi - \varphi\sin\varphi),$$

$$\frac{d\upsilon}{d\varphi}\Big|_{\varphi=\gamma} = -A\sin\gamma + B\cos\gamma - \frac{PR^3}{2EI}(\cos\gamma - \gamma\sin\gamma) = 0. \qquad \text{(c)}$$

Solution of (b) and (c) are

$$B = \frac{P\pi R^3}{2EI}\cos^2\gamma,$$
$$A = -\frac{PR^3}{2EI}(\cos\gamma\sin\gamma - \gamma).$$

Radial displacement becomes

$$\upsilon(\varphi) = \frac{PR^3}{2EI}\left[(\gamma - \cos\gamma\sin\gamma)\cos\varphi + \cos^2\gamma\sin\varphi - \varphi\cos\varphi\right]. \qquad \text{(d)}$$

Special Cases

1. For free end of the bar $\varphi = 0$. In this case formula (d) becomes

$$\upsilon(0) = \frac{PR^3}{2EI}(\gamma - \cos\gamma\sin\gamma) = \frac{PR^3}{EI}\left(\frac{\gamma}{2} - \frac{\sin 2\gamma}{4}\right).$$

2. Let the central angle be $\gamma = \pi/2$. In this case for any φ we get

$$\upsilon(\varphi) = \frac{P\pi R^3}{4EI}\left(1 - 2\frac{\varphi}{\pi}\right)\cos\varphi.$$

3. Let $\gamma = \pi/2$ and $\varphi = 0$. In this case $\Delta_{\text{vert}} = (P\pi R^3/4EI)(\downarrow)$.

We can derive expressions for displacement in a similar manner when we are dealing with different types of loads. The most important types of loads and the corresponding results are presented in Tables A.6 and A.7.

Discussion

It is evident that the differential relations allow us to find the displacement equation for an arbitrary section, while the Mohr integral allows us to determine displacement at a fixed section.

Detailed analysis of uniform two-hinged circular arch subjected to a single force is presented in [Tim72]. Numerous reference data, related to circular and elliptical rings, subjected to different loads is presented in [Roa75], [You89], [Bir68, vol. 1].

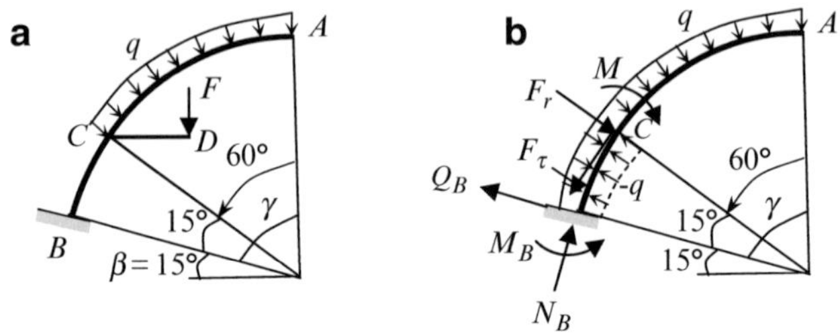

Fig. 1.29 Design diagram of circular bar and its equivalent presentation

Example 1.8. The circular rod *AB* of radius $R = 24$ m and central angle $\gamma = 75^{\circ}$ together with cantilever *CD* of length 6 m is loaded as shown in Fig. 1.29a. The structure is subjected to a concentrated force $F = 10$ kN and radial uniformly distributed load $q = 2$ kN/m. (a) Determine the reactions of support *B*; (b) determine the displacements at point *C* and at the free end *A*.

Solution. We first perform some preliminary operations. Transfer the force *F* to the point *C*; additional clockwise couple is $M = F \times CD = 60\,\text{kN m}$. Resolve a force *F* at point *C* into the radial and axial components $F_{\text{rad}} = F \sin(15 + \beta) = 5\,\text{kN}$ and $F_\tau = -F \cos(15 + \beta) = -8.66\,\text{kN}$. The negative sign corresponds to the data in Table A.6. Figure 1.29b presents the initial design diagram in the equivalent form: the load *q* acts within the entire rod *AB*, and load $(-q)$ within the portion *BC*; also, the axial force F_τ, radial force F_{rad}, and couple *M* act at point *C*.

Reactions at the support B. According to the principle of superposition and data in Tables A.6 and A.7 we get (A.6, 1–1 means Appendix Table A.6, column 1, row 1)

$$\begin{aligned}Q_B &= Q(q) + Q(-q) + Q(F_{\text{rad}}) + Q(F_\tau) + Q(M)\\ &= \underbrace{qR\sin 75^\circ}_{\text{A.6,1-1}} - \underbrace{qR\sin 15^\circ}_{\text{A.6,1-1}} + \underbrace{F_{\text{rad}}\cos 15^\circ}_{\text{A.7,1-1}} + \underbrace{F_\tau \sin 15^\circ}_{\text{A.7,2-1}} + \underbrace{0}_{\text{A.7,3-1}} = 36.53\,\text{kN},\end{aligned}$$

$$\begin{aligned}N_B &= N(q) + N(-q) + N(F_{\text{rad}}) + N(F_\tau) + N(M)\\ &= \underbrace{-qR(1-\cos 75^\circ)}_{\text{A.6,1-2}} + \underbrace{qR(1-\cos 15^\circ)}_{\text{A.6,1-2}} - \underbrace{F_{\text{rad}}\sin 15^\circ}_{\text{A.7,2-2}} + \underbrace{F_\tau \cos 15^\circ}_{\text{A.7,3-2}} + \underbrace{0}_{\text{A.7,3-3}}\\ &= -43.6\,\text{kN},\end{aligned}$$

$$\begin{aligned}M_B &= \underbrace{qR^2(1-\cos 75^\circ)}_{\text{A.6,1-3}} - \underbrace{qR^2(1-\cos 15^\circ)}_{\text{A.6,1-3}} + \underbrace{F_{\text{rad}}R\sin 15^\circ}_{\text{A.7,1-3}} + \underbrace{F_\tau R(1-\cos 15^\circ)}_{\text{A.7,2-3}} + \underbrace{M}_{\text{A.7,3-3}}\\ &= 898.5\,\text{kNm}.\end{aligned}$$

Displacements at point C. Displacement components of the point *C* caused by loads which act on the portion *BC* $\left(F_{\text{rad}},\ F_\tau,\ M,\ -q\right)$ are presented in Table 1.8; for computation of displacements we use the principle of superposition and data in Tables A.6 and A.7.

Table 1.8 Computation of displacements at point C

	Radial $\lambda(15°)$	Axial $\xi(15°)$	Slope $\psi(15°)$
Influence F_{rad}	$F_{rad} \times \delta_{11}(15°)$	$F_{rad} \times \delta_{21}(15°)$	$F_{rad} \times \delta_{31}(15°)$
	$= F_{rad} \times R^2 \frac{R}{EI}\left(\frac{1}{2}\frac{\pi}{12} - \frac{\sin 30°}{4}\right)$	$= F_{rad} \times R^2 \frac{R}{EI}\frac{(1-\cos 15°)^2}{2}$	$= F_{rad} \times R\frac{R}{EI}(1-\cos 15°)$
	$= \frac{407.8}{EI}$ (From A.7, 1 – 4)	$= \frac{40.1}{EI}$ (From A.7, 1 – 5)	$= \frac{98.1}{EI}$ (From A.7, 1 – 6)
Influence F_{axial}	$F_\tau \times \delta_{12}(15°)$	$F_\tau \times \delta_{22}(15°)$	$F_\tau \times \delta_{32}(15°)$
	$= F_\tau \times R^2 \frac{R}{EI}\frac{(1-\cos 15°)^2}{2}$	$= F_\tau \times R^2 \frac{R}{EI}\left(\frac{3}{2}\frac{\pi}{12} + \frac{\sin 30°}{4} - 2\sin 15°\right)$	$= F_\tau \times R\frac{R}{EI}\left(\frac{\pi}{12} - \sin 15°\right)$
	$= -\frac{69.5}{EI}$ (From A.7, 2 – 4)	$= -\frac{7.411}{EI}$ (From A.7, 2 – 5)	$= -\frac{14.9}{EI}$ (From A.7, 2 – 6)
Influence M	$M \times \delta_{13}(15°)$	$M \times \delta_{23}(15°)$	$M \times \delta_{33}(15°)$
	$= M \times R\frac{R}{EI}(1-\cos 15°)$	$= F_\tau \times R\frac{R}{EI}\left(\frac{\pi}{12} - \sin 15°\right)$	$= M \times R\frac{R}{EI}\frac{\pi}{12}$
	$= \frac{1,177.6}{EI}$ (From A.7, 3 – 4)	$= \frac{103}{EI}$ (From A.7, 3 – 5)	$= \frac{377}{EI}$ (From A.7, 3 – 6)
Influence $(-q)$	$-qR^3\frac{R}{EI}\frac{(1-\cos 15°)^2}{2}$	$-qR^3\frac{R}{EI}\left(\frac{3}{2}\frac{\pi}{12} - 2\sin 15° + \frac{\sin 30°}{4}\right)$	$-qR^2\frac{R}{EI}\left(\frac{\pi}{12} - \sin 30°\right)$
	$= -\frac{385.2}{EI}$ (From A.6, 1 – 4)	$= -\frac{41.1}{EI}$ (From A.6, 1 – 5)	$= -\frac{82.4}{EI}$ (From A.6, 1 – 6)
Total	$\lambda_{rad}(15°) = \lambda^C_{rad}$	$\lambda_{rad}(15°) = \lambda^C_{rad}$	$\psi(15°) = \psi^C = \frac{377.8\,\text{kN m}^2}{EI}$
	$= \frac{1,307\,\text{kN m}^3}{EI}$	$= \frac{94.635\,\text{kN m}^3}{EI}$	

From A.7, 1–4 = Appendix, Table A.7, column 1, row 4. Unit "kN m³" is related to the numerator

Displacements at the free end. ($\gamma_{AB} = 75° = 5\pi/12; \quad \gamma_{AC} = 60° = \pi/3$).
For calculation of displacements at the free end we use the transfer matrix (1.40)

$$\begin{aligned}\lambda_A &= \lambda_C \cos \gamma_{AC} - \xi_C \sin \gamma_{AC} + \psi_C R \sin \gamma_{AC} + F_1(q)\\ \xi_A &= \lambda_C \sin \gamma_{AC} + \xi_C \cos \gamma_{AC} + \psi_C R(1 - \cos \gamma_{AC}) + F_2(q)\\ \psi_A &= \psi_C + F_3(q).\end{aligned}$$

Functions $F_i(q)$ take into account load q within all length AB. In our case

$$\begin{aligned}\lambda_A =& \frac{1130.7}{EI} \cos 60° - \frac{94.63}{EI} \sin 60° + \frac{377.8}{EI} R \sin 60°\\ &+ \underbrace{qR^3 \frac{R}{EI} \frac{(1 - \cos 75°)^2}{2}}_{\text{A.6, 1–4}} = \frac{190,597\,\text{kNm}^3}{EI}\end{aligned}$$

$$\begin{aligned}\xi_A =& \frac{1130.7}{EI} \sin 60° + \frac{94.63}{EI} \cos 60° + \frac{377.8}{EI} R(1 - \cos 60°)\\ &+ \underbrace{qR^3 \frac{R}{EI} \left(\frac{3}{2} \times \frac{5\pi}{12} - 2 \sin 75° + \frac{\sin 150°}{4}\right)}_{\text{A.6,1–5}} = \frac{109,504\,\text{kNm}^3}{EI}\end{aligned}$$

$$\psi_A = \frac{377.8}{EI} + \underbrace{qR^2 \frac{R}{EI} \left(\frac{5\pi}{12} - \sin 75°\right)}_{\text{A.6,1–6}} = \frac{9,863\,\text{kNm}^2}{EI}.$$

These values will be used for analysis of symmetrical circular uniform arch in Sect. 3.10.2.

Chapter 2
Three-Hinged Arches

This chapter is devoted to the analysis of statically determinate three-hinged arches, subjected to fixed and moving loads. Analysis of an arch in the case of fixed loads implies determination of reactions of supports and construction of internal force diagrams. Analysis of an arch in the case of moving load implies construction of influence lines for reactions, thrust, and internal forces.

Some important concepts are discussed. Among them are a reference beam, thrust, nil points of influence lines, etc. Analytical formulas for computation of internal forces as well as for construction of influence lines for reactions and internal forces are developed. Special types of arches are considered; among them are arches with simple and complex ties, arches with support points on different levels. Analysis of the multispan arched structure and truss enforced by arched chain are discussed.

Fundamental investigation in the area of static analysis of arches is attributed to Bresse [Bre59], Kirchhoff [Kir76], and Winkler [Tim53] to name a few.

2.1 General

Idealized design diagram of the arch without overarched members is shown in Fig. 2.1a. This diagram contains two curvilinear members which are hinged together at the crown; connections of curvilinear members with abutment are also hinged. These three hinges are distinguishing features of the three-hinged arch. Design diagram also contains information about the shape of the neutral line of the arch. Usually, this shape is given by an expression of the form $y = f(x)$. Expressions for some characteristic shapes are presented in Tables A.1 and A.2.

Degrees of freedom of the arch in Fig. 2.1a, according to Chebushev formula [Kar10], are determined by the formula

$$W = 3D - 2H_0 - S_0 = 3 \times 2 - 2 \times 1 - 4 = 0, \tag{2.1}$$

I.A. Karnovsky, *Theory of Arched Structures: Strength, Stability, Vibration*, DOI 10.1007/978-1-4614-0469-9_2, © Springer Science+Business Media, LLC 2012

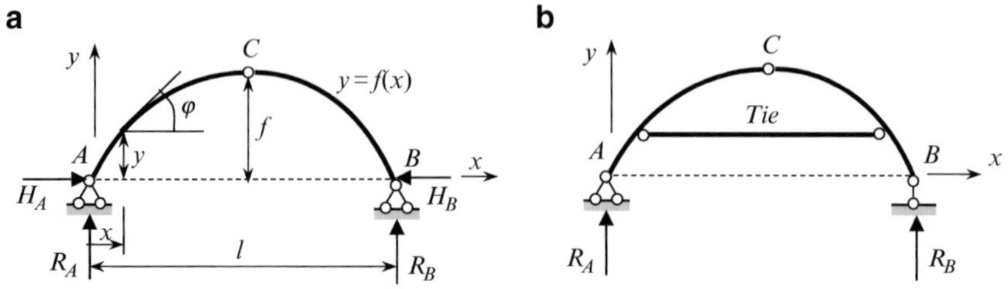

Fig. 2.1 (**a**, **b**) Design diagram of three-hinged arch without tie and with elevated tie

where D, H_0, and S_0 are the number of rigid discs, the number of simple hinges, and the number of constraints of support, respectively. Since $W = 0$, this structure does not have redundant constraints, while all existing constraints constitute the geometrically unchangeability. Indeed, two rigid discs AC and BC are connected with the ground by two hinges A and B and line AB does not pass through the intermediate hinge C.

This structure has four unknown reactions, i.e., two vertical reactions R_A, R_B and two horizontal reactions H_A, H_B. For their determination, three equilibrium equations can be formulated considering the structure in whole. Since bending moment at the hinge C is zero, this provides additional equilibrium equation. It means that the sum of the moments of all external forces, which are located on the right (or on the left) part of the structure with respect to hinge C is zero

$$\sum_{\text{left}} M_C = 0 \quad \text{or} \quad \sum_{\text{right}} M_C = 0 \tag{2.2}$$

These four equations of equilibrium determine all four reactions at the supports. Therefore, three-hinged arch is a geometrically unchangeable and statically determinate structure.

The fundamental feature of arched structure is that horizontal reactions appear even if the structure is subjected to vertical load only. These horizontal reactions $H_A = H_B = H$ called as a *thrust*; such types of structures are often called as thrusted structures.

It will be shown later that at any cross section of the arch, the bending moments, shear, and axial forces arise. However, the bending moments and shear forces are considerably smaller than corresponding internal forces in a simply supported beam covering the same span and subjected to the same load. This is the fundamental property of the arch thanks to thrust. Thrusts in both supports are oriented toward each other and reduce the bending moments that would arise in beams of the same span and load. Therefore, the height of the cross section of the arch can be much less then the height of a beam to resist the same loading. So the three-hinged arch is more economical than simply supported beam, especially for large-span structures.

Introducing a tie into the system increases the number of constraints by one and therefore, in order for the arch with a tie to remain statically determinate, one of the

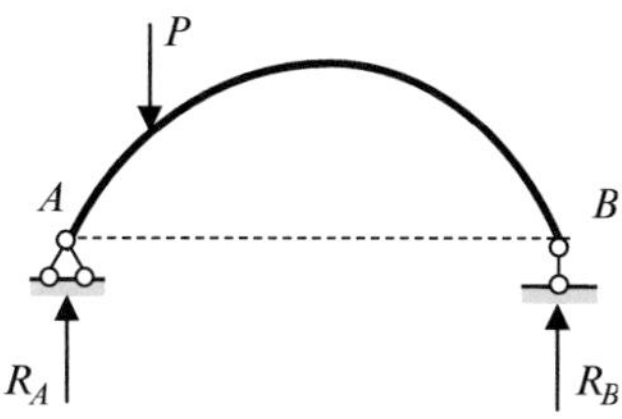

Fig. 2.2 Simply supported thrustless curvilinear member

pinned support must be replaced by a rolled support. A tie changes the distribution of internal forces in arch. The tie may be located at the level of the supports or above them. Arch with an elevated tie is shown in Fig. 2.1b. If tie is connected with arch by means of hinges, then the tie is subjected only to a tensile internal force.

In the case of vertical loads, which act on the arch with a tie, the horizontal reactions of supports equals zero while an extended force (thrust) arises in a tie.

Let us have a quick look at the structure shown in Fig. 2.2. Is this an arch? The arch is characterized by two fundamental markers such as a curvilinear axis and appearance of the thrust. Therefore, the structure in Fig. 2.2 presents the curvilinear trustless simply supported element, i.e., this is just a member with a curvilinear axis, but not an arch.

It is obvious that, unlike the beam, in this structure the axial compressed forces arise; however, the distribution of bending moments for this structure and for a beam of the same span and load will not differ, while the shear forces are less in this structure than that in beam. Thus, the fundamental feature of the arch (decreasing of the bending moments due to appearance of the thrust) for structure in Fig. 2.2 is not observed.

2.2 Reactions of Supports and Internal Forces

Let us consider a three-hinged symmetrical arch with intermediate hinge C at the highest point of the arch and with supports A and B at one elevation. Design diagram of the corresponding three-hinged arch is presented in Fig. 2.3; the span and rise of the arch are labeled as l and f, respectively. Equation of central line of the arch is $y = y(x)$.

Reactions of Supports

The stress analysis, and especially, construction of influence lines for internal forces of the three-hinged arch may be easily and elegantly performed if the conception of the “reference (or substitute) beam” is introduced. The reference

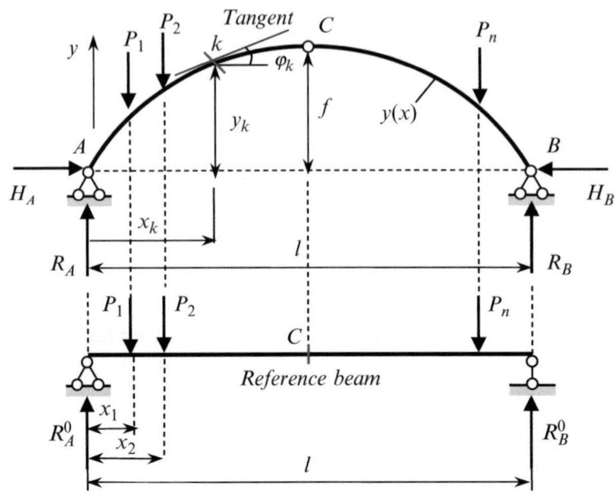

Fig. 2.3 Three-hinged arch. Design diagram and reference beam

beam is a simply supported beam of the same span as the given arch and subjected to the same loads, which act on the arch (Fig. 2.3).

The following reactions arise in arch: $R_A,\ R_B,\ H_A,\ H_B$. The vertical reactions of three-hinged arches carrying the vertical loads have same values as the reactions of the reference beam

$$R_A = R_A^0; \quad R_B = R_B^0. \tag{2.3}$$

The horizontal reactions (thrust) at both supports of three-hinged arches subjected to the vertical loads are equal in magnitude and opposite in direction

$$H_A = H_B = H. \tag{2.4}$$

Bending moment at the hinge C of the arch is zero. Therefore, by definition of the bending moment

$$M_C = \underbrace{R_A\frac{l}{2} - P_1\left(\frac{l}{2} - x_1\right) - P_2\left(\frac{l}{2} - x_2\right)}_{M_C^0} - H_A \times f = 0.$$

Underlined set of terms is the bending moment acting over section C of the reference beam (this section is located under the hinge of the arch). Therefore, last equation may be rewritten in the form

$$M_C^0 - H_A \times f = 0,$$

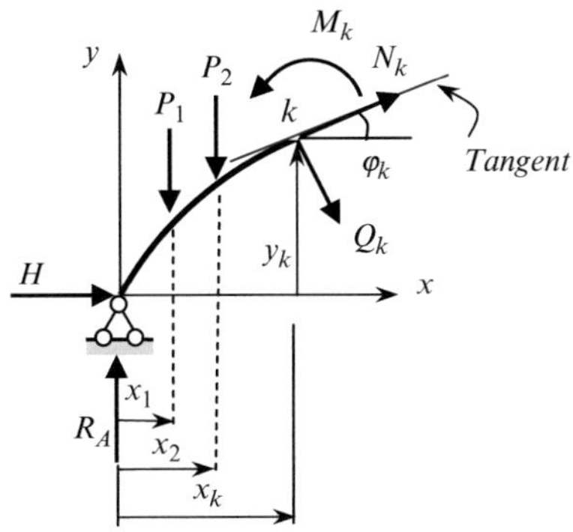

Fig. 2.4 Positive internal forces at any section k

which immediately allows us to calculate the thrust

$$H = \frac{M_C^0}{f}. \tag{2.5}$$

Thus, the thrust of the arch equals to bending moment at section C of the reference beam divided by the rise of the arch.

Internal Forces

In any section k of the arch, the following internal forces arise: the bending moment M_k, shear Q_k, and axial force N_k. The positive directions of internal forces are shown in Fig. 2.4.

Internal forces acting over a cross section k may be obtained considering the equilibrium of free body diagram of the left or right part of the arch. It is convenient to use the left part of the arch. By definition

$$M_k = R_A x_k \quad \sum_{\text{left}} P_i(x_k - x_i) - Hy_k,$$

$$Q_k = \left(R_A - \sum_{\text{left}} P\right) \cos\varphi_k - H \sin\varphi_k,$$

$$N_k = -\left(R_A - \sum_{\text{left}} P\right) \sin\varphi_k - H \cos\varphi_k,$$

where P_i are forces which are located at the left side of section k; x_i are corresponding abscises of the points of application; x_k and y_k are coordinates of point k; and φ_k is the angle between the tangent to the center line of the arch at point k and a horizontal.

These equations may be represented in the following convenient form

$$\begin{aligned} M_k &= M_k^0 - Hy_k, \\ Q_k &= Q_k^0 \cos\varphi_k - H \sin\varphi_k, \\ N_k &= -Q_k^0 \sin\varphi_k - H \cos\varphi_k, \end{aligned} \tag{2.6}$$

where expressions

$$M_k^0 = R_A x - \sum_{\text{left}} P_i(x - x_i), \text{ and } Q_k^0 = R_A - \sum_{\text{left}} P,$$

represent the bending moment and shear force at section k for the reference beam (beam's bending moment and beam's shear).

Analysis of (2.5) and (2.6)

1. Thrust of the arch is inversely proportional to the rise of the arch.
2. In order to calculate the bending moment in any cross section of the three-hinged arch, the bending moment at the same section of the reference beam should be decreased by the value Hy_k. Therefore, the bending moment in the arch less than that of in the reference beam. This is the reason why the three-hinged arch is more economical than simply supported beam, especially for large-span structures.

   In order to calculate shear force in any cross section of the three-hinged arch, the shear force at the same section of the reference beam should be multiplied by $\cos\varphi_k$ and this value should be decreased by $H \sin\varphi_k$.
3. Unlike beams loaded by vertical loads only, there are axial forces, which arise in arches loaded by vertical loads only. These axial forces are always compressed.

Example 2.1. Design diagram of the three-hinged circular arch subjected to fixed loads is presented in Fig. 2.5a. The forces $P_1 = 10$ kN, $P_2 = 8$ kN, $q = 2$ kN/m. It is necessary to construct the internal force diagrams M, Q, N.

Solution. Reference beam. The reactions are determined from the equilibrium equations of all the external forces:

$$\sum M_B = 0 \rightarrow -R_A^0 \times 32 + P_1 \times 24 + q \times 8 \times 12 + P_2 \times 4 = 0 \rightarrow R_A^0 = 14.5\,\text{kN},$$
$$\sum M_A = 0 \rightarrow R_B^0 \times 32 - P_1 \times 8 - q \times 8 \times 20 - P_2 \times 28 = 0 \rightarrow R_B^0 = 19.5\,\text{kN}.$$

The bending moment M^0 and shear Q^0 diagrams for reference beam are presented in Fig. 2.5b. At point C ($x = 16$ m), the bending moment is $M_C^0 = 152$ kN m.

Three-hinged arch. The vertical reactions and thrust of the arch are

$$R_A = R_A^0 = 14.5\,\text{kN}, \qquad R_B = R_B^0 = 19.5\,\text{kN}, \quad H = \frac{M_C^0}{f} = \frac{152}{8} = 19\,\text{kN}.$$

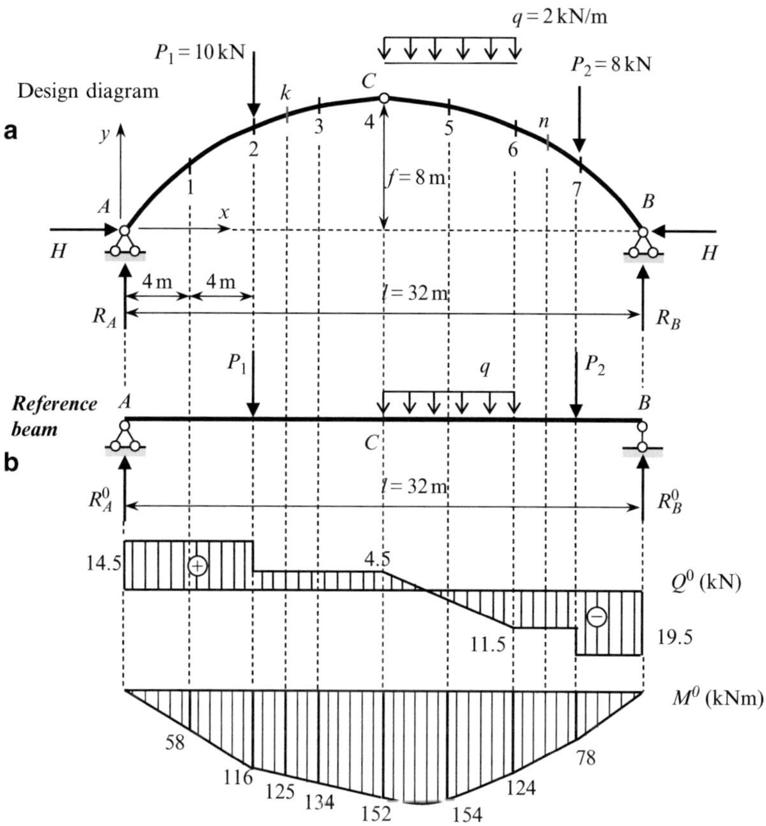

Fig. 2.5 (**a**) Design diagram of three-hinged circular arch and (**b**) reference beam and corresponding internal forces diagrams

For construction of internal forces diagrams of the arch, a set of sections has to be considered and for each section internal forces should be calculated. All computations concerning geometrical parameters and internal forces of the arch are presented in Table 2.1. The column 0 contains the numbers of sections. For specified sections A, 1–7, and B, the abscissa x and corresponding ordinate y (in meters) are presented in columns 1 and 2, respectively. Radius of curvature of the arch is

$$R = \frac{f}{2} + \frac{l^2}{8f} = \frac{8}{2} + \frac{32^2}{8 \times 8} = 20\,\text{m}.$$

Coordinates y are calculated using the following expression

$$y(x) = \sqrt{R^2 - \left(\frac{l}{2} - x\right)^2} - R + f = \sqrt{400 - (16 - x)^2} - 12\,(\text{m}).$$

Table 2.1 Internal forces in three-hinged circular arch (Fig. 2.6) ($R_A = 14.5\,\text{kN}$; $R_B = 19.5\,\text{kN}$; $H = 19\,\text{kN}$)

Section	x (m)	y (m)	sin φ	cos φ	M_x^O (kNm)	Hy (kNm)	M_x (kNm)	Q_x^O (kN)	Q_x (kN)	N_x (kN)
0	1	2	3	4	5	5′	6	7	8	9
A	0.0	0.0	0.8	0.6	0	0.0	0	14.5	−6.5	−23
1	4.0	4.0	0.6	0.8	58	76	−18	14.5	0.2	−23.9
2	8.0	6.330	0.4	0.9165	116	120.27	−4.27	14.5/4.5[a]	5.6892/−3.4757	−23.213/−19.213
k	10	7.0788	0.3	0.9539	125	134.497	−9.497	4.5	−1.4074	−19.474
3	12	7.596	0.2	0.9798	134	144.324	−10.324	4.5	0.6091	−19.516
4 (***C***)	16	8.0	0.0	1.0	152	152	0.0	4.5	4.5	−19.00
5	20	7.596	−0.2	0.9798	154	144.324	9.676	−3.5	0.3707	−19.316
6	24	6.330	−0.4	0.9165	124	120.27	3.73	−11.5	−2.9397	−22.013
n	26	5.3205	−0.5	0.8660	101	101.089	−0.089	−11.5	−0.459	−22.204
7	28	4	−0.6	0.8	78	76	2	−11.5/−19.5	2.2/−4.2	−22.1/−26.9
B	32	0.0	−0.8	0.6	0	0.0	0	−19.5	3.5	−27

Note the values of discontinuity due to concentrated load equal to P cos φ and P sin φ in shear and normal force diagrams, respectively

[a] Values in numerator and denominator (columns 7–9) mean value of the force to the left and to the right of corresponding section

Columns 3 and 4 contain values of sin φ and cos φ, which are calculated by the formula

$$\sin\varphi = \frac{l-2x}{2R} = \frac{32-2x}{40}, \quad \cos\varphi = \frac{y+R-f}{R} = \frac{y+12}{20}.$$

Values of bending moment and shear for reference beam, which are presented in columns 5 and 7, are taken directly from the corresponding diagrams in Fig. 2.5b. Values for Hy are contained in column 5′. Columns containing separate terms for $Q^0 \cos\varphi$, $Q^0 \sin\varphi$, $H\cos\varphi$, $H\sin\varphi$ are not presented. Values of bending moment, shear, and normal forces for three-hinged arch are tabulated in columns 6, 8, and 9. They have been computed using (2.6). For example, for section A we have

$$\begin{aligned} Q_A &= Q_A^0 \cos\varphi_A - H \sin\varphi_A = 14.5 \times 0.6 - 19 \times 0.8 = -6.5\,\text{kN}, \\ N_A &= -Q_A^0 \sin\varphi_A - H \cos\varphi_A = -14.5 \times 0.8 - 19 \times 0.6 = -23\,\text{kN}. \end{aligned}$$

The final internal force diagrams for the arch are presented in Fig. 2.6. Bending moment diagram is shown on the side of the extended fibers, thus the signs of bending moments are omitted. As for beam, the bending moment and shear diagrams satisfy to Schwedler's differential relationships. In particularly, if at any point a shear changes its sign, then a slope of the bending moment diagram equals zero, i.e., at this point the bending moment has local extreme (e.g., points 2, 7, etc.). It can be seen that the bending moments which arise in cross sections of the arch are much less than that of in a reference beam.

It is obvious that for supports $R_A^2 + H^2 = Q_A^2 + N_A^2$ and $R_B^2 + H^2 = Q_B^2 + N_B^2$.

2.3 Rational Shape of the Arch

The shape of the arch, which is subjected to a given fixed load, is called rational if the bending moments in the cross section of the arch equal to zero. An example of a rational arch could be in the form of a circular arch which is loaded by uniform radial (hydrostatic) load [Rzh82].

2.3.1 Vertical Load Does Not Depend on the Shape of the Arch

In this case, the reactions of the arch and bending moments for reference beam do not depend on the shape of the arch. Thus, for a rational arch, we have the condition

$$M_k = M_k^0 - Hy_k = 0, \tag{2.7}$$

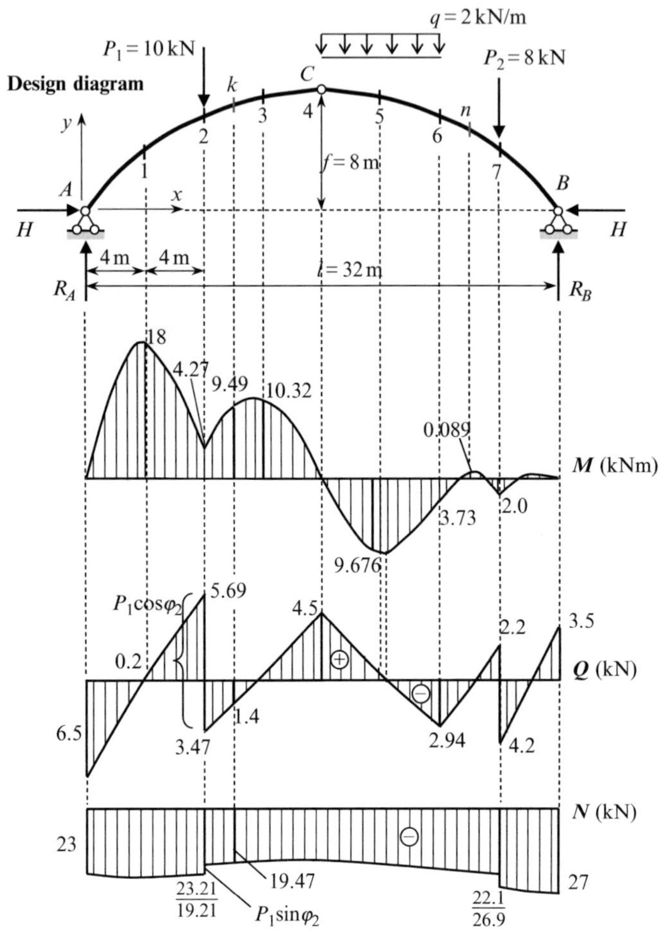

Fig. 2.6 Design diagram of three-hinged circular arch. Internal forces diagrams

where M_k^0 is a bending moment in the reference simply supported beam; H is a thrust of the arch; y_k is a vertical coordinate of the point on the axis of the arch. Therefore, the shape of the rational arch is determined by its y coordinate

$$y_k = \frac{M_k^o}{H}. \tag{2.8}$$

It is easy to prove the following statement: if a three-hinged arch is subjected to a vertical load and the vertical ordinates y of the arch, measured from the support line AB, are proportional to corresponding ordinates of the bending moment diagram of the reference beam, then the bending moments at all sections of the arch are equal to zero. This statement is true for any position of the intermediate hinge C [Rab60].

Indeed, let for any section k of the arch, the y-ordinate of the axis and bending moment of the reference beam be related by the formula $y_k = nM_k^0$, where n is an arbitrary number. Bending moment at section k is

$$M_k = M_k^0 - Hy_k = M_k^0 - HnM_k^0 = M_k^0(1 - nH).$$

For crown hinge C, the bending moment $M_C = M_C^0(1 - nH) = 0$. Since $M_C^0 \neq 0$, then $(1 - nH) = 0$.

Thus, the bending moment at any section equals to zero.

Example 2.2. Three-hinged symmetric arch of span l and rise f is loaded by a uniformly distributed load q within the entire span. Origin is placed on the left support and the axis x is directed to right. Expression for bending moment of the reference beam is $M_x^0 = qx(l - x)/2$.

The thrust of the arch is $H = M_C^0/f = ql^2/(8f)$. Therefore, the required equation of the axis of the arch becomes

$$y(x) = \frac{M_x^0}{H} = \frac{4f}{l^2}x(l - x).$$

Thus, if a uniformly distributed vertical load acts within the entire span of the three-hinged parabolic arch, then the bending moments do not arise in the arch.

Note, if a given load is governed by the law $q(x) = q_0 + kx$, then the bending moment diagram and the rational axis of the arch are characterized by third-order polynomials [Kis60].

2.3.2 Vertical Load Depends on Arch Shape

Let us consider a three-hinged arch load as shown in Fig. 2.7. We can see that a shape of the arch determine the value of load. According to the definition, in the case of a rational arch, only axial forces arise in the cross sections.

Free body diagram for infinitesimal element i–j is shown in Fig. 2.7; horizontal projection of this element is dx. Equilibrium equation

$$\sum X = N\cos\varphi - (N + dN)\cos(\varphi + d\varphi) = 0,$$

leads to $d(N\cos\varphi) = 0$. It means that

$$N\cos\varphi = \text{const} = H, \tag{2.9}$$

where H is the thrust of the arch.

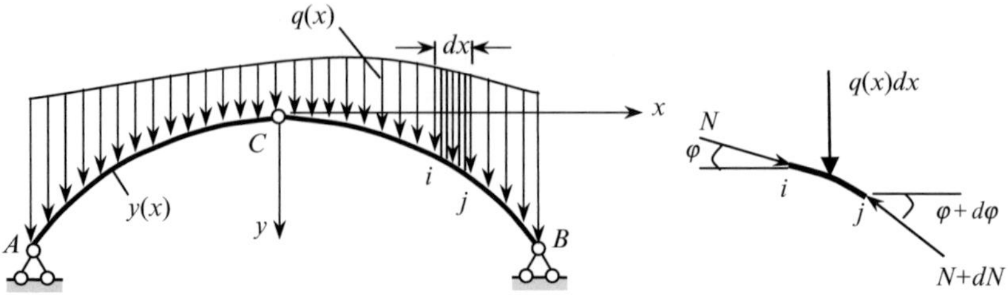

Fig. 2.7 Three-hinged arch subjected to load which depends on the shape of the arch

Equilibrium equation

$$\sum Y = N\sin\varphi + q(x)\mathrm{d}x - (N+\mathrm{d}N)\sin(\varphi+\mathrm{d}\varphi) = 0 \text{ leads to } \frac{\mathrm{d}}{\mathrm{d}x}(N\sin\varphi) = q(x). \tag{2.9a}$$

Since $N = H/\cos\varphi$, (2.9a) can be rewritten as follows

$$\frac{\mathrm{d}}{\mathrm{d}x}(H\tan\varphi) = q(x) \text{ or } \frac{\mathrm{d}}{\mathrm{d}x}\left(H\frac{\mathrm{d}y}{\mathrm{d}x}\right) = H\frac{\mathrm{d}^2y}{\mathrm{d}x^2} = q(x).$$

Thus, the equation of the rational axis of the arch in the case of a load, which depends on the shape of the arch obeys the differential equation [Kis60]

$$\frac{\mathrm{d}^2y}{\mathrm{d}x^2} = \frac{q(x)}{H}. \tag{2.10}$$

For each specified load, the problem of determining the rational shape of the arch comes down to integration of (2.10).

Example 2.3. Symmetrical three-hinged arch of span l and rise f is subjected to vertical load $q(x)$, which consists of two parts. One part of load, q_0, is uniformly distributed within the entire span of the arch. The second part of load depends on the shape of the arch. Assume that this part of the load is proportional to coordinate y. Thus, the total load becomes $q(x) = q_0 + \gamma \times y$. Design diagram of right-hand part of the arch and location of the x and y axis are shown in Fig. 2.8.

Differential equation (2.10) becomes

$$\frac{\mathrm{d}^2y}{\mathrm{d}x^2} = \frac{q_0 + \gamma\times y}{H} \text{ or } \frac{\mathrm{d}^2y}{\mathrm{d}x^2} - k^2y = \frac{q_0}{\gamma}k^2, \quad k^2 = \frac{\gamma}{H}.$$

Its solution and first derivative are

$$y = A\sinh kx + B\cosh kx - \frac{q_0}{\gamma};\ \frac{\mathrm{d}y}{\mathrm{d}x} = Ak\cosh kx + Bk\sinh kx.$$

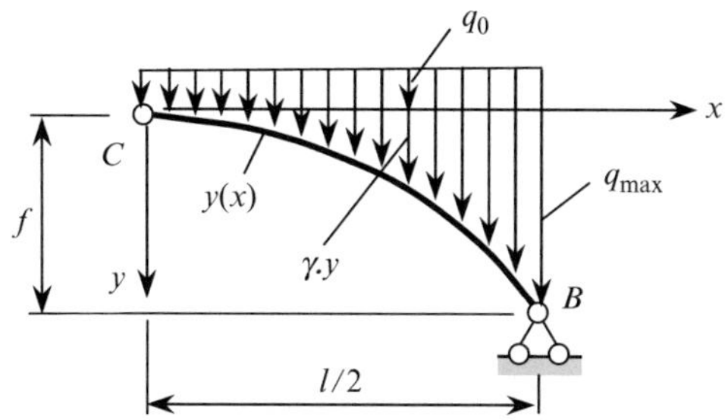

Fig. 2.8 Load change according to the shape of the arch, $q(x) = q_0 + \gamma y$

Constants of integration are found from the boundary conditions for symmetrical arch:

1. At $x = 0$ (indeterminate hinge C), $dy/dx = 0$. This condition leads to $A = 0$.
2. At $x = 0$ $y = 0$, so $B = q_0/\gamma$.

Equation of the axis of the rational shape of the arch becomes

$$y(x) = \frac{q_0}{\gamma}(\cosh kx - 1).$$

This curve is called a catenary [Kis80]. Some data for catenary arch with the given span l and rise f and parameter of the load $\delta = q_{max}/q_0$ are presented below.

Equation of the shape of the arch is

$$y = \frac{f}{\delta - 1}(\cosh kx - 1),$$

where relationship between parameters k and δ is

$$\delta = \cosh \frac{kl}{2}, \quad \text{so } k = \frac{2}{l} \text{arc} \cosh \delta.$$

The slope of the axis of the arch is

$$\tan \varphi = \frac{f}{\delta - 1} k \sinh kx.$$

The thrust H of the arch, axial force N in any cross section of the arch, and maximum axial force N_{max} are:

$$H = \frac{q_0(\delta - 1)}{fk^2}, \quad N = H\sqrt{1 + \tan^2\varphi}, \quad N_{max} = H\sqrt{1 + k^2 f^2 \frac{\delta + 1}{\delta - 1}}.$$

Vertical component of the reaction of support is

$$V = H \frac{f}{\delta - 1} k \sinh \frac{kl}{2}.$$

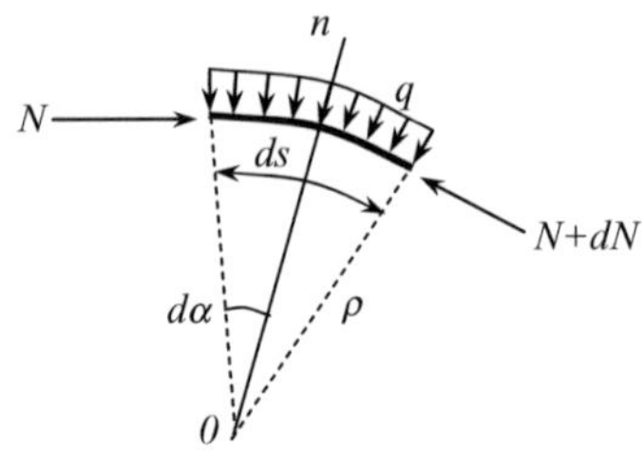

Fig. 2.9 Infinitesimal element subjected to radial load and axial force

2.3.3 Radial Load

Let us consider an arch with arbitrary equation for the central axis. The arch is loaded by a radial load. It means that the load is directed along the radius of curvature at each infinitesimal element of the arch. Design diagram of such element of length ds, central angle $2d\alpha$, and radius of curvature ρ is shown in Fig. 2.9. The load q is directed to the center of curvature; the load q should be treated as uniformly distributed within the portion ds. Since the arch is rational, then bending moments are absent.

From the equilibrium equations

$$\sum M_O = N\rho - (N + \mathrm{d}N)\rho = 0,$$

we get d$N = 0$. It means that in the case of a radial load, the axial force in arch is constant.

Since $\sin d\alpha \cong d\alpha$ and $\mathrm{d}s = \rho \times 2d\alpha$, then the equilibrium equation in projection of all forces onto the normal axis

$$\sum n = N \sin \mathrm{d}a + (N + \mathrm{d}N) \sin \mathrm{d}a - q\mathrm{d}s = 0,$$

leads to the following expression for the radius of curvature $\rho = N/q$. Curvature of the axis of the rational arch is proportional to the intensity q of the external load. In the case of a uniformly distributed radial load (q = const), the axis of the rational arch presents a circle [Kis60].

The simplest problems of optimal three-hinged and redundant uniform arches are presented in [Gol80]: in these problems, it is necessary to find the shape of the arch which minimize its volume. Different types of loading are considered. Among them are fixed, moving, and wind loads.

2.4 Influence Lines for Reactions and Internal Forces

This section is devoted to construction of influence line for reactions, thrust, and internal forces. Three precise approaches are considered. They are the analytical approach, the nil points of influence lines, and fictitious beam methods. Influence lines method for structural analysis was developed by Winkler (1835–1888) and independently by Mohr (1835–1918) in 1868.

2.4.1 Analytical Approach

Equations (2.3), (2.5), and (2.6) can be used for deriving the equations for influence lines. The equations for influence lines for vertical reactions of the arch are derived from (2.3). Therefore, the equations for influence lines become

$$\mathrm{IL}(R_A) = \mathrm{IL}\left(R_A^O\right); \qquad \mathrm{IL}(R_B) = \mathrm{IL}\left(R_B^O\right). \tag{2.11}$$

The equation of influence lines for thrust is derived from (2.5). Since for a given arch, a rise f is a *fixed* number, then the equations for influence lines becomes

$$\mathrm{IL}(H) = \frac{1}{f} \times \mathrm{IL}\left(M_C^O\right). \tag{2.12}$$

Thus, influence line for trust H may be obtained from the influence line for bending moment at section C of the reference beam, if all ordinates of the latter will be divided by parameter f.

The equations for influence lines for internal forces at any section k may be derived from (2.6). Since for a given section k, the parameters y_k, sin φ_k, and cos φ_k are *fixed* numbers, then the equations for influence lines become

$$\begin{aligned} \mathrm{IL}(M_k) &= \mathrm{IL}\left(M_k^O\right) - y_k \times \mathrm{IL}(H), \\ \mathrm{IL}(Q_k) &= \cos\varphi_k \times \mathrm{IL}\left(Q_k^O\right) - \sin\varphi_k \times \mathrm{IL}(H), \\ \mathrm{IL}(N_k) &= -\sin\varphi_k \times \mathrm{IL}\left(Q_k^O\right) - \cos\varphi_k \times \mathrm{IL}(H). \end{aligned} \tag{2.13}$$

In order to construct the influence line for bending moment at section k, it is necessary to sum two graphs: one of them is influence line for bending moment at section k for reference beam and second is influence line for thrust H with all ordinates of which have been multiplied by a constant factor $(-y_k)$.

Equation of influence lines for shear also has two terms. The first term presents influence line for shear at section k in the reference beam, all the ordinates of which have been multiplied by a constant factor cos φ_k. The second term presents the influence line of the thrust of the arch, all the ordinates of which have been multiplied by a constant factor $(-\sin\varphi_k)$. Summation of these two graphs leads to the required influence line for shear force at section k. Similar procedure should be applied for the construction of influence line for axial force. Note that both terms for axial force are negative.

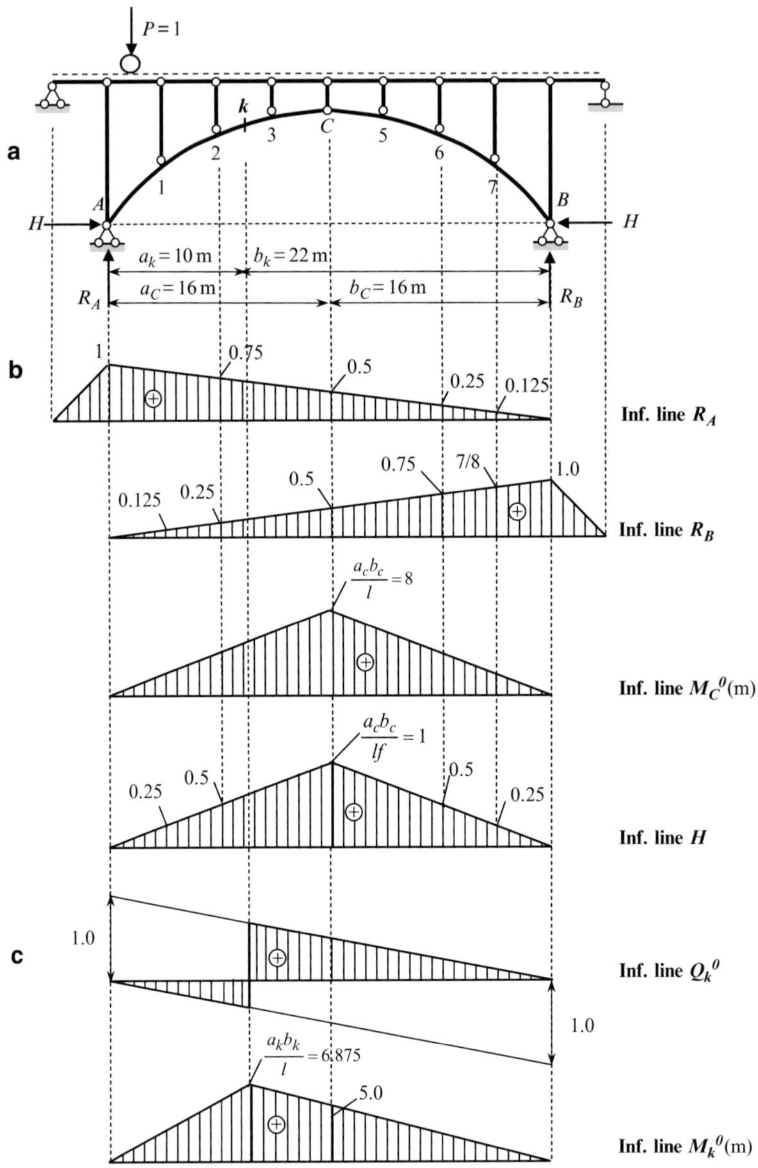

Fig. 2.10 Three-hinged arch. (**a**) Design diagram; (**b**) influence lines for reactions of the arch; and (**c**) influence lines for internal forces at section *k* for reference beam

Figure 2.10a presents the arched structure consists of the arch itself and overarched construction, which includes the set of simply supported beams and vertical posts with hinged ends. Unit load, which moves along the horizontal beams, is transmitted over the posts on the arch at discrete points. Thus, this design diagram corresponds to indirect load application. Parameters of the arch are same as in Fig. 2.5a.

It is required to construct the influence lines for vertical reactions, thrust and for bending moment M_k, shear Q_k, and normal force N_k for section k.

Influence Lines for Reactions

According to (2.11), influence lines for vertical reactions R_A and R_B of the arch do not differ from influence lines for reaction of supports of a simply supported beam.

Influence line for thrust may be constructed according to (2.12); the maximum ordinate of influence line for bending moment at section C of the reference beam equals to $a_c b_c / l = 8\,\text{m}$. Therefore, the maximum ordinate of influence line for thrust H of the arch becomes $(1/f) \times (a_c b_c / l) = l/4f = 32/4 \times 8 = 1$. Influence lines for reactions of supports of the arch and internal forces for reference beam are shown in Fig. 2.10b, c. Indirect load application is taken into account [Kar10].

Influence Lines for Internal Forces at Section *k*

Section k is characterized by the following parameters: $a_k = 10\,\text{m}$, $b_k = 22\,\text{m}$, $y_k = 7.0788$ m, sin $\varphi = 0.30$, cos $\varphi = 0.9539$ (Table 2.1). Algorithms for the construction of influence lines of internal forces for arch are described in Sect. 2.4.1.

Bending moment. Influence line for M at section k may be constructed according to (2.13).

$$\text{IL}(M_k) = \text{IL}\left(M_k^0\right) - y_k \times \text{IL}(H). \tag{2.13a}$$

Step 1. Influence line for bending moment at section k of reference beam M_k^0 presents the triangle with maximum ordinate $a_k b_k / l = 10 \times 22/32 = 6.875\,\text{m}$ at sections k and 5.0 m at section C (Figs. 2.10 and 2.11).

Step 2. Influence line for thrust H presents triangle with maximum ordinate $l/(4f) = 1$ at section C. Term $y_k \times \text{IL}(H)$ presents the similar graph; the maximum ordinate is $y_k \times 1 = 7.0788\,\text{m}$. So the specified ordinates of graph $y_k \times \text{IL}(H)$ at section k and C are 4.42425 and 7.0788 m, respectively (Fig. 2.11).

Step 3. Procedure (2.13a) is presented in Fig. 2.11, construction of influence line M_k. Since both terms in (2.13a) has *different* signs, then both graphs, $\text{IL}\left(M_k^0\right)$ and $y_k \times \text{IL}(H)$ should be plotted on the *one side* on the basic line. The ordinates of required $\text{IL}(M_k)$ will be located *between* these both graphs. Specified ordinates of final influence line (2.13a) at section k and C are

$$6.875 - 4.42425 = 2.45075\,\text{m and } 5.0 - 7.0788 = -2.0788\,\text{m}.$$

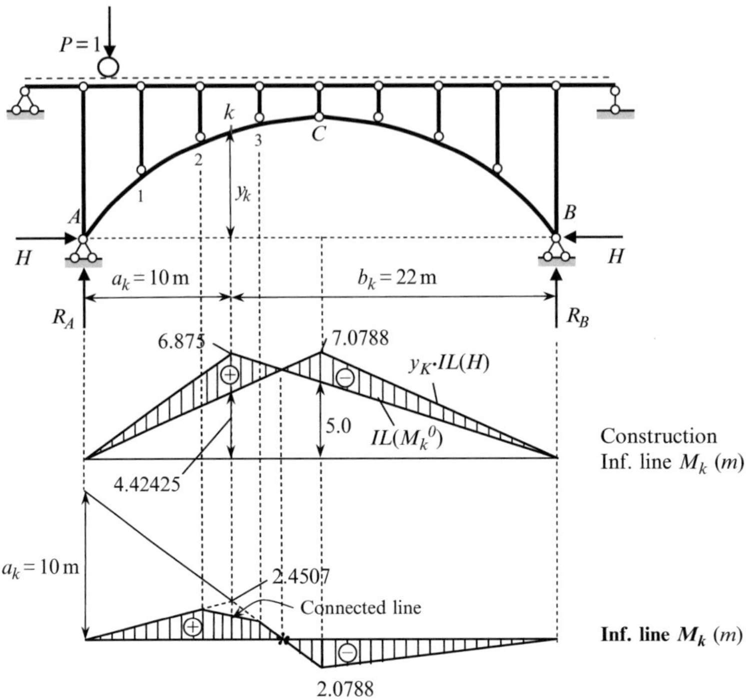

Fig. 2.11 Three-hinged arch. Design diagram and construction of influence line for bending moment at section k of the arch

Step 4. Influence line between joins 2 and 3 presents a straight line because of indirect load application [Kar10]; this connected line is shown by solid line. Final influence line $\mathrm{IL}(M_k)$ is shown in Fig. 2.11.

Shear force. This influence line may be constructed according to equation

$$\mathrm{IL}(Q_k) = \cos\varphi_k \times \mathrm{IL}\left(Q_k^0\right) - \sin\varphi_k \times \mathrm{IL}(H). \tag{2.13b}$$

Step 1. Influence line for shear at section k for the reference beam is shown in Fig. 2.10c; the specified ordinates at supports A and B equal to 1.0. The first term $\cos\varphi_k \times \mathrm{IL}\left(Q_k^0\right)$ of (2.13b) presents a similar graph with specified ordinates $\cos\phi_k = 0.954$ at supports A and B, so ordinates at the left and right of section k are -0.298 and 0.656, while at crown C is 0.477.

Step 2. Influence line for thrust is shown in Fig. 2.10b; the specified ordinates at crown C equals to 1.0. The second term $\sin\varphi_k \times \mathrm{IL}(H)$ of (2.13b) presents a similar graph with specified ordinates $0.3 \times 1.0 = 0.3$ at crown C. Specified ordinate at section k is 0.1875.

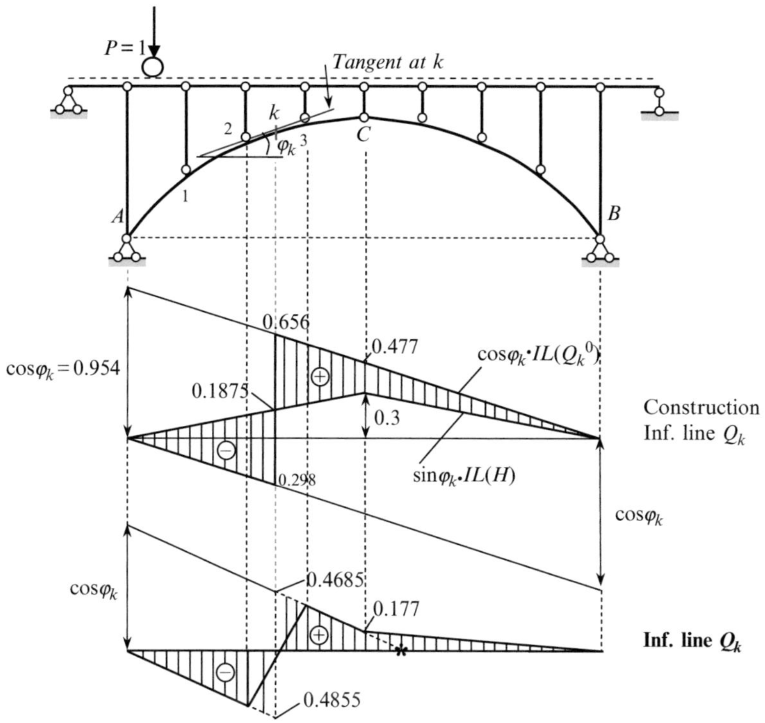

Fig. 2.12 Three-hinged arch. Design diagram and construction of influence line for shear at section k of the arch

Step 3. Procedure (2.13b) is presented in Fig. 2.12. As in case for bending moment, both terms in (2.13b) has *different* signs, therefore both graphs $\cos\varphi_k \times \mathrm{IL}\left(Q_k^0\right)$ and $\sin\varphi_k \times \mathrm{IL}(H)$ should be plotted on the *one side* on the basic line. Ordinates *between* both graphs present the required ordinates for influence line for shear. Specified ordinates of final influence line (2.13b) at left and right of section k are

$$0.298 + 0.1875 = 0.4855 \text{ and } 0.656 - 0.1875 = 0.4685.$$

At crown C, ordinate of influence line Q_k is $0.477 - 0.3 = 0.177$.

Step 4. Influence line between joins 2 and 3 presents a straight line; this connected line is shown by a solid line. Final influence line $\mathrm{IL}(Q_k)$ is shown in Fig. 2.12.

Axial force. This influence line may be constructed according to the following equation

$$\mathrm{IL}(N_k) = -\sin\varphi_k \times \mathrm{IL}\left(Q_k^0\right) - \cos\varphi_k \times \mathrm{IL}(H). \tag{2.13c}$$

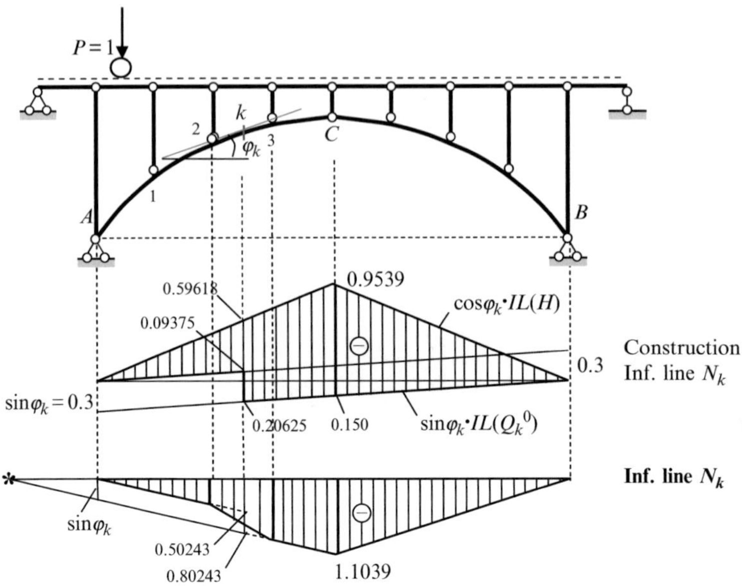

Fig. 2.13 Three-hinged arch. Design diagram and construction of influence lines for axial force at section k of the arch

Step 1. Influence line for shear at section k for the reference beam is shown in Fig. 2.10c. The first term $\sin\varphi_k \times \mathrm{IL}\left(Q_k^0\right)$ of (2.13c) presents a similar graph with specified ordinates $\sin\varphi_k = 0.30$ at supports A and B, so at the left and right of section k ordinates are 0.09375 and $-$ 0.20625, while at crown C is $-$ 0.15.

Step 2. Influence line for thrust is shown in Fig. 2.10b; the specified ordinates at crown C equals to 1.0. The second term $\cos\varphi_k \times \mathrm{IL}(H)$ of (2.13c) presents a similar graph with specified ordinates $0.9539 \times 1.0 = 0.9539$ at crown C. Specified ordinate at section k is 0.59618.

Step 3. Procedure (2.13c) is presented in Fig. 2.13. Both terms in (2.13c) has *same* signs; therefore, both graphs, $\sin\varphi_k \times \mathrm{IL}\left(Q_k^0\right)$ and $\cos\varphi_k \times \mathrm{IL}(H)$, should be plotted on the *different sides* on the basic line. Ordinates for required $\mathrm{IL}(N_k)$ are located *between* these both graphs. Specified ordinates of final influence line (2.13c) at left and right of section k are $-(0.59618 - 0.09375) = -0.50243$ and $-(0.59618 + 0.20625) = -0.80843$.
At crown C, ordinate of influence line N_k is $-(0.9539 + 0.15) = -1.1039$.

Step 4. Influence line between joins 2 and 3 presents a straight line; this connected line is shown by a solid line. Final influence line $\mathrm{IL}(N_k)$ is shown in Fig. 2.13.

Properties of the Influence Lines for Internal Forces

1. Influence line for bending moment has significantly less ordinates than for reference beam. This influence line contains the positive and negative ordinates. It means that at section k, extended fibers can be located below or above the neutral line depending on where the load is placed.

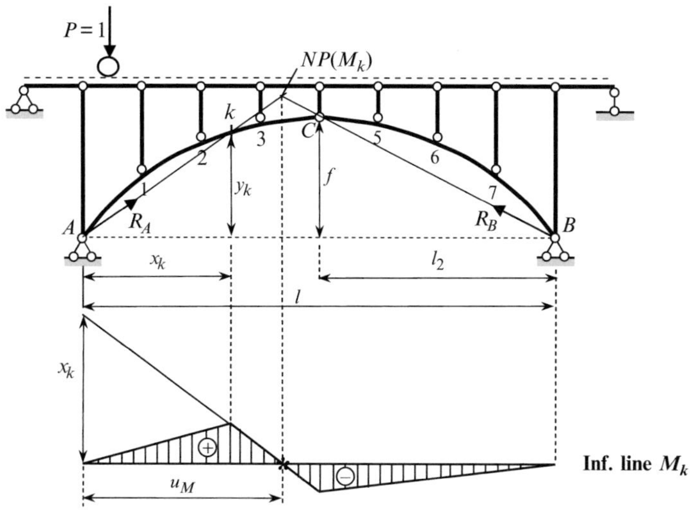

Fig. 2.14 Construction of influence line M_k using the nil point method

2. Influence line for shear, as in the case of reference beam, has two portions with positive and negative ordinates; all ordinates are significantly less than that of in the reference beam.
3. Influence line for axial force has only negative ordinates. So in case of arbitrary load, the axial forces in arch are always compressed.

2.4.2 Nil Points Method

Each influence lines shown in Figs. 2.11–2.13 has the specified point labeled as (*). These points are called as nil (or neutral) point of corresponding influence line. Such points of influence lines indicate a position of the concentrated load on the arch, so internal forces M, Q, and N in the given section k would be zero. Nil points may be used as simple procedure for the construction of influence lines for internal forces and checking the influence lines which were constructed by the analytical approach. This procedure for three-hinged arch of span l is discussed below.

Bending Moment

Step 1. Find nil point (NP) of influence line M_k. If load P is located on the left half of the arch, then reaction of the support B pass through crown C. Bending moment at section k equals zero, if reaction of support A pass through point k. Therefore, NP (M_k) is the point of intersection of line BC and Ak (theorem about three concurrent forces). The nil point (*) is always located between the crown C and section k (Fig. 2.14).

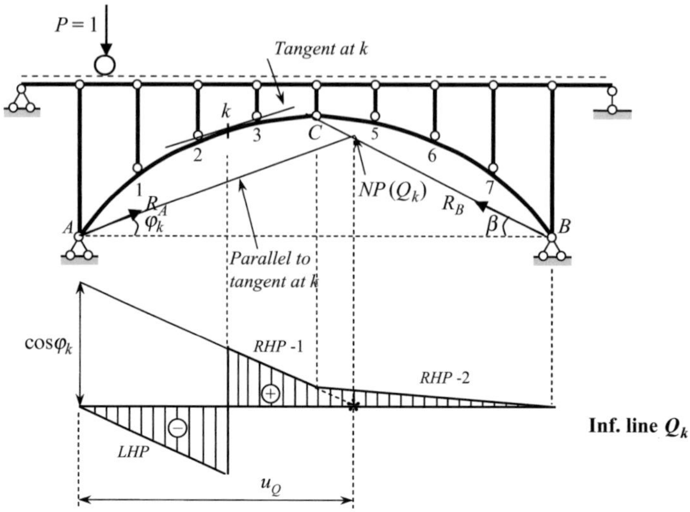

Fig. 2.15 Symmetrical three-hinged arch. Construction of influence line Q_k using the nil point method. The case of fictitious nil point

Step 2. Lay off along the vertical passing through the support *A*, the abscissa of section *k*, i.e., x_k.

Step 3. Connect this ordinate with nil point and continue this line till a vertical passing through crown C and then connect this point with support *B*.

Step 4. Take into account indirect load application; connecting line between joints 2 and 3 is not shown.

Location of NP(M_k) may be computed by the formula

$$u_M = \frac{lf\,x_k}{y_k l_2 + x_k f}. \tag{2.14}$$

Shear Force

Step 1. Find nil point (NP) of influence line Q_k. If load *P* is located on the left half of the arch, then reaction of the support *B* pass through crown *C*. Shear force at section *k* equals zero, if reaction of support *A* will be parallel to tangent at point *k*. Therefore, NP(Q_k) is point of intersection of line *BC* and line which is parallel to tangent at point *k*. For a given design diagram and specified section *k*, the nil point (*) is *fictitious* one (Fig. 2.15).

Step 2. Lay off along the vertical passing through the support *A*, the value $\cos\varphi_k$.

Step 3. Connect this ordinate with nil point. A working zone of influence line is portion between section *k* and vertical passing through crown C – right-hand portion

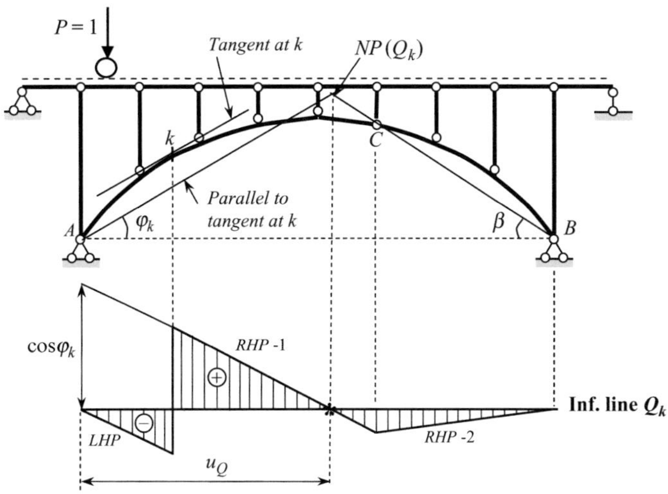

Fig 2.16 Nonsymmetrical three-hinged arch. Construction of influence line Q_k using the nil point method. The case of real nil point

1 (RHP-1). Then connect the point under crown C with support *B* – right-hand portion 2 (RHP-2).

Step 4. Left-hand portion (LHP) is parallel to right-hand portion 1 and connects two points: zero ordinate at support *A* and point under section *k*.

Figure 2.16 presents a nonsymmetrical three-hinged arch with *real* nil point for influence line Q_k; this point is located within the span of the arch. Therefore, we have one portion with positive shear and two portions with negative shear.

Location of NP(Q_k) for cases in Figs. 2.15 and 2.16 may be computed by the formula

$$u_Q = \frac{l \tan \beta}{\tan \beta + \tan \varphi_k}. \tag{2.15}$$

Axial Force

The nil point of influence line N_k is point of intersection of line *BC* and line passing through support *A* perpendicular to tangent at section *k*.

Step 1. Find nil point (NP) of influence line N_k. If load *P* is located on the left half of the arch, then reaction of the support *B* pass through crown C. Axial force at section *k* equals zero, if reaction of support *A* will be perpendicular to tangent at point *k*. The nil point (*) is located beyond the arch span (Fig. 2.17).

Step 2. Lay off along the vertical passing through the support *A*, the value $\sin \varphi_k$.

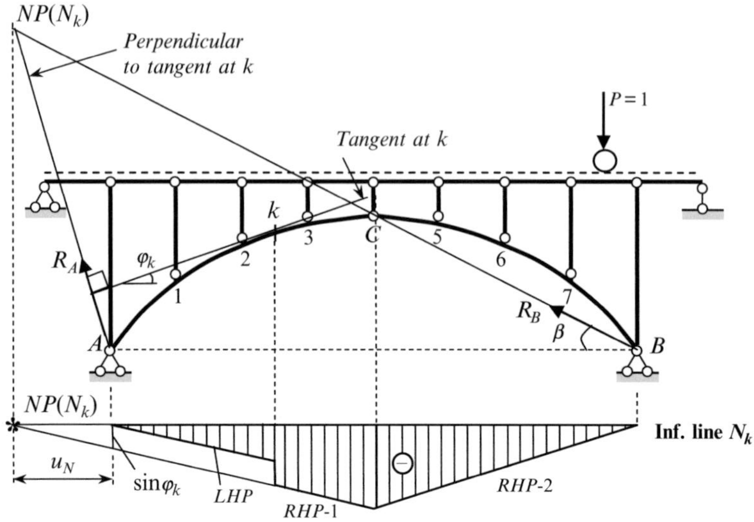

Fig. 2.17 Construction of influence line N_k using the nil point method

Step 3. Connect this ordinate with nil point and continue this line till vertical passes through crown C. A working zone is portion between section k and vertical passing through crown C (first right-hand portion RHP-1). Then connect the point under crown C with support B (second right-hand portion – RHP-2).

Step 4. LHP is parallel to RHP-1 and connects two points: zero ordinate at support A and point under section k.

Location of NP(N_k) may be computed by the formula

$$u_N = \frac{l \tan\beta}{\tan\beta - \cot\varphi_k}. \tag{2.16}$$

2.4.3 Fictitious Beam Method

Influence lines for internal forces of the three-hinged arch may be constructed as the bending moment diagram for the fictitious beam subjected to the special type of loads [Uma72-73].

Influence Line for M_k

Fictitious beam is loaded by two forces $P_k^f = 1$ at section k and $V_C^f = y_k/f$ at section C (Fig. 2.18). For arch in Fig. 2.5a and Table 2.1, we get $V_C^f = y_k/f = 7.0788/8 = 0.88485$.

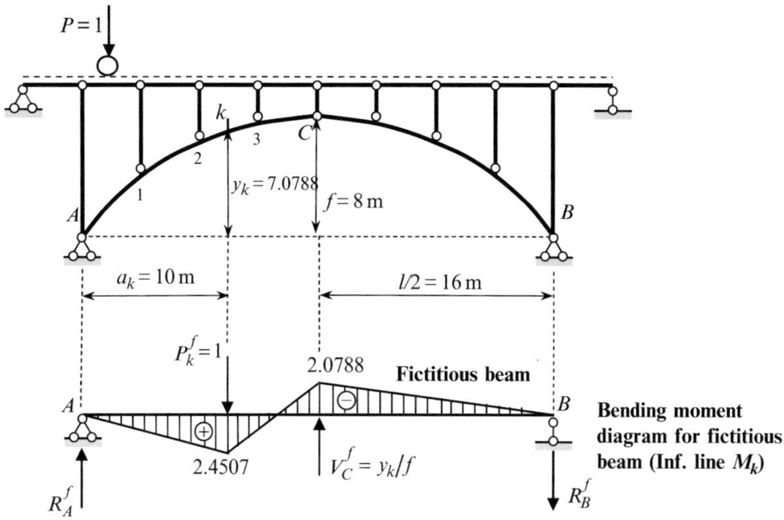

Fig. 2.18 Three-hinged arch. Fictitious beam for M_k is loaded by two forces $P_k^f = 1$ at section k and $V_C^f = y_k/f$; the bending moment diagram presents the influence line for M_k for the entire arch

Reactions of fictitious beam are

$$R_A^f = \frac{1 \times 22 - 0.88485 \times 16}{32} = 0.245075\,(\uparrow), \quad R_B^f = 0.129925\,(\downarrow).$$

All forces and reactions are dimensionless. Bending moment diagram is shown on the extended fibers (positive ordinates are placed below the neutral line). Bending moments at specified points of the fictitious beam are

$$M_k^f = R_A^f \times a_k = 0.245075 \times 10 = 2.4507\,\text{m},$$

$$M_C^f = -R_B^f \times \frac{l}{2} = -0.129925 \times 16 = -2.0788\,\text{m}.$$

These ordinates of influence line for M_k have been obtained earlier and presented in Fig. 2.11.

Influence Line for Q_k

Fictitious beam is loaded by the couple $M_k^f = \cos\varphi_k = 0.9539$ (clockwise) at section k and force $V_C^f = \sin\varphi_k/f\,(1/\text{m})$ (upwards) at section C (Fig. 2.19). For arch in Fig. 2.5a and Table 2.1, we get $V_C^f = \sin\varphi_k/f = 0.3/8 = 0.0375\,(1/\text{m})$.

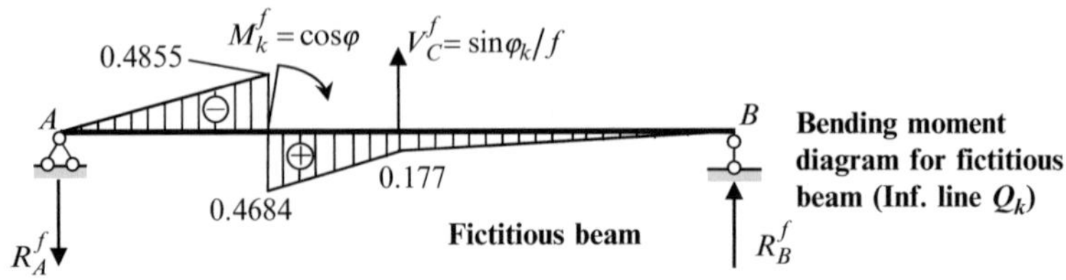

Fig. 2.19 Fictitious beam for Q_k. Bending moment diagram for fictitious beam presents the influence line Q_k for the entire arch

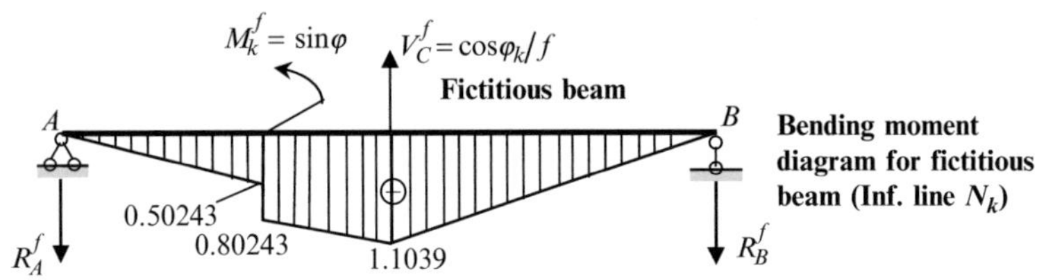

Fig. 2.20 Fictitious beam for N_k. Bending moment diagram for the fictitious beam presents the influence line N_k for the entire arch

Reactions of fictitious beam are

$$R_A^f = 0.048559\ (1/\mathrm{m})(\downarrow) \text{ and } R_B^f = 0.011059\ (1/\mathrm{m})(\uparrow).$$

Bending moments at specified points of the fictitious beam are

$$M_k^{f,\mathrm{left}} = -R_A^f \times a_k = -0.048559 \times 10 = -0.4855,$$

$$M_k^{f,\mathrm{right}} = -R_A^f \times a_k + 0.9539 = -0.048559 \times 10 + 0.9539 = 0.4684,$$

$$M_C^f = R_B^f \times \frac{l}{2} = 0.011059 \times 16 = 0.177.$$

Fictitious bending moments are dimensionless. These ordinates of influence line for Q_k have been obtained earlier and presented in Fig. 2.12.

Influence Line for N_k

Fictitious beam is loaded by the couple $M_k^f = \sin\varphi_k = 0.30$(counterclockwise) at section k and force $V_C^f = \cos\varphi_k/f = 0.9539/8 = 0.11924\ (1/\mathrm{m})$ (upwards) at section C (Fig. 2.20).

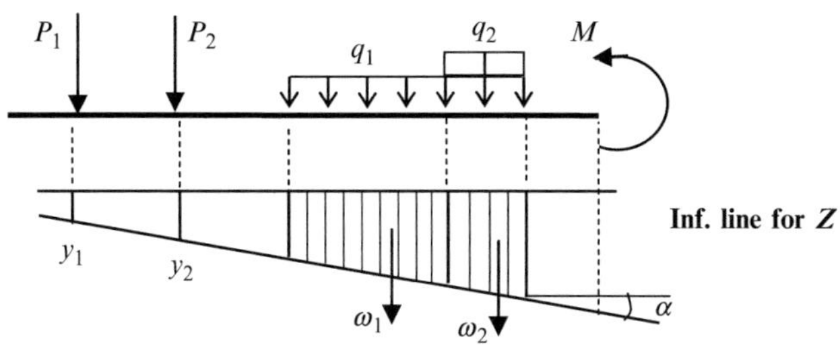

Fig. 2.21 Application of influence line for fixed loads

Reactions of fictitious beam are

$$R_A^f = 0.05024\,(1/\text{m})(\downarrow) \text{ and } R_B^f = 0.06899\,(1/\text{m})(\downarrow).$$

Bending moments at specified points of the fictitious beam are

$$M_k^{f,\text{left}} = -R_A^f \times a_k = -0.050243 \times 10 = -0.50243,$$

$$M_k^{f,\text{right}} = -R_A^f \times a_k - \sin\varphi_k = -0.50243 - 0.30 = -0.80243,$$

$$M_C^f = -R_B^f \times \frac{l}{2} = -0.06899 \times 16 = -1.1039.$$

Fictitious bending moments are dimensionless. These ordinates of influence line for N_k have been obtained earlier and presented in Fig. 2.13.

2.4.4 *Application of Influence Lines*

Influence lines, which describe the variation of any function Z (reaction, bending moment, shear, etc.) in the fixed section due to moving concentrated unit load $P = 1$ may be effectively used for calculation of this function Z due to *arbitrary fixed* and *moving* loads [Dar89], [Kar10].

Fixed load. Three types of fixed loads will be considered: concentrated loads P_i, uniformly distributed loads q_j, and couples M_k (Fig. 2.21).

Any function Z as a result of application of these loads may be calculated by the formula

$$Z = \pm\sum P_i y_i \pm \sum q_j \omega_j \pm \sum M_k \tan\alpha_k, \tag{2.17}$$

where y is the ordinates of influence line for function Z at the point where force P is applied; ω is the area of influence line graph for function Z within the portion where load q is applied; α_k is the angle between the x-axis and the portion of influence line for function Z within which M is applied.

The sign of Z due by load P depends on the sign of ordinate y of influence line. The sign of the area ω coincides with sign of ordinates of influence line; if the influence line within the load limits has different signs, then the areas must be taken with appropriate signs. If couple tends to rotate influence line toward base line through an angle less than 90°, then the sign is positive.

Formula (2.17) reflects the superposition principle and may be applied for any type of statically determinate and redundant structures.

Example 2.4. Assume that arch is subjected to fixed loads as shown in Fig. 2.5a. Calculate the reactions and internal forces of the arch at section k using influence lines.

Solution. Reactions of supports. Ordinates of influence line for R_A at the points of application the loads P_1 and P_2 are 0.75 and 0.125, respectively (Fig. 2.10b). The area of the influence line under the uniformly distributed load is

$$\omega = \frac{0.5 + 0.25}{2} \times 8 = 3.0\,(\mathrm{m}).$$

Therefore, the reaction $R_A = P_1 \times 0.75 + q \times 3 + P_2 \times 0.125 = 14.5\,\mathrm{kN}$.
The thrust H of the arch, using influence line (Fig. 2.10b) equals

$$H = P_1 \times 0.5 + q\frac{1 + 0.5}{2} \times 8 + P_2 \times 0.25 = 19\,\mathrm{kN}.$$

Internal forces in section k. The internal forces can be found in a similar way, using the relevant influence lines (Figs. 2.11–2.13). They are following:

$$M_k = P_1 \times 1.96 - q\frac{2.0788 + 1.0394}{2} \times 8 - P_2 \times 0.5194 = -9.500\,\mathrm{kN\,m}$$

$$Q_k = -P_1 \times 0.3883 + q\frac{0.177 + 0.0885}{2} \times 8 + P_2 \times 0.04425 = -1.405\,\mathrm{kN},$$

$$N_k = -P_1 \times 0.40194 - q\frac{1.1039 + 0.5519}{2} \times 8 - P_2 \times 0.2759 = -19.473\,\mathrm{kN}.$$

The magnitudes of just found internal forces M_k, Q_k, and N_k coincide with those computed in Example 2.1 and presented in Table 2.1.

These values of reactions coincide with those computed previously (Example 2.1).

Moving loads. Influence line for any function Z allows us to calculate Z for *any position* of a moving load, and that is very important, the most unfavorable position of the moving loads and corresponding value of the relevant function. Unfavorable (or dangerous) position of a moving load is such position, which leads to the maximum (positive or negative) value of the function Z. The following types of moving loads will be considered: one concentrated load, a set of loads, and a distributed load.

The set of connected moving loads may be considered as a model of moving truck. Specifications for truck loading may be found in various references, for example, in the American Association of State and Highway Transportation Officials (AASHTO). This code presents the size of the standard truck and the

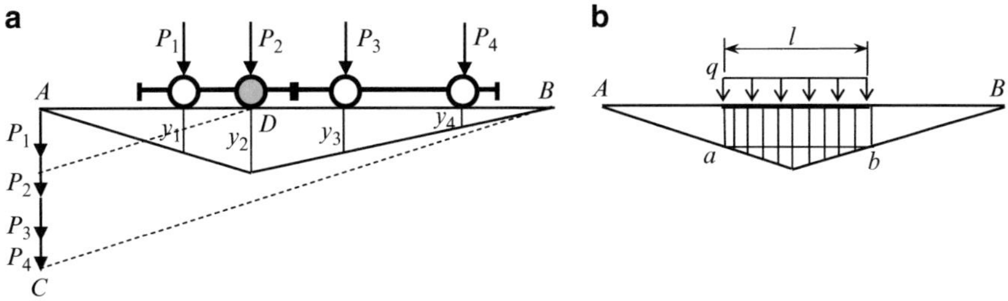

Fig. 2.22 Graphical definition of the unfavorable position of load for triangular influence line. (**a**) Set of concentrated load and (**b**) uniformly distributed load of fixed length l

distribution of its weight on each axle. The moving distributed load may be considered as a model of a set of containers which may be placed along the loading counter of the arch at arbitrary position.

Note that the term "moving load" with respect to influence line concept implies only that position of the load is arbitrary, i.e. this is a *static* load, which may have different positions along the beam. The time, velocity of the moving load, and any dynamic effects are not taken into account. Thus, for convenience, in this section we will use notion of "moving" or "traveling" load for static load, which may have *different* position along the structure.

The most unfavorable position of a single concentrated load is its position at a section with maximum ordinate of influence line. If influence line has positive and negative signs, then it is necessary to calculate corresponding maximum of the function Z using the largest positive and negative ordinates of influence line.

In case of set of concentrated moving loads, we assume that some of loads may be connected. This case may be applicable for moving cars, bridge cranes, etc. We will consider different forms of influence line.

Influence Line Forms a Triangle

A dangerous position occurs when one of the loads is located *over the vertex* of an influence line; this load is called a critical load. (The term "critical load" for problems of elastic stability, Chaps. 4 and 5, has a different meaning.) The problem is to determine *which* load among the group of moving loads is critical. After a critical load is known, all other loads are located according to the given distances between them.

The critical load may be easily defined by a graphical approach. Let the moving load be a model of two cars, with loads P_i on the each axle (Fig. 2.22). All distance between forces are given.

Step 1. Trace the influence line for function Z. Plot all forces P_1, P_2, P_3, P_4 in order using arbitrary scale from the left-most point A of influence line; the last point is denoted as C.

Step 2. Connect the right most point B with point C.

Step 3. On the base line show point D, which corresponds to the vertex of influence line and from this point draw a line, which is parallel to the line CB until it intersect with the vertical line AC.

Step 4. The intersected force (in our case P_2) presents a critical load; unfavorable location of moving cars presented in Fig. 2.22a.

Step 5. Maximum (or minimum) value of relative function is $Z = \sum P_i \times y_i$.

Influence Line Forms a Polygon

A dangerous position of the set of moving concentrated loads occurs when one or more loads stand over vertex of the influence line. Both the load and the apex of the influence line over which this load must stand to induce a maximum (or minimum) of the function under consideration are called critical. The critical apex of the influence line must be convex.

In case of uniformly distributed moving load, the maximum value of the function Z corresponds to the location of a distributed load q, which covers maximum one-sign area of influence line. The negative and positive portions of influence line must be considered in order to obtain minimum and maximum of function Z.

The special case of uniformly distributed moving load happens, if load is distributed within the *fixed length* l. In case of triangular influence line, the most unfavorable location of such load occurs when the portion $ab = l$ and base AB will be parallel (Fig. 2.22b).

Example 2.5. Simply supported beam with two overhangs is presented in Fig. 2.23. Determine the most unfavorable position of load, which leads to maximum (positive and negative) values of the bending moment and shear at section k. Calculate corresponding values of these functions. Consider the following loads: uniformly distributed load q and two connected loads P_1 and P_2 (a twin-axle cart with different wheel loads).

Solution. Influence lines for required functions Z are presented in Fig. 2.23.

Action of a uniformly distributed load $q = 1.6$ kNm. Distributed load leads to maximum value of the function if the area of influence lines within the distributed load is maximum. For example, the positive shear at section k is peaked if load q covers all portions of influence line with positive ordinates; for minimum shear in the same section the load q must be applied within portion with negative ordinates.

$$Q_{k(\max +)} = 1.6 \times \frac{1}{2}(0.3 \times 3 + 0.4 \times 4) = 2\,\text{kN};$$

$$Q_{k(\max -)} = -1.6 \times \frac{1}{2}(0.6 \times 6 + 0.3 \times 3) = -3.6\,\text{kN};$$

$$M_{k(\max +)} = 1.6 \times \frac{1}{2} 10 \times 2.4 = 19.2\,\text{kNm};$$

$$M_{k(\max -)} = -1.6 \times \frac{1}{2}(1.2 \times 3 + 1.8 \times 3) = -7.2\,\text{kNm}.$$

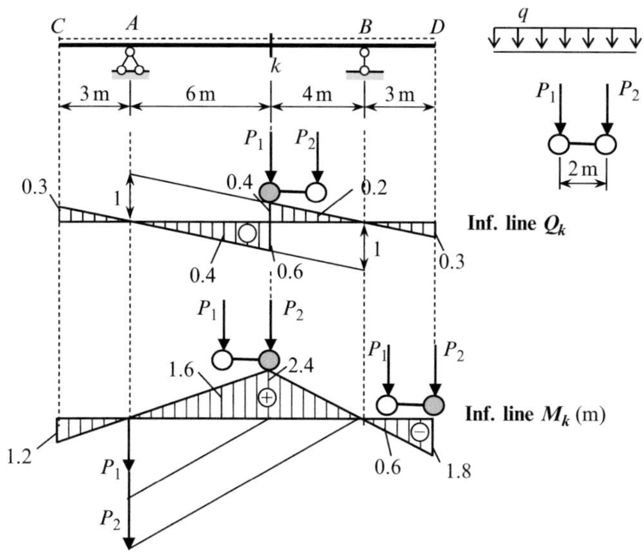

Fig. 2.23 Design diagram of the beam, influence lines, and most unfavorable positions of two connected loads

Positive value of $M_{k\ \max}$ means that if load is located between AB, the tensile fibers of the beam at section k are located below longitudinal axis of the beam. If load is located within the overhangs, then bending moment at section k is negative, i.e., the tensile fibers at section k are located above the longitudinal axis of the beam.

Action of the set of loads $P_1 = 5$ kN and $P_2 = 8$ kN. Unfavorable locations of two connected loads are shown in Fig. 2.23. Critical load for bending moment at section k (triangular influence line) is defined by the graphical method; the load P_2 is a critical one and it should be placed over the vertex of influence line.

$$Q_{k(\max +)} = 5 \times 0.4 + 8 \times 0.2 = 3.6\,\text{kN},$$
$$Q_{k(\max -)} = -(5 \times 0.4 + 8 \times 0.6) = -6.8\,\text{kN},$$
$$M_{k(\max)+} = 5 \times 1.6 + 8 \times 2.4 = 27.2\,\text{kNm},$$
$$M_{k(\max -)} = -(5 \times 0.6 + 8 \times 1.8) = -17.4\,\text{kNm}.$$

If a set of loads P_1 and P_2 modeling a crane bridge, then the *order* of loads is fixed and cannot be changed. If a set of loads P_1 and P_2 is a model of a moving car, then we need to consider the case when a car moves in opposite direction. In this case, the order of forces from left to right becomes P_2 and P_1.

2.5 Core Moments and Normal Stresses

This paragraph is devoted to simplifying the procedure to calculate the normal stresses caused by the simultaneous action of M and N. The concept of the "core moments" is introduced and their influence lines are constructed. We discuss the most unfavorable loading of the influence line.

2.5.1 Normal Stresses

Let us consider an arbitrary section nm of the arch. Assume that the load acts in the one of the main planes of the cross section. The point of application of the resultant R is shifted from the axial line of the arch by a length e; magnitude of this force, its direction, and point of application may be determined using a concept "curve of pressure" as explained in Appendix "Pressure curve". This force is resolved into the normal N and shear force Q (Fig. 2.24a).

In the case of an eccentrically loaded bar, the maximum normal stresses, caused by the bending moment M and compressed force N, arises at the extreme fibers of the cross section

$$\sigma = -\frac{N}{A} \pm \frac{M}{W}, \tag{2.18}$$

where N is the normal component of a force R and the bending moment $M = Ne$; A, W, I_x are the area, elastic section modulus, and moment of inertia of the cross section of the arch, respectively. In the case of a nonsymmetrical section, the elastic section moduli are $W_n = I_x/a_1$ and $W_m = I_x/a_2$, where a_1 and a_2 are the distances from the neutral line to an extreme fibers.

For determining the maximum normal stresses due to moving load, it is necessary to load the influence lines for M and N. These influence lines have different shapes and the influence lines for M can alternate in sign. Therefore, this procedure becomes cumbersome. However, the two-termed formula (2.18) may be simplified.

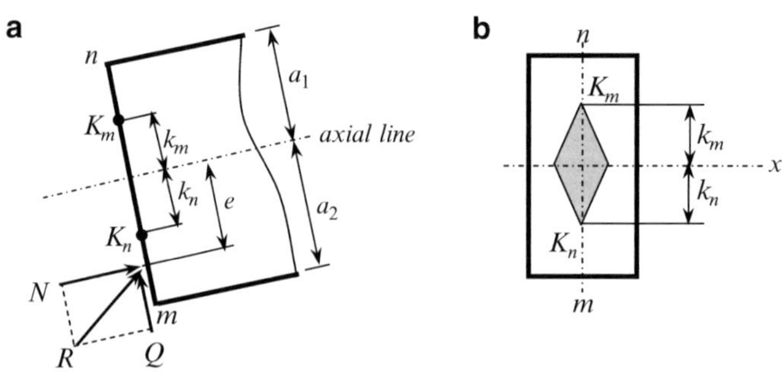

Fig. 2.24 (**a**) Internal forces at section n–m and (**b**) core of the cross section

Figure 2.24b presents the core (kern) for rectangular cross section; determination of its shapes and dimensions for arbitrary cross section may be found in the strength of materials textbooks. The concept of the core of the cross section was introduced by Bresse [Bre54], [Tim53], [Tod60]. The top and bottom points of the core are denoted by K_m and K_n.

If a force is applied at the bottom point K_n of the core, then $M = N \times k_n$ and normal stresses at the top fibers n equals zero

$$\sigma_n = -\frac{N}{A} + \frac{M}{W_n} = -\frac{N}{A} + \frac{Nk_n}{W_n} = 0. \tag{2.18a}$$

This equation leads to the formula $k_n = W_n/A$. Similarly, if a force is applied at the top point K_m of the core, then normal stresses at the bottom fibers m equals to zero and we get $k_m = W_m/A$.

If the compressed force N is applied as shown in Fig. 2.24a, then the normal stress at the bottom point m is

$$\sigma_m = -\frac{N}{A} - \frac{M}{W_m} = -\frac{N}{A} - \frac{Ne}{W_m} = -\frac{N}{W_m}\left(\frac{W_m}{A} + e\right) = -\frac{N}{W_m}(k_m + e).$$

The core moment presents the moment of the force N about the top core point K_m

$$M_{K_m}^{\text{core}} = N(e + k_m). \tag{2.19}$$

This moment differs from the usual bending moment by a term Nk_m. Finally, for normal stress in the bottom fibers of the cross section, we get the formula

$$\sigma_m = -\frac{M_{K_m}^{\text{core}}}{W_m}. \tag{2.20}$$

This formula shows that the maximal normal stresses caused by the moment M and force N equal to the normal stress caused by the core moment only. Similarly, the normal stress at the upper fibers n may be calculated by the formula $\sigma_n = M_{K_n}^{\text{core}}/W_n$, where core moment

$$M_{K_n}^{\text{core}} = N(e - k_n),$$

presents the moment of the force N about the bottom core point K_n.

2.5.2 *Influence Lines for Core Moments*

For construction of the influence line for core moments at section k, we will use the nil point method. This procedure will be the same as for construction of influence

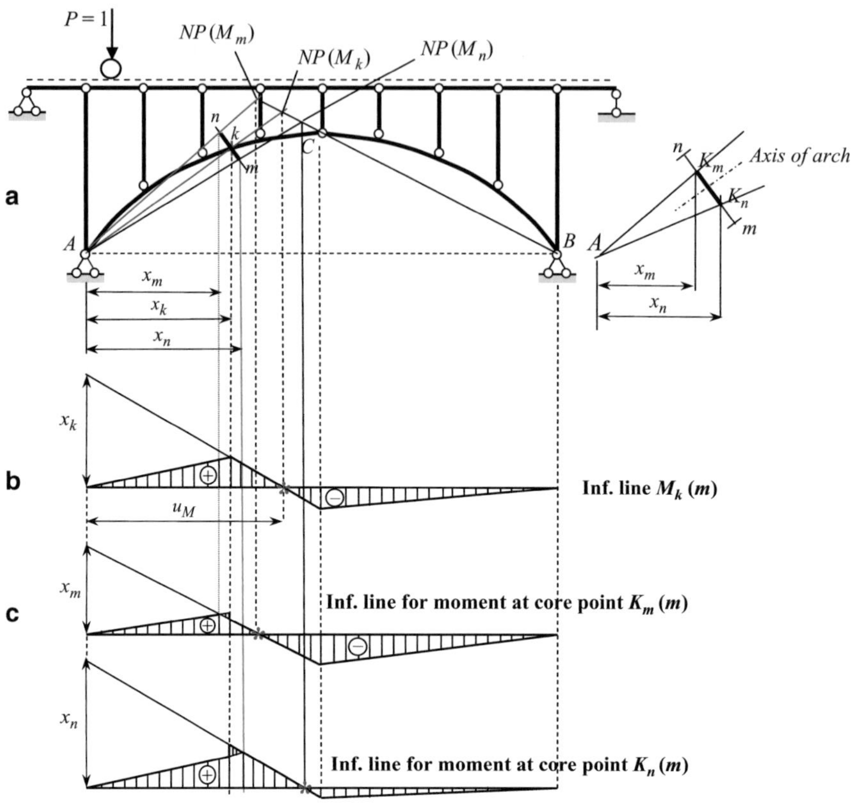

Fig. 2.25 Three-hinged arch. (**a**) Design diagram; (**b**) influence lines for bending moment M_k; and (**c**) influence lines for core moments at section k

line for bending moment at section k (Figs. 2.14 and 2.25a, b); indirect load application is not taken into account.

We show the top and bottom fiber points n and m at section k and denote the top and bottom core points by K_m and K_n (Fig. 2.25). These core points have coordinates x_m and x_n. Influence lines for core moments contain additional areas which are placed between two vertical lines; one of these lines passes through point k laying on the axis of the arch, and other vertical line passes over the core point (Fig. 2.25c). Additional areas of influence lines arise because in this section of the arch the influence line of axial force has a jump. Ordinates of this additional area of influence line are small and they may be neglected [Dar89]. However, it is important that the location of the nil points for core moments do not coincides with nil point for bending moment.

Influence lines for core moments allow us to answer the following question: which part of influence lines should be loaded by a uniformly distributed load (or any live load) in order for the tensile normal stresses at extrados (top) fibers of section k to be maximum.

The stresses at the top fibers n will be *tensile* if a resultant of all external left-hand (or right-hand) forces will passes below the bottom core point K_n. Given this, the moment about the core point K_n will be negative. Therefore, the load should be placed over the *negative* ordinates of the influence line for bending moment at core point K_n. If load will be placed over the positive ordinates of the same influence line, then a compressed stresses at extrados fibers n of section k will arise.

2.6 Special Types of Three-Hinged Arches

This paragraph contains analysis of the special types of three-hinged arch subjected to fixed and moving loads. Among them are the circular arch with elevated simple tie, parabolic arch with complex tie, and askew arch.

2.6.1 Arch with Elevated Simple Tie

Three-hinged arch with tie may be obtained from an ordinary three-hinged arch without a tie, if the horizontal constraint at support B (or A), which prevents horizontal displacement of the abutment hinge, is replaced by a tie. The tie may be located on the level of the supported points (Fig. 2.26a) or above them (elevated tie) (Fig. 2.26b). Application of complex tie is also possible. One type of an arch with a complex tie is shown in Fig. 2.26c. Three-hinged arches with ties represent geometrically unchangeable statically determinate structures and have certain peculiarities of their analysis, which are presented below.

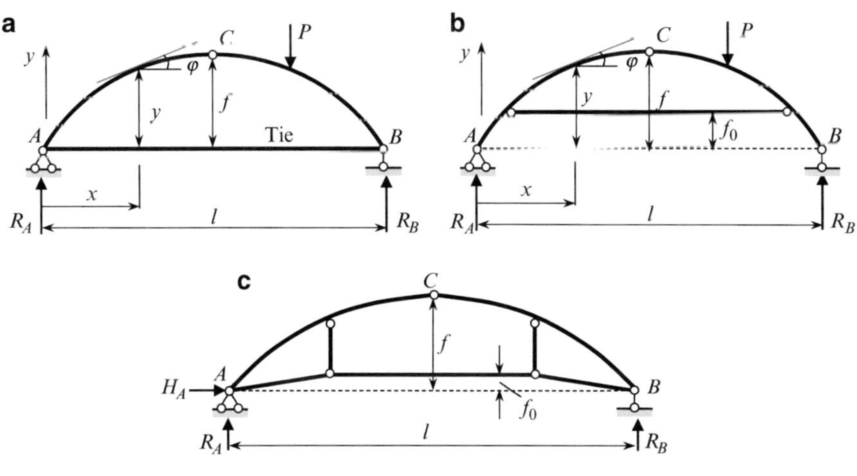

Fig. 2.26 Design diagrams of three-hinged arches with tie. (**a**) Simple tie on support level; (**b**) arch with elevated simple tie; and (**c**) arch with complex tie

In case (2.26a), the tensile force in the tie (thrust) is $H = M_C^O/f$, where M_C^O represent the bending moment at section C for the reference beam. Two forces H act at points A and B, as for an ordinary three-hinged arch without tie. Therefore, internal forces in cross sections of the given arch will be exactly the same as for arch without tie and may be calculated using (2.6). However, support B of the arch with tie has a horizontal displacement due to the elastic properties of the tie, while a three-hinged arch without tie has no a horizontal displacement.

In case (2.26b), the thrust in the tie is

$$H = \frac{M_C^O}{f - f_0}. \tag{2.21}$$

Two forces H act above points A and B. Internal forces in cross sections of the arch are obtained from modified (2.6); they depend on location of the section on the arch (below or above the tie). If sections are located below the tie level then

$$M_x = M_x^0, \quad Q_x = Q_x^0 \cos\varphi, \quad N_x = -Q_x^0 \sin\varphi. \tag{2.22}$$

If sections are located above the tie level, then

$$\begin{aligned} M_x &= M_x^0 - H(y - f_0), \\ Q_x &= Q_x^0 \cos\varphi - H \sin\varphi, \\ N_x &= -Q_x^0 \sin\varphi - H \cos\varphi, \end{aligned} \tag{2.23}$$

where M_x^0, Q_x^0 are bending moment and shear force at section x for the reference beam.

In the case of a complex tie, it is necessary to determine a thrust in the tie, then internal forces in all the members of the tie and finally, internal forces in the arch itself. The complex tie of the arch allows us not only to increase the strength of the arch structure but also to distribute internal forces in the arch as required.

Example 2.6. Design diagram of three-hinged circular arch with elevated tie is presented in Fig. 2.27. Geometrical parameters of the arch and loads are the same as for a three-hinged arch without tie (Fig. 2.5a). We need to compute the internal forces in the arch and compare results obtained for the same arch without tie.

Solution. The vertical reactions of supports, as in Example 2.1, are $R_A = R_A^0 = 14.5\,\text{kN}$, $\quad R_B = R_B^0 = 19.5\,\text{kN}$.

Horizontal reaction H_A at the support A may be calculated from the equation $\sum X = 0 \rightarrow H_A = 0$.

The force H in the tie may be determined using equilibrium condition for left (or right) part of the arch (section 1–1)

$$H \rightarrow \sum M_C^{\text{left}} = 0 \rightarrow R_A \times 16 - P_1 \times 8 - H(f - f_0) = 0 \rightarrow H = \frac{M_C^0}{f - f_0} = \frac{152}{8 - 2} = 25.33\,\text{kN}.$$

Computations of the geometrical parameters and internal forces of the arch are presented in Table 2.2.

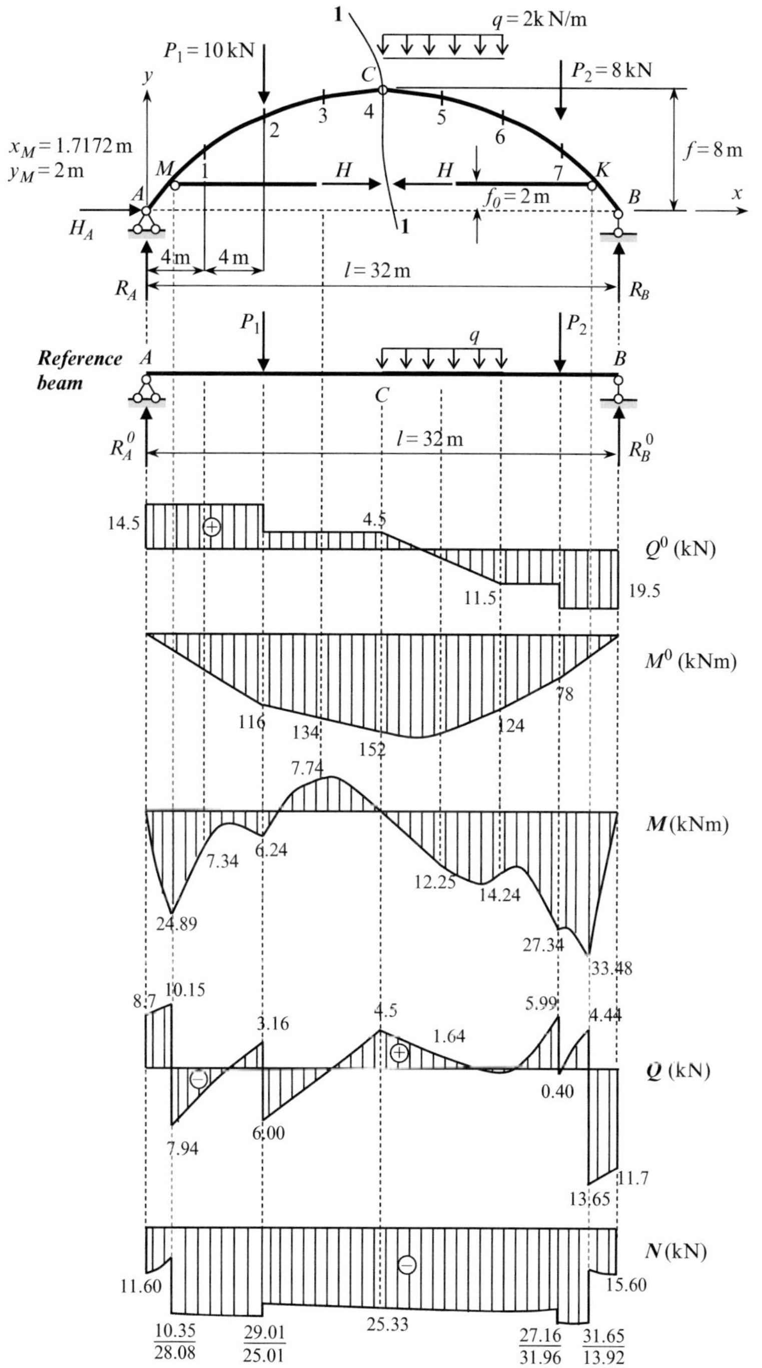

Fig. 2.27 Three-hinged arch with simple elevated tie. Design diagram, reference beam, and internal force diagrams

Table 2.2 Internal forces in three-hinged circular arch with simple elevated tie ($R_A = 14.5\,\text{kN}$; $R_B = 19.5\,\text{kN}$; $H = 25.33\,\text{kN}$)

Section	x (m)	y (m)	$\sin\varphi$	$\cos\varphi$	M_x^O (kNm)	$H(y - f_0)$ (kNm)	M_x (kNm)	Q_x^O (kN)	Q_x (kN)	N_x (kN)
0	1	2	3	4	5	5′	6	7	8	9
A	0.0	0.0	0.8	0.6	0	–	0	14.5	8.7	−11.6
M	1.7171	2.0	0.7141	0.70	24.8975	0.0	24.8975	14.5	10.15/−7.938	−10.3544/−28.0854
1	4	4.0	0.6	0.8	58	50.66	7.34	14.5	−3.598	−28.964
2	8	6.333	0.4	0.9165	116	109.755	6.245	14.5/4.5[a]	3.1572/−6.0	−29.0149/−25.0149
3	12	7.596	0.2	0.9798	134	141.747	−7.7467	4.5	−0.6569	−25.718
4 (C)	16	8.0	0.0	1.0	152	152	0.0	4.5	4.5	−25.33
5	20	7.596	−0.2	0.9798	154	141.746	12.2533	−3.5	1.6367	−25.5183
6	24	6.333	−0.4	0.9165	124	109.755	14.245	−11.5	−0.4077	−27.8149
7	28	4	−0.6	0.8	78	50.66	27.34	−11.5/−19.5	5.998/−0.402	−27.164/−31.964
K	30.2828	2.0	−0.7141	0.70	33.4834	0.0	33.4834	−19.5	4.4381/−13.65	−31.6559/−13.9249
B	32	0.0	−0.8	0.6	0	–	0	−19.5	−11.7	−15.6

Columns containing separate terms $Q^0\cos\varphi$, $Q^0\sin\varphi$, $H\cos\varphi$, $H\sin\varphi$ are not presented

Note: Discontinuity of shear and axial force diagrams at sections 2 and 7 is caused by vertical forces P_1 and P_2, while discontinuity of same diagrams at sections M and K is caused by thrust H

[a] Values in numerator and denominator (columns 7–9) are the mean value of the force to the left and to the right of corresponding section

Radius of the circle, according to (1) is $R = (f/2) + l^2/8f = (8/2) + [32^2/(8 \times 8)] = 20\,\text{m}$. Columns 1 and 2 contain ordinate x and corresponding ordinate y (in meters) for specified sections. Ordinate $y(x)$ is

$$y = \sqrt{R^2 - \left(\frac{l}{2} - x\right)^2} - R + f = \sqrt{400 - (16 - x)^2} - 12\,(\text{m}).$$

Columns 3 and 4 contain values of $\sin\varphi = (l - 2x)/2R = (32 - 2x)/40$ and $\cos\varphi = (\mathrm{y} + \mathrm{R} - \mathrm{f})/\mathrm{R} = (\mathrm{y} + 12)/20$.

Values of bending moment and shear for reference beam are tabulated in columns 5 and 7 and taken directly from corresponding diagrams, which are presented in Fig. 2.27. Values of $H(y - f_0)$ are given in column 5′. Sections A and B have no entries for column 5′, which means that force in the tie does not effect on the bending moment at corresponding section of the arch. Values of bending moment, shear, and normal forces for three-hinged arch are tabulated in columns 6, 8, and 9. They have been computed using (2.22) for sections which are located below the tie. For example, for section A, we get

$$\begin{aligned} Q_A &= Q_A^0 \cos\varphi_A = 14.5 \times 0.6 = 8.7\,\text{kN} \\ N_A &= -Q_A^0 \sin\varphi_A = -14.5 \times 0.8 = -11.6\,\text{kN}. \end{aligned}$$

For sections above the tie, we need to use (2.23). For example, for section 3, we get

$$\begin{aligned} M_x &= M_x^0 - H(y - f_0) = 134 - 25.33 \times (7.596 - 2) = -7.7467\,\text{kNm}, \\ Q_x &= Q_x^0 \cos\varphi - H \sin\varphi = 4.5 \times 0.9798 - 25.33 \times 0.2 = -0.6569\,\text{kN}, \\ N_x &= -Q_x^0 \sin\varphi - H \cos\varphi = -4.5 \times 0.2 - 25.33 \times 0.9798 = -25.718\,\text{kN}. \end{aligned}$$

Corresponding diagrams are presented in Fig. 2.27. Bending moment diagrams for beam and arch are shown on the extended fibers; therefore, the signs of bending moments are omitted. For convenience, different scales have been adopted for different diagrams.

Verification. The vertical concentrated force P leads to value of discontinuity $P\cos\varphi$ and $P\sin\varphi$ for diagram Q and N, respectively; the horizontal force H leads to value of discontinuity $H\sin\varphi$ and $H\cos\varphi$ for same diagrams Q and N.

Values of discontinuity on shear and normal force diagrams due to concentrated forces H and P_i are:

Shear force diagram point at M: $10.15 - (-7.938) = 18.088 = H\sin\varphi$,
Point 2: $3.1572 - (-6.0) = 9.1572 = P_1\cos\varphi$.

Normal force diagram point at M: $-10.35 - (-28.08) = 17.73 = H\cos\varphi$,
Point 2: $-25.01 - (-29.01) = 4.0 = P_1\sin\varphi$.

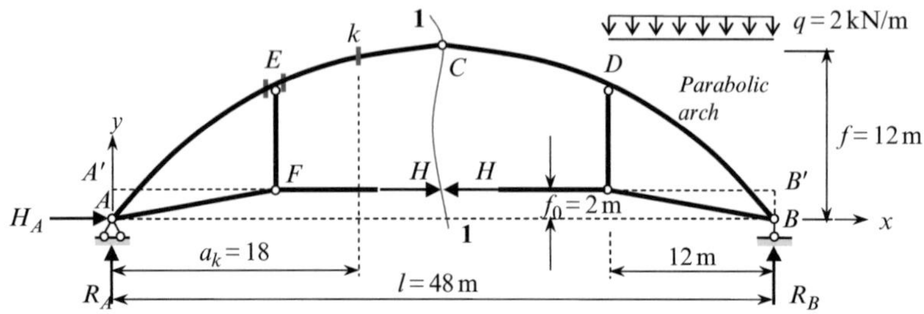

Fig. 2.28 Design diagram of the arch with complex tie

Values of discontinuity on shear and normal force diagrams for points 7 and K are verified in a similar manner.

Now we will compare the internal force diagrams for the arch without tie (Fig. 2.6) and the arch with the elevated tie (Fig. 2.27). Unlike the arch without tie, two horizontal forces H act at points M and K. Therefore, the shear and axial force diagrams at points M and K have abrupt changes $H \sin\varphi$ for the Q diagram and $H \cos\varphi$ for the N diagram. The axial force N for both arches remains compressed.

The fundamental change occurs in the distribution of bending moments. For example, for all sections of the left part of the arch without tie, the extended fibers are located above the neutral line (Fig. 2.6), while for arch with the tie, the extended fibers are located below the neutral line (Fig. 2.27) (portion A-2 and slightly further). For the right part of the arch without tie, the bending moment diagram changes the sign three times: in the neighborhood of point n and 7, the extended fibers are located above and below the neutral line, respectively, while for arch with tie, the entire right part of the arch has extended fibers below the neutral line.

2.6.2 *Arch with Complex Tie*

Analysis of such structure subjected to fixed and moving load has some features. Design diagram of the symmetrical parabolic arch with complex tie is presented in Fig. 2.28. The arch is loaded by vertical uniformly distributed load $q = 2$ kN/m. We need to determine the reactions of the supports, thrust, and internal forces at section k ($a_k = 18$ m, $y_k = 11.25$ m, $\tan\varphi_k = 0.25$, $\cos\varphi_k = 0.970$, $\sin\varphi_k = 0.2425$) as well as to construct the influence line for above-mentioned factors.

Reactions and Internal Forces at Section *k*

The vertical reactions are determined from the equilibrium equations of all the external forces acting on the arch

$$R_A \to \sum M_B = 0: \quad -R_A \times 48 + q \times 12 \times 6 = 0 \to R_A = 3\,\text{kN},$$
$$R_B \to \sum M_A = 0: \quad R_B \times 48 - q \times 12 \times 42 = 0 \to R_B = 21\,\text{kN}.$$

Horizontal reaction at support A is $H_A = 0$.

The thrust H in the tie (section 1–1) is determined from the following equation

$$H \to \sum M_C^{\text{left}} = 0: \quad -R_A \frac{l}{2} + H(f - f_0) = 0 \to H = M_C^0/(f - f_0) = 7.2\,\text{kN}. \tag{2.24a}$$

Equilibrium equations of joint F lead to the axial forces at the members of AF and EF of the tie.

Internal forces at section k for a reference simply supported beam are as follows:

$$M_k^0 = R_A \times x_k = 3 \times 18 = 54\,\text{kNm},$$
$$Q_K^0 = R_A = 3\,\text{kN}.$$

Internal forces at point k for three-hinged arch are determined as follows

$$\begin{aligned} M_k &= M_k^0 - H(y_k - f_0) = 54 - 7.2 \times (11.25 - 2) = -12.6\,\text{kNm}, \\ Q_k &= Q_k^0 \cos\varphi_k - H \sin\varphi_k = 3 \times 0.970 - 7.2 \times 0.2425 = 1.164\,\text{kN}, \\ N_k &= -\left(Q_k^0 \sin\varphi_k + H \cos\varphi_k\right) = -(3 \times 0.2425 + 7.2 \times 0.970) = -7.711\,\text{kN}. \end{aligned} \tag{2.24b}$$

Note, that the discontinuity of the shear and normal forces at section E left and right of the vertical member EF is $N_{EF} \times \cos\varphi$ and $N_{EF} \times \sin\varphi$, respectively.

Influence Lines for Thrust and Internal Forces (*M*, *Q*, *N*) at Section *k*

Influence lines for vertical reactions R_A and R_B for arch and for reference simply supported beam coincide, i.e.,

$$\text{IL}(R_A) = \text{IL}\left(R_A^0\right), \qquad \text{IL}(R_B) = \text{IL}\left(R_B^0\right).$$

According to (2.24a), the equation of influence line for thrust becomes

$$\text{IL}(H) = \frac{1}{f - f_0} \times \text{IL}\left(M_C^0\right).$$

The maximum ordinate of influence line for H at crown C is $1/(f - f_0)\times (l/4) = 48/[4 \times (12 - 2)] = 1.2$. Influence line for thrust H may be considered as a key influence line (Fig. 2.29).

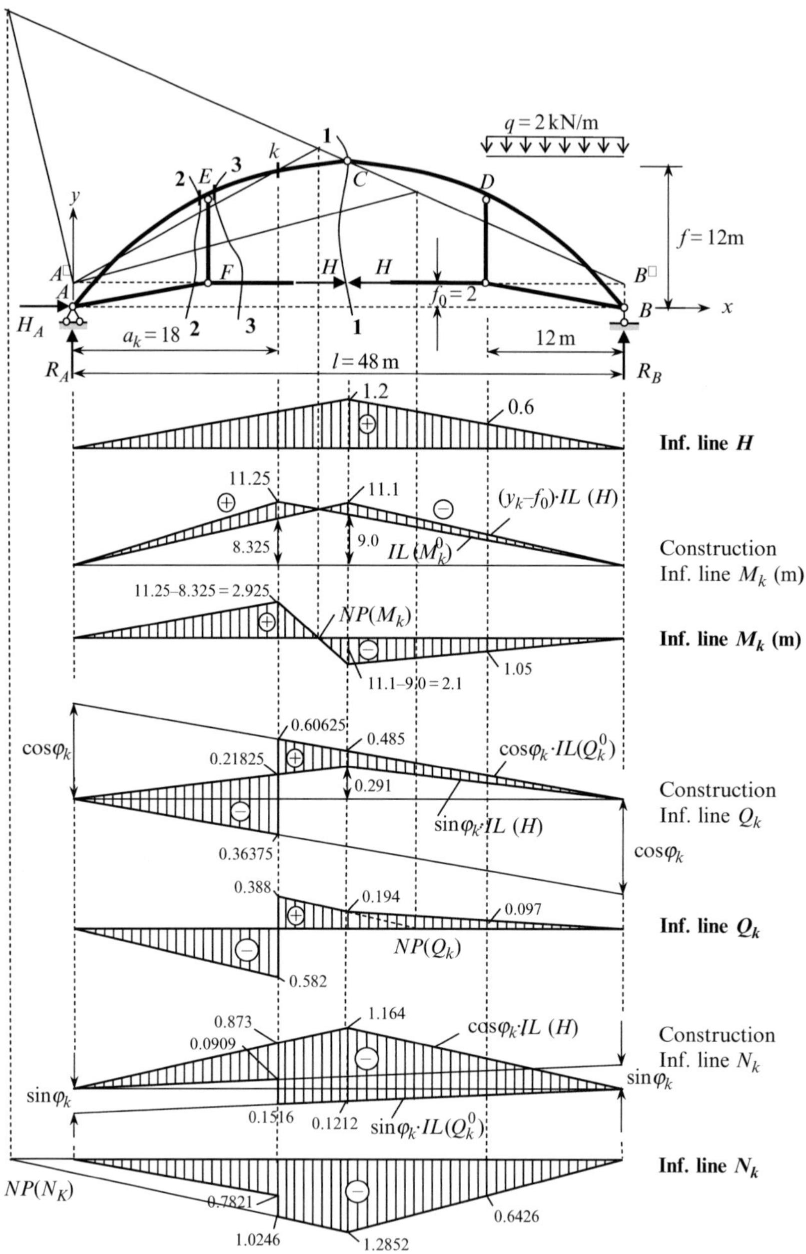

Fig. 2.29 Three-hinged arch with complex tie. Design diagram and influence lines

Bending Moment

According to (2.24b) for bending moment at any section, the equation of influence line for bending moment at section k is

$$\mathrm{IL}(M_k) = \mathrm{IL}\left(M_k^0\right) - (y_k - f_0) \times \mathrm{IL}(H) = \mathrm{IL}\left(M_k^0\right) - 9.25 \times \mathrm{IL}(H). \qquad (2.24\text{c})$$

Influence line M_K^0 presents a triangle with maximum ordinate $a_k b_k / l = 18 \times 30/48 = 11.25\,\mathrm{m}$ at section k, so the ordinate at crown C equals to 9 m. Influence line for thrust H presents the triangle with maximum ordinate 1.2 at crown C. Ordinate of the graph $(y_k - f_0) \times \mathrm{IL}(H)$ at crown C equals $(11.25 - 2) \times 1.2 = 11.1\,\mathrm{m}$, so ordinate at section k equals 8.325 m. Detailed construction of influence line M_k is shown in Fig. 2.29. Since both terms in (2.24c) has *different* signs, they should be plotted on the one side on the basic line and the final ordinates of influence line are located *between* two graphs $\mathrm{IL}\left(M_k^0\right)$ and $9.25 \times \mathrm{IL}(H)$.

Shear Force

According to (2.24b) for shear at any section, the equation of influence line for shear at section k is

$$\begin{aligned}\mathrm{IL}(Q_k) &= \cos\varphi_k \times \mathrm{IL}\left(Q_k^0\right) - \sin\varphi_k \times \mathrm{IL}(H)\\ &= 0.970 \times \mathrm{IL}\left(Q_k^0\right) - 0.2425 \times \mathrm{IL}(H).\end{aligned} \qquad (2.24\text{d})$$

Ordinates of the graph $0.970 \times \mathrm{IL}\left(Q_k^0\right)$ are 0.36375 and 0.60625 to the left and to the right at section k, so ordinate at crown C is 0.485. Maximum ordinate of the graph $0.2425 \times \mathrm{IL}(H) = 0.2425 \times 1.2 = 0.291$ is located at crown C, so ordinate at section k is 0.21825.

Ordinate of influence line for shear at crown C equals $0.485–0.291 = 0.194$; the left and the right of section k ordinates of influence line become

$$-(0.36375 + 0.21825) = -0.582 \text{ and } 0.60625 - 0.21825 = 0.388.$$

Detailed construction of influence line Q_k is shown in Fig. 2.29.

Normal Force

According to (2.24b) for normal force at any section, the equation of influence line for normal force at section k is

$$\begin{aligned}\mathrm{IL}(N_k) &= -\sin\varphi_k \times \mathrm{IL}\left(Q_k^0\right) - \cos\varphi_k \times \mathrm{IL}(H)\\ &= -\left[0.2425 \times \mathrm{IL}\left(Q_k^0\right) + 0.970 \times \mathrm{IL}(H)\right]\end{aligned} \qquad (2.24\text{e})$$

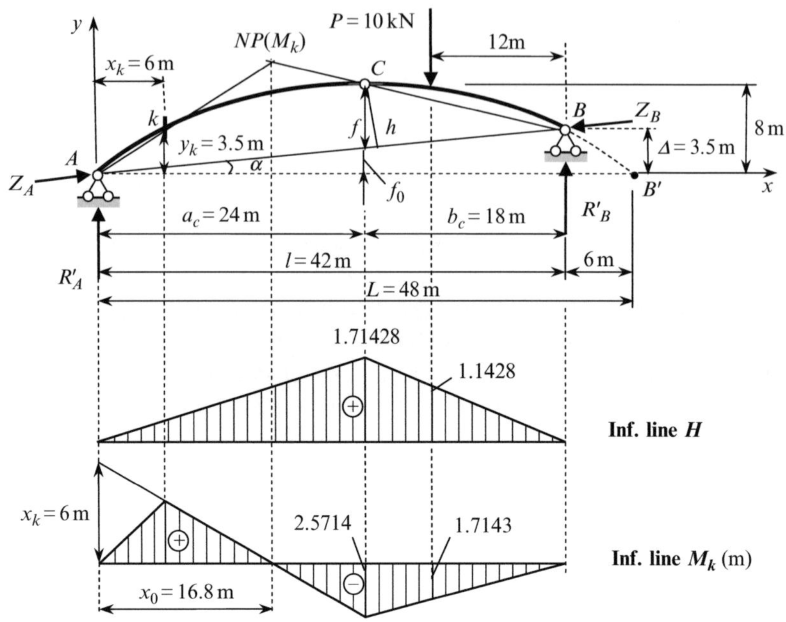

Fig. 2.30 Three-hinged askew arch. Design diagram and influence lines

Maximum ordinate of the graph $0.970 \times \mathrm{IL}(H)$ is $0.970 \times 1.2 = 1.164$; this ordinate is located at crown C. Specific ordinates of the graph $0.2425 \times \mathrm{IL}(Q_k^0)$ are 0.09094 and 0.1516 and located to the left and to the right of section k.

Detailed construction of influence line N_k is shown in Fig. 2.29. This figure also represents the construction of influence lines using nil point method; note that construction of the nil points must be done on the basis of conventional supports A' and B'.

2.6.3 Askew Arch

The arch with support points located on the different levels is called askew (or rising) arch. Three-hinged askew arch is geometrically unchangeable and statically determinate structure. Analysis of askew arch subjected to fixed and moving loads has some features.

Design diagram of three-hinged askew arch is presented in Fig. 2.30. Let the shape of the arch is parabola, span of the arch $l = 42$ m and support B is $\Delta = 3.5$ m higher than support A. The total height of the arch at hinge C is 8 m. The arch is loaded by force $P = 10$ kN. It is necessary to calculate the reactions and bending moment at section k, construct the influence lines for thrust and bending moment M_k, and apply influence lines for calculation of bending moment and reactions due to fixed load.

Equation of the axis of parabolic arch is

$$y = 4(f + f_0) \times (L - x) \times \frac{x}{L^2},$$

where span for arch A–C–B' with support points on the same level is $L = 48$ m. For $x = 42$ m (support B), ordinate $y = 3.5$ m, so

$$\tan\alpha = \frac{\Delta}{l} = \frac{3.5}{42} = 0.0833 \rightarrow \cos\alpha = 0.9965 \rightarrow \sin\alpha = 0.08304.$$

Other geometrical parameters are

$$f_0 = 24\tan\alpha = 2.0\,\text{m} \rightarrow f = 8 - 2 = 6\,\text{m} \rightarrow h = f\,\cos\alpha = 6 \times 0.9965 = 5.979\,\text{m}.$$

For $x = 6$ m (section k), the ordinate $y_k = 3.5$ m.

Reactions and Bending Moment at Section *k*

It is convenient to resolve total reaction at point A into two components. One of them, R'_A, has vertical direction and other, Z_A, is directed along the line AB. Similarly resolve the reaction at support B. These components are R'_B and Z_B. The vertical forces R'_A and R'_B represent a *part* of the total vertical reactions. These vertical forces may be computed as for the reference beam

$$R'_A \rightarrow \sum M_B = 0: \qquad -R'_A \times 42 + P \times 12 = 0 \rightarrow R'_A = 2.857\,\text{kN},$$
$$R'_B \rightarrow \sum M_A = 0: \qquad R'_B \times 42 - P \times 30 = 0 \rightarrow R'_B = 7.143\,\text{kN}.$$

Since a bending moment at crown C is zero then

$$Z_A \rightarrow \sum M_C^{\text{left}} = 0: \quad Z_A \times h - M_C^0 = 0 \rightarrow Z_A = \frac{M_C^0}{h} = \frac{2.857 \times 24}{5.979} = 11.468\,\text{kN},$$
$$Z_A = Z_B = Z,$$

where M_C^0 is a bending moment at section C for the reference beam.

Thrust H represents the horizontal component of the Z, i.e., $H = Z\cos\alpha = 11.468 \times 0.9965 = 11.428\,\text{kN}$.

The total vertical reactions may be defined as follows

$$R_A = R'_A + Z\sin\alpha = 2.857 + 11.468 \times 0.08304 = 3.809\,\text{kN},$$
$$R_B = R'_B - Z\sin\alpha = 7.143 - 11.468 \times 0.08304 = 6.191\,\text{kN}.$$

Bending moment at section k:

$$M_k = M_k^0 - Hy = 3.809 \times 6 - 11.428 \times 3.5 = -17.144\,\text{kN}.$$

Influence Lines for Thrust and Bending Moment M_k

Thrust. Since $H = Z\cos\alpha = \left(M_C^0/h\right)\cos\alpha$, then equation of influence line for thrust becomes

$$\text{IL}(H) = \frac{\cos\alpha}{h} \times \text{IL}\left(M_C^0\right).$$

The maximum ordinate of influence line occurs at crown C and equals

$$\frac{\cos\alpha}{h} \times \frac{a_C b_C}{l} = \frac{0.9965}{5.979} \times \frac{24 \times 18}{42} = 1.71428.$$

Bending moment M_k. Since $M_k = M_k^0 - Hy_k$, then equation of influence line for bending moment at section k becomes

$$\text{IL}(M_k) = \text{IL}\left(M_k^0\right) - y_k \times \text{IL}(H).$$

Influence line may be easily constructed using the nil point method. Equation of the line Ak is

$$y = \frac{3.5}{6}x = 0.5833x.$$

Equation of the line BC is

$$y - y_C = m(x - x_C) \rightarrow y - 8 = -\frac{4.5}{18}(x - 24) \rightarrow y = 14 - 0.25x,$$

where m is a slope of the line BC.

The nil point NP(M_k) of influence line for M_k is the point of intersection of lines Ak and BC. Solving these equations leads to $x_0 = 16.8$ m. Influence lines for H and M_k are presented in Fig. 2.30. Maximum positive and negative bending moment at section k occurs if load P is located at section k and hinge C, respectively. If load P is located within portion x_0, then extended fibers at section k are located below the neutral line of the arch.

The thrust and bending moment at section k may be calculated using the relevant influence lines

$$H = Py = 10 \times 1.1428 = 11.428\,\text{kN}$$
$$M_k = Py = 10 \times (-1.7143) = -17.143\,\text{kNm}.$$

These values coincide exactly with those calculated previously.

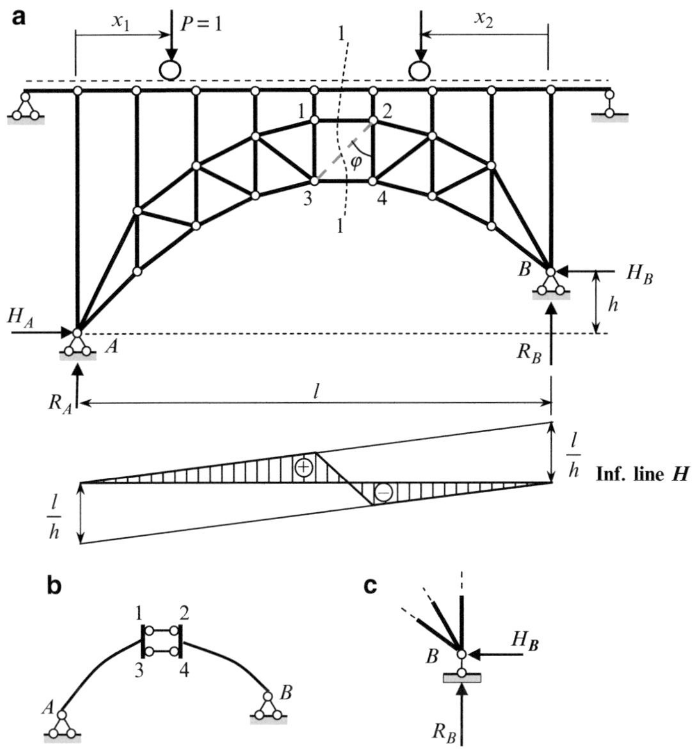

Fig. 2.31 Modified askew arched structure

As before, the influence line for thrust constructed *once* may be used for its computation for different cases of arbitrary loads. Then, knowing the vertical reactions and thrust, the internal forces at any point of the arch may be calculated by definition without using influence line for that particular internal force.

2.6.4 Latticed Askew Arch

Design diagram of the modified askew arched structure with over-arch construction is presented in Fig. 2.31a. Pinned supports A and B are located at different elevations. Each half-arch itself (A-1–3 and B-2–4) represents the structure with webbed members. Panel block 1–2–3–4 has no diagonal member, thus both half-arches are connected by means of two parallel rods 1–2 and 3–4. Therefore, the vertical relative displacement of two half-arches is possible (Fig. 2.31b), while in the classic three-hinged arch only angular relative displacement of two half-arches is possible. The vertical posts are used only to transmit loads directly to the upper chord of the structure.

Degree of freedom equals

$$W = 2J - S - S_0 = 2 \times 27 - 47 - 7 = 0,$$

where J, S, and S_0 are the number of hinged joints, members of structure and constraints of supports, respectively [Kar10]. Though the both part of arch represent the simplest truss (or rigid disc), they are connected in a specific way, mainly by members 1–2 and 3–4 as well as an imagine member AB (ground). These members are not parallel. The structure is statically determinate and geometrically unchangeable.

For analysis of this structure, we will apply the following procedure:

1. Replace the constraint of the support B, which prevents horizontal displacement, by a diagonal member 2–3 (dotted line in Fig. 2.31a) and apply external force H_B at point B (Fig. 2.31a, c). Such a substitution does not change the number of degree of freedom.
2. Consider two positions x_1 and x_2 of a moving load P and determine thrust $H_A = H_B = H$ in terms of x, l, and h, when the internal force in the substitute member 2–3 is zero.

Force $P = 1$ is located at the left part of the structure. Thrust $H \rightarrow \sum M_A = 0$:

$$R_B l + Hh - Px = 0 \rightarrow R_B = \frac{1}{l}(Px - Hh).$$

Internal force in substitute member D_{23} section 1–1 is determined as follows

$$\sum Y^{\text{right}} = 0: \quad R_B - D_{23} \cos \varphi = 0.$$

Taking into account the previous result for reaction R_B, internal force in diagonal member becomes

$$\frac{1}{l}(Px - Hh) - D_{23} \cos \varphi = 0.$$

However, diagonal member is absent, therefore $D_{23} = 0$ and the expression for thrust is

$$H = \frac{Px}{h}, \text{ so } \mathrm{IL}(H) = \frac{x}{h}.$$

Force $P = 1$ is located at the right part of the structure. Thrust $H \rightarrow \sum M_A = 0$:

$$R_B l + Hh - P(l - x) = 0 \rightarrow R_B = \frac{P(l - x) - Hh}{l}$$

Internal force in the substitute member $D_{23} \rightarrow \sum Y^{\text{right}} = 0 : R_B - D_{23} \times \cos \varphi - P = 0$.

Taking into account the previous result for reaction R_B, equation for internal force in diagonal member becomes

$$-D_{23} \cos \varphi - \frac{Px}{l} - \frac{Hh}{l} = 0.$$

However, $D_{23} = 0$ so the expression for thrust becomes

$$H = -\frac{Px}{h}, \text{ so IL}(H) = -\frac{x}{h}.$$

Influence line for H represents two parallel lines with ordinates l/h at the support points and connecting line within the panels 1–2. The sign of thrust H depends on location of the moving load (unlike previously considered arched structures).

Influence line H is a fundamental characteristic of the system. Knowing the influence line H allows us to calculate this reaction for any type of loadings. Calculation of all other reactions and internal forces in any members presents no difficulties.

Note if supports A and B will be located at the same level, then the system becomes instantaneously changeable. Indeed, in this case, two rigid discs (the left and right parts of the structure) are connected by three parallel members, mainly 1–2, 3–4 and AB [Kar10].

2.7 Complex Arched Structures

This paragraph contains analysis of the complex arched structures subjected to fixed and moving load. Among them are the multispan three-hinged arched structure and trusses with arched hinged chain.

2.7.1 Multispan Three-Hinged Arched Structure

Multispan three-hinged arched structure is a geometrically unchangeable structure, which consists of three-hinged arches connected together by means of hinges. Figure 2.32a presents the multispan arched structure which contains three-hinged arch ACB with overhang BG, arch DIF with overhang HD, and central three-hinged arch GEH, which is connected with left and right arches by means of hinges G and H.

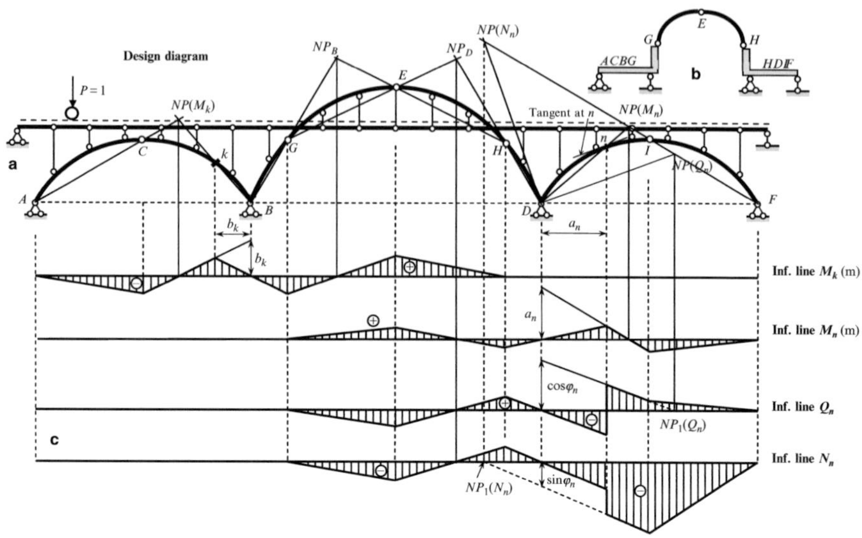

Fig. 2.32 Multispan arched structure. (**a**) Design diagram, (**b**) interaction scheme, and (**c**) influence lines for internal forces at sections *k* and *n*

It is necessary to construct the influence lines for bending moment, shear, and normal forces at sections *k* and *n*, using the nil point method. Indirect application of the load on the arch system should not be taken into account.

Kinematical Analysis

Degrees of freedom of this arch structure, according to Chebushev formula, are determined as $W = 3D - 2H_0 - S_0 = 3 \times 6 - 2 \times 5 - 8 = 0$, where D, H_0, and S_0 are number of rigid discs, number of simple hinges, and number of constraints of support, respectively [Kar10].

The whole structure may be presented as two main arched structures *ACBG* and *HDIF* and a suspended arch *GEH*; corresponding interaction diagram is shown in Fig. 2.32b. Each arched structures *ACBG* and *HDIF* present two rigid discs, connected with the ground. Two curvilinear members *GE* and *EH* are connected by hinge *E* and supported by two unmovable rigid discs, which can be considered as ground. Thus, the entire structure is statically determinate and geometrically unchangeable.

Influence line for bending moment M_k. There exist two nil points of influence line for M_k as the points of intersection of two lines:

1. Lines *AC* and *Bk*: their intersection point is $NP(M_k)$.
2. Lines *BG* and *HE*: their intersection point is NP_B.

The nil point NP_B possesses interesting properties. If moving load is traveling along the horizontal portion *GE*, then reaction at *H* is passing through hinge *E* and reaction at G might have various directions in accordance to the theorem about three concurrent forces R_H, $P = 1$, and R_G. The last reaction R_G is transformed as active force on the arch $ACBG$, in which reactions R_A and R_B arise. Reaction R_G is an active force for arch $ACBG$, passing through support B. This force is perceived by support B and reaction R_A is zero. Therefore, at all sections of the arch ACB, all internal forces are zero. Thus, if load P is located at NP_B, then *all* internal forces of the arch ACB are zero.

Since arch GEH is suspended, the bending moment M_k does not arise if load P is traveling along the portion HF.

Influence line for M_n. There exist two nil points of IL(M_n) as the points of intersection of two lines:

1. Lines FI and Dn: their intersection point is $NP(M_n)$.
2. Lines DH and GE: their intersection point is NP_D.

It is evident that point NP_D possesses the same properties for arch DIF as point NP_B for arch ACB: if moving load is located on the vertical passing through point NP_D, then at all sections of the arch DIF all internal forces are zero.

Influence line for shear force Q_n. There exist two nil points of influence line for Q_n. They are the point of intersection of line FI and the line which is parallel to tangent at section n and point NP_D.

Influence line for axial force N_n. There exist two nil points of influence line for N_n. They are the point of intersection of line FI and line which is perpendicular to tangent at section n and point NP_D. Specific ordinates and positions of the nil points allow us to easily construct the influence lines. Some of them are presented in Fig. 2.32c.

Note that the nil points $NP_1(Q_n)$ and $NP_1(N_n)$ are not real; they only facilitate the construction of influence lines Q_n and N_n, respectively.

It is left to reader to construct influence lines of shear and normal force in section k; construction of influence lines for internal forces for any section of central arch GEH should present no challenge.

2.7.2 Arched Combined Structures

Some examples of arches combined structures are presented in Fig. 2.33. In all cases, these systems consist of two trusses, AC and CB, connected by hinge C and stiffened by additional structures called a hinged (or arched) chain. The hinged chain may be located above or below the trusses. Vertical members connect the hinged chain with the trusses. The connections between the members of the arched chain and the hangers or posts are hinged. In case (c), all the hinges of the hinged chain are located on one line. In cases (a) and (b), a load is applied to the truss directly, while in case (c), the load is applied to the joint of the hinged chain and then transmitted to the truss.

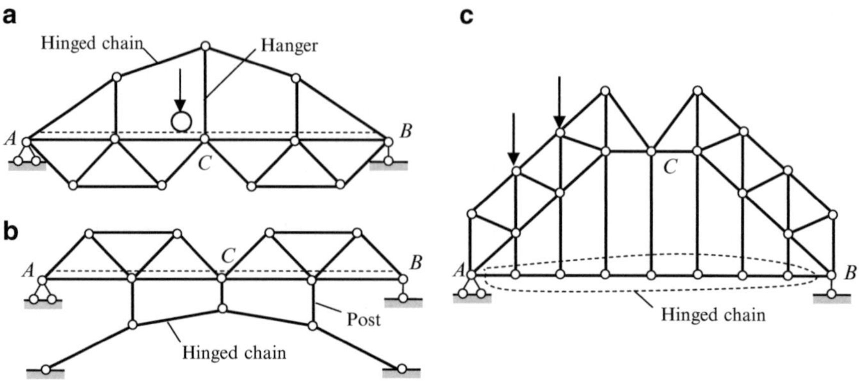

Fig. 2.33 Trusses with hinged chain

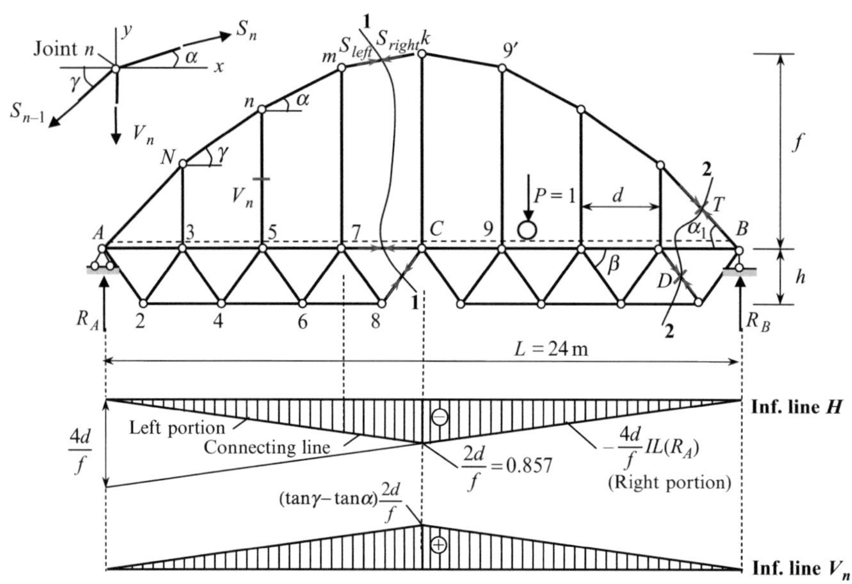

Fig. 2.34 Truss with over-truss arched chain. Design diagram and influence lines

Truss with Over-Truss Arched Chain

The typical truss with a hinged chain located above the truss is shown in Fig. 2.34. Assume that the parameters of the structure are as follows: $d = 3$ m, $h = 2$ m, $f = 7$ m, $L = 24$ m. We need to construct the influence lines for the reactions and the internal forces in hanger, V_{n-1}.

As usual we start with the kinematical analysis of the structure. Since the structure consist only members with hinges at the ends, then degrees of freedom of this complex arched structure is determined as $W = 2J - S - S_0 = 2 \times 24 - 45 - 3 = 0$, so the structure is geometrically unchangeable and statically determinate.

Reaction of Supports and Internal Forces

Reactions R_A and R_B for any load can be calculated using following equilibrium conditions:

$$R_A \to \sum M_B = 0: \quad R_B \to \sum M_A = 0.$$

For calculation of the internal forces that arise in the members of the hinged chain, we need to show the free body diagram for any joint n (Fig. 2.34). The equilibrium condition $\sum X = 0$ leads to relationship

$$S_n \cos\alpha = S_{n-1} \cos\gamma = H. \tag{2.25}$$

Thus, for any vertical load acting on the given structure, the horizontal component of the forces, which arise in all the members of the hinged chain, is equal. The horizontal component of the forces S_n, S_{n-1} is called a thrust.

Now we will provide an analysis for the case of a moving load. The influence lines for reactions R_A and R_B are the same as for a simply supported beam. However, the construction of an influence line for thrust H has some special features. Let us consider them.

Thrust H (section 1–1, the sectioned panel of the load contour SPLC is panel 7-*C*; Ritter's point is *C*). Internal force S, which arises in the element *m*–*k* of the hinged chain, is denoted as S_{left} and S_{right}. The meaning of the subscript notation is clear from Fig. 2.34.

If load $P = 1$ is located to the left of joint 7, then thrust H can be calculated by considering the *right* part of the structure. The active forces are reaction R_B and internal forces $S_{7\text{-}C}$, $S_{8\text{-}C}$, and S_{right}. The last force S_{right} can be resolved into two components: a horizontal component, which is the required thrust H, and a vertical component, which acts along the vertical line *C*–*k*. Now we form the sum of the moment of all forces acting on the right part of the structure around point C, i.e., $H \to \sum M_C^{\text{right}} = 0$. In this case, the vertical component of force S_{right} produces no moment, while the thrust produces moment Hf.

If load $P = 1$ is located right at joint C, then thrust H can be calculated by considering the *left* part of the structure. The active forces are reaction R_A and internal forces $S_{7\text{-}C}$, $S_{8\text{-}C}$, and S_{left}. The force S_{left}, which is applied at joint m, can be resolved into a horizontal component H and a vertical component. The latter component acts along vertical line *m*–7. Now we find the sum of the moment of all the forces, which act on the *left* part of the truss, around point C. In this case, the vertical component of force S_{right} produce the nonzero moment around joint C and thrust H has a new arm (*m*–7) around the center of moments C. In order to avoid these difficulties, we translate the force S_{left} along the line of its action from joint m into joint k. After that we resolve this force into its vertical and horizontal components. This procedure allows us to eliminate the moment due to the vertical component of S, while the moment due to the horizontal component of S is easily calculated as Hf.

Construction of the influence line for H is presented in the table below.

$P = 1$ left at SPLC	$P = 1$ right at SPLC
$H \to \sum M_C^{\text{right}} = 0: \quad R_B 4d + Hf = 0$	$H \to \sum M_C^{\text{left}} = 0: \quad R_A 4d + Hf = 0$
$H = -\frac{4d}{f} R_B \to \text{IL}(H) = -\frac{4d}{f} \text{IL}(R_B)$	$H = -\frac{4d}{f} R_A \to \text{IL}(H) = -\frac{4d}{f} \text{IL}(R_A)$

The left portion of the influence line for H (portion A-7) presents the influence line for R_B multiplied by coefficient $-4d/f$ and the right-hand portion (portion C–B) presents the influence line for R_A multiplied by the same coefficient. The connecting line is between points 7 and C (Fig. 2.34). The negative sign for thrust indicates that all members of the arched chain are in compression.

Force V_n. Equilibrium condition for joint n leads to the following result:

$$\sum Y = 0: \quad -V_n + S_n \sin\alpha - S_{n-1}\sin\gamma = 0 \to V_n = H(\tan\alpha - \tan\gamma).$$

Therefore,

$$\text{IL}(V_n) = (\tan\alpha - \tan\gamma) \times \text{IL}(H).$$

Since $\alpha < \gamma$ and H is negative, then all hangers are in tension. The corresponding influence line is shown in Fig. 2.34.

The influence line for thrust H can be considered as the key influence line, since thrust H always appears in any cut-section for the entire structure. This influence line allows us to calculate thrust for an arbitrary load. After that, the internal force in any member can be calculated simply by considering all the external loads, the reactions, and the thrust as an additional external force.

Discussion

For any location of a load, the hangers are in tension and all members of the chain are compressed. The maximum internal force at *any* hanger occurs if load P is placed at joint C.

To calculate the internal forces in different members caused by an arbitrary fixed load, the following procedure is recommended:

1. Construct the influence line for the thrust.
2. Calculate the thrust caused by a fixed load.
3. Calculate the required internal force considering thrust as an additional external force.

This algorithm combines both approaches: the methods of fixed and of moving loads and so provides a very powerful tool for the analysis of complex structures.

Example 2.7. The structure in Fig. 2.34 is subjected to a uniformly distributed load q within the entire span L. Calculate the internal forces T and D in the indicated elements.

Solution. The thrust of the arch chain equals $H = q\omega_H = -q(1/2)L(2d/f) = -(qLd/f)$, where ω_H is area of the influence line for H under the load q. After that, the required force T according to (a) is

$$T = \frac{H}{\cos\alpha_1} = -\frac{qLd}{f\cos\alpha_1}.$$

We can see that in order to decrease the force T, we must increase the height f and/or decrease the angle α_1.

To calculate force D, we can use section 2–2 and consider the equilibrium of the right part of the structure:

$$D \to \sum Y = 0: \quad D\sin\beta + R_B + T\sin\alpha_1 = 0 \to$$
$$D = -\frac{1}{\sin\beta}\left(\frac{qL}{2} - \frac{qLd}{f}\tan\alpha_1\right).$$

Thus, this problem is solved using the fixed and moving load approaches: thrust H is determined using corresponding influence lines, while internal forces D and T are computed using H and the classical method of through sections.

Arched Chain with Over-Arch Trussed Structure

The typical arched chain with a truss located above the arched chain is shown in Fig. 2.35. Assume that the parameters of the structure are as follows: $d = 2$ m, $h = 2$ m, $f = 8$ m, $l = 12d = 24$ m, $a_K = 6$ m. We need to construct the influence lines for the reactions, thrust, and the internal forces in indicated members U_4 and D_4.

Kinematical analysis shows that degree of freedom is $W = 2J - S - S_0 = 2 \times 34 - 61 - 7 = 0$, so the structure is statically determinate and geometrically unchangeable. The structure has the four support points: A_1, A_2, B_1, and B_2 and the following reactions: $R_{A1},\ R_{A2}, R_{B1},\ R_{B2},\ H_A,\ H_B$.

Reactions of Support and Internal Forces

Total vertical reactions *of a structure as a whole are* $R_A = R_{A1} + R_{A2}$, $R_B = R_{B1} + R_{B2}$, *where*

$$R_A \to \sum M_B = 0: \quad -R_A l + P(l-x) = 0 \to R_A = \frac{P(l-x)}{l} \to \mathrm{IL}(R_A) = \frac{l-x}{l},$$
$$R_B \to \sum M_A = 0: \quad R_B l - Px = 0 \to R_B = \frac{Px}{l} \to \mathrm{IL}(R_B) = \frac{x}{l}.$$

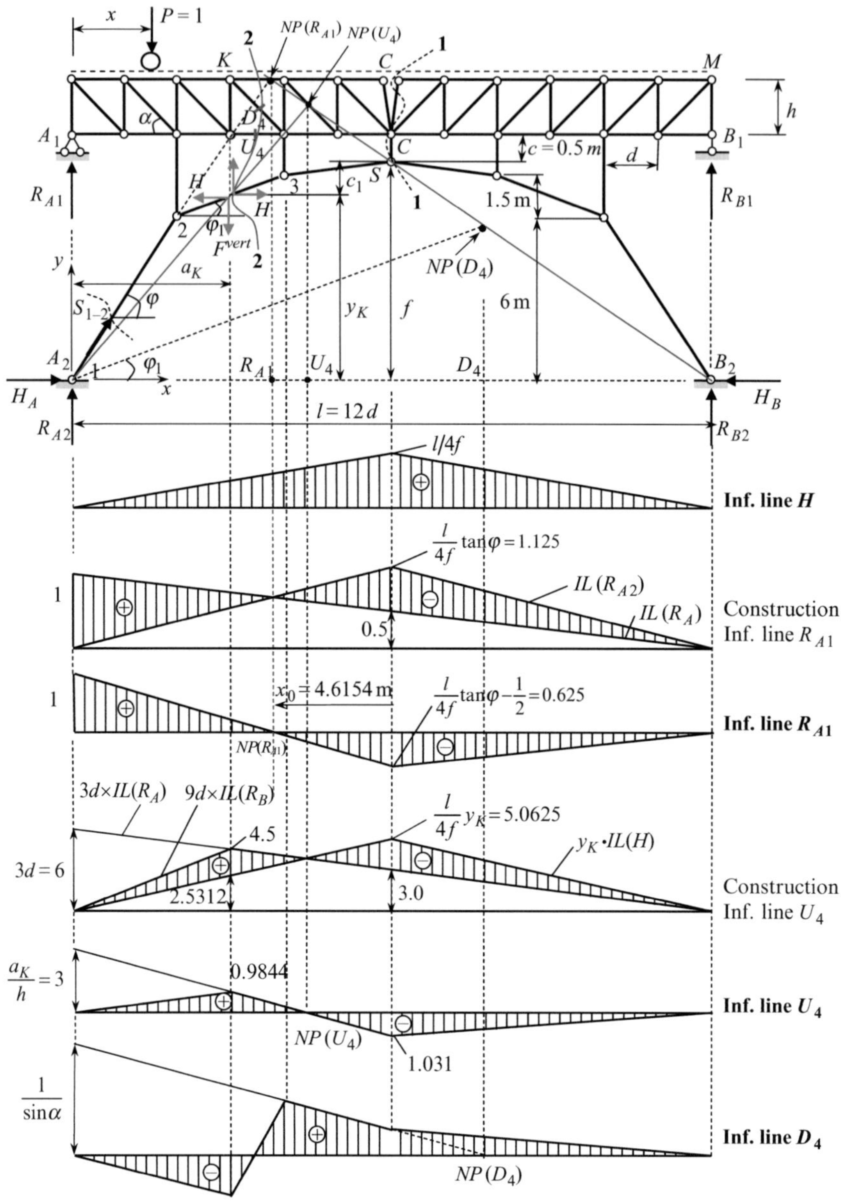

Fig. 2.35 Arched chain with over-arch trussed structure. Design diagram and influence lines

Influence lines for total vertical reactions of support are the same as for a simply supported beam.

Thrust. For the entire structure, the equilibrium condition $\sum X = 0$ leads to relationship $H_A = H_B = H$. Section 1–1 passes through joints S and C.

$P = 1$ left at joint C'	$P = 1$ right at joint C'
$H \to \sum M_S^{\text{right}} = 0: \quad -Hf + R_B \frac{l}{2} = 0$	$H \to \sum M_S^{\text{left}} = 0: \quad Hf - R_A \frac{l}{2} = 0$
$H = \frac{l}{2f} R_B \to \text{IL}(H) = \frac{l}{2f} \times \text{IL}(R_B)$	$H = \frac{l}{2f} R_A \to \text{IL}(H) = \frac{l}{2f} \times \text{IL}(R_A)$

The maximum ordinate under the joint C is equal to $l/4f$.

Vertical components of reactions. Equilibrium conditions for joint A_2 are

$$\sum X = 0: \quad S_{1-2} \cos\varphi + H = 0 \to S_{1-2} = -H/\cos\varphi,$$
$$\sum Y = 0: \quad S_{1-2} \sin\varphi + R_{A2} = 0 \to R_{A2} = -S_{1-2} \sin\varphi.$$

So the vertical component of reaction at point A_2 becomes $R_{A2} = H \tan\varphi$; corresponding influence line is

$$\text{IL}(R_{A2}) = \tan\varphi \times \text{IL}(H).$$

Similarly, $\text{IL}(R_{B2}) = \tan\varphi \times \text{IL}(H)$.

Influence lines for R_{A2} and R_{B2} may be obtained by multiplying all ordinates of influence line for H by a constant factor tan φ. The maximum ordinate under joint C is equal to $(l/4f)\tan\varphi$.

Reaction at point A_1. Since total reaction $R_A = R_{A1} + R_{A2}$, then

$$R_{A1} = R_A - R_{A2} \to \text{IL}(R_{A1}) = \text{IL}(R_A) - \text{IL}(R_{A2}) = \text{IL}(R_A) - \tan\varphi \times \text{IL}(H).$$

Construction and final influence line for vertical component R_{A1} is presented in Fig. 2.35. The nil point of influence line for R_{A1} is point of intersection of lines B_2–S and 1–2. The location of this nil point is defined by the formula $x_0 = (l/2) \times [(l\tan\varphi - 2f)/(l\tan\varphi + 2f)]$. For the entire structure, we get $\tan\varphi = (3/2)$ and $x_0 = 4.6154$ m.

Ordinate of influence line R_{A1} at point C equals to $(l/4f)\tan\varphi - (1/2) = [24/(4 \times 8)] \times (3/2) - (1/2) = 0.625$.

Note that reaction R_{A1} may be directed upward and downward as well.

Force U_4. Section 2–2 passes across the fourth panel of the truss and arch member 2–3 just under joint K; the vertical line passing through joint K intersects the member 2–3 at $y_K = 6.75$ m. The internal force F_{2-3} in the arch chain is resolved into vertical F^{vert} and horizontal H components. Obviously, the horizontal component equals to thrust H.

$P = 1$ left at SPLC	$P = 1$ right at SPLC
$U_4 \to \sum M_K^{\text{right}} = 0$	$U_4 \to \sum M_K^{\text{left}} = 0$
$-U_4 h + R_B 9d - H(f + c + h) + H(c_1 + c + h) = 0$	$U_4 h - R_A 3d + H(f + c + h) - H(c_1 + c + h) = 0$
$-U_4 h + R_B 9d - H \times y_K = 0$	$U_4 h - R_A 3d + H \times y_K = 0$
$U_4 = \frac{1}{h}(R_B 9d - H \times y_K) \to$	$U_4 = \frac{1}{h}(R_A 3d - H \times y_K) \to$
$\text{IL}(U_4) = \frac{1}{h}[9d \times \text{IL}(R_B) - y_K \times \text{IL}(H)]$	$\text{IL}(U_4) = \frac{1}{h}[3d \times \text{IL}(R_A) - y_K \times \text{IL}(H)]$

The term $H(c_1 + c + h)$ presents the moment with respect to point K due to thrust, which arise in member 2–3; the moment with respect to the same point K due to F^{vert} (vertical component of force $F_{2\text{–}3}$) is zero.

The nil point of influence line for U_4 is the point of intersection of lines B_2–S and the line which originates from joint 1 and passes through member 2–3 under point K. This point is real.

Ordinate of influence line U_4 at point C equals to $1/h\,(5.0625 - 3.0) = 1.031$.

Force D_4 (section 1–1). Assume that internal force $S_{2\text{–}3}$ is tensile.

$P = 1$ left at SPLC	$P = 1$ right at SPLC
$D_4 \to \sum Y^{\text{right}} = 0$	$D_4 \to \sum Y^{\text{right}} = 0$
$D_4 \sin\alpha + R_B - S_{2\text{–}3}\sin\varphi_1 = 0, S_{2\text{–}3} = -\dfrac{H}{\cos\varphi_1}$	$R_A - D_4\sin\alpha + S_{2\text{–}3}\sin\varphi_1 = 0, S_{2\text{–}3} = -\dfrac{H}{\cos\varphi_1}$
$D_4 \sin\alpha + R_B + H\tan\varphi_1 = 0$	$R_A - D_4\sin\alpha - H\tan\varphi_1 = 0$
$D_4 = -\dfrac{1}{\sin\alpha}(R_B + H \times \tan\varphi_1) \to$	$D_4 = \dfrac{1}{\sin\alpha}(R_A - H \times \tan\varphi_1) \to$
$\text{IL}(D_4) = -\dfrac{1}{\sin\alpha}[\text{IL}(R_B) + \tan\varphi_1 \times \text{IL}(H)]$	$\text{IL}(D_4) = \dfrac{1}{\sin\alpha}[\text{IL}(R_A) - \tan\varphi_1 \times \text{IL}(H)]$

Construction and final influence line for D_4 is presented in Fig. 2.35. The nil point of influence line for D_4 is point of intersection of lines B_2–S and the line which originates from joint 1 and passes parallel to the member 2–3. This point for given φ_1 is fictitious.

Influence line for thrust H of the structure is very useful for the calculation of internal force in any member of the truss. Let the structure be subjected to uniformly distributed load q along the entire span l of the truss. In this case, the thrust of the arch chain equals to $H = q\omega_H = q(1/2)l(l/4f) = (ql^2/8f)$, where ω_H is the area of influence line for H under the load q. Positive sign indicates that shown direction for the thrust at points A_2 and B_2 coincides with actual direction of thrust. Knowing the thrust allows us to perform an analysis of the structure. For example, the force $S_{1-2} = -(H/\cos\varphi) = -(ql^2/8f\cos\varphi)$. Negative sign indicates that the member 1–2 is compressed.

In the case of a fixed concentrated force P at joint C' and uniformly distributed load q within C'–M, we get:

$$U_4 = -1.031P - \frac{1}{2} \times 1.031 \times 6d \times q = -1.031P - 6.186q\,(\text{kN}).$$

2.8 Deflection of Three-Hinged Arches Due to External Loads

This section presents computation of displacement of three-hinged arch. Different approaches are applied: Maxwell–Mohr integral and graph multiplication method using Simpson–Kornouhkov rule.

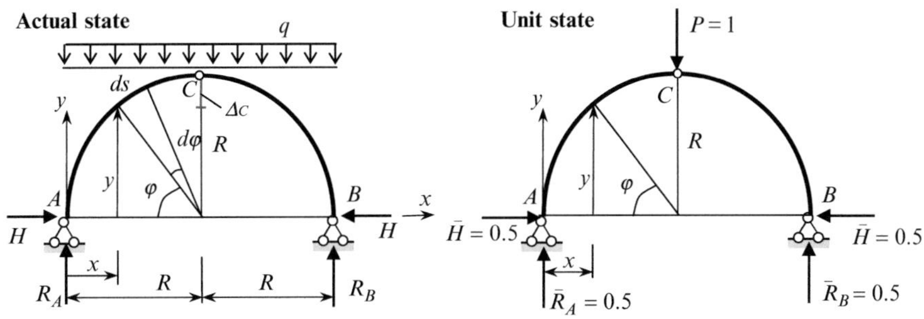

Fig. 2.36 Design diagram of the arch and unit state

2.8.1 Uniform Circular Arch: Exact Solution

Three-hinged semicircular uniform arch of radius R carrying uniformly distributed load q is shown in Fig. 2.36. The flexural stiffness is EI. For calculating the vertical displacement of the hinge C, we assume that influence of axial and shear forces on displacement is negligible. The expression for displacement for this problem takes into account only the bending moments

$$\Delta_C = \int_0^s \frac{M_P \bar{M}}{EI} \mathrm{ds},$$

where M_P denotes the bending moment due to actual load. Now we will consider two states, the actual and unit ones, and form the expressions for bending moments for both of them.

Actual State

The vertical reactions of supports and thrust are:

$$R_A = R_B = \frac{ql}{2} = qR; \quad H = \frac{M_C^0}{f} = \frac{q(2R)^2}{8R} = \frac{qR}{2},$$

where $l = 2R$ is the span of the arch; M_C^0 is the bending moment at point C for reference beam; f is the rise of the arch, $f = R$. The magnitude of the bending moment induced at any section by the given load q is

$$M_P = R_A x - Hy - \frac{qx^2}{2} = q\left(Rx - \frac{Ry}{2} - \frac{x^2}{2}\right).$$

Unit State

This state presents the same arch subjected to unit vertical force P at hinge C. The vertical reactions of supports and thrust are:

$$\bar{R}_A = \bar{R}_B = \frac{1}{2}; \quad \bar{H} = \frac{M_C^0}{f} = \frac{1 \times l}{4R} = \frac{1}{2}.$$

The magnitude of the bending moment induced at any section by the unit load P is

$$\bar{M} = \bar{R}_A x - \bar{H} y = \frac{1}{2} x - \frac{1}{2} y.$$

Now, the vertical displacement at point C may be presented as:

$$\Delta_C = 2 \int_0^{\pi R/2} \frac{M_P \bar{M}}{EI} \mathrm{d}s = \frac{2q}{EI} \int_0^{\pi R/2} \left(Rx - \frac{Ry}{2} - \frac{x^2}{2} \right) \times \left(\frac{x}{2} - \frac{y}{2} \right) \mathrm{d}s \qquad (2.26a)$$

Let us change to polar coordinates: $\mathrm{d}s = R\mathrm{d}\varphi$, $y = R \sin \varphi$, $x = R - R \cos \varphi = R(1 - \cos \varphi)$. The upper limit $s = \pi R/2$ should be changed to $\varphi = \pi/2$. In this case, (2.26a) becomes

$$\begin{aligned} \Delta_C &= \frac{q}{EI} \int_0^{\pi/2} \left[R^2(1 - \cos\varphi) - \frac{R^2}{2} \sin\varphi - \frac{R^2}{2} (1 - \cos\varphi)^2 \right] \\ &\quad \times [R(1 - \cos\varphi) - R \sin\varphi] R\mathrm{d}\varphi \\ &= \frac{qR^4}{EI} \int_0^{\pi/2} \left[1 - \cos\varphi - \frac{1}{2} \sin\varphi - \frac{1}{2} (1 - \cos\varphi)^2 \right] \times (1 - \cos\varphi - \sin\varphi)\, \mathrm{d}\varphi. \end{aligned}$$

Integrating procedure is cumbersome, but elementary. On rearrangement, the final result for vertical displacement at C can be written as

$$\Delta_C = \frac{qR^4}{4EI} (\pi - 3). \qquad (2.26b)$$

In case of concentrated force P at point C, the vertical displacement at C is $\Delta_C = (PR^3/2EI)(\pi - 3)$.

2.8.2 *Nonuniform Arch of Arbitrary Shape: Approximate Solution*

In general case of the nonuniform arch and arbitrary shape, the general idea for computation of displacement remains the same – it is necessary to "multiply" the

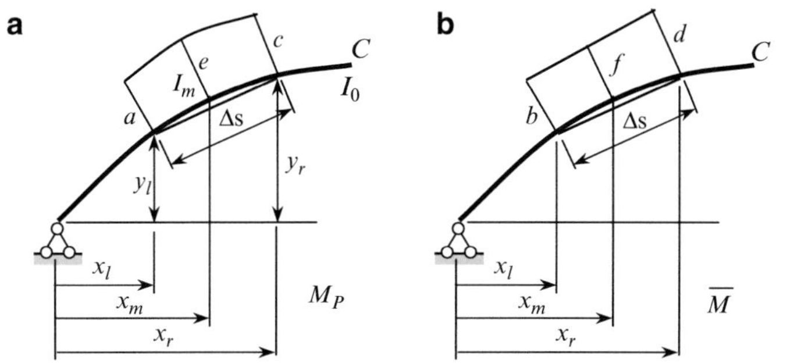

Fig. 2.37 Notation of ordinates of the bending moment diagrams within the one straight segment; M_P and $\overline{M}$ are bending moment diagrams in the actual and unit states

bending moments diagrams in the entire and the unit states. However, the Vereshchagin rule becomes none applicable, since the basic line of both diagrams is curvilinear. Therefore, it is only possible to determine the displacement in the general case of the arch numerically. For this, a curvilinear axis of the arch should be presented as a set of straight elements (usually 8–10), followed by a multiplication procedure of two bending moment diagrams. As before, the normal and shear forces will be neglected.

Let us subdivide the arch into segments with equal horizontal projections. The length of the ith chord between two nodal points equals $\Delta s = \sqrt{(x_r - x_l)^2 + (y_r - y_l)^2}$. Ordinates of the left and right ends of the portion, x_l, y_l and x_r, y_r, should be calculated according to the equation $y = f(x)$ of the axis of the arch. Now Mohr integral may be presented in approximate form

$$\Delta_{iP} = \int \frac{\overline{M}_i M_P}{EI} \mathrm{d}s \cong \frac{1}{EI_0} \sum_n \overline{M}_i M_P \times \frac{I_0}{I_m} \Delta s, \tag{2.27}$$

where M_P and $\overline{M}_i$ are bending moment diagrams in the entire and unit states, respectively; n is the total number of segments, I_0 and I_m are the moment of inertia of the cross section at the crown C and at the middle of the segment Δs. The moment of inertia I_m should be calculated according to the law $I = I(x)$, or as half-sum of the moments of inertia at the ends of a segment. Simpson's formula [Dar89]

$$EI_0 \Delta_{iP} = \sum_n \frac{\Delta s'}{6}(ab + 4ef + cd), \Delta s' = \Delta s \frac{I_0}{I_m} \tag{2.28}$$

is applied to each straight segment and is subsequently summed over all the segments. Ordinates a, e, and c of the bending moment diagram M_P in the loading state relate to the left end, the middle point, and the right end of the ith segment (Fig. 2.37a); ordinates b, f, d of the bending moment diagram $\bar{M}$ in the unit state relate to the same points (Fig. 2.37b).

Figure 2.38 presents a nonuniform parabolic arch and its approximate model.

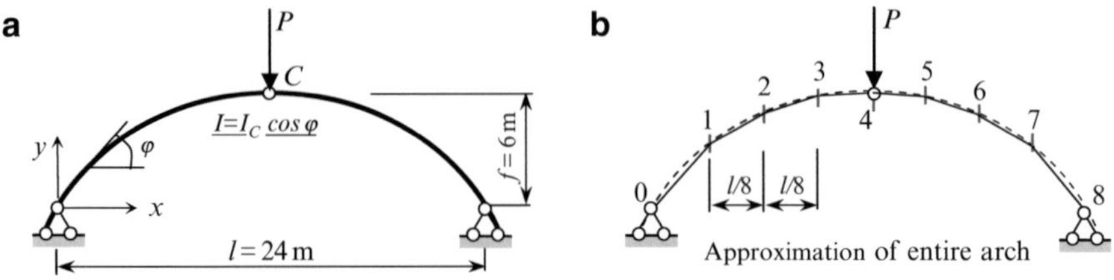

Fig. 2.38 Parabolic arch. Design diagram and approximation of entire arch

Table 2.3 Geometrical parameters of parabolic arch

	Coordinates (m)			
Points	x	y	$\tan\varphi$	$\cos\varphi$
0	0	0.0	1.00	0.7070
1	3	2.625	0.75	0.800
2	6	4.500	0.50	0.8944
3	9	5.625	0.25	0.9701
4	12	6.000	0.0	1.0
5	15	5.625	−0.25	0.9701
6	18	4.500	−0.5	0.8944
7	21	2.625	−0.75	0.800
8	24	0.0	−1.00	0.7070

Table 2.4 The chord lengths of each straight segment

Portion	0–1	1–2	2–3	3–4	4–5	5–6	6–7	7–8
Length (m)	3.9863	3.5377	3.2040	3.0233	3.0233	3.2040	3.5377	3.9863

Table 2.5 Geometrical parameters at specified sections of nonuniform arch ($I = I_C \cos\varphi$)

	Geometrical parameters (m)								
Portion	x_l	x_m	x_r	y_l	y_m	y_r	Δs	I_m (factor I_C)	$\Delta s' = \Delta s \frac{I_C}{I_m}$
0–1	0	1.5	3	0	1.4062	2.625	3.9863	0.7535	5.2904
1–2	3	4.5	6	2.625	3.6562	4.500	3.5377	0.8475	4.1743
2–3	6	7.5	9	4.500	5.1562	5.625	3.2040	0.9323	3.4366
3–4	9	10.5	12	5.625	5.9062	6.000	3.0233	0.9850	3.0693

Specified Points of the Arch

The *span* of the arch is divided into eight equal parts; the specified points are labeled 0–8. Parameters of the arch for these sections are presented in Table 2.3; the following formulas for calculation of trigonometric functions of the angle φ between the tangent to the arch and x-axis have been used: $\tan\varphi = y' = [4f(l-2x)]/l^2$, $\cos\varphi = 1/\sqrt{1+\tan^2\varphi}$.

The lengths of each straight segment are presented in Table 2.4. Table 2.5 presents the geometrical parameters at specified sections of the arch, and computation of the conventional length $\Delta s'$ for each segment.

Table 2.6 Bending moments at specified sections and computation of deflection

		M_P, factor $(-P)$			$\overline{M}$ factor (-1)			$\frac{\Delta s'}{6}(ab + 4ef + cd)$
Portion	$\Delta s'$	a	e	c	b	f	d	
0–1	5.2904	0.0	0.65625	1.125	0.0	0.65625	1.125	2.6348P
1–2	4.1743	1.125	1.40625	1.50	1.125	1.40625	1.50	7.9491P
2–3	3.4366	1.50	1.40625	1.125	1.50	1.40625	1.125	6.5443P
3–4	3.0693	1.125	0.65625	0.0	1.125	0.65625	0.0	1.5286P

The moment of inertia $I_m = 0.5(I_l + I_r)$. For example, for segment 0–1 we get

$$I_{0-1} = 0.5\,(0.707 + 0.800)\,I_C = 0.7535 I_C.$$

Table 2.6 contains the bending moments at specified sections for loaded and unit states. These moments are calculated by the formula $M_k = M_k^0 - H y_k$, where $H = M_C^0/f = Pl/4f = 1 \times P$. For each segment, the section at the left end has ordinates a and b, at the middle section the ordinates are e and f and at the right end ordinates are c and d.

For example, in P-condition $R_A = P/2$ and $H = P$, so for points 1 and 2 - (portion 1–2) we get

$$M_1 = \frac{P}{2} \times 3 - P \times 2.625 = -1.125P, \quad M_2 = \frac{P}{2} \times 6 - P \times 4.5 = -1.5P.$$

Data for the right half-arch is not presented due to the symmetry of structure.

Required displacement of point C is equals to twice the sum of the members of the last column. In our case,

$$\Delta_{iP} - 2(2.6348 + 7.9491 + 6.5443 + 1.5286)\frac{P}{EI_C} - 37.3136\frac{P}{EI_C}. \tag{2.26c}$$

The above-discussed procedure is very effective for computation of displacement of any nonsymmetrical three-hinged arches. If it is necessary to take into account the shear and axial forces, the corresponding terms of Maxwell Mohr integral (1.8) should be included and Table 2.6 to be expanded [Rab54a].

2.9 Displacement Due to Settlement of Supports and Errors of Fabrication

Settlement of supports and errors of fabrication often occur in engineering practice. If this happens in a statically determinate structure, the internal stresses in the members of the structures are not induced. So computation of displacement of any point of statically determinate structures reflects the kinematical nature of a problem.

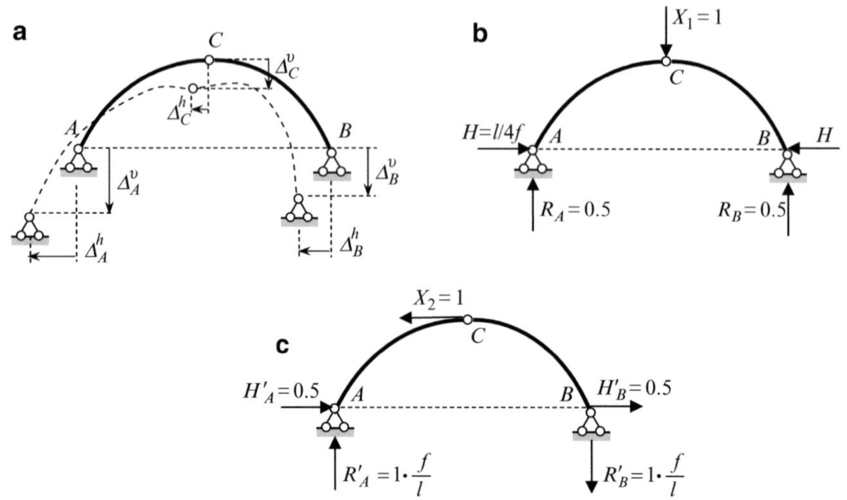

Fig. 2.39 (**a**) Settlement of supports *A* and *B*; (**b**) unit state for calculation of Δ_C^v; and (**c**) unit state for calculation of Δ_C^h

2.9.1 *Settlements of Supports*

Let us consider a three-hinged arch of span l and rise f; supports A and B settles in vertical and horizontal directions as shown in Fig. 2.39a. The new position of the arch, in an exaggeration scale, is shown by a dotted line. It is necessary to calculate the vertical Δ_C^v and horizontal Δ_C^h displacements of the hinge C. Unit state presents the same structure subjected to unit force X, which corresponds to the required displacement.

An effective method for solution of this type of problem is the principle of virtual displacements

$$\sum \delta W_{\text{act}} = 0. \tag{2.29}$$

According to this principle, *the elementary work done by all active forces on any virtual displacements, which are compatible with constraints, is zero.*

Procedure for Computation of Displacement Caused by the Settlement of Support

1. At point K where displacement should be determined, we need to apply a unit generalized force $X = 1$, corresponding to the required displacement.
2. Show reactions R at the settled support, caused by unit generalized force $X = 1$, and compute these reactions.

3. Calculate the total work (2.29) done by unit force and all reactions on the displacements of the supports.
4. Solve this equation with respect to required displacement.

Vertical displacement of the hinge C. Let us apply $X_1 = 1$ in vertical direction; this force corresponds to the required vertical displacement Δ_C^{v}. Reactions at the supports A and B are shown in Fig. 2.39b. These reactive forces should be considered as active, and (2.29) becomes

$$X_1 \times \Delta_C^{v} - R_A \times \Delta_A^{v} - R_B \times \Delta_B^{v} - H \times \Delta_A^{\mathrm{h}} + H \times \Delta_B^{\mathrm{h}} = 0.$$

Since $X = 1$, then

$$\Delta_C^{v} = -\sum R \times \Delta = R_A \times \Delta_A^{v} + R_B \times \Delta_B^{v} + H \times \Delta_A^{\mathrm{h}} - H \times \Delta_B^{\mathrm{h}}. \qquad (2.29a)$$

Formula (2.29a) may be generalized for the case of displacements caused by settlements of *several* supports

$$\Delta_{ks} = -\sum R\Delta, \qquad (2.30)$$

where Δ_{ks} is the displacement in kth direction due to settlement of supports, Δ is the given settlement of support; R are the reactions in the support which is settled; this reaction caused by unit load which corresponds to the required displacement. Summation covers all supports.

Horizontal displacement of the hinge C. Horizontal force $X_2 = 1$ corresponds to the required horizontal displacement Δ_C^{h}. Reactions at the supports A and B are shown in Fig. 2.39b. Equation (2.30) leads to the following result:

$$\Delta_C^{\mathrm{h}} = -\sum R \times \Delta = R_A' \times \Delta_A^{v} - R_B' \times \Delta_B^{v} + H_A' \times \Delta_A^{\mathrm{h}} + H_B' \times \Delta_B^{\mathrm{h}}.$$

Discussion

1. Equation (2.30) reflects a *kinematical* nature of problem; it means that displacements of any point of a statically determinate structure are determined by the geometrical parameters of a structure without taking into account the deformations of its elements. Any settlement of support of such structure does not depend on the stiffness of the structure, and therefore leads to displacement of its separate parts as rigid discs.
2. The positive results for required Δ_i means that unit load X on the displacement Δ_i performs positive work.
3. Assume that $\Delta_A^{v} \neq 0$, while all other displacements are zero. Thus, in case of vertical displacement of only one of the support, the crown hinge C has the vertical and horizontal displacements.

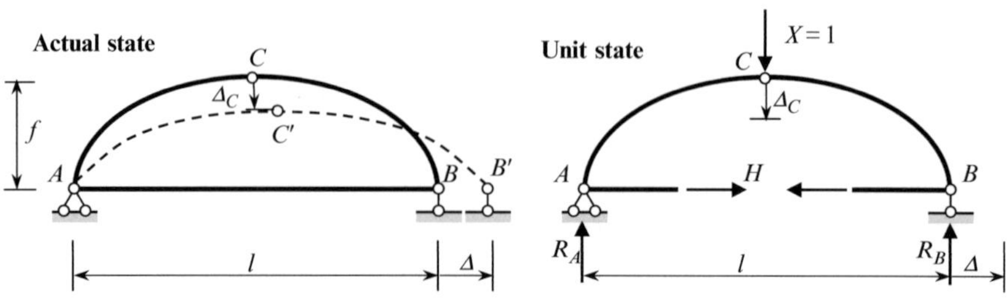

Fig. 2.40 Design diagram of the arch (error fabrication) and unit state

2.9.2 Errors of Fabrication

Deflections of the structural members may occur as a result of the geometric misfit. This topic is sometimes referred to as geometric incompatibility.

The following procedure may be applied for this type of problems:

1. At point K, where displacement should be determined, we need to apply a unit generalized force $X = 1$ corresponding to the required displacement.
2. Compute all reactions caused by the unit generalized force $X = 1$.
3. Calculate the work done by these reactions on the displacements.

Example 2.8. The tie AB of the arch ACB in Fig. 2.40 is $\Delta = 0.02$ m longer then required length l. Find the vertical displacement at point C, if $l = 48$ m, $f = 6$ m.

Solution. The actual position of the tie is AB' instead of required AB position. For computation of the vertical displacement Δ_C we have to apply a unit vertical force at C. Reactions of the three-hinged arch and thrust in tie caused by the force $P = 1$ equal

$$R_A = R_B = 0.5, H = M_C^0/f = l/(4f) = 2.$$

Application of principle of virtual displacements leads to the following expression

$$X \times \Delta_C - H \times \Delta = 0.$$

Since $X = 1$, then the required displacement becomes

$$\Delta_C = +H \times \Delta = +0.04\,\text{m (downward)}$$

It is obvious that the effect of geometric incompatibility may be useful for the regulation of stresses in the structure. Let us consider a three-hinged arch which is loaded by any fixed load. The bending moments are $M(x) = M^0 - Hy$, where M^0 is the bending moment in the reference beam. If a tie is fabricated longer than is required, then the thrust becomes $H = H_1 + H_2$ where H_1 and H_2 are thrust due to fixed load and errors of fabrication, respectively.

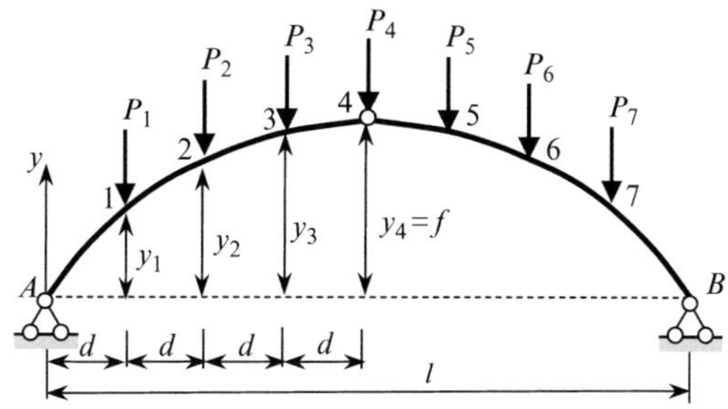

Fig. 2.41 Design diagram of three-hinged parabolic arch

Discussion

For computation of displacement due to the settlement of supports and errors of fabrication, we use the principle of virtual work. This principle and the Maxwell–Mohr integral method have the general concept of generalized coordinate and corresponding generalized unit force in common.

2.10 Matrix Form Analysis of Arches Subjected to Fixed and Moving Load

This paragraph presents the matrix analysis of three-hinged arch subjected to fixed and moving load.

Design diagram of three-hinged parabolic arch subjected to fixed load is shown in Fig. 2.41. The span of the arch is divided into n equal portions, so $d = l/n$; in the case of Fig. 2.41, $n = 8$.

Let the span and rise of the arch be $l = 16$ m and $f = 4$ m, respectively.

If the equation of the arch obeys formula (3), then

$$y_1 = y_7 = 1.75\,\text{m}, \quad y_2 = y_6 = 3.0\,\text{m}, \quad y_3 = y_5 = 3.75\,\text{m}, \quad y_4 = f = 4.0\,\text{m}.$$

Vector of bending moments at the nodal points 1–7 is

$$\vec{\mathbf{M}} = \mathbf{L}_m^* \mathbf{L}_m \vec{\mathbf{P}}, \tag{2.31}$$

where the influence matrix of bending moments is

$$\mathbf{L}_m^* = \begin{bmatrix} 1 & 0 & 0 & m_1^* & 0 & 0 & 0 \\ 0 & 1 & 0 & m_2^* & 0 & 0 & 0 \\ 0 & 0 & 1 & m_3^* & 0 & 0 & 0 \\ 0 & 0 & 0 & 0 & 0 & 0 & 0 \\ 0 & 0 & 0 & m_5^* & 1 & 0 & 0 \\ 0 & 0 & 0 & m_6^* & 0 & 1 & 0 \\ 0 & 0 & 0 & m_7^* & 0 & 0 & 1 \end{bmatrix}$$

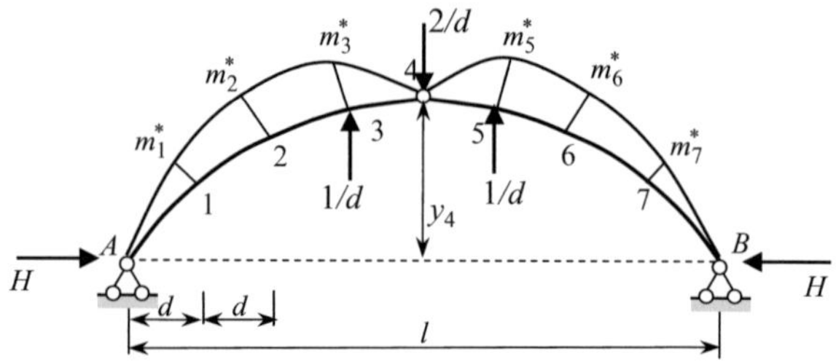

Fig. 2.42 Bending moment diagram due to self-balanced load $1/d$–$2/d$–$1/d$

To find the entries m_i^* we need to construct a bending moment diagram for three-hinged arch subjected to self-balanced load, which acts as shown in Fig. 2.42.

The trust is $H = 1/f$, so the bending moments at the nodal points are

$$m_1^* = m_7^* = -\frac{y_1}{y_4} = -\frac{1.75}{4}; \quad m_2^* = m_6^* = -\frac{y_2}{y_4} = -\frac{3}{4};$$

$$m_3^* = m_5^* = -\frac{y_3}{y_4} = -\frac{3.75}{4}.$$

Therefore, matrix L_m^* becomes

$$\mathbf{L}_m^* = \begin{bmatrix} 1 & 0 & 0 & -0.4375 & 0 & 0 & 0 \\ 0 & 1 & 0 & -0.75 & 0 & 0 & 0 \\ 0 & 0 & 1 & -0.9375 & 0 & 0 & 0 \\ 0 & 0 & 0 & 0 & 0 & 0 & 0 \\ 0 & 0 & 0 & -0.9375 & 1 & 0 & 0 \\ 0 & 0 & 0 & -0.75 & 0 & 1 & 0 \\ 0 & 0 & 0 & -0.4375 & 0 & 0 & 1 \end{bmatrix}.$$

Influence matrix of bending moments for the arch coincides with influence matrix for bending moments for simply supported beam of the same span

$$\mathbf{L}_m = \frac{d}{n}\mathbf{I}_{(n-1)},$$

where $\mathrm{I}_{(n-1)}$ is a matrix of order $n-1$ and has the following special form

$$\mathbf{I}_{(n-1)} = \begin{bmatrix} n-1 & n-2 & n-3 & \ldots & 1 \\ n-2 & \ldots & \ldots & \ldots & 2 \\ \ldots & \ldots & \ldots & \ldots & \ldots \\ 2 & 4 & 6 & \ldots & n-2 \\ 1 & 2 & 3 & \ldots & n-1 \end{bmatrix}.$$

If $n = 8$ then L_m becomes

$$\mathbf{L}_m = \frac{d}{8}\begin{bmatrix} 7 & 6 & 5 & 4 & 3 & 2 & 1 \\ 6 & 12 & 10 & 8 & 6 & 4 & 2 \\ 5 & 10 & 15 & 12 & 9 & 6 & 3 \\ 4 & 8 & 12 & 16 & 12 & 8 & 4 \\ 3 & 6 & 9 & 12 & 15 & 10 & 5 \\ 2 & 4 & 6 & 8 & 10 & 12 & 6 \\ 1 & 2 & 3 & 4 & 5 & 6 & 7 \end{bmatrix}.$$

This matrix is symmetric with respect to both diagonals. The entries of the last row and last column, as well as the entries of the first column (from bottom to top) and first row (from right to left) present the natural numbers 1,2,..., $(n-1)$. Any entry m_{ki}, which is located on the main diagonal or above, is determined as a product of the k-th entry of the very first row and the number i of the row .

Vector of bending moments is the result of multiplication of the following matrices

$$\vec{\mathbf{M}} = \begin{bmatrix} M_1 \\ M_2 \\ M_3 \\ M_4 \\ M_5 \\ M_6 \\ M_7 \end{bmatrix} = \begin{bmatrix} 1 & 0 & 0 & -0.4375 & 0 & 0 & 0 \\ 0 & 1 & 0 & -0.75 & 0 & 0 & 0 \\ 0 & 0 & 1 & -0.9375 & 0 & 0 & 0 \\ 0 & 0 & 0 & 0 & 0 & 0 & 0 \\ 0 & 0 & 0 & -0.9375 & 1 & 0 & 0 \\ 0 & 0 & 0 & -0.75 & 0 & 1 & 0 \\ 0 & 0 & 0 & -0.4375 & 0 & 0 & 1 \end{bmatrix}$$

$$\times \frac{2}{8}\begin{bmatrix} 7 & 6 & 5 & 4 & 3 & 2 & 1 \\ 6 & 12 & 10 & 8 & 6 & 4 & 2 \\ 5 & 10 & 15 & 12 & 9 & 6 & 3 \\ 4 & 8 & 12 & 16 & 12 & 8 & 4 \\ 3 & 6 & 9 & 12 & 15 & 10 & 5 \\ 2 & 4 & 6 & 8 & 10 & 12 & 6 \\ 1 & 2 & 3 & 4 & 5 & 6 & 7 \end{bmatrix} \times \begin{bmatrix} P_1 \\ P_2 \\ P_3 \\ P_4 \\ P_5 \\ P_6 \\ P_7 \end{bmatrix}.$$

If we assume that the vector of external loads is $\vec{P} = \lfloor 1 \ \ 4 \ \ 2 \ \ 0 \ \ 0 \ \ 2.5 \ \ 0 \rfloor^T$ [Kle80], then the vector of bending moments at the nodal points 1–7 becomes

$$\vec{\mathbf{M}} = \lfloor 2.75 \quad 6 \quad 3.75 \quad 0 \quad -1.25 \quad 0 \quad -1.25 \rfloor^{\mathrm{T}}.$$

This matrix approach may be effectively used for the construction of influence lines for bending moments. If force $P = 1$ is placed only at joint 1, then the vector of external load becomes

$$\vec{P} = \lfloor 1 \quad 0 \quad 0 \quad 0 \quad 0 \quad 0 \quad 0 \rfloor^{\mathrm{T}}$$

and procedure (2.31) gives us the bending moments at the nodal points 1–7.

In order to calculate all ordinates of influence lines for bending moments at sections 1–7, the vector of loads $\vec{P}$ should be replaced by an identity matrix **P**; if $n = 8$, then this matrix is of order 7.

The final result of multiplication of the three squared matrices is

$$\mathbf{M} = \begin{bmatrix} \mathrm{IL}(M_1) \\ \mathrm{IL}(M_2) \\ \mathrm{IL}(M_3) \\ \mathrm{IL}(M_4) \\ \mathrm{IL}(M_5) \\ \mathrm{IL}(M_6) \\ \mathrm{IL}(M_7) \end{bmatrix} = \begin{bmatrix} 1.3125 & 0.6250 & -0.0625 & -0.7500 & -0.5625 & -0.3750 & -0.1875 \\ 0.7500 & 1.5000 & 0.2500 & -1.0000 & -0.7500 & -0.5000 & -0.2500 \\ 0.3125 & 0.6250 & 0.9375 & -0.7500 & -0.5625 & -0.3750 & -0.1875 \\ 0 & 0 & 0 & 0 & 0 & 0 & 0 \\ -1.1875 & -0.3750 & -0.5625 & -0.7500 & 0.9375 & 0.6250 & 0.3125 \\ -0.2500 & -0.5000 & -0.7500 & -1.0000 & 0.2500 & 1.5000 & 0.7500 \\ -1.1875 & -0.3750 & -0.5625 & -0.7500 & -0.0625 & 0.6250 & 1.3125 \end{bmatrix}.$$

The ith row of this matrix represents the influence line of bending moment at the ith nodal point.

It is easy to verify that each influence line for bending moment consist of the strength portions; this means that the structure under consideration is indeed statically determined.

Chapter 3
Redundant Arches

This chapter is devoted to the analysis of statically indeterminate arches. Numerous examples of application of this method for the analysis of different arches are presented. Among them are uniform and nonuniform arches with and without ties, arches with elastic supports, etc., subjected to external fixed and moving loads, temperature changes, concrete shrinkage, settlements of supports, and errors of fabrication.

3.1 Types, Forms, and Peculiarities of Redundant Arches

Typical statically indeterminate arches are shown in Fig. 3a–c. A two-hinged arch (Fig. 3b) has one redundant constraint. A hingeless arch (arch with fixed ends) (Fig. 3a) has three redundant constraints. One-hinged arch (Fig. 3c) has two redundant constraints. All of these arches may be symmetrical or nonsymmetrical.

In the general case, the following internal forces arise in an arch: bending moments M, shear Q, and axial force N. In the statically indeterminate arches, internal forces are the result of the action of external loads, as well as settlements of supports, change of temperature, errors of fabrication, and shrinkage. Since a two-hinged arch is more flexible than a hingeless arch, the internal forces are smaller in a two-hinged arch when compared to the hingeless counterpart. Distribution of internal forces and displacements in an arch depends on the shape of the arch. The shape of the arch is described by the equation $y = f(x)$ of the axis of the arch, as well as the law of change of moment of inertia of the cross section of the arch along its axis.

The choice of the arch shape is determined by the engineer on the basis of a number of requirements. The first important requirement is that the distribution of material along the axis of the arch must corresponds to distribution of internal forces. For example, in the pinned support, the bending moment equals to zero; therefore, the moment of inertia of the cross-section should be minimal.

I.A. Karnovsky, *Theory of Arched Structures: Strength, Stability, Vibration*,
DOI 10.1007/978-1-4614-0469-9_3, © Springer Science+Business Media, LLC 2012

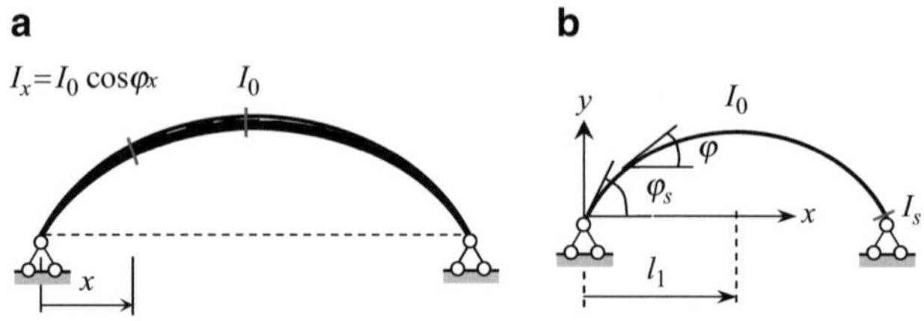

Fig. 3.1 Two-hinged nonuniform arch

The second requirement concerns the ability to obtain accurate solutions, which will in turn allow the engineer to perform an analytical analysis of results. Several laws of the change of moment of inertia of the cross section for symmetrical arches are presented below.

3.1.1 Two-Hinged Arch

In the case of an arch of variable cross-section, the following law for moment of inertia I_x of cross-section may be assumed $I_x = I_0 \cos \varphi_x$, where I_0 is the moment of inertia at the crown, φ_x is the angle between tangent to the axis of the arch and the horizontal line [Kle80], [Dar89]. This expression describes a decrease in flexural stiffness of the arch from crown to supports (Fig. 3.1a) and corresponds to insignificant bending moments in the vicinity of the supports of the arch subjected to a uniformly distributed load. This law also satisfies the esthetic requirements.

It is possible to assume that $A_x = A_0 \cos \varphi_x$, where A_0 is the area of the cross-section at the crown [Dar89].

For two-hinged symmetrical arches, the following relationships are also possible (origin is placed at the left support) (Fig. 3.1b) [Kis60]:

$$I = \frac{I_0}{\cos\varphi} \times \frac{1}{1 + (n-1)\frac{x}{l_1}}, \qquad I = \frac{I_0}{\cos\varphi} \times \frac{1}{1 + (n-1)\left(\frac{x}{l_1}\right)^5}, \tag{3.1}$$

where l_1 is a half-span of the arch; $n = I_0/(I_s \cos \varphi_s)$, subscript "s" refers to the cross-section at the support of the arch.

If $n = 1$, then $I_s \cos \varphi_s = I_0$; this case leads to the change of moment of inertia by the law $I = I_0/\cos\varphi$ [Sni66], [Dar89]. According to this formula, the flexural stiffness decreases from support to the crown. In the case of a parabolic arch, this relationship allows us to perform the integration procedure in a close form.

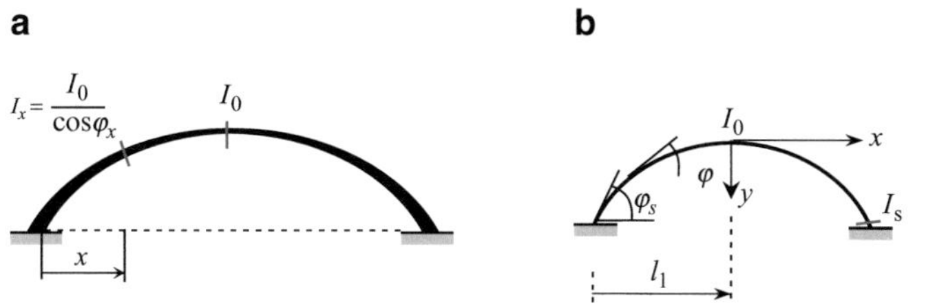

Fig. 3.2 Hingeless nonuniform arch

3.1.2 Hingeless Arch

For such arches, the maximum bending moments arise at supports (for typical loads, such as a uniformly distributed load along half-span, or nonuniformly distributed loads, etc.). For these cases, the following law for moment of inertia of cross-section may be assumed: $I_x \cos\varphi_x = I_C$. This expression corresponds to an increase of flexural stiffness of the arch from crown to supports (Fig. 3.2a).

For hingeless symmetrical arches, it is also possible to assume the following relationships (origin is placed at the highest point of the arch (Fig. 3.2b) [Kis60], [Sni66])

$$I = \frac{I_0}{\cos\varphi} \times \frac{1}{1-(1-n)\frac{x}{l_1}}, \quad I = \frac{I_0}{\cos\varphi} \times \frac{1}{1-(1-n)\frac{y}{f}}, \tag{3.2}$$

where $n = I_0/(I_s \cos\varphi_s)$, I_0 and I_s are moment of inertia at the crown and at support; l_1, f are half of the span and rise of the arch, respectively.

If $n = 1$, then $I_s \cos\varphi_s = I_0$. This case leads to the change of moment of inertia by the law $I = I_0/\cos\varphi$ (Fig. 3.2a).

Thus, it can be observed that the shape in Fig. 3.2a is not wise to use for pinned supports, while the shape in Fig. 3.1a is dangerous to use in case of clamped supports. It is obvious that the laws for moment of inertia of cross-section in real structures are not limited to the cases considered above. There are other laws of variation of moments of inertia listed in the books [Mor35], [Mel31], [Str27], and [Ric99].

If an axial force should be taken into account, then we need to know how to change the area A_x of the cross section of the arch along the axis of the arch. For rectangular cross section of the arch with width $b =$ constant and variable thickness $h(x)$, the moment of inertia at any section and crown are $I_x = bh_x^3/12, I_0 = bh_0^3/12$, respectively. Since $A_x = bh_x$, $A_0 = bh_0$ then in case of $n = 1$ we get $A_x = A_0/\sqrt[3]{\cos\varphi_x}$. This formula may be replaced by the simpler approximate form $A_x \approx A_0/\cos\varphi_x$. This formula leads to an accurate approximation of the thrust and bending moments [Dar89].

In the case of flat arches ($f<l/8$), we can assume $\cos\varphi_x \approx 1$, the length of the elementary segment ds and its horizontal projection dx are equals, and $A_x = A_0 = \text{const}$ for all cross sections of the arch.

3.2 Force Method

The Force method presents a powerful method for analyzing linear elastic statically indeterminate structures; this method also has a wide application in problems of stability and dynamics of structures. The method is very attractive because it has clear physical meaning, which is based on a convenient and well-ordered procedure of calculation of displacements of deformable structures, and presently, this method has been brought to elegant simplicity and perfection.

3.2.1 Primary System and Primary Unknowns

Degree of redundancy, or statical indeterminacy, equals to the number of redundant constraints whose elimination leads to the new geometrically unchangeable and statically determinate structure. Thus, degree of statical redundancy is the difference between the number of constraints and the number of independent equilibrium equations that can be written for a given structure.

Primary unknowns represent *reactions* (forces and/or moments), which arise in redundant constraints. Unknown *internal* forces also may be treated as primary unknowns. Primary system is such structure, which is obtained from the given one by eliminating redundant constraints and replacing them by primary unknowns [Kar10].

3.2.2 Canonical Equations of the Force Method

A two-hinged arch (Fig. 3.3a) presents a structure with one redundant constraint. Let the primary unknown X_1 be the horizontal reaction of the right support. The primary system is shown in Fig. 3.3b; this structure is subjected to given loads as well as the force X_1.

The canonical equation of the Force method and primary unknown is given as follows:

$$\delta_{11}X_1 + \Delta_{1P} = 0, \quad X_1 = -\frac{\Delta_{1P}}{\delta_{11}}. \tag{3.3}$$

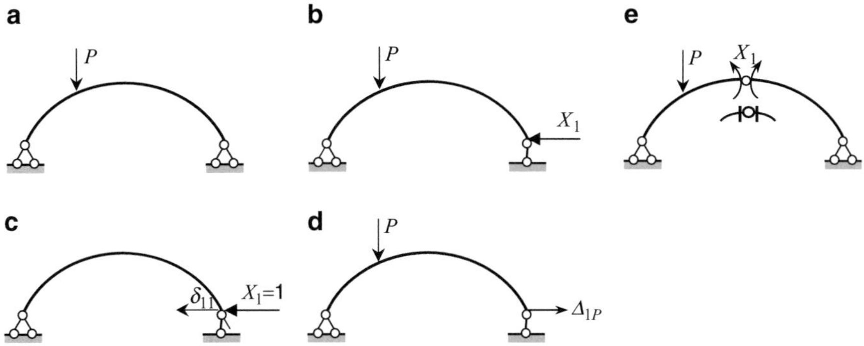

Fig. 3.3 Design diagram and primary systems of two-hinged arch

Coefficient δ_{11} of canonical equation represents the *displacement* of the primary structure along the direction of unknown X_1 due to *unit* primary unknowns (Fig. 3.3c), and therefore this coefficient is called *unit displacement*. Unit displacement is always strictly positive, $\delta_{11} > 0$. The term $\delta_{11}X_1$ represents the displacement along the direction of unknown X_1 due to the action of real unknown X_1. Free term Δ_{1P} represents displacement in the primary system along the direction of unknown X_1 due to the action of actual load. Displacement caused by applied loads Δ_{1P} is called *load term* (Fig. 3.3d).

Left part of (3.3), $\delta_{11}X_1 + \Delta_{1P}$, represents the total displacement along the direction of unknown X_1 due to its action and a given load. Total displacement, which occurs in the primary structure in the direction of eliminated restriction caused by primary unknown and applied load, equals to zero. In this case, the difference between the given and primary structures vanishes.

The form of presentation of the canonical equation $\delta_{11}X_1 + \Delta_{1P} = 0$ is always the same for a two-hinged arch; it does not depend on its peculiarities (uniform/nonuniform arch, symmetrical/nonsymmetrical arch), and type of external actions (forces, support settlements, temperature change, fabrication error).

It is possible to adopt the primary system as a three-hinged arch (Fig. 3.3e). In this case, the primary unknown will be the moment at a crown. Coefficient δ_{11} is a *mutual* angle of rotation at the crown caused by unit moment $X_1 = 1$, and Δ_{1P} is the *mutual* angle of rotation of two sections – left and right at the crown – caused by a given load. In this case, the canonical equation means that a *mutual* angle of rotation at the crown caused by primary unknown X_1 and given load equals to zero.

A two-hinged arch with tie presents a structure with one redundant constraint. Internal force in a tie may be treated as the primary unknown X_1. In this case, the primary system represents a simply supported curvilinear rod. Coefficient δ_{11} is *mutual* linear displacement due to the unit force $X_1 = 1$, and Δ_{1P} is *mutual* linear displacement in the tie due to the given load. The canonical equation means that *mutual* linear displacement of any two sections, which belongs to the tie, caused by primary unknown X_1 and given load equals to zero.

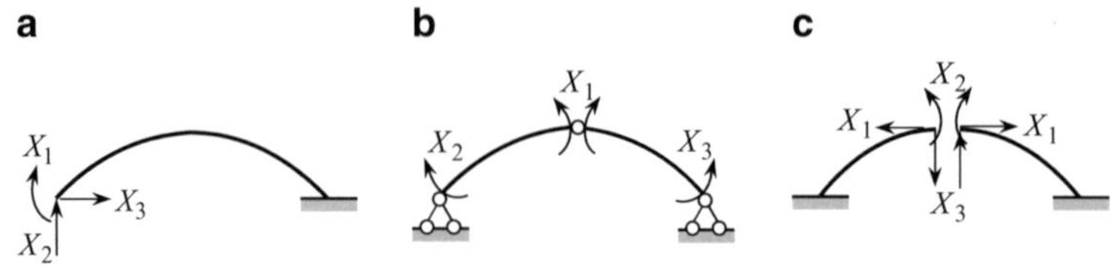

Fig. 3.4 Primary systems for hingeless arch

Hingeless arch present a structure with three redundant constraints. The different primary systems are shown in Fig. 3.4.

Nonsymmetrical primary system is shown in Fig. 3.4a. The redundant constraints are all constraints of one of the support. Primary unknowns are three reactions of the eliminated constraints.

Primary system as a three-hinged symmetric arch is shown in Fig. 3.4b. Primary unknowns are *pair wise* moments at the crown and moments at supports.

Primary system may be chosen as two symmetric curvilinear fixed-free bars (Fig. 3.4c). Primary unknowns are internal forces which arise at the axis of symmetry; they are *pair wise* axial forces X_1, moments X_2, and shear force X_3. It is obvious that other versions of the primary systems are possible.

For any primary system in Fig. 3.4, the canonical equation of Force method is written as follows

$$\begin{aligned} \delta_{11}X_1 + \delta_{12}X_2 + \delta_{13}X_3 + \Delta_{1P} &= 0, \\ \delta_{21}X_1 + \delta_{22}X_2 + \delta_{23}X_3 + \Delta_{2P} &= 0, \\ \delta_{31}X_1 + \delta_{32}X_2 + \delta_{33}X_3 + \Delta_{3P} &= 0. \end{aligned} \tag{3.4}$$

The form of presentation of the canonical equation (3.4) is always the same for hingeless arches; it does not depend on the peculiarities of the arch, and type of external exposures (forces, support settlements, temperature change, fabrication, error). Canonical equations for an n-times redundant structure may be written in a similar form.

All coefficients δ_{ik} of canonical equations represent a *displacement* of the primary structure due to *unit* primary unknowns; these coefficients are called *unit displacements*.

Coefficient δ_{ik} is the displacement along the direction of unknown X_i due to action of unit unknown X_k; term $\delta_{ik}X_k$ presents displacement along the direction of unknown X_i due to action of real unknown X_k. Coefficients δ_{ik}, which are located on the principal diagonal ($i = k$), are called principal (main) displacements. All other displacements $\delta_{ik}(i \neq k)$ are called the secondary unit displacements.

Free term Δ_{iP} presents displacement along the direction of unknown X_i due to the action of actual load on primary system. Displacements Δ_{iP} caused by applied loads are called the *loaded terms or free terms*. Loaded displacements Δ_{iP} may be positive, negative, or equals zero.

Physical meaning of the canonical equations: The left part of the *i*th equation presents the total displacement along the direction of unknown X_i due to action of all real unknowns X_k as well as applied load. Total displacement of the primary structure in directions of eliminated constraints caused by primary unknowns and applied load equals zero. In this case, the difference between the given and primary structures is vanished.

3.2.3 Unit and Loading Displacements

Computation of coefficients and free terms of canonical equations presents significant and very important part of analysis of any statically indeterminate structure. For their calculation, any methods can be applied. The graph multiplication method is the best suited for arched structures. For this, it is necessary in primary system to construct bending moment diagrams $\bar{M}_1, \bar{M}_2, ..., \bar{M}_n$ due to unit primary unknowns X_i, $i = 1,\ldots,n$ and diagram M_{P}^0 due to given load. Unit displacements and loaded terms are calculated by Maxwell–Mohr formulas. If the shear and axial forces are neglected, then

$$\delta_{ik} = \sum \int \frac{\overline{M}_i \times \overline{M}_k}{EI} \mathrm{d}s, \quad \Delta_{i\mathrm{P}} = \sum \int \frac{\overline{M}_i \times M_P^0}{EI} \mathrm{d}s. \tag{3.5}$$

Computation of these displacements may be performed using the graph multiplication method. Book [Kar10] presents different types of verifications for unit and loaded displacements.

Properties of unit coefficients are as follows:

1. Main displacements are strictly positive ($\delta_{ii}>0$).
2. Secondary displacements δ_{ik}, $i \neq k$ may be positive, negative, or zero.
3. Secondary displacements satisfy the reciprocal displacement theorem

$$\delta_{ik} = \delta_{ki}. \tag{3.6}$$

It means that unit displacements symmetrically placed with respect to the principal diagonal of canonical equations are equal.

The unit of displacements δ_{ik} presents the ratio of unit for displacement according to index *i* and units for force according to index *k*. For example, for primary system in Fig. 3.4a, we get

$$\delta_{11}(\mathrm{rad}/(\mathrm{kNm})), \quad \delta_{12}(\mathrm{rad}/\mathrm{kN}), \quad \delta_{21}(\mathrm{m}/(\mathrm{kNm}) = 1/\mathrm{kN}), \quad \delta_{22}(\mathrm{m}/\mathrm{kN}).$$

3.2.4 Procedure for Analysis

Solution of canonical equation for structure with n redundant constraints is the primary unknowns X_i, $i = 1,\ldots,n$. After that the primary system may be loaded by primary unknowns and given load, internal forces may be computed as for usual statically determinate structure. However, the following way allows once again an effective use of the bending moment diagrams in primary system. The final bending moment diagram M_{P} may be constructed by the formula given below

$$M_{\mathrm{P}} = \bar{M}_1 X_1 + \bar{M}_2 X_2 + \cdots + \bar{M}_n X_n + M_{\mathrm{P}}^0. \tag{3.7}$$

Thus, in order to compute the ordinates of the resulting bending moment diagram, it is necessary to multiply each unit bending moment diagrams $\overline{M}_k$ by the corresponding primary unknown X_k and summing up with bending moment diagram due to applied load in the primary system M_{P}^0. This formula expresses the superposition principle. Advantage of (3.6) is that it may be effectively presented in tabulated form.

Shear and axial forces may be calculated by formulas

$$Q_{\mathrm{P}} = \bar{Q}_1 X_1 + \bar{Q}_2 X_2 + \cdots + \bar{Q}_n X_n + Q_{\mathrm{P}}^0,$$

$$N_{\mathrm{P}} = \bar{N}_1 X_1 + \bar{N}_2 X_2 + \cdots + \bar{N}_n X_n + N_{\mathrm{P}}^0, \tag{3.7a}$$

where $\bar{Q}_i$, and $\bar{N}_i$ are shear and axial force diagram due to unit ith primary unknown $X_i = 1$.

Note, for framed structure, shear forces may be easily calculated on the basis of bending moment diagram using Schwedler theorem and axial forces may be calculated on the basis of shear force diagram by considering equilibrium of joints of the structure [Kar10]. Finally, having internal force diagrams, all reactions are easy to determine.

Procedure for the analysis of statically indeterminate arches is as follows:

1. Choose the primary system of the Force method.
2. Accept the simplified model of the arch, i.e., the arch is divided into several portions and each curvilinear portion is changed by straight member. Calculate the geometrical parameters of the arch at specified points.
3. Calculate the unit and loaded displacements; (3.5) may be applied.
4. Find the primary unknown using canonical equation of the Force method.
5. Construct the internal force diagrams; (3.7)–(3.7a) may be applied.
6. Calculate the reactions of supports and provide their verifications.

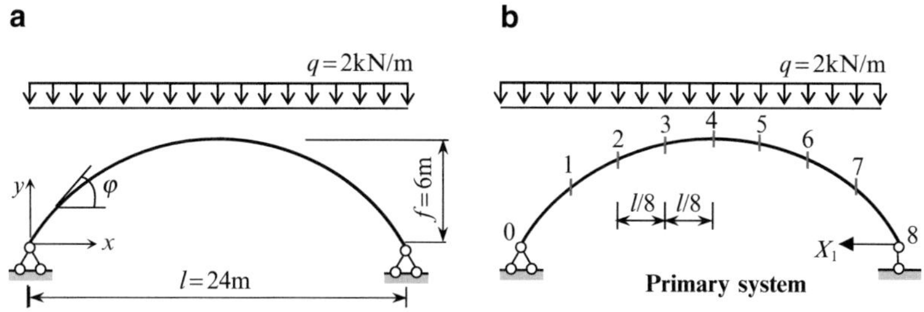

Fig. 3.5 Parabolic arch. Design diagram and primary system

3.3 Arches Subjected to Fixed Loads

This section contains analysis of parabolic two-hinged uniform arch, subjected to uniformly distributed load q within all span. It is shown that this arch is rational.

3.3.1 Parabolic Two-Hinged Uniform Arch

Design diagram of two-hinged uniform arch subjected to uniformly distributed load q within all span is shown in Fig. 3.5a. The flexural stiffness of the cross section of the arch is EI. The equation of the neutral line of the arch is $y = 4fx(l - x)/l^2$. It is necessary to find the distribution of internal forces.

The arch under investigation is statically indeterminate of the first degree. The primary system is shown in Fig. 3.5b; the primary unknown X_1 is the horizontal reaction of the right support. Canonical equation of the Force method is $\delta_{11}X_1 + \Delta_{1P} = 0$.The primary unknown $X_1 = -\Delta_{1P}/\delta_{11}$.

Specified Points of the Arch

The *span* of the arch is divided into eight equal parts; the specified points are labeled 0–8. Parameters of the arch for these sections are presented in Table 3.1; the following formulas for the calculation of trigonometric functions of the angle between the tangent to the arch and x-axis have been used:

$$\tan\varphi = y' = \frac{4f(l-2x)}{l^2}, \quad \cos\varphi = \frac{1}{\sqrt{1+\tan^2\varphi}}, \ \sin\varphi = \cos\varphi\tan\varphi.$$

The length of the chord between points n and $n - 1$ equals

$$l_{n,n-1} = \sqrt{(x_n - x_{n-1})^2 + (y_n - y_{n-1})^2}.$$

The chord lengths of each portion of the arch are presented in Table 3.2.

Table 3.1 Geometrical parameters of parabolic arch

Points	Coordinates (*m*) x	y	$\tan\varphi$	$\cos\varphi$	$\sin\varphi$
0	0	0.0	1.00	0.7070	0.7070
1	3	2.625	0.75	0.800	0.6000
2	6	4.500	0.50	0.8944	0.4472
3	9	5.625	0.25	0.9701	0.2425
4	12	6.000	0.0	1.0	0.0
5	15	5.625	−0.25	0.9701	−0.2425
6	18	4.500	−0.5	0.8944	−0.4472
7	21	2.625	−0.75	0.800	−0.6000
8	24	0.0	−1.00	0.7070	−0.7070

Table 3.2 Chord length for portions of the arch

Portion	Chord length (m)
0–1	3.9863
1–2	3.5377
2–3	3.2040
3–4	3.0233
4–5	3.0233
5–6	3.2040
6–7	3.5377
7–8	3.9863

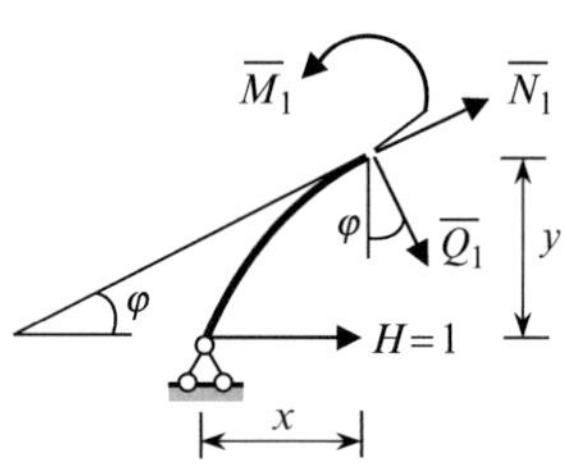

Fig. 3.6 Positive directions of internal forces

Internal Forces in the Unit State

The arch is subjected to unit primary unknown $X_1 = 1$ (Fig. 3.5b). Horizontal reaction $H = 1$ and the positive directions of internal forces are shown in Fig. 3.6.

$$\begin{aligned} \overline{M}_1 &= -1 \times y, \\ \overline{Q}_1 &= -1 \times \sin\varphi, \\ \overline{N}_1 &= -1 \times \cos\varphi. \end{aligned} \tag{3.8a}$$

Table 3.3 Internal forces of the arch in the unit state

Points	$\overline{M}_1$	$\overline{Q}_1$	$\overline{N}_1$
0	0.0	−0.7070	−0.7070
1	−2.625	−0.6000	−0.8000
2	−4.50	−0.4472	−0.8944
3	−5.625	−0.2425	−0.9701
4	−6.00	0.0	−1.0000
5	−5.625	0.2425	−0.9701
6	−4.50	0.4472	−0.8944
7	−2.625	0.6000	−0.8000
8	0.0	0.7070	−0.7070

Internal forces at specified points in the unit state according to (3.8a) are presented in Table 3.3.

The unit displacement caused by primary unknown $X = 1$ equals

$$\delta_{11} = \int_{(s)} \frac{\overline{M}_1 \times \overline{M}_1}{EI} \mathrm{d}s. \tag{3.8b}$$

Thus, the *only* column $\overline{M}_1$ (Table 3.3) will be used for the calculation of unit displacement; the columns $\overline{Q}_1$ and $\overline{N}_1$ will be used for computation of final shear and axial forces as indicated in step 5 of the procedure.

Internal Forces in the Loaded State

Displacement in the primary system caused by the applied load is given as follows

$$\Delta_{1\mathrm{P}} = \int_{(s)} \frac{\overline{M}_1 \times M_{\mathrm{P}}^0}{EI} \mathrm{d}s, \tag{3.8c}$$

where M_{P}^0 is bending moments in the arch in the primary system due to the given load q. Thus, as in the case of unit displacement, for the computation of loaded displacement, we will take into account only bending moment.

The reactions of supports of the primary system in the loaded state are $R_A^0 = R_B^0 = 24$ kN; this state is not shown. Expressions for internal forces are follows $(0 \leq x \leq 24)$

$$M_{\mathrm{P}}^0 = R_A^0 x - q\frac{x^2}{2} = 24x - x^2,$$

$$Q_{\mathrm{P}}^0 = \left(R_A^0 - qx\right) \cos\varphi = (24 - 2x) \cos\varphi,$$

Table 3.4 Internal forces of the arch in the loaded state

Points	M_{P}^0	Q_{P}^0	N_{P}^0
0	0.0	16.968	−16.968
1	63	14.400	−10.800
2	108	10.7328	−5.178
3	135	5.8206	−1.455
4	144	0.0	0.0
5	135	−5.8206	−1.455
6	108	−10.7328	−5.178
7	63	−14.400	−10.800
8	0.0	−16.968	−16.968

$$N_{\mathrm{P}}^0 = -\left(R_A^0 - qx\right)\sin\varphi = -(24 - 2x)\sin\varphi.$$

Internal forces at specified points in the loaded state of the primary system are presented in Table 3.4.

Computation of Unit and Loaded Displacements

For the calculation of displacement, the Simpson formula is applied. Unit and loaded displacements are

$$\delta_{11} = \frac{\overline{M}_1 \times \overline{M}_1}{EI} = \sum_1^n \frac{l_i}{6EI}\left(a_1^2 + 4c_1^2 + b_1^2\right),$$

$$\Delta_{1\mathrm{P}} = \frac{\overline{M}_1 \times M_{\mathrm{P}}^0}{EI} = \sum_1^n \frac{l_i}{6EI}\left(a_1 a_{\mathrm{P}} + 4c_1 c_{\mathrm{P}} + b_1 b_{\mathrm{P}}\right),$$

where l_i is the length of the ith straight portion of the arch (Table 3.2); n is the number of straight portions of the arch (in our case, $n = 8$); a_1, a_P are the ordinates of the bending moment diagrams $\overline{M}_1$ and M_{P}^0 at the extreme left end of the portion; b_1, b_P are the ordinates of the same bending moment diagrams at the extreme right end of the portion; c_1, c_P are the ordinates of the same bending moment diagrams at the middle point of the portion.

Calculation of the unit and loaded displacements is presented in Table 3.5. In columns a_1, c_1, and b_1 of columnhead "Unit state" contain data from column $\overline{M}_1$ of Table 3.3. In columns a_P, c_P, and b_P of columnhead "Loaded state" contain data from column M_{P}^0 of Table 3.4. As an example for portion 1–2, the entries of columns 6 and 10 are obtained by the following way

Table 3.5 Calculation of coefficient and free term of canonical equation

Portion	$\frac{l}{6EI}$	Unit state a_1	c_1	b_1	$\frac{\overline{M}_1 \times \overline{M}_1}{EI}$	Loaded state a_P	c_P	b_P	$\frac{\overline{M}_1 \times M_P^0}{EI}$
1	2	3	4	5	6	7	8	9	10
0–1	0.6644	0.0	−1.3125	−2.625	9.1563	0.0	31.5	63	−219.75
1–2	0.5896	−2.625	−3.5625	−4.500	**45.9335**	63	85.5	108	**−1102.40**
2–3	0.5340	−4.500	−5.0625	−5.625	82.4529	108	120.5	135	−1978.87
3–4	0.5039	−5.625	−5.8125	−6.000	102.1815	135	139.5	144	−2452.35
4–5	0.5039	−6.000	−5.8125	−5.625	102.1815	144	139.5	135	−2452.35
5–6	0.5340	−5.625	−5.0625	−4.500	82.4529	135	121.5	108	−1978.87
6–7	0.5896	−4.500	−3.5625	−2.625	45.9335	108	85.5	63	−1102.40
7–8	0.6644	−2.625	−1.3125	0.0	9.1563	63	31.5	0.0	219.40

$$\frac{3.5377}{6EI}\times[(-2.625)\times(-2.625)+4(-3.5625)\times(-3.5625)+(-4.50)\times(-4.50)]$$
$$=\frac{45.9335}{EI},$$
$$\frac{3.5377}{6EI}\times[(-2.625)\times 63+4(-3.5625)\times 85.5+(-4.50)\times 108]=-\frac{1,102.40}{EI}.$$

Sum of the columns 6 and 10 are:

$$\delta_{11}^{\text{arch}}=\frac{479.4484}{EI}(\text{m/kN}),$$

$$\Delta_{1\text{P}}=-\frac{11,506.74}{EI}(\text{m}).$$

Canonical equation and primary unknown (thrust) become

$$\frac{479.4484}{EI}X_1-\frac{11,506.74}{EI}=0 \rightarrow X_1=24.00\text{ kN}.$$

Construction of Internal Force Diagrams

Internal forces, which arise in the entire structure, may be calculated by the formulas given below

$$M=\overline{M}_1X_1+M_{\text{P}}^0,$$
$$Q=\overline{Q}_1X_1+Q_{\text{P}}^0,$$
$$N=\overline{N}_1X_1+N_{\text{P}}^0.$$

Table 3.6 Calculation of internal forces at specified points of the arch

Points	$\overline{M}_1X_1$	$\overline{Q}_1X_1$	$\overline{N}_1X_1$	M_P^0	Q_P^0	N_P^0	M	Q	N
0	0.0	−16.968	−16.968	0.0	16.968	−16.968	0.0	0.0	−33.936
1	−63	−14.400	−19.2	63	14.400	−10.80	0.0	0.0	−30.0
2	−108	−10.733	−21.466	108	10.733	−5.178	0.0	0.0	−26.644
3	−135	−5.820	−23.282	135	5.820	−1.455	0.0	0.0	−24.737
4	−144	0.0	−24.00	144	0.0	0.0	0.0	0.0	−24.0
5	−135	5.820	−23.282	135	−5.820	−1.455	0.0	0.0	−24.737
6	−108	10.733	−21.466	108	−10.733	−5.178	0.0	0.0	−26.674
7	−63	14.40	−19.20	63	−14.40	−10.800	0.0	0.0	−30.0
8	0.0	16.968	−16.968	0.0	−16.968	−16.968	0.0	0.0	−33.936

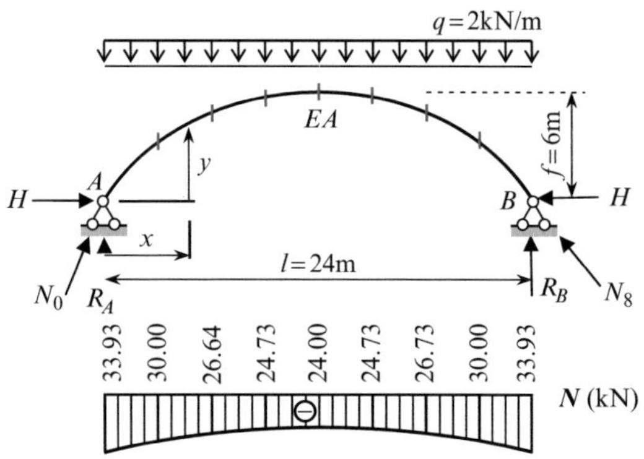

Fig. 3.7 Two-hinged parabolic arch. Axial force diagram and reaction of supports

Calculation of internal forces in the arch due to the given fixed load is presented in Table 3.6; internal forces $\overline{M}_1$, $\overline{Q}_1$, and $\overline{N}_1$ due to unit primary unknown $X_1 = 1$ are presented earlier in Table 3.3.

Corresponding axial force diagram N is presented in Fig. 3.7.

Knowing the internal forces at points 0 and 8, we can calculate the reactions at supports A and B. Axial force $N_0 = 33.936$ kN is shown at support A (Fig. 3.7). Reactions of this support are

$$R_A = N_0 \sin \varphi_0 = 33.936 \times 0.707 = 24 \text{ kN}, \quad H = N_0 \cos \varphi_0 = 24 \text{ kN}.$$

For primary unknown X_1, we have obtained previously the same result.

Discussion

1. If two-hinged uniform parabolic arch is subjected to uniformly distributed load within all span, then this arch is rational since the bending moments and shear forces equal to zero in all sections of the arch. In this case, only axial compressed forces arise in all section of the arch. Such a conclusion can be made if the effect

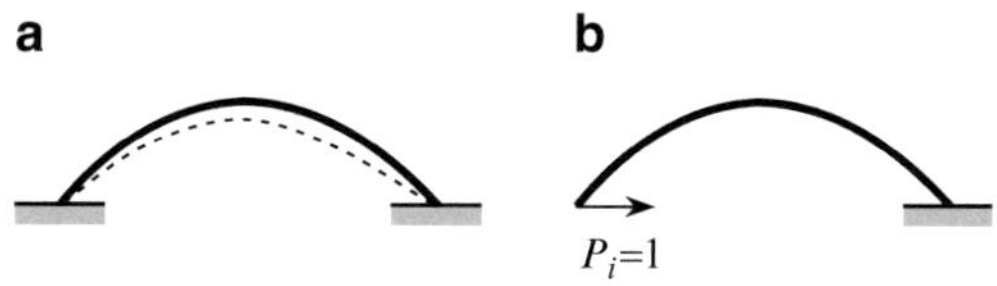

Fig. 3.8 Design diagram of hingeless arch and unit state for calculation of horizontal displacement of the left support

of axial and shear forces on the unit and loaded displacements are not taken into account.

If the axial forces are taken into account, then the change of the primary unknown is insignificant. However, the condition of the arch has changed fundamentally since the arch has ceased to be rational. Detailed analysis is presented in Sect. 3.11.2.

2. Procedure for the analysis of nonuniform arch remains the same. However, in this case, Table 3.1 must contain an additional column with parameter EI for *each point* 0–8, Table 3.2 must contain the parameter EI_i for the *middle point* of each portion, and column 2 of the Table 3.5 should be replaced by the column $l_i/6\mathrm{EI}_i$.

3.3.2 Some Comments About Rational Axis

In the previous section, we showed that uniform parabolic two-hinged arch, subjected to uniform vertical load, is a rational arch. Taking into account the axial forces which arise in the arch, this wonderful feature of the arch immediately disappears. It is possible to show that if a hingeless arch is shaped according to pressure polygon of three-hinged arch (Appendix, Pressure Polygon), then inevitably bending moments arise in the arch if the axial forces are taken into account.

Let us show that it is essentially impossible to construct a rational hingeless arch (i.e., an arch wherein all the sections of which the bending moments are zero).

Assume that it is possible to construct a rational hingeless arch. For this arch, $M_\mathrm{P} = Q_\mathrm{P} = 0$ and the axial force in the arch is $N_\mathrm{P} = H/\cos\varphi$. Let us determine the horizontal displacement of the left support (Fig. 3.8a). Unit state presents *any statically determinate system* subjected to unit load, which corresponds to required displacement [Kar10].

One version of primary system and corresponding unit state is shown in Fig. 3.8b.

In this state, internal forces $\overline{M}_i \neq 0, \overline{Q}_i \neq 0$, and $\overline{N}_i = \cos\varphi$. Therefore, required displacements becomes

$$\Delta_{\mathrm{hor}}^{\mathrm{left}} = \int_s \frac{\overline{N}_i N_\mathrm{P}}{EA} \mathrm{d}s = \int_x \cos\varphi \times \frac{H}{\cos\varphi} \frac{1}{EA} \times \frac{\mathrm{d}x}{\cos\varphi}.$$

It is obvious that this result does not equal to zero, so the initial assumption about zero bending moments is false. Thus, any change in the axis of the arch cannot lead to the zero bending moments [Kis60]. Strictly speaking, even for rational axis of the three-hinged arch, deflection of the arch caused by axial forces leads to the change of the curvature of the arch, and as a result leads to appearance of bending moments.

Note that it is possible to achieve a zero bending moment by controlling the stresses in the arch. This problem is considered in ref. [Kis60].

3.4 Symmetrical Arches

This section describes properties of symmetrical structures, introduces the concept of elastic center, and contains analysis of parabolic nonuniform arch and semicircular uniform arch.

3.4.1 Properties of Symmetrical Structures

Symmetrical structures mean their geometrical symmetry, symmetry of supports, and symmetry for stiffness of the members. Symmetrical arch subjected to vertical load P is shown in Fig. 3.9a. This load may be presented as a sum of symmetrical and antisymmetrical components (Fig. 3.9b, c). In general, any load may be presented as the sum of symmetrical and antisymmetrical components.

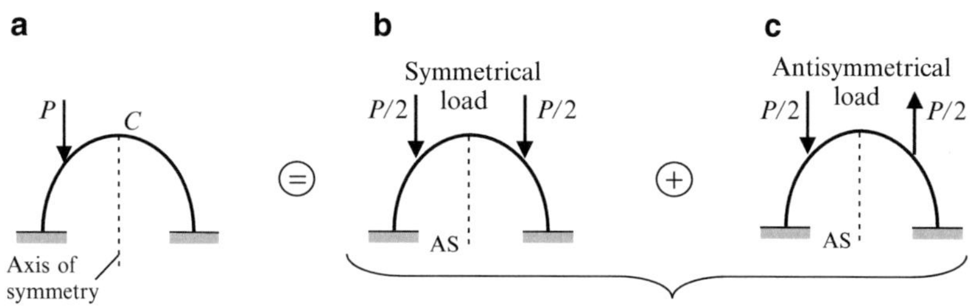

Fig. 3.9 Presentation of the load P as a sum of the symmetrical and antisymmetrical components

At any section of a member, the following *internal forces* arise: *symmetrical* unknowns, such as bending moment M and axial force N and *antisymmetrical* unknown shear force Q (Fig. 3.10).

At the point of the axis of symmetry, the following *displacements* arise: the vertical Δ_v, horizontal Δ_h, and angular φ. In the case of symmetrical load, the horizontal and angular displacements at the point C are zero, while the vertical displacement occurs. In the case of antisymmetrical load, the horizontal and angular displacements at the point A exist, while a vertical displacement is zero.

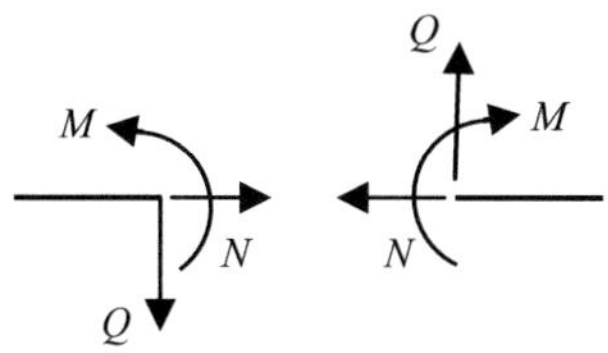

Fig. 3.10 Symmetrical and antisymmetrical internal forces

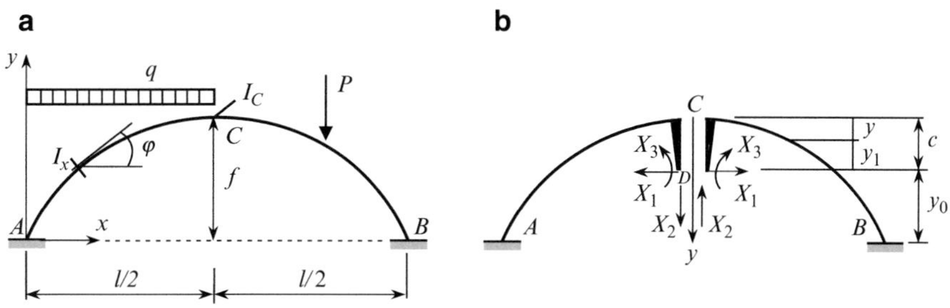

Fig. 3.11 Primary system using the concept of elastic center

Fundamental properties of internal force diagrams for symmetrical structures:

1. In the case of symmetrical loading, the internal force diagrams for symmetrical unknowns (M, N) are symmetrical and the diagram for antisymmetrical unknown (Q) is antisymmetrical.
2. In the case of antisymmetrical loading, the internal force diagrams for symmetrical unknowns (M, N) are antisymmetrical and the diagram for antisymmetrical unknown (Q) is symmetrical.

More detail of the properties for the symmetrical structures is presented in ref. [Kar10].

3.4.2 *Elastic Center*

Let us consider a hingeless arch subjected to arbitrary load (Fig. 3.11a). For such arch, the system of three canonical equations is written in the form of (3.4). Elastic center is a special point, where primary unknowns should be placed so that a system of coupled canonical equations of the Force method would be presented as three independent equations [Dar89], [Bro06].

Let us consider symmetric arch and find location of the elastic center. Let the primary unknowns be the axial force X_1, shear X_2, and bending moment X_3; they act at the crown C of the arch. Since the bending moments diagrams $\overline{M}_1$ and $\overline{M}_3$ are symmetrical while $\overline{M}_2$ is antisymmetrical, then

$$\delta_{12} = \delta_{21} = \int \frac{\overline{M}_1 \overline{M}_2}{EI} \mathrm{d}s = 0 \quad \text{and} \quad \delta_{23} = \delta_{32} = \int \frac{\overline{M}_2 \overline{M}_3}{EI} \mathrm{d}s = 0.$$

In this case, the dependent equations (3.4) are reduced to the system of two coupled equations with respect to symmetrical unknowns X_1 and X_3

$$\begin{aligned} \delta_{11}X_1 + \delta_{13}X_3 + \Delta_{1\mathrm{P}} &= 0, \\ \delta_{31}X_1 + \delta_{33}X_3 + \Delta_{3\mathrm{P}} &= 0, \end{aligned} \tag{3.9}$$

and one independent equation with respect to antisymmetrical unknown X_2.

$$\delta_{22}X_2 + \Delta_{2\mathrm{P}} = 0. \tag{3.9a}$$

Note, that in case of one-hinged symmetrical arch, the bending moment X_3 at the crown equals zero.

Now the secondary unit displacements of the system (3.9) may be reduced to zero. For this, we introduce two absolutely rigid cantilevers *CD* of length *c*, and the primary unknowns are located at point *D* (Fig. 3.11b). The origin of coordinates coincides with the crown C and axis *y* is directed downward. If three unknowns are placed at the point *D*, then the bending moments due to unit primary unknowns are $\overline{M}_1 = -1 \times y_1, \overline{M}_2 = -1 \times x$ and $\overline{M}_3 = 1$, where y_1 is ordinate of the axis of the arch with respect to line of the unknown X_1. Let us find the length *c* so that unit displacement be zero: $\delta_{13} = 0$. In this case, two coupled equations break down into two independent equations. It happens if

$$\delta_{13} = \int \frac{\overline{M}_1 \overline{M}_3}{EI} \mathrm{d}s = -\int_A^C \frac{(y_1) \times 1}{EI} \mathrm{d}s = \int_A^C \frac{y_1 \mathrm{d}s}{EI} = 0.$$

Thus, initial system of (3.4) is transformed into three independent equations:

$$\delta_{11}X_1 + \Delta_{1\mathrm{P}} = 0,$$

$$\delta_{33}X_3 + \Delta_{3\mathrm{P}} = 0,$$

$$\delta_{22}X_2 + \Delta_{2\mathrm{P}} = 0.$$

The *i*th canonical equation means that a *mutual* displacement of two sections at point *D* caused by primary unknown X_i and external load is zero. Unit displacement δ_{ii} presents a mutual displacements of two sections at point *D* due to primary unknown $X_i = 1$, while $\Delta_{i\mathrm{P}}$ is a mutual displacement of two sections at same point *D* due to the given load.

The primary unknowns are $X_i = -\Delta_{i\mathrm{P}}/\delta_{ii}$. The point *D* is called the elastic center (Fig. 3.11b).

The coordinate of elastic center is

$$c = \frac{S_1}{S_2}, \quad S_1 = \int_0^{l/2} y \frac{\mathrm{d}s}{EI}; \quad S_2 = \int_0^{l/2} \frac{\mathrm{d}s}{EI}. \tag{3.10}$$

Indeed, an integral $\int_A^C y_1 \mathrm{d}s/EI = 0$ may be presented as

$$\int_A^C \frac{(c-y)\mathrm{d}s}{EI} = c\int_A^C \frac{\mathrm{d}s}{EI} - \int_A^C \frac{y\mathrm{d}s}{EI} = 0,$$

and this expression immediately leads to (3.10). Positions of elastic center for different shapes of arches are presented below.

3.4.2.1 Hingeless Symmetrical Parabolic Nonuniform Arch

Assume that $y = 4f/l^2(lx - x^2)$ and $I_x = I_\mathrm{C}/\cos\varphi$.

Since $\mathrm{d}s = \mathrm{d}x/\cos\varphi$, then $\mathrm{d}s/EI = \mathrm{d}x/EI_\mathrm{C}$, so $S_1 = 4f/(l^2EI_\mathrm{C})\int_0^{l/2}(lx - x^2)\mathrm{d}x$, $S_2 = \int_0^{l/2}\mathrm{d}x/EI_\mathrm{C}$ and for coordinate of elastic center we get $y_0 = 2f/3, c = f/3$.

3.4.2.2 Hingeless Symmetrical Cubic Nonuniform Arch

Let equation of the axis and the moment of the cross section be [Sni66].

$$y = \left[\mu\frac{x^2}{l_1^2} + (1-\mu)\frac{x^2}{l_1^2}\right]f \quad \text{(origin is placed at the crown)}$$

$$I_\varphi = \frac{I_0}{\cos\varphi} \times \frac{1}{1 + \beta(x/l_1)},$$

where $2l_1$ is a span of the arch, μ and β are any parameters, $0 \le \mu \le 1.4$, and $-1 \le \beta \le 0$. In this case,

$$c - f\frac{15 + 5\mu + \beta(12 + 3\mu)}{30(2+\beta)}.$$

3.4.2.3 Hingeless Symmetrical Circular Uniform Arch [Kle80]

In this case, the elastic center coincides with a center of gravity of the thin semicircular line, i.e., $y_0 = 2R/\pi$.

Table 3.7 General form for calculation of unit and loaded displacements

Portion	Δs	I	$\Delta s'$	x	y	$y^2\Delta s'$	$x^2\Delta s'$	M_P^0	$yM_P^0\Delta s'$	$xM_P^0\Delta s'$	$M_P^0\Delta s'$
1											
...											

3.4.2.4 Any Symmetrical Hingeless Nonuniform Arch [Bro06]

In case of symmetrical arch of arbitrary shape and with any law of the moment of inertia I of the cross section, the coordinate of elastic center may be determined by numerical method with the following steps.

1. The span of the arch is divided into 8–12 equal portions and each curvilinear segment of the arch is replaced by a straight member of length Δs.
2. For each nodal point of the arch, calculate the moment of inertia; for middle point of the straight member, the moment of inertia may be calculated by the formula $I = 0.5(I_l + I_r)$, where I_l and I_r are the moments of inertia at the left and right end points of the straight member.
3. Determine conventional length of each member: $\Delta s' = (I_C/I)\Delta s$ (I_C is a moment of inertia at the crown).
4. Determine coordinates x and y for middle point of each straight member.
5. For middle point of each straight member, determine the bending moment M_P^0 in the primary system caused by the given load.

For calculation of unit and loaded displacements, we obtain the following formulas

$$
\begin{aligned}
EI_C\delta_{11} &= \int \overline{M}_1^2 ds' = \sum_A^C y^2\Delta s'; \quad EI_C\Delta_{1P} = \int \overline{M}_1 M_P^0 ds' = \sum_A^C yM_P^0\Delta s',\\
EI_C\delta_{22} &= \int \overline{M}_2^2 ds' = \sum_A^C x^2\Delta s'; \quad EI_C\Delta_{2P} = \int \overline{M}_2 M_P^0 ds' = -\sum_A^C xM_P^0\Delta s',\\
EI_C\delta_{33} &= \int \overline{M}_3^2 ds' = \sum_A^C \Delta s'; \quad EI_C\Delta_{3P} = \int \overline{M}_3 M_P^0 ds' = \sum_A^C M_P^0\Delta s'.
\end{aligned}
\tag{3.11}
$$

All calculations are presented in the tabulated form (Table 3.7).

If it is necessary, the normal and shear forces may be easily taken into account [Rab54a].

The primary unknowns are $X_i = -\Delta_{iP}/\delta_{ii}, \quad i = 1, 2, 3$. The reactions of supports may be calculated considering equilibrium of the left and right half-arch subjected to given loads and primary unknowns.

In case of any nonsymmetrical hingeless arch, for full separation of the coupled equations we need to determine two coordinates $(x_0,\ y_0)$ and angle of rotation (α) of the axis (Fig. 3.12) [Rab54], [Bro06]. These axis are called the principal axes.

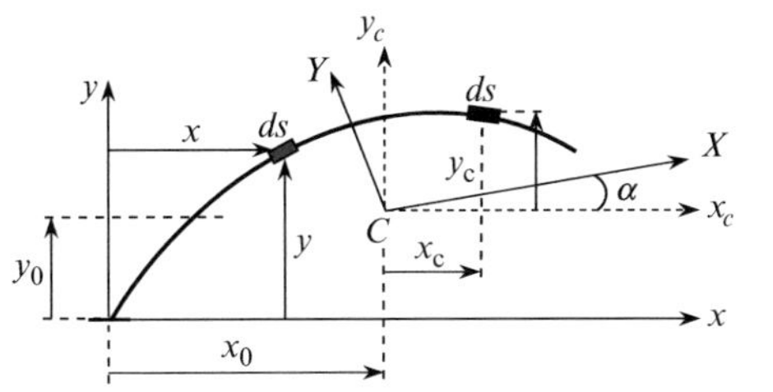

Fig. 3.12 Elastic center concept for nonsymmetrical hingeless arch

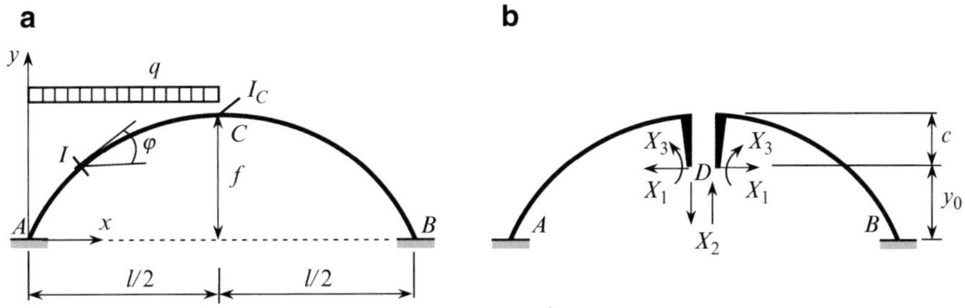

Fig. 3.13 (**a, b**) Design diagram and primary system for hingeless arch

Position of elastic center and direction of principal axis may be calculated by the formulas presented in ref. [Bro06]. However, the numerical procedure to calculate the position of the elastic center turns out to be very cumbersome, and thus the advantage of using the elastic center in all further calculations becomes questionable.

The physical meaning of the concept *"elastic center"*: in the middle point of each straight segment apply a load ds/I, which is called the elementary fictitious elastic load. The center of gravity of these loads represents the elastic center [Dar89].

3.4.3 *Parabolic Hingeless Nonuniform Arch*

Symmetrical parabolic nonuniform arch is subjected to uniformly distributed load q (Fig. 3.13a). Equation of the central line of the arch is given by $y = 4fx(l - x)/l^2$. The moment of inertia of the cross section of the arch changes according to the law $I = I_C/\cos\varphi$. The primary system is shown in Fig. 3.13b.

Position of the elastic center D is $c = f/3$. In this case, the primary unknowns are $X_i = -\Delta_{iP}/\delta_{ii}$, $i = 1, 2, 3$. If the axial and shear forces are neglected, then expressions for unit and loaded displacements become

$$\delta_{ii} = \int \frac{\overline{M}_i^2 \mathrm{d}s}{EI_x} = \frac{1}{EI_C} \int \overline{M}_i^2 \mathrm{d}x, \quad \Delta_{iP} = \int \frac{\overline{M}_i M_P \mathrm{d}s}{EI_x} = \frac{1}{EI_C} \int \overline{M}_i M_P \mathrm{d}x.$$

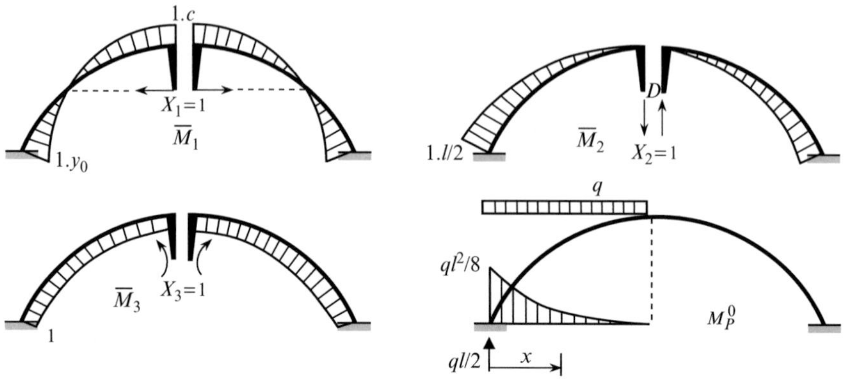

Fig. 3.14 Bending moment diagrams in the primary system caused by unit primary unknowns and external load

Bending moment diagrams in a primary system caused by unit primary unknowns and given load are shown in Fig 3.14.

Construct expressions for bending moments in the primary system caused by unit primary unknowns $X_i = 1$ and external load q. Let the origin be placed at support A for left half-arch and at support B for right-half arch; the positive direction of the axis x for right-half arch is directed from right to left.

$$\overline{M}_1^{\text{left}} = \overline{M}_1^{\text{right}} = \frac{2}{3}f - y,$$

$$\overline{M}_2^{\text{left}} = -\frac{l}{2} + x; \quad \overline{M}_2^{\text{right}} = \frac{l}{2} - x,$$

$$\overline{M}_3^{\text{left}} = \overline{M}_3^{\text{right}} = 1,$$

$$M_{\text{P}}^{\text{left}} = \frac{ql}{2}x - \frac{ql^2}{8} - qx\frac{x}{2}; \quad M_{\text{P}}^{\text{right}} = 0.$$

For unit displacements, we get

$$\delta_{11} = \int \frac{\overline{M}_1^2 \mathrm{d}s}{EI_x} = \frac{2}{EI_{\text{C}}} \int_0^{l/2} \left(\frac{2}{3}f - y\right)^2 \mathrm{d}x = \frac{4f^2 l}{45EI_{\text{C}}},$$

$$\delta_{22} = \int \frac{\overline{M}_2^2 \mathrm{d}s}{EI_x} = \frac{2}{EI_{\text{C}}} \int_0^{l/2} \left(\frac{l}{2} - x\right)^2 \mathrm{d}x = \frac{l^3}{12EI_{\text{C}}},$$

$$\delta_{33} = \int \frac{\overline{M}_3^2 ds}{EI_x} = \frac{2}{EI_{\text{C}}} \int_0^{l/2} \mathrm{d}x = \frac{l}{EI_{\text{C}}}.$$

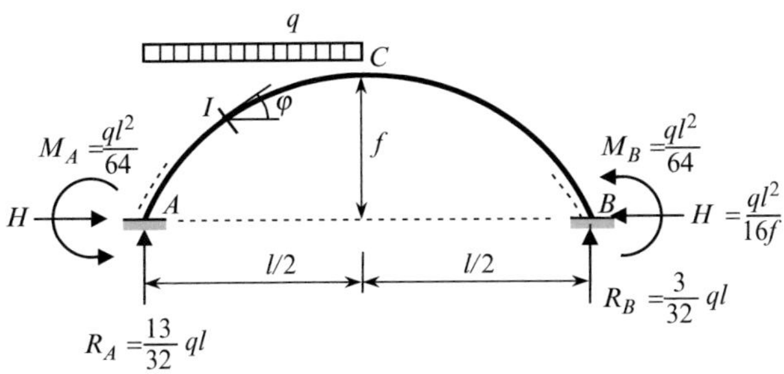

Fig. 3.15 Reactions of hingeless parabolic arch; $I = I_C/\cos\varphi$. Location of expanded fibers in the vicinity of supports are shown by *dotted lines*

The loaded displacements become

$$\Delta_{1P} = \int \frac{\overline{M}_1 M_P ds}{EI_x} = \frac{1}{EI_C}\int_0^{l/2}\left(\frac{2}{3}f - y\right)\left(\frac{ql}{2}x - \frac{ql^2}{8} - qx\frac{x}{2}\right)dx$$

$$= \frac{1}{EI_C}\int_0^{l/2}\left[\frac{2}{3}f - \frac{4f}{l^2}\left(lx - x^2\right)\right]\left(\frac{ql}{2}x - \frac{ql^2}{8} - qx\frac{x}{2}\right)dx = -\frac{qfl^3}{180EI_C}$$

$$\Delta_{2P} = \int \frac{\overline{M}_2 M_P ds}{EI_x} = \frac{1}{EI_C}\int_0^{l/2}\left(-\frac{l}{2} + x\right)\left(\frac{ql}{2}x - \frac{ql^2}{8} - qx\frac{x}{2}\right)dx = \frac{ql^4}{2 \times 64EI_C},$$

$$\Delta_{3P} = \int \frac{\overline{M}_3 M_P ds}{EI_x} = \frac{1}{EI_C}\int_0^{l/2} 1 \times \left(\frac{ql}{2}x - \frac{ql^2}{8} - qx\frac{x}{2}\right)dx = -\frac{ql^3}{48EI_C}.$$

Primary unknowns are

$$X_1 = -\frac{\Delta_{1P}}{\delta_{11}} = \frac{ql^2}{16f}, \quad X_2 = -\frac{\Delta_{2P}}{\delta_{22}} = -\frac{3}{32}ql, \quad X_3 = -\frac{\Delta_{3P}}{\delta_{33}} = \frac{ql^2}{48}.$$

Now we can calculate the reaction of supports (Fig. 3.15): thrust and vertical reactions are

$$H = X_1 = \frac{ql^2}{16f}, \quad R_A = \frac{ql}{2} + X_2 = \frac{13}{32}ql(\uparrow), \quad R_B = X_2 = \frac{3}{32}ql(\uparrow).$$

Bending moment at the crown and at the supports are

$$M_C = X_3 - X_1 c = \frac{ql^2}{48} - \frac{ql^2}{16f} \times \frac{f}{3} = 0.$$

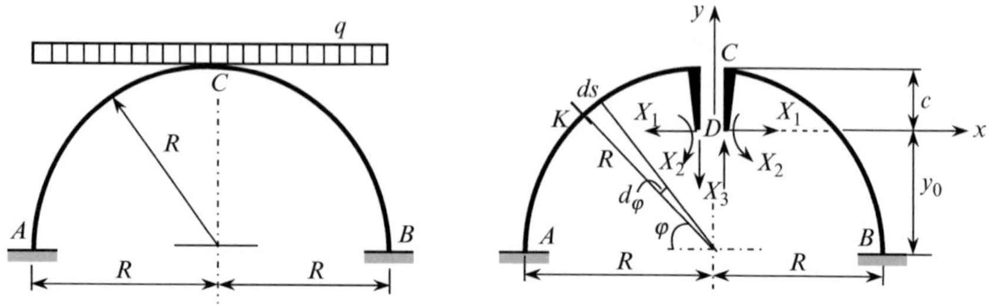

Fig. 3.16 Design diagram of the arch and primary system

$$M_A = X_1 y_0 - X_2 \frac{l}{2} + X_3 - \frac{ql}{2} \times \frac{l}{4} = \frac{ql^2}{16f} \times \frac{2}{3} f + \frac{3}{32} ql \times \frac{l}{2} + \frac{ql^2}{48} - \frac{ql^2}{8} = -\frac{ql^2}{64}$$

(extrados fibers are extended).

$$M_B = X_1 y_0 + X_2 \frac{l}{2} + X_3 = \frac{ql^2}{16f} \times \frac{2}{3} f - \frac{3}{32} ql \times \frac{l}{2} + \frac{ql^2}{48} = \frac{ql^2}{64}$$

(intrados fibers are extended).

These results are presented in Table A.31, line 3 for case $v = 0, k = 1$.

Now we can calculate the reactions of supports in the case of a uniform loading of parabolic arch within the *entire* span; they are $R_A = R_B = ql/2, H = ql^2/8f$, $M_A = M_B = 0$. Bending moment at crown $M_C = 0$.

3.4.4 Circular Hingeless Uniform Arch

Semicircular uniform arch is subjected to a uniformly distributed load q (Fig. 3.16). The primary system contains the absolutely rigid cantilevers of length c. Primary unknowns X_i are placed in the elastic center D at the axis of symmetry of the arch, $y_0 = 2R/\pi$. The origin coincides with center D.

Since loading of the arch is symmetrical, the antisymmetrical unknown $X_3 = 0$. Canonical equations become

$$\delta_{11} X_1 + \Delta_{1P} = 0,$$
$$\delta_{22} X_2 + \Delta_{2P} = 0$$

and the primary unknowns are $X_1 = -\Delta_{1P}/\delta_{11}, \quad X_2 = -\Delta_{2P}/\delta_{22}$. For calculation of unit and loaded displacements, the shear and axial forces are ignored.

Coordinates of arbitrary section K of the arch are

$$x = R\cos\varphi,$$
$$y = R\sin\varphi - y_0 = R\sin\varphi - \frac{2R}{\pi}.$$

Bending moment at section K caused by unit primary unknowns $X_1 = 1, X_2 = 1$ and external load are

$$\overline{M}_1 = -1 \times y = -R\sin\varphi + \frac{2R}{\pi}; \quad \overline{M}_2 = -1; \quad M_P^0 = -\frac{qx^2}{2} = -\frac{qR^2}{2}\cos^2\varphi.$$

During the calculation of unit and loaded displacements, integration along the length of the arch, ds, is replaced by integration of its polar counterpart dφ, d$s = R$dϕ.

Unit displacements are

$$EI\delta_{11} = 2\int_0^{\pi/2}\left(-R\sin\varphi + \frac{2R}{\pi}\right)^2 R\mathrm{d}\varphi = \left(\frac{\pi}{2} - \frac{4}{\pi}\right)R^3,$$
$$EI\delta_{22} = 2\int_0^{\pi/2} R\mathrm{d}\varphi = \pi R.$$

Displacements along X_1 and X_2 due the external load are

$$EI\Delta_{1P} = 2\int_0^{\pi/2}\left(-\frac{qR^2}{2}\cos^2\varphi\right)\left(-R\sin\varphi + \frac{2R}{\pi}\right)R\mathrm{d}\varphi = -\frac{qR^4}{6},$$

$$EI\Delta_{2P} = 2\int_0^{\pi/2}\left(-\frac{qR^2}{2}\cos^2\varphi\right)(-1)R\mathrm{d}\varphi = \frac{\pi qR^3}{4}.$$

Primary unknowns are

$$X_1 = -\frac{\Delta_{1P}}{\delta_{11}} = \frac{qR}{6(\pi/2 - 4/\pi)} = 0.5602qR, \quad X_2 = -\frac{\Delta_{2P}}{\delta_{22}} = -\frac{qR^2}{4}.$$

Bending moments at supports and at the crown are

$$M_A = M_B = X_1 y_0 - X_2 - \frac{qR^2}{2} = \frac{qR}{6(\pi/2 - 4/\pi)} \times \frac{2R}{\pi} + \frac{qR^2}{4} - \frac{qR^2}{2} = 0.108qR^2,$$
$$M_C = -X_1(R - y_0) - X_2 = \frac{qR}{6(\pi/2 - 4/\pi)}\left(R - \frac{2R}{\pi}\right) + \frac{qR^2}{2} = 0.046qR^2.$$

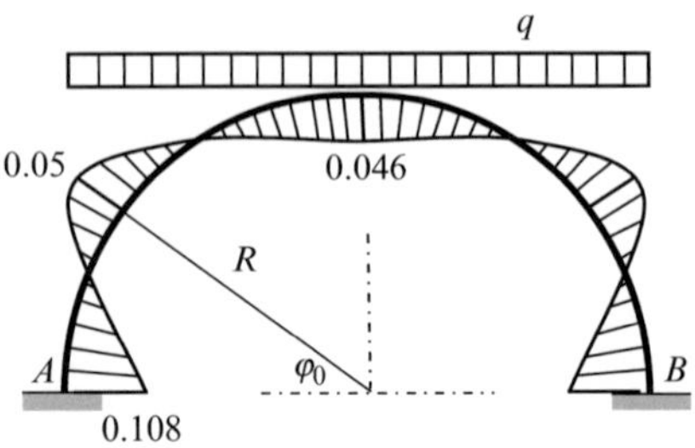

Fig. 3.17 Bending moment diagram; factor qR^2

Bending moment at any section of the arch is

$$M_x = -X_1 y - X_2 - \frac{qx^2}{2} = -\frac{qR}{6(\pi/2 - 4/\pi)}\left(R\sin\varphi - \frac{2R}{\pi}\right) + \frac{qR^2}{4} - \frac{qR^2\cos^2\varphi}{2}.$$

Corresponding bending moment diagram is shown in Fig. 3.17.

We find a section of the arch where the maximal bending moment arises. Condition $\mathrm{d}M/\mathrm{d}\varphi = 0$ leads to $\sin\varphi_0 = 0.562$, $\cos\varphi_0 = 0.827$, and $\varphi_0 = 34^\circ 10'$. The maximal negative moment (extrados fibers are extended) becomes $M_{\max} = -0.05qR^2$.

3.5 Settlements of Supports

If there is a relative displacements of supports of the redundant arches, then internal forces arises in the arch. This section is devoted to the analysis of two-hinged and hingeless arches subjected to the settlements of supports. Settlements of rigid supports occur regardless of a presence of external load.

3.5.1 Two-Hinged Arch

Different settlements of support of two-hinged arch are shown in Fig. 3.18. Horizontal reaction (thrust) will constitute the primary unknown X_1. Canonical equation is $\delta_{11}X_1 + \Delta_{1\Delta} = 0$, where the free term $\Delta_{1\Delta}$ is displacement in the primary system along the primary unknown caused by the given displacement of support. This term is determined by the formula

$$\Delta_{1\Delta} = -\overline{R}_1\Delta, \tag{3.12}$$

where $\overline{R}_1$ is a component of reaction (due to $X_1 = 1$) which coincides with the direction of displacement Δ [Kar10]. Let us consider two cases.

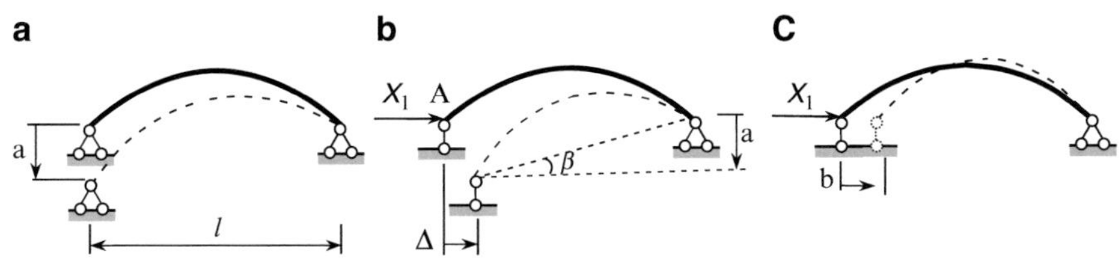

Fig. 3.18 Different settlements of support of two-hinged arches

1. *Vertical displacement of support* (Fig. 3.18a). Assume that displacement a is small. Reaction $\overline{R}_1$ (caused by $X_1 = 1$) along displacement (i.e., the vertical reaction) is zero and therefore $\Delta_{1\Delta} = -0 \times a = 0$. Thus, in the case of a small vertical displacement, the thrust of the arch is equal to zero.

   If a vertical displacement a of support A is large, then this support also has a horizontal displacement Δ (Fig. 3.18b). Approximate expression for this displacement is given by

$$\Delta = l(1 - \cos\beta) \cong l\left(1 - \sqrt{1 - \frac{a^2}{l^2}}\right).$$

   Along the displacement Δ, the reaction is $\overline{R}_1 = -1$. The negative sign means that this reaction and primary unknown $X_1 = 1$ have opposite directions. Free term of canonical equation becomes

$$\Delta_{1\Delta} = -\overline{R}_1\Delta = -(-1) \times \Delta = l\left(1 - \sqrt{1 - \frac{a^2}{l^2}}\right).$$

2. *Horizontal displacement of support* (Fig. 3.18c). Assume that displacement b occurs inside of the arch. Along the displacement b, the reaction is $\overline{R}_1 = -1$. According to (3.12), we get

$$\Delta_{1\Delta} = -\overline{R}_1\Delta = -(-1) \times b - b.$$

   The thrust $H = X_1 = -\Delta_{1P}/\delta_{11}$. In both cases, unit displacement δ_{11} is calculated as usual. Computation of internal forces does not pose a problem when thrust is known.

3.5.2 Hingeless Arch

Symmetrical hingeless arch and settlements of its right support is shown in Fig. 3.19a.

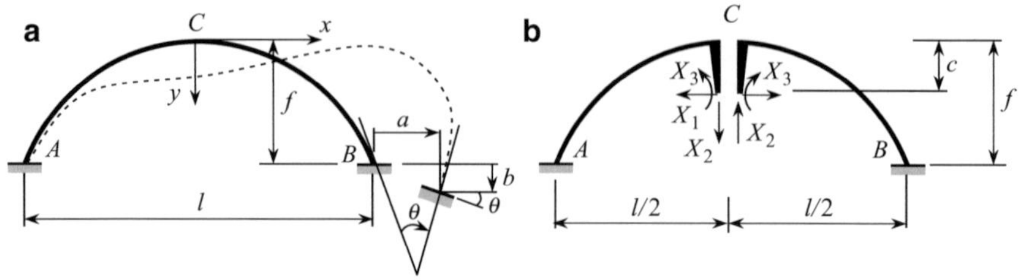

Fig. 3.19 (**a, b**) Hingeless arch. Settlement of support and primary system

If the primary unknowns are placed in the elastic center (Fig. 3.19b), then (3.4) will be separated

$$\delta_{ii}X_1 + \Delta_{i\Delta} = 0. \tag{3.13}$$

Primary unknowns are $X_i = -\Delta_{i\Delta}/\delta_{ii}, \quad i = 1, 2, 3.$

Unit displacements are $\delta_{ii} = \int \overline{M}_i^2 \mathrm{d}s/EI_x$; all of them may be presented in the form $\delta_{ii} = K_i/EI_0$, where K_i are some strictly positive numbers, and EI_0 is a flexural rigidity of a certain section of the arch. It means that the primary unknowns in case of the settlements of supports are proportional to the flexural rigidity of the arch. Increasing the flexural rigidity leads to increasing of primary unknowns [Kis60].

Free terms $\Delta_{i\Delta}(i = 1, 2, 3)$ of the system (3.13) represent displacements of the primary system in the direction of primary unknowns X_i due to displacements of supports; subscript Δ refers to displacements caused by *settlements* of supports. For calculation of these terms, we need to use the theorem of reciprocal unit displacements and reactions (1.39).

It is easy to show [Kis60], [Kar10] that the formula for calculation of the free terms of (3.13) is

$$\Delta_i = -\sum_i \overline{R}_{ik} \times d_i. \tag{3.14}$$

In this formula, the subscript Δ –"settlements" – at $\Delta_{i\Delta}$ is omitted; $\overline{R}_{ik}$ is reaction of the constraint in the direction of a given displacement d_i due to unit primary unknown $X_k = 1$. In other words, $\overline{R}_{i1}$ and $\overline{R}_{i2}$ are found as reactions in the primary system due to primary unknowns $X_1 = 1$ and $X_2 = 1$, respectively; these reactions must be determined in the supports, which are subjected to settlement.

Each primary unknown leads to appearance of the unit reactions in the displaced support; these reactions in direction of *the settlement* of support are shown in Fig. 3.20.

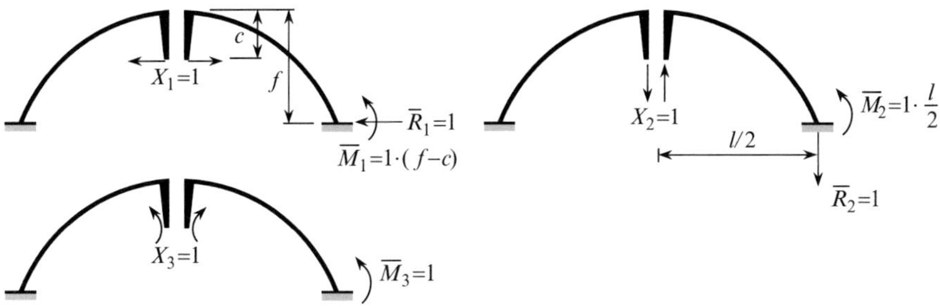

Fig. 3.20 Reactions along the settlements of support due to unit primary unknowns X_i

Formula (3.14) leads to the following results

$$\Delta_{1\Delta} = -\sum \overline{R}_{1i} d_1 = -\left(\overline{R}_1 a + \overline{M}_1 \theta\right) = -[-1 \times a - 1 \times (f-c)\theta] = a + (f-c)\theta,$$

$$\Delta_{2\Delta} = -\sum \overline{R}_{2i} d_2 = -\left(\overline{R}_2 b + \overline{M}_2 \theta\right) = -\left[+1 \times b - 1 \times \frac{l}{2}\theta\right] = -b + \frac{l}{2}\theta,$$

$$\Delta_{3\Delta} = -\sum \overline{R}_{3i} d_2 = -\left(\overline{M}_3 \theta\right) = -(-1 \times \theta) = \theta.$$

After calculation of the primary unknowns X_i, construction of internal forces is performed as usual.

3.6 Arches with Elastic Supports

If the arch is supported by the elastic-yielding structure or insufficiently robust foundation, then the elastic properties of supports should be taken into account. In this paragraph, we are dealing with the deformation of supports caused by external loads unlike settlements of supports without action of any loads as it is discussed in Sect. 3.5. Taking into account the compliance of support enables us to find a more accurate distribution of internal forces in the arch.

Let us consider a hingeless arch with elastic supports. Figure 3.21 presents the support A subjected to the unit forces X_a, X_b, and a moment X_c. New position of flexible support is A'. The stiffness of a support is characterized by three principal positive displacements δ_{aa}, δ_{bb}, δ_{cc}. The first subscript represents direction and second represents cause. The second displacements δ_{ab}, δ_{ac}, δ_{bc} may also be negative.

Primary unknowns (shear force X_1, normal force X_2, and moment X_3) are applied at the elastic center D (Fig. 3.22).

Now we will consider action of each primary unknown separately (Fig. 3.23).

Primary unknown $X_1 = 1$ leads to the following *pressure* on the support A: the vertical force $X_a = 1$ and a clockwise moment $1 \times l/2$. Therefore, each

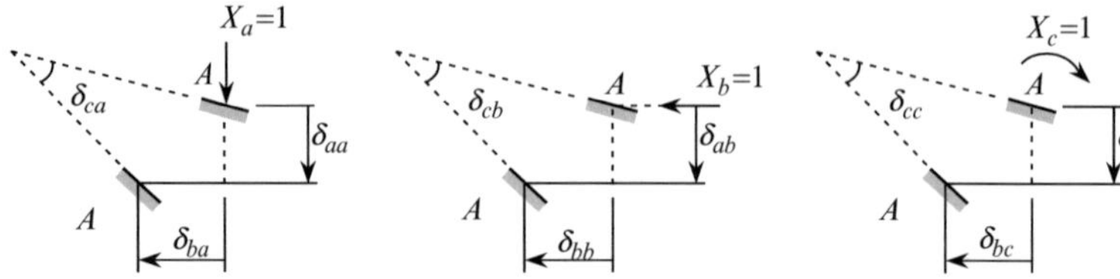

Fig. 3.21 Notation of displacements due to action of each unit load at support *A*

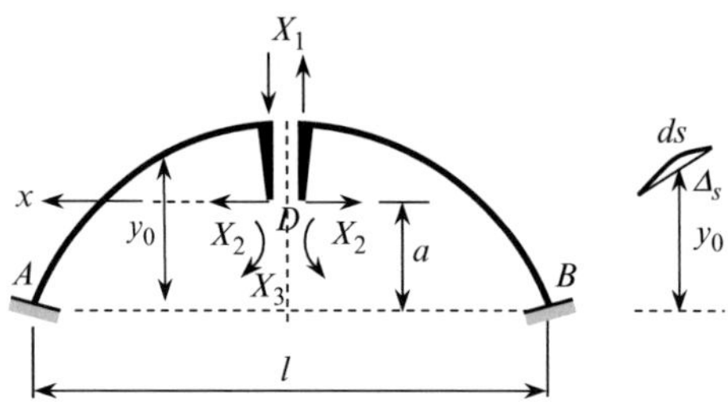

Fig. 3.22 Primary system and primary unknowns (supports *A* and *B* are elastic)

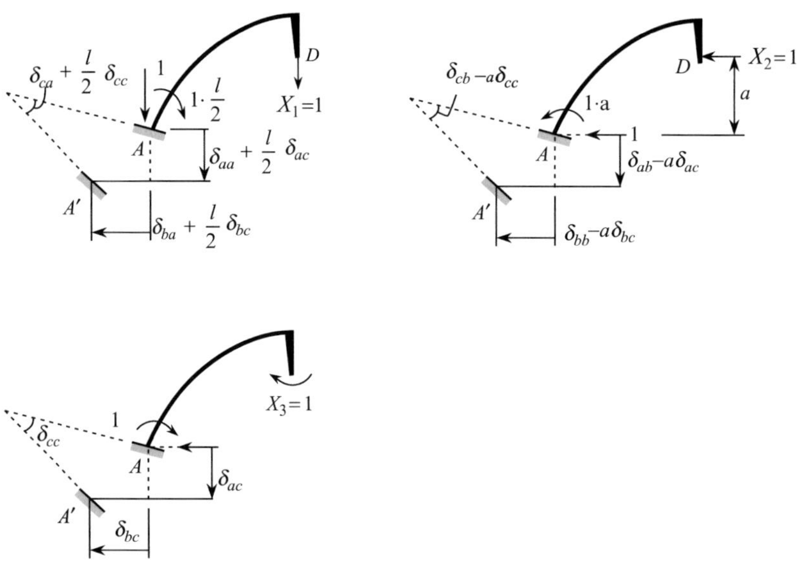

Fig. 3.23 Computation of displacements at flexible support *A* due to action of unit primary unknowns

displacement consists of two terms. The first term $(\delta_{aa}, \delta_{ba}, \delta_{ca})$ corresponds to unit force $X_a = 1$, while the second terms $((l/2)\delta_{ac}, (l/2)\delta_{bc}, (l/2)\delta_{cc})$ corresponds to the moment $1 \times l/2$ as shown in Fig. 3.21.

Similarly unit displacements may be calculated in case of $X_2 = 1$. If the primary system is loaded by $X_3 = 1$, then a unit displacement contains only one term since the action on the support A is a unit moment, which coincides with loading $X_c = 1$ as shown in Fig. 3.21.

Since point D is the elastic center, we get three uncoupled equation. Note that the position of the elastic center in case of elastic supports does not coincide with the elastic center for arch with absolutely rigid supports. Position of the elastic center may be calculated by the formula [Rab60]

$$a = \frac{K + EI_0\delta_{bc}}{\int_A^C \frac{I_0}{I}\mathrm{d}s + EI_0\delta_{cc}}, \quad K = \int_A^C y_0 \frac{I_0}{I}\mathrm{d}s \cong \sum_A^C y_0 \frac{I_0}{I}\Delta s, \tag{3.15}$$

where I_0 is a moment of inertia of the cross section of the arch in crown; displacements δ_{bc} and δ_{cc} according to Fig. 3.21 are elastic characteristics of the abutment.

Procedure for Analysis

Curvilinear axis line of the arch is presented as a set of straight members of length Δs. For the middle point of each member, calculate coordinate y_0 (Fig 3.22) and a moment of inertia I of a cross section of the arch. Then find position of the elastic center by (3.15). Primary unknowns are $X_i = -\Delta_{i\mathrm{P}}/\delta_{ii}, \quad i = 1, 2, 3$.

Unit displacements which depend on deflections of arch itself are denoted by δ_{11}^0, δ_{22}^0, δ_{33}^0. Taking into account elastic properties of supports, the unit displacements become

$$\begin{aligned} &EI_0\delta_{11} - EI_0\delta_{11}^0 + 2EI_0\left(\delta_{aa} + l\delta_{ac} + \frac{l^2}{4}\delta_{cc}\right), \\ &EI_0\delta_{22} = EI_0\delta_{22}^0 + 2EI_0\left(\delta_{bb} - 2a\delta_{bc} + a^2\delta_{cc}\right), \\ &EI_0\delta_{33} = EI_0\delta_{33}^0 + 2EI_0\delta_{cc}. \end{aligned} \tag{3.16}$$

Analysis of properties of these expressions is presented in ref. [Rab60].

The external load in the primary system leads to the pressure and moments which act on the supports A and B; they are denoted by F^{left}, F^{right} and M_A^P, M_B^P, respectively. The forces are positive, if they are directed downwards. The moments are positive if intrados fibers are extended.

Table 3.8

	Problem 1	Problem 2
	Redundant arch with rigid supports (hinged of fixed)	Redundant arch with elastic supports
Settlements of supports	Does not depend on the external load	Depend on the external loads
Unit displacement of canonical equations	Depend on the elastic properties of the arch itself and do not depend on the settlements of supports	Depend on the elastic properties of the arch and elastic supports
Loaded terms	Depend only the settlements of supports	Depend on the elastic properties of the arch and elastic supports
Elastic center	Depend on the elastic properties of the arch itself and do not depend on the settlements of supports	Depend on the elastic properties of the arch and elastic supports

Loaded displacements which depend on deflections of arch itself are denoted by $\Delta^0_{1\mathrm{P}}$, $\Delta^0_{2\mathrm{P}}$, Δ^0_{P}. Taking into account elastic properties of supports, the loaded displacements become [Rab60]

$$\begin{aligned}
EI_0\Delta_{1\mathrm{P}} &= EI_0\Delta^0_{1\mathrm{P}} + EI_0\left(F^{\mathrm{left}} - F^{\mathrm{right}}\right)\left(\delta_{aa} + \frac{l}{2}\delta_{ca}\right) + EI_0\left(M^P_B - M^P_A\right)\left(\delta_{ac} + \frac{l}{2}\delta_{cc}\right),\\
EI_0\Delta_{2\mathrm{P}} &= EI_0\Delta^0_{2\mathrm{P}} + EI_0\left(F^{\mathrm{left}} + F^{\mathrm{right}}\right)(\delta_{ba} - a\delta_{ca}) - EI_0\left(M^P_B + M^P_A\right)(\delta_{bc} - a\delta_{cc}),\\
EI_0\Delta_{3\mathrm{P}} &= EI_0\Delta^0_{3\mathrm{P}} + EI_0\left(F^{\mathrm{left}} + F^{\mathrm{right}}\right)\delta_{ca} - EI_0\left(M^P_B + M^P_A\right)\delta_{cc}.
\end{aligned} \tag{3.17}$$

In case of an absolutely rigid abutment, all displacements in Fig. 3.21 are zero; therefore all unit and loaded terms depend only on the deflections of the arch itself. Knowing the primary unknowns, computation of internal forces should be carried out as usual.

Difference between analysis of the redundant arch in case of the settlements of supports (Problem 1, Sect. 3.5), and analysis of the redundant arch in case of the elastic supports (Problem 2) is presented in Table 3.8.

3.7 Arches with Elastic Tie

This section presents an exact solution for circular uniform arch with elastic tie and approximate solution for nonuniform arch of arbitrary shape with elastic tie.

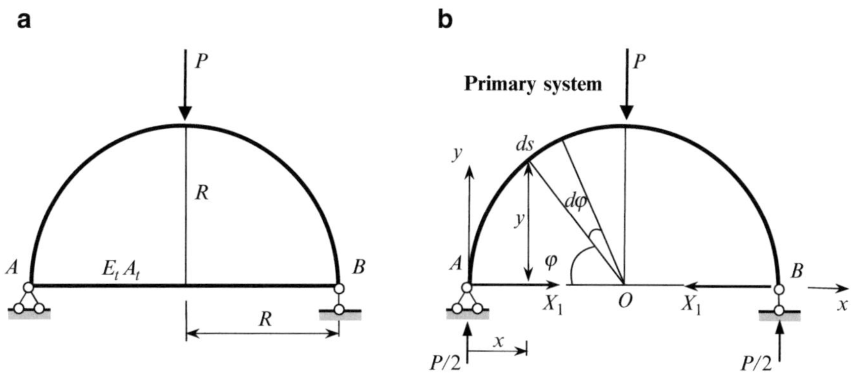

Fig. 3.24 Two-hinged arch with tie

3.7.1 Semicircular Uniform Arch

Design diagram of the circular arch with tie is shown in Fig. 3.24a. The flexural stiffness of the arch $EI = \text{const}$. The stiffness of the tie is E_tA_t.

The structure has one unknown of the Force method. Canonical equation is $\delta_{11}X_1 + \Delta_{1P} = 0$. Let the primary unknown X_1 be the force which arises in the tie (Fig. 3.24b). Canonical equation means that a *mutual* displacement of two sections of the tie caused by primary unknown X_1 and external load is zero. Unit displacement δ_{11} represents the mutual displacements of two sections of a tie due to the primary unknown $X_1 = 1$, while Δ_{1P} is the mutual displacement of two sections of a tie due to a given load.

In the polar coordinates, we get $x = R(1 - \cos\varphi)$; $y = R\sin\varphi$; $ds = Rd\varphi$.

The bending moments in primary system due to P and $X_1 = 1$ are

$$M_P = \frac{Px}{2} = \frac{PR(1-\cos\varphi)}{2},$$
$$\overline{M}_1 = -1 \times y = -R\sin\varphi.$$

In calculating the unit displacement δ_{11}, we take into account the bending of the arch itself and extension of the tie

$$\delta_{11} = \int_0^{\pi R} \frac{\overline{M}_1 \times \overline{M}_1}{EI} ds + \int_0^l \frac{\overline{N}_1 \times \overline{N}_1}{E_tA_t} dx$$
$$= \int_0^{\pi} \frac{R^2\sin^2\varphi}{EI} Rd\varphi + \int_0^{l=2R} \frac{1 \times 1}{E_tA_t} dx = \frac{\pi R^3}{2EI} + \frac{2R}{E_tA_t}. \qquad (3.18)$$

In calculating the loaded displacement Δ_{1P}, we take into account the bending of the arch itself

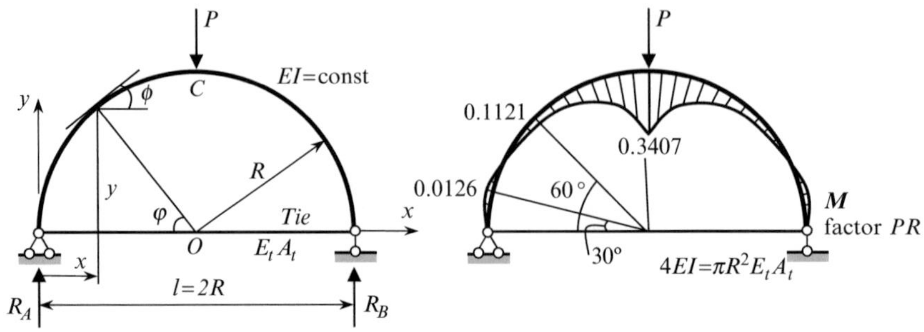

Fig. 3.25 Bending moment diagrams for the arch; $4EI/E_tA_t = \pi R^2$

$$\Delta_{1P} = \int_0^{\pi R} \frac{\overline{M}_1 \times M_P^0}{EI} ds = -\frac{2}{EI}\int_0^{\pi/2} \frac{PR}{2}(1-\cos\varphi)R\sin\varphi \times Rd\varphi$$
$$= -\frac{PR^3}{2EI}. \tag{3.19}$$

The force in the tie (thrust) becomes

$$X_1 = H = -\frac{\Delta_{1P}}{\delta_{11}} = \frac{PR^2}{F}, \quad \text{where } F = \pi R^2 + \frac{4EI}{E_tA_t}.$$

It can be seen that increasing the stiffness of the tie leads an increase of internal force in the tie.

Knowing the thrust, the internal forces may be calculated by the formulas

$$M_k = M_k^O - Hy_k,$$
$$Q_k = Q_k^O \cos\varphi_k - H\sin\varphi_k,$$
$$N_k = -\left(Q_k^O \sin\varphi_k + H\cos\varphi_k\right),$$

where M_k^O and Q_k^O related to the simple supported reference beam.

The bending moment in arch depends on the stiffness of the tie. As in the case of three-hinged arch, the presence of a tie leads to a decrease of the bending moments in comparison with reference beam. Increasing the stiffness of the tie leads to a decrease of the positive moments in the arch.

Let $4EI/E_tA_t = \pi R^2$. In this case, the thrust is $H = P/2\pi$. Bending moments at specific sections are

$$M(x = 0.134R,\ \varphi = 30^\circ) = \frac{P}{2}R(1-\cos 30^\circ) - \frac{P}{2\pi}R\sin 30^\circ = -0.0126PR,$$
$$M(x = 0.5R,\ \varphi = 60^\circ) = 0.1121PR, \quad M(x = R,\ \varphi = 90^\circ) = 0.3407PR.$$

Final bending moment diagram is shown in Fig. 3.25.

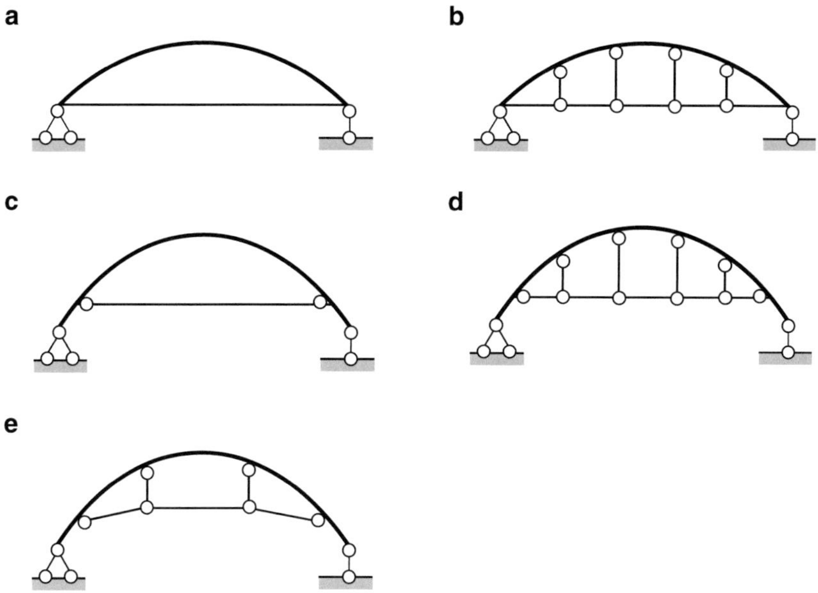

Fig. 3.26 Two-hinged arches with tie

If $E_tA_t = \infty$, then thrust becomes $H = P/\pi$. Bending moments at specific sections are

$$M(x = 0.134R,\ \varphi = 30^\circ) = \frac{P}{2}R(1 - \cos 30^\circ) - \frac{P}{\pi}R\sin 30^\circ = -0.092PR,$$
$$M(x = 0.5R,\ \varphi = 60^\circ) = 0.02565PR, \quad M(x = R,\ \varphi = 90^\circ) = 0.1817PR.$$

If a tie presents an absolutely rigid member, then thrust in the tie is more than the thrust which arises in a two-hinged arch without a tie. This result may be explained by taking into account the expression for X and δ_{11}.

3.7.2 Nonuniform Arch of Arbitrary Shape

Design diagrams of two-hinged arches with ties are shown in Fig. 3.26. Among them are arches with ties on elevated supports (a, b), arches with elevated ties (c, d), and an arch with a complex tie (e).

Let us provide the structural analysis of the nonuniform arch of arbitrary shape with tie at the level of supports (Fig. 3.26a). Assume that the flexural stiffness of the arch EI along its axis is variable. The axis of the arch is described by an arbitrary equation $y = f(x)$. The tie has a constant cross section along its length; the modulus of elasticity and area of the cross section of the tie are E_t and A_t, respectively. Let us

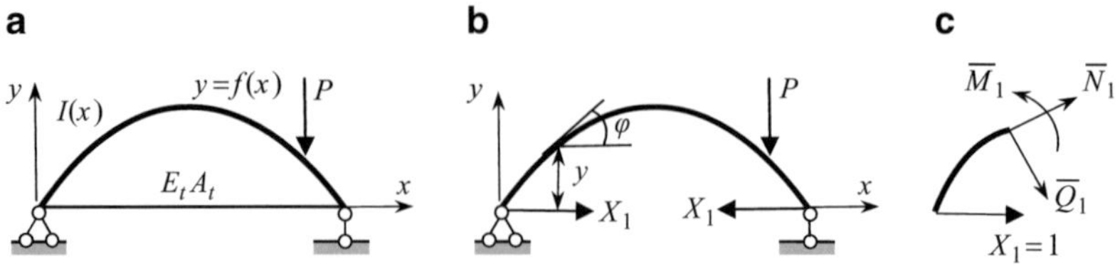

Fig. 3.27 Design diagram and primary system

find expression for thrust, taking into account all components of the Mohr integral and the stiffness of the tie.

Canonical equation of the Force method is $\delta_{11}X_1 + \Delta_{1P} = 0$. Primary unknown is the force X_1 in a tie (thrust H). Primary system is shown in Fig. 3.27.

For computation of displacements δ_{11} and Δ_{1P}, we will take into account not only the bending moments but also the axial and shear forces.

The unit primary unknown $X_1 = 1$ and external forces lead to the following internal forces

$$\overline{M}_1 = -1 \times y, \quad \overline{N}_1 = -\cos\varphi, \quad \overline{Q}_1 = -\sin\varphi,$$

$$M_P = M_0, \quad N_P = -Q_0 \sin\varphi, \quad Q_P = Q_0 \cos\varphi,$$

where M_0 and Q_0 are bending moment and shear of the reference beam; the positive signs of internal forces are shown in Fig. 3.27c.

Let I_0 be the moment of inertia of the cross section of the arch at the specified section, for example, at the crown. Then, Maxwell–Mohr integral for unit and loaded displacements becomes

$$EI_0\delta_{11} = \int y^2 \frac{I_0}{I} \mathrm{d}s + \int \cos^2\varphi \frac{I_0}{A} \mathrm{d}s + \int \sin^2\varphi \frac{\mu EI_0}{GA} \mathrm{d}s + \int_0^l 1^2 \frac{EI_0}{E_tA_t} \mathrm{d}s, \quad (3.20)$$

$$EI_0\Delta_{1P} = -\int yM_0 \frac{I_0}{I} \mathrm{d}s + \int Q_0 \sin\varphi \cos\varphi \frac{I_0}{A} \mathrm{d}s - \int Q_0 \sin\varphi \cos\varphi \frac{\mu EI_0}{GA} \mathrm{d}s. \quad (3.21)$$

With the exception of last term for δ_{11}, integration of all the terms should be performed along the axis of the arch. The last term in (3.20) evaluates to $(EI_0/E_tA_t)l$. The precise integrating may be done only in the specified cases. In the general case, numerical integration must be used: the axis of the arch should be divided into separate straight portions of equal length Δs, for middle point of the portion the quantities y, I_0/I, $\sin\varphi$, $\cos\varphi$, etc., are computed. For the given equation of the axis line of the arch, $y = f(x)$, trigonometric functions of the angle φ are

Table 3.9 General form for calculation of unit displacement; $\Delta s' = (I_0/I)\Delta s$

Member	y	A	I	$\sin\varphi$	$\cos\varphi$	$\Delta s'$	$y^2\Delta s'$	$\cos^2\left(\varphi\frac{I_0}{A}\right)$	$\sin^2\left(\varphi\frac{\mu EI_0}{GA}\right)$
1	2	3	4	5	6	7	8	9	10
1									
...									
							Σ_1	Σ_2	Σ_3

Table 3.10 General form for calculation of the loaded displacement Δ_{1P}; $\Delta s' = (I_0/I)\Delta s$

Member	y	A	I	$\sin\varphi$	$\cos\varphi$	$\Delta s'$	M_0	Q_0	$yM_0\Delta s'$	$Q_0\dfrac{I_0}{2A}\sin 2\varphi$	$Q_0\dfrac{\mu EI_0}{2GA}\sin 2\varphi$
1	2	3	4	5	6	7	8	9	10	11	12
1											
...											
									Σ_4	Σ_5	Σ_6

$$\tan\varphi = \frac{\mathrm{d}y}{\mathrm{d}x}, \quad \cos\varphi = \pm\frac{1}{\sqrt{1+\tan^2\varphi}}; \quad \sin\varphi = \tan\varphi\cos\varphi.$$

For the right-hand half-arch, the functions $\sin\varphi<0$ and $\cos\varphi>0$. Calculation of unit δ_{11} and loaded Δ_{1P} displacements is shown in Tables 3.9 and 3.10, respectively.

Expression for unit displacement becomes

$$EI_0\delta_{11} = \Sigma_1 + (\Sigma_2+\Sigma_3)\Delta s + \frac{EI_0}{E_t A_t}l. \tag{3.22}$$

The term Σ_1 takes into account the bending moments in the arch, while terms $\Sigma_2\Delta s$ and $\Sigma_3\Delta s$ take into account the axial and shear forces in the arch; the last term in (3.22) takes into account the axial force in the tie.

Similarly, expression for loaded displacement (3.21) may be presented as follows

$$EI_0\Delta_{1P} = -\Sigma_4 + (\Sigma_5 - \Sigma_6)\Delta s. \tag{3.23}$$

Computations of Σ_4, Σ_5, and Σ_6 are presented in Table 3.10. Note that the tie is not subjected to any external forces in the primary system.

Primary unknown is

$$X = H = -\frac{\Delta_{1P}}{\delta_{11}} = -\frac{EI_0\Delta_{1P}}{\Sigma_1 + (\Sigma_2+\Sigma_3)\Delta s + \frac{EI_0}{E_t A_t}l}. \tag{3.24}$$

This is the precise expression for thrust which arises in the tie. Ignoring the axial force (terms Σ_2 and Σ_5) or/and shear force (terms Σ_3 and Σ_6) leads to the approximate expression for thrust. Influence of the axial forces is significant for shallow (gently) arches. The greater the parameter h/l(h is a height of the cross section of the arch), the greater is the influence of the shear forces. As shown in Table 3.10 (columns 10–12), the relative influence of different terms depends not only on the shape of the arch and dimensions of the cross sections, but also on the external load [Rab54a].

If $E_tA_t = 0$, then the force in the tie is zero, and arch reduces into a simply supported curvilinear *beam*. In the case of absolutely rigid (nondeformable) tie, the thrust H becomes

$$H_{\text{lim}} = -\frac{EI_0\Delta_{1P}}{\Sigma_1 + (\Sigma_2 + \Sigma_3)\Delta s}.$$

In this case, the elongation of tie is zero; this leads to the usual two-hinged arch without tie.

All internal forces in the arch may be calculated by (2.6) if the force in the tie is known.

Note that a rational axis of the two-hinged arch with tie may be obtained if and only if the tie is nondeformable and influence of the shear and axial forces on the deflection of the arch is ignored [Rab54a].

3.8 Special Effects

This section is devoted to analysis of redundant arches in case of change of temperature and shrinkage of concrete. Two types of arches are considered. They are two-hinged arch with tie and a hingeless nonuniform arch.

3.8.1 Change of Temperature

If statically indeterminate arch is subjected to a change of temperature, then internal forces arise in the arch. Corresponding structural analysis of the arches may be effectively carried out by the Force method. Free terms of canonical equations present the temperature displacement in primary system. For their computation, (1.10) should be applied. This formula takes into account the uniform and nonuniform thermal effect.

3.8.1.1 Two-Hinged Arch with Tie

Let us consider the uniform arch with tie; the span of the arch is l and height of the cross section is h_0. The temperature of extrados and intrados fibers of the arch has been increased by t_1 and t_2 degrees, respectively. Coefficients of thermal expansion of the material of the arch and tie are α and α_t. Let the primary unknown be the force H in tie. Canonical equation and primary unknown are

$$\delta_{11}X_1 + \Delta_{1t} = 0, \quad X = H = -\frac{\Delta_{1t}}{\delta_{11}}. \tag{3.25}$$

The unit displacement δ_{11} should be calculated as usual. Free term of canonical equation may be presented as

$$\Delta_{1t} = \int \overline{N}_1 \alpha t \mathrm{d}s + \int \frac{\overline{M}_1 \alpha t'}{h_0} \mathrm{d}s, \tag{3.26}$$

where $t = (t_1 + t_2)/2$ and $\Delta t = t' = t_1 - t_2$ are average temperature and temperature gradient; forces caused by unit primary unknown $X_1 = 1$ are $\overline{N}_1 = -\cos\varphi, \quad \overline{M}_1 = -y$ for arch and $\overline{N}_1 = 1, \quad \overline{M}_1 = 0$ for tie.

Assume that temperature along the arch and tie is constant. In this case,

$$\Delta_{1t} = -\alpha t \int_s \cos\varphi \mathrm{d}s - \alpha t' \int_s \frac{y}{h_0} \mathrm{d}s + \alpha_t t l. \tag{3.27a}$$

The first term in (3.27a) corresponds to the heating of the arch itself. The negative sign means that heating of the arch leads to an increase of the distance between two points of the tie, while the positive unknown X_1 leads to a decrease of the distance. Since $\cos\varphi \mathrm{d}s = \mathrm{d}x$, then

$$-\alpha t \int_s \cos\varphi \mathrm{d}s = -\alpha t \int_l \mathrm{d}x = -\alpha t l.$$

The second term in (3.27a) corresponds to the nonuniform heating of the arch. The temperature gradient is positive if it leads to the convexity of the arch being in same direction as a positive unknown X_1, i.e., upward. The third term corresponds to a uniform heating of the tie. In the case of an arch without a tie, we let $\alpha_t = 0$.

In case of uniform heating, the gradient $t' = t_1 - t_2 = 0$ and the free term for arch with tie becomes

$$\Delta_{1t} = -\alpha t l + \alpha_t t l = (\alpha_t - \alpha) t l. \tag{3.27b}$$

The primary unknown is

$$H = -\frac{\Delta_{1t}}{\delta_{11}} = -\frac{(\alpha_t - \alpha) t l}{\delta_{11}}.$$

In the case of an arch without tie, the free term of (3.25) is $\Delta_{1t} = -\alpha t l$. Internal forces in the arch itself are

$$M = -Hy; \quad N = -H\cos\varphi; \quad Q = -H\sin\varphi. \tag{3.27c}$$

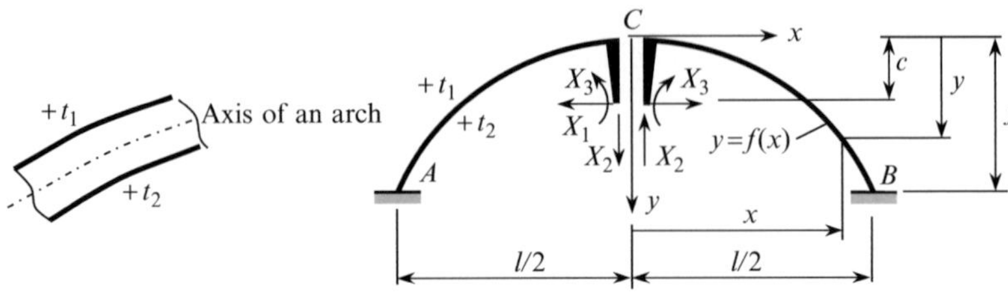

Fig. 3.28 Hingeless arch subjected to thermal effect

If $t > 0$, then in an arch without a tie, the thrust is directed into the span. In an arch with a tie, the direction of the H depends on the relationship between α and α_t. If $\alpha > \alpha_t$, then tie is extended. If $\alpha = \alpha_t$, then $H = 0$. If compressed temperature force in a tie ($\alpha < \alpha_t$) is more than extended force caused by external load, then a loss of stability of the tie may happen [Rab54a].

3.8.1.2 Symmetrical Hingeless Arch

Let us consider an arch with an arbitrary shape of the axis line; the flexural stiffness EI_x of the cross section of the arch varies by an arbitrary law. The temperature of extrados and intrados fibers of the arch has been increased by t_1 and t_2 degrees, respectively. Coefficient of thermal expansion of the arch's material is α. As before, the gradient and average temperature are denoted as $\Delta t = t_1 - t_2$ and $t = (t_1 + t_2)/2$.

The primary unknowns are placed in the elastic center (Fig 3.28). Canonical equations become

$$\delta_{11}X_1 + \Delta_{1t} = 0, \quad \delta_{22}X_2 + \Delta_{2t} = 0, \quad \delta_{33}X_3 + \Delta_{3t} = 0.$$

Since the deformation of the arch is symmetrical, the antisymmetrical unknown is $X_2 = 0$. Unit displacements are

$$\begin{aligned} \delta_{11} &= \int_s (y-c)^2 \frac{ds}{EI_x} + \int_s \cos^2\varphi_x \frac{ds}{EA_x}, \\ \delta_{33} &= \int_s \frac{ds}{EI_x}. \end{aligned} \tag{3.28}$$

In expression for δ_{11}, the second term takes into account the axial forces. Free terms of canonical equations are

$$\begin{aligned} \Delta_{1t} &= \alpha(t_1 - t_2)\int_s \overline{M}_1 \frac{ds}{h} + \alpha\frac{t_1 + t_2}{2}\int_s \overline{N}_1 ds = -\alpha\Delta t\int_s (y-c)\frac{ds}{h} - \alpha t\int_s \cos\varphi_x ds, \\ \Delta_{3t} &= \alpha(t_1 - t_2)\int_s \overline{M}_3 \frac{ds}{h} = -\alpha\Delta t\int_s \frac{ds}{h}, \end{aligned} \tag{3.29}$$

where h is a height of the cross section of the arch.

Primary unknowns are $X_1 = -(\Delta_{1t}/\delta_{11})$, $X_3 = -(\Delta_{3t}/\delta_{33})$. The bending moment at the support of the arch due to the change of temperature is $M_{\text{sup}} = X_3 + X_1(f - c)$.

Symmetrical Parabolic Arch

If the origin is placed in the crown C, then equation of the axis of the arch is $y = 4fx^2/l^2$ The moment of inertia of cross section and the height of the cross section at the crown C are denoted by I_C and h_C. Let the moment of inertia of the cross section varies by the law $I_x = I_C/\cos\varphi_x$. If a cross section is a rectangle, then the height of the cross section becomes $h_x = h_C/\sqrt[3]{\cos\varphi_x}$. However, for shallow arches with a sufficient degree of accuracy, we can assume $h_x = h_C/\cos\varphi_x$. These assumptions lead to the following expressions for primary unknowns

$$X_1 = \frac{45\alpha t E I_C}{4(1+\mu)f^2},$$
$$X_3 = \frac{\alpha \Delta t E I_C}{h_C}.$$

Parameter which takes the axial force into account is

$$\mu = \frac{J_1}{J_2}, \quad J_1 = \int_s \cos^2\varphi_x \frac{ds}{A_x}, \quad J_2 = \int_s (y-c)^2 \frac{ds}{I_x}.$$

The expression for X_1 reflects the following fact: the thrust due to a change in temperature increases together with the rigidity of the arch and with reduction of its rise [Dar89].

Knowing the primary unknowns, the internal forces in the arch may be calculated by formulas

$$M_x = -X_1 y + X_3,$$
$$N_x = -X_1 \cos\varphi.$$

Special Cases

1. If axial force is neglected, then parameter $\mu = 0$.
2. In case of uniform change of temperature, the temperature gradient $\Delta t = 0$ and primary unknown $X_3 = 0$.

3. If conditions 1 and 2 hold simultaneously, then bending moment at the crown and supports are

$$M_C = X_1 c = \frac{15\alpha t E I_C}{4f} \quad \text{(extrados fibers extended)},$$

$$M_A = M_B = X_1(f - c) = \frac{15\alpha t E I_C}{2f} \quad \text{(intrados fibers extended)}.$$

3.8.2 *Shrinkage of Concrete*

Shrinkage of concrete effect of the arch may be equivalent presented as a change of temperature. Indeed, if the coefficient of thermal expansion of the concrete is α, then αt is the elongation of the unit length of the concrete member, if a temperature is changed by t°. During shrinkage of concrete, the linear dimensions of the member decrease by 0.025%. If we assume $\alpha = 0.00001$ for concrete, then shrinkage of concrete is equivalent to a decrease in temperature by 25°C. For practical purposes this value is taken to be between 10 and 15°C. Thus, analysis of the arch in the case of shrinkage of concrete boils down to analysis of the arch subjected to a uniform decrease of temperature by 10–15°C. Therefore, the bending moment diagram caused by the shrinkage of concrete will be the same as in the case of a uniform change of temperature [Kis60]. Note that the effect of shrinkage of concrete may be particularly suppressed by controlling internal forces using the tie [Dar89], [Kis60].

3.9 Influence Lines

This section is devoted to analytical construction of influence lines for redundant arches. The Force method is applied. The following arches are considered: two-hinged parabolic nonuniform arch, two-hinged circular uniform arch with tie, and hingeless parabolic nonuniform arch.

Construction of influence lines for internal forces in redundant arches starts from construction of influence lines for primary unknowns. The following procedure is recommended.

1. Adopt the primary unknowns and show the corresponding primary system. Write the canonical equations of the Force method.
2. Compute the coefficients of canonical equations. These unit displacements depend on the primary system, distribution of stiffnesses, and therefore present some *values*.

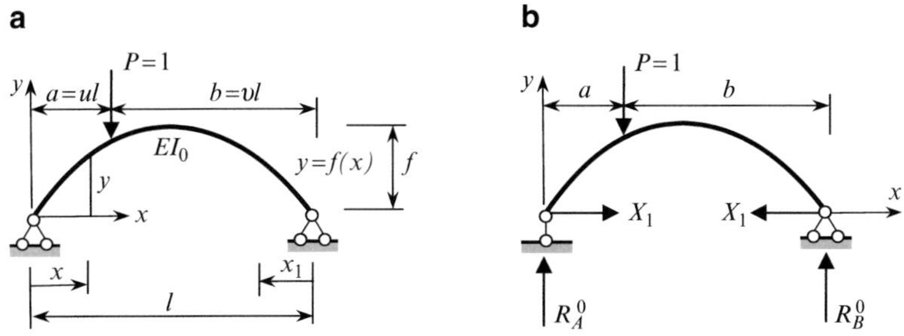

Fig. 3.29 Two-hinged arch subjected to moving load $P = 1$

3. Compute the free terms of canonical equations. These displacements depend on the *location* of the unit load and therefore present some *functions*.
4. Solve the canonical equations; since the free terms are functions, then the primary unknowns will also present *functions* of location of unit load.
5. Construct the influence lines for reactions and internal forces at the specified section of the arch.

3.9.1 Two-Hinged Parabolic Nonuniform Arch

Design diagram of symmetrical parabolic arch of span l and rise f is shown in Fig. 3.29a. Equation of the axial line is $y = 4f\,x(l - x)/l^2$. The moment of inertia of the cross section of the arch varies by the law $I = I_0/\cos\varphi$, where I_0 is a moment of inertia of the cross section of the arch at the crown, φ is the angle between tangent at the given section of the arch and a horizontal line.

Canonical equation of the Force method is $\delta_{11}X_1 + \Delta_{1P} = 0$. The primary system is shown in Fig. 3.29b. The primary unknown is $X_1 = \Delta_{1P}/\delta_{11}$. In this formula, the unit displacement δ_{11} is a *number*, while the loaded displacement Δ_{1P} presents a *function* of the location of the force $P = 1$. Therefore, the equation for the required thrust should be rewritten as follows

$$\mathrm{IL}(X_1) = \mathrm{IL}(H) = -\frac{\mathrm{IL}(\Delta_{1P})}{\delta_{11}}. \tag{3.30}$$

In the Mohr–Maxwell integral, we will only take bending moments into account. Bending moment in the primary system caused by unit primary unknown $X_1 = 1$ is

$$\overline{M}_1 = -1 \times y.$$

Bending moments in the primary system caused by unit single force P are $M_P^0 = R_A^0 x = (Pb/l)x$ for left portion of the arch $(0 \le x \le a)$, $M_P^0 = R_B^0 x_1 = (Pa/l)x_1$ for right portion of the arch $(0 \le x_1 \le b)$.

Taking into account the expressions for moment of inertia, the unit displacement becomes

$$\delta_{11} = \int_s \frac{\overline{M}_1 \times \overline{M}_1}{EI} \mathrm{d}s = \int_0^l \frac{\overline{M}_1 \times \overline{M}_1}{EI_0} \mathrm{d}x = \int_0^l \frac{y^2}{EI_0} \mathrm{d}x = \frac{16f^2}{l^4} \int_0^l x^2(l-x)^2 \frac{\mathrm{d}x}{EI_0} = \frac{8f^2 l}{15EI_0}. \tag{3.31}$$

The loaded displacement, taking into account two portions of the arch, is

$$\Delta_{1P} = \int_s \frac{\overline{M}_1 \times M_P^0}{EI} \mathrm{d}s = -\int_0^a y \frac{Pbx}{l} \frac{\mathrm{d}x}{EI_0} - \int_0^b y \frac{Pax_1}{l} \frac{\mathrm{d}x_1}{EI_0}. \tag{3.32}$$

Substitution of the given equation $y = f(x)$ into (3.32) leads to the following expression

$$\begin{aligned} \Delta_{1P} &= -\int_0^a \frac{4f}{l^2} x(l-x) \frac{Pbx}{l} \frac{\mathrm{d}x}{EI_0} - \int_0^b \frac{4f}{l^2} x(l-x) \frac{Pax_1}{l} \frac{\mathrm{d}x_1}{EI_0} \\ &= -\frac{4Pf}{l^3 EI_0} \left[b \int_0^a (l-x)x^2 \mathrm{d}x - a \int_0^b (l-x)x_1^2 \mathrm{d}x_1 \right]. \end{aligned} \tag{3.33a}$$

After integration, we get

$$\Delta_{1P} = -\frac{4Pf}{12 l^3 EI_0} \left[ba^3(4l - 3a) + ab^3(4l - 3b) \right].$$

Assume $P = 1$, dimensionless parameter $u = a/l$, then for loaded displacement, we get

$$\Delta_{1P} = -\frac{l^2 f}{3EI_0} u(1-u)\left(1 + u - u^2\right). \tag{3.33b}$$

Thus, the equation of the influence line for thrust becomes

$$\mathrm{IL}(H) = \frac{5}{8} \frac{l}{f} u(1-u)\left(1 + u - u^2\right). \tag{3.34}$$

It can be seen that this curve is fourth order. Influence line for thrust is shown in Fig. 3.30a.

Now we can construct the influence lines for internal forces in the arch. As for a three-hinged arch, we have

Fig. 3.30 Influence lines for thrust H and internal forces at section k

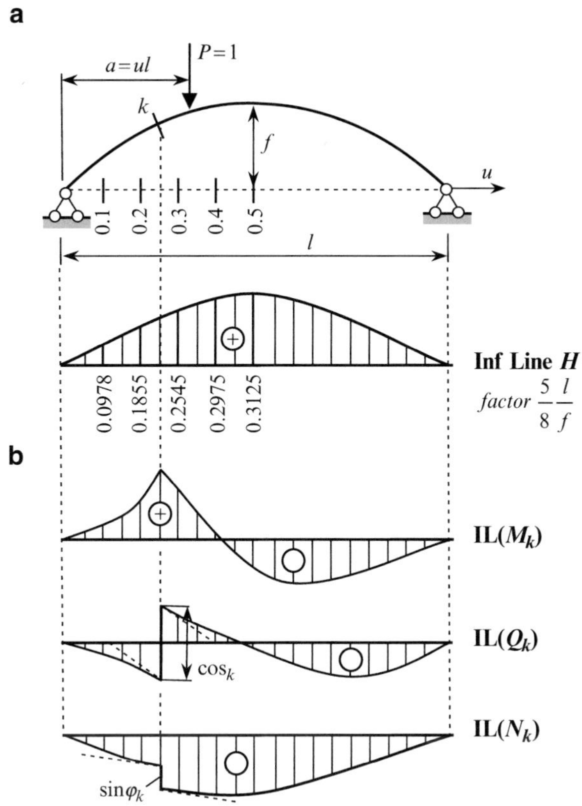

$$\mathrm{IL}(M_k) = \mathrm{IL}\left(M_k^O\right) - y_k \mathrm{IL}(H),$$

$$\mathrm{IL}(Q_k) = \cos\varphi_k \mathrm{IL}\left(Q_k^O\right) - \sin\varphi_k \mathrm{IL}(H),$$

$$\mathrm{IL}(N_k) = -\sin\varphi_k \mathrm{IL}\left(Q_k^O\right) - \cos\varphi_k \mathrm{IL}(H).$$

Here $\mathrm{IL}\left(M_k^O\right)$ and $\mathrm{IL}\left(Q_k^O\right)$ are influence lines for bending moment and shear at the section k of the simply supported beam; y_k and φ_k are ordinate of point k and angle between the tangent at the same point and horizontal line, respectively. Corresponding influence lines are shown in Fig. 3.30b.

Influence lines for shear and axial force at section k has discontinuity $\cos\varphi_k$ and $\sin\varphi_k$, respectively. Two tangents at the left and right portions at section k of these influence lines are parallel; these tangents are shown by dotted lines.

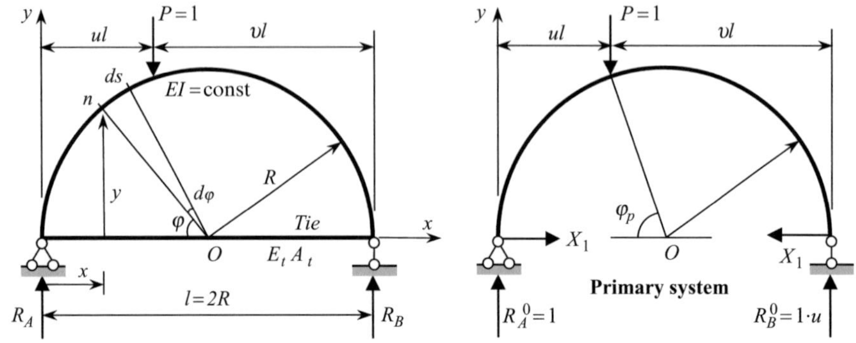

Fig. 3.31 Circular uniform arch with elastic tie subjected to moving load $P = 1$

3.9.2 Two-Hinged Circular Uniform Arch with Elastic Tie

The uniform semicircular two-hinged arch with elastic tie is shown in Fig. 3.31. Flexural stiffness of the arch is EI, axial stiffness of the tie is E_tA_t. Let the primary unknown be the internal force in the tie.

Canonical equation of the Force method is $\delta_{11}X_1 + \Delta_{1P} = 0$. Primary unknown is $X_1 = -\Delta_{1P}/\delta_{11}$.

Assumption

We will take into account deflection of the arch due to bending of the arch itself and deflection of the tie due to tension.

Unit displacement is

$$\delta_{11} = \int_0^{\pi R} \frac{\overline{M}_1 \times \overline{M}_1}{EI} ds + \int_0^{l} \frac{\overline{N}_1 \times \overline{N}_1}{E_tA_t} dx, \tag{3.35}$$

where $\overline{M}$, $\overline{N}$ are bending moment in the arch and normal force in the tie due to the primary known $X_1 = 1$.

Coordinates of the arbitrary point n are $x = R(1 - \cos\varphi)$, $y = R\sin\varphi$. Since $ds = Rd\varphi$, $\overline{N} = 1$ and $\overline{M}_1 = -1 \times y = -R\sin\varphi$, then

$$\delta_{11} = \int_0^{\pi} \frac{R^2\sin^2\varphi}{EI} d\varphi + \int_0^{l=2R} \frac{1 \times 1}{E_tA_t} dx = \frac{\pi R^3}{2EI} + \frac{2R}{E_tA_t}. \tag{3.35a}$$

Free term of canonical equation is

$$\Delta_{1P} = \int_0^{\pi R} \frac{\overline{M}_1 \times M_P^0}{EI} ds, \tag{3.36}$$

where M_P^0 is the bending moment in the primary system due to the given load $P = 1$.

Location of the moving load P is defined by central angle φ_P, and therefore

$$ul = R(1 - \cos\varphi_P) \to u = \frac{1}{2}(1 - \cos\varphi_P) \to \upsilon = 1 - u = \frac{1}{2}(1 + \cos\varphi_P).$$

The vertical reaction and bending moment in the primary system are

$$\begin{aligned}
R_A^0 &= 1 \times \upsilon = \frac{1}{2}(1 + \cos\varphi_P),\\
M_P^0 &= R_A^0 x = \frac{1}{2}(1 + \cos\varphi_P)R(1 - \cos\varphi), \quad x \le ul,\\
M_P^0 &= R_A^0 x - P(x - ul) = \frac{1}{2}(1 + \cos\varphi_P)R(1 - \cos\varphi) - [R(1 - \cos\varphi)\\
&\quad -R(1 - \cos\varphi_P)],\ x \ge ul.
\end{aligned}$$

Free term of canonical equation becomes

$$\begin{aligned}
\Delta_{1P} = \frac{1}{EI}\int_0^{\varphi_P} &\underbrace{\frac{1}{2}(1 + \cos\varphi_P)R(1 - \cos\varphi)}_{M_P^0} \times \underbrace{(-R\sin\varphi)}_{\overline{M}_1}\underbrace{Rd\varphi}_{ds}\\
&+ \int_{\varphi_P}^{\pi}\underbrace{\left[\frac{1}{2}(1 + \cos\varphi_P)R(1 - \cos\varphi) - R(\cos\varphi_P - \cos\varphi)\right]}_{M_P^0} \times \underbrace{(-R\sin\varphi)}_{\overline{M}_1}\underbrace{Rd\varphi}_{ds}.
\end{aligned}$$

After integration, this formula may be presented as follows

$$\Delta_{1P} = -\frac{R^3}{EI}f(\varphi_P), \quad f(\varphi_P) = 2\left(\frac{3}{4} - \cos^2\varphi_P + \frac{1}{4}\cos 2\varphi_P\right). \tag{3.36a}$$

The primary unknown caused by arbitrary load P becomes

$$X_1 = -\frac{\Delta_{1P}}{\delta_{11}} = \frac{PR^2}{\pi R^2 + \frac{4EI}{E_t A_t}}f(\varphi_P). \tag{3.37}$$

Ordinates of influence line for thrust H, according to (3.37), are presented in Table 3.11.

Table 3.11 Influence line for thrust H, factor $R^2/(\pi R^2+k)$, $k=4EI/E_tA_t$

φ_P (degrees)	0.0	15	20	30	45	60	70	75	90
$f(\varphi_P)$	0.0	0.0670	0.116980	0.25	0.5	0.75	0.883	0.993	1.0

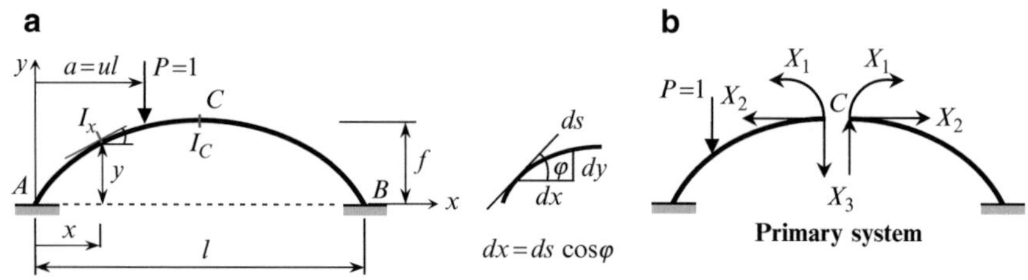

Fig. 3.32 Parabolic arch with clamped ends. (**a**) Design diagram; (**b**) primary system

Verification

If load P is located at the highest point of the arch, then axial load in the tie equals

$$X_1 = H = \frac{1.0\times PR^2}{\pi R^2 + \frac{4EI}{E_tA_t}}.$$

Parameter 1.0 is taken from Table 3.11 for the given position of a load P. This result has been obtained in Sect. 3.7.1.

Special Cases

1. Let $E_tA_t=\infty$. In this case, $X_1=H=(P/\pi)f(\varphi_P)$; this case corresponds to a two-hinged arch without tie.
2. If $4EI/E_tA_t=\pi R^2$, then $X_1=H=(P/2\pi)f(\varphi_P)$.
3. Let $E_tA_t=0$. In this case, $X_1=0$; this case corresponds to a simply supported curvilinear rod.

3.9.3 Hingeless Nonuniform Parabolic Arch

Design diagram of a symmetrical parabolic arch with clamped ends is shown in Fig. 3.32a. The equation of the neutral line is $y=4f\,x(l-x)/l^2$. The cross-sectional moments of inertia varies by law $I_x=I_C/\cos\varphi_x$, where I_C corresponds to the highest point C of the arch; this law corresponds to increasing the moment of inertia from crown to supports. It is necessary to construct the influence lines for reactions of the support A and bending moment at the crown C.

Table 3.12 Bending moments due to unit primary unknowns and given unit load

	$\overline{M}_1$	$\overline{M}_2$	$\overline{M}_3$	$M^0_{\mathrm{P}=1}$
Bending moment expression	1	$1 \times (f - y)$	$-1 \times \left(\frac{l}{2} - x\right)$	$M^0_{\mathrm{P}} = -1 \times (a - x)$

This arch is statically indeterminate to the third degree. Let us accept the primary system presented in Fig. 3.32b, so the primary unknowns are the bending moment X_1, the normal force X_2, and shear X_3 at the crown C of the arch.

Canonical equations of the Force method are written in the form of (3.4).

The unit coefficients are $\delta_{ik} = \int_0^l (\overline{M}_i \times \overline{M}_k / EI_x)\mathrm{d}s$. Since X_1 and X_2 are symmetrical unknowns, the unit bending moment diagrams $\overline{M}_1$ and $\overline{M}_2$ are symmetrical, while $\overline{M}_3$ diagram is antisymmetrical. It is obvious that all displacements computed by multiplying symmetrical diagram by antisymmetrical ones equal to zero. Therefore, $\delta_{13} = \delta_{31} = 0, \ \delta_{23} = \delta_{32} = 0$, and the canonical equations fall into two independent systems

$$\begin{cases} \delta_{11}X_1 + \delta_{12}X_2 + \Delta_{1\mathrm{P}} = 0, \\ \delta_{21}X_1 + \delta_{22}X_2 + \Delta_{2\mathrm{P}} = 0, \end{cases} \quad \text{and} \quad \delta_{33}X_3 + \Delta_{3\mathrm{P}} = 0.$$

Note, we do not use the concept of the elastic center.

Coefficients and free terms of canonical equations will be calculated taking into account only bending moments, which arise in the arch. The expression for bending moments in the left part of the primary system for unit and loaded states (the force $P = 1$ is located within the left part of the arch) are presented in Table 3.12.

Unit Coefficients

$$\delta_{11} = \int_0^l \frac{\overline{M}_1 \times \overline{M}_1}{EI_x}\mathrm{d}s = \int_0^l \frac{1 \times 1 \times \cos\varphi_x}{EI_\mathrm{C}} \times \frac{\mathrm{d}x}{\cos\varphi_x} = \int_0^l \frac{\mathrm{d}x}{EI_\mathrm{C}} = \frac{l}{EI_\mathrm{C}},$$

$$\delta_{12} = \int_0^l \frac{\overline{M}_1 \times \overline{M}_2}{EI_x}\mathrm{d}s = \int_0^l 1 \times (f - y)\frac{\mathrm{d}x}{EI_\mathrm{C}} = \int_0^l \left[f - \frac{4f}{l^2}(l - x)x\right]\frac{\mathrm{d}x}{EI_\mathrm{C}} = \frac{fl}{3EI_\mathrm{C}},$$

$$\delta_{22} = \int_0^l \frac{\overline{M}_2 \times \overline{M}_2}{EI_x}\mathrm{d}s = \int_0^l (f - y)^2\frac{\mathrm{d}x}{EI_\mathrm{C}} = \int_0^l \left[f - \frac{4f}{l^2}(l - x)x\right]^2\frac{\mathrm{d}x}{EI_\mathrm{C}} = \frac{f^2 l}{5EI_\mathrm{C}},$$

$$\delta_{33} = 2\int_0^{l/2} \frac{\overline{M}_3 \times \overline{M}_3}{EI_x}\mathrm{d}s = 2\int_0^{l/2} \left(\frac{l}{2} - x\right)^2\frac{\mathrm{d}x}{EI_\mathrm{C}} = \frac{l^3}{12EI_\mathrm{C}}.$$

Free Terms

Since $P = 1$, then the free terms are denoted through $\delta_{i\mathrm{P}}$

$$\delta_{1\mathrm{P}} = \int_0^a \frac{\overline{M}_1 \times M_\mathrm{P}^0}{EI_x} \mathrm{d}s = -\int_0^a 1 \times (a-x) \frac{\mathrm{d}x}{EI_\mathrm{C}} = -\frac{a^2}{2EI_\mathrm{C}},$$

$$\delta_{2\mathrm{P}} = \int_0^a \frac{\overline{M}_2 \times M_\mathrm{P}^0}{EI_x} \mathrm{d}s = -\int_0^a 1 \times (f-y) \times 1 \times (a-x) \frac{\mathrm{d}x}{EI_\mathrm{C}}$$
$$= -\int_0^a 1 \times \left[f - \frac{4f}{l^2}(l-x)x \right] (a-x) \frac{\mathrm{d}x}{EI_\mathrm{C}} = \frac{a^2 f}{EI_\mathrm{C}} \left(-\frac{1}{2} + \frac{2}{3}u - \frac{1}{3}u^2 \right),$$

$$\delta_{3\mathrm{P}} = \int_0^a \frac{\overline{M}_3 \times M_\mathrm{P}^0}{EI_x} \mathrm{d}s = \int_0^a \left(\frac{l}{2} - x \right)(a-x) \frac{\mathrm{d}x}{EI_\mathrm{C}} = \frac{l^3}{EI_\mathrm{C}} u^2 \left(\frac{1}{4} - \frac{u}{6} \right).$$

Canonical equations become

$$lX_1 + \frac{fl}{3} X_2 = \frac{a^2}{2},$$
$$\frac{fl}{3} X_1 + \frac{1}{5} f^2 l X_2 = -fa^2 \left(-\frac{1}{2} + \frac{2}{3}u - \frac{1}{3}u^2 \right) \text{ and } \frac{l^3}{12} X_3 + l^3 u^2 \left(\frac{1}{4} - \frac{u}{6} \right) = 0.$$

The solution of these equations leads to the following expressions for the primary unknowns in terms of dimensionless parameter $u = a/l$, which defines the location of the unit force P:

$$X_1 = u^2 \left(-\frac{3}{4} + \frac{5}{2}u - \frac{5}{4}u^2 \right) l,$$
$$X_2 = \frac{15}{4} u^2 (1-u)^2 \frac{l}{f}, \tag{3.38}$$
$$X_3 = 12u^2 \left(-\frac{1}{4} + \frac{u}{6} \right).$$

These formulae should be applied for $0 \le u \le 0.5$. Since X_1 and X_2 are symmetrical unknowns, the expressions for these unknowns for the right part of the arch $(0.5 \le u \le 1)$ may be obtained from (3.38) by substitution $u \to 1 - u$. Since X_3 is antisymmetrical unknown, the sign for X_3 should be changed and parameter u should be substituted by $1 - u$. Influence lines for the primary unknowns X_1, X_2, and X_3 may be constructed easily.

After computation of the primary unknowns, we can calculate the reaction and internal forces at any section of the arch.

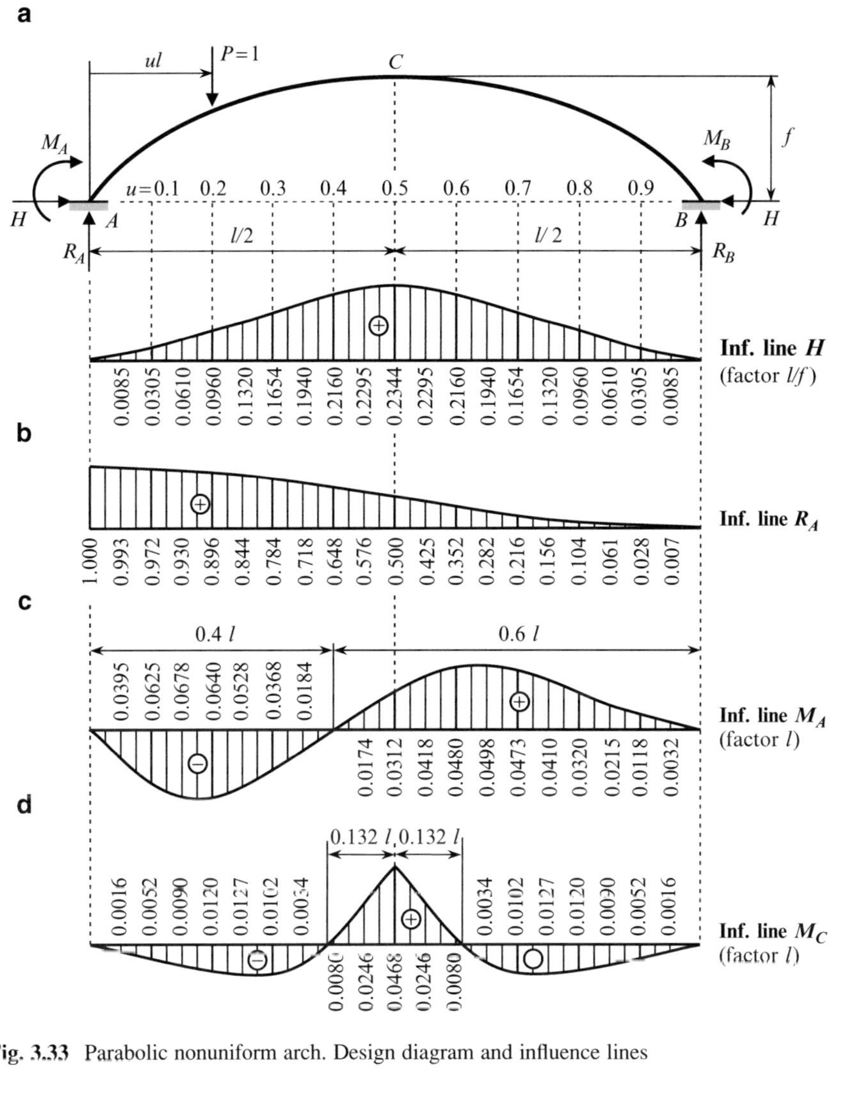

Fig. 3.33 Parabolic nonuniform arch. Design diagram and influence lines

Reactions of Support A

The following reactions should be calculated: thrust, vertical reaction, and moment.

Thrust: $H = X_2 = (15/4)u^2(1-u)^2(l/f)$ for $0 \le u \le 1.0$

This formula presents the thrust of the arch as the function of the dimensionless parameter u, i.e., this expression is the influence line for H (Fig. 3.33a).

Maximum thrust is $H_{\max} = 0.2344Pl/f$ and it occurs when the force P is located at the crown C. This formula shows that decreasing of the rise f leads to increasing of the thrust H.

Vertical reaction: $R_A = X_3 + 1 = 12u^2(-1/4 + u/6) + 1$ for $0 \le u \le 0.5$

Since X_3 is antisymmetrical unknown, for the right part of the arch it is necessary to change sign on the opposite and make the change $u \to 1 - u$. Therefore, if unit load P is located on the right part of the arch, then reaction R_A is

$$R_A = X_3 = -12(1-u)^2\left(-\frac{1}{4}+\frac{1-u}{6}\right) \quad \text{for } 0.5 \le u \le 1.0.$$

Corresponding influence line is presented in Fig. 3.33b.

Moment at Support A

$$M_A = -1 \times ul + X_1 + X_2 \times f - X_3 \times \frac{l}{2} = u\left(-1 + \frac{9}{2}u - 6u^2 + \frac{5}{2}u^3\right)l$$

$$\text{for } 0 \le u \le 0.5,$$

$$M_A = X_1 + X_2 f - X_3 \frac{l}{2} = (1-u)^2\left(\frac{5}{2}u^2 - u\right)l \quad \text{for } 0.5 \le u \le 1.0.$$

Corresponding influence line is presented in Fig. 3.33c.

Bending Moment at Crown C

$$M_C = X_1 = u^2\left(-\frac{3}{4} + \frac{5}{2}u - \frac{5}{4}u^2\right)l \quad \text{for } 0 \le u \le 0.5.$$

Since X_1 is symmetrical unknown, for the right part of the arch, it is necessary to make the change $u \to 1 - u$. Therefore, if unit load P is located on the right part of the arch, then bending moment at crown is

$$M_C = X_1 = (1-u)^2\left(-\frac{3}{4} + \frac{5}{2}(1-u) - \frac{5}{4}(1-u)^2\right)l \quad \text{for } 0.5 \le u \le 1.0.$$

Corresponding influence line is presented in Fig. 3.33d.

Conclusions

If load P is placed in the portion of $0.132l$ in both sides from crown C, then the extended fibers at C are located below the neutral line of the arch. The direction of the support moment M_A depends on the location of the load: if load P is placed within $0.4l$ from the left support, then extended fibers in vicinity of the A are extrados fibers.

Discussion

1. For the given parabolic nonuniform arch, we obtained the precise results. It happens because the area moment of inertia of a cross section of the arch varies according to formula $I_x = I_C/\cos\varphi_x$. Since $dx = ds \cos \varphi_x$, $ds/EI_x = dx/EI_C$ and all integrals are presented in exact form.

   For numerical construction of influence lines, the elastic loads method may be recommended [Dar89].
2. In arch with clamped supports subjected to distributed load along half-span, the maximum bending moments arise at supports. For this case, the law for moment of inertia of cross-section $I_x \cos\varphi_x = I_C$ corresponds to increasing of flexural rigidity of the arch from crown to supports.
3. In arch with pinned supports, the zeros bending moments arise at supports. For these cases, the following law for moment of inertia of cross-section may be taken as: $I_C \cos\varphi_x = I_x$. This expression corresponds to decreasing of flexural rigidity of the arch from crown to supports.

Thus, it can be observed that it is not wise to use shape $I_x \cos\varphi_x = I_C$ for arches with pinned supports; and it is dangerous to use the shape $I_C \cos\varphi_x = I_x$ in case of clamped supports [Kar10]. It is obvious that the laws for moment of inertia of cross-section in real structures are not limited to two considered cases above [Kis60]. Note that influence lines for the primary unknowns of a nonuniform catenary arch are presented in ref. [Kis60].

Influence Lines for Internal Forces at Arbitrary Section

Expressions (3.38) for primary unknowns are used for construction of influence lines for internal forces at arbitrary section. The specified section k of the arch has the following parameters $x_k,\ y_k,\ \sin\varphi_k,\ \cos\varphi_k$ (Fig. 3.34a). First of all, we need to construct the expressions for internal forces at section k and after that transform these expressions into an equation of the corresponding influence lines.

For calculation of internal forces at section k, we will consider the left part of the arch (part AC, Fig. 3.34b) and take into account the forces which are located right hand at section k, i.e., on the portion $k - C$.

Bending moment at section k is

$$M_k = X_1 + X_2(f - y_k) - X_3\left(\frac{l}{2} - x_k\right) - \underline{P(ul - x_k)}.$$

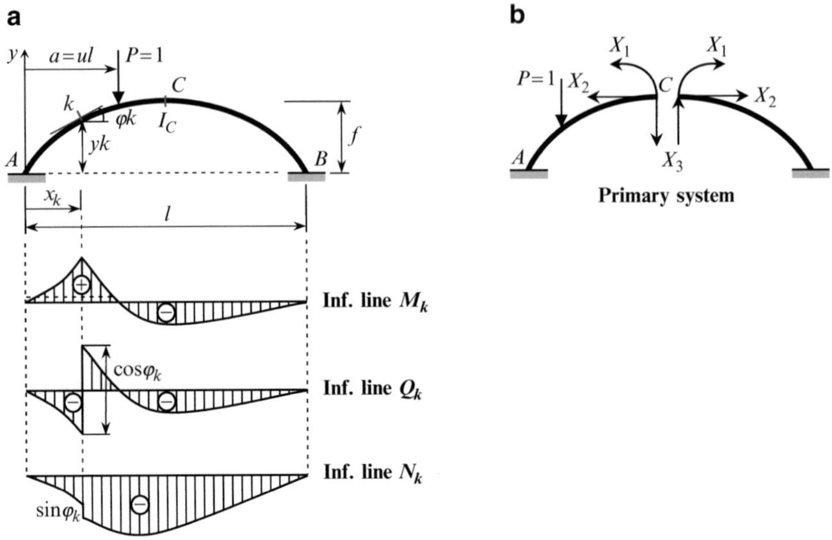

Fig. 3.34 Parabolic arch with clamped supports. (**a**) Design diagram; (**b**) primary system; (**c**) shape of influence lines

Underlined terms should be taken into account only if load P is located within the portion $k - C$ $(ul - x_k \geq 0)$

Now equation of influence line for bending moment at section k becomes

$$\text{IL}(M_k) = \text{IL}(X_1) + (f - y_k)\text{IL}(X_2) - \left(\frac{l}{2} - x_k\right)\text{IL}(X_3) - \underline{1 \times (ul - x_k)}. \quad (3.39\text{a})$$

Similarly, shear force at section k and corresponding influence line are

$$\begin{aligned} Q_k &= -X_2 \sin\varphi_k + X_3 \cos\varphi_k + \underline{P\cos\varphi_k}, \\ \text{IL}(Q_k) &= -\sin\varphi_k \text{IL}(X_2) + \cos\varphi_k \text{IL}(X_3) + \underline{1 \times \cos\varphi_k}. \end{aligned} \quad (3.39\text{b})$$

For axial force at section k and its influence line, we get

$$\begin{aligned} N_k &= -(X_2 \cos\varphi_k + X_3 \sin\varphi_k) - \underline{P\sin\varphi_k}, \\ \text{IL}(N_k) &= -\cos\varphi_k \text{IL}(X_2) - \sin\varphi_k \text{IL}(X_3) - \underline{1 \times \sin\varphi_k}. \end{aligned} \quad (3.39\text{c})$$

General shape of influence lines are shown in Fig. 3.34c. Maximum positive bending moment at section k arises if load P is placed at this section; position of force P, which leads to the extended intrados fibers at section k, is shown by a dotted line.

3.9.4 Application of Influence Lines

Let us show the application of the obtained influence lines for computation of reactions for support A and internal forces (shear and bending moment) at the crown C. Design diagram of symmetric parabolic nonuniform arch with clamped ends is presented in Fig. 3.35. The cross-sectional moment of inertia varies by law $I_x = I_C/\cos\varphi_x$, as considered above. The arch is subjected to concentrated load $P = 30$ kN and uniformly distributed load $q = 2$ kN/m.

Influence lines for reactions and bending moment at the crown C are presented in Figs. 3.33 and 3.35.

Internal forces may be defined using the corresponding influence lines by formula $S = Py + q\Omega$, where y is the ordinate of influence line under concentrated force, Ω is the area of influence line within acting distributed load. The area of curvilinear influence line may be calculated approximately by replacing curvilinear segments between two neighboring ordinates by straight lines (Fig. 3.36).

If a horizontal distance h, which separates these ordinates, remains constant, then the area bounded by two ordinates y_n and y_m will be given by the formula

$$\Omega_n^m = h\left(\frac{y_n}{2} + y_{n+1} + y_{n+2} + \cdots + y_{m-1} + \frac{y_m}{2}\right). \qquad (3.40)$$

Ordinates of influence lines in Fig. 3.35 are presented over $0.05l = 1.2$ m. Now we can calculate reactions of support A due to fixed force P and q.

Thrust:

$$H = \frac{l}{f}\left[P \times 0.1320 + q\left(\frac{0.2344}{2} + 0.2295 + \frac{0.2160}{2}\right)1.2\right] = 20.20 \text{ kN}.$$

Vertical reaction:

$$R_A = \left[P \times 0.844 + q\left(\frac{0.5}{2} + 0.425 + \frac{0.352}{2}\right)1.2\right] = 27.36 \text{ kN}.$$

Moment at support:

$$M_A = l\left[-P \times 0.0528 + q\left(\frac{0.0312}{2} + 0.0418 + \frac{0.048}{2}\right)1.2\right] = -33.32 \text{ kNm}.$$

Obtained values of reactions at support A (as well as the influence lines for primary unknowns X_i) allow us to calculate all internal forces at any section of the arch. For this, it is necessary to eliminate all constraints at the left end of the arch and replace them by the reactive forces just found, i.e., to consider the given arch as a statically determinate structure clamped at B only and subjected to given load and reactions at support A. For example, bending moment at crown C, by definition, equals

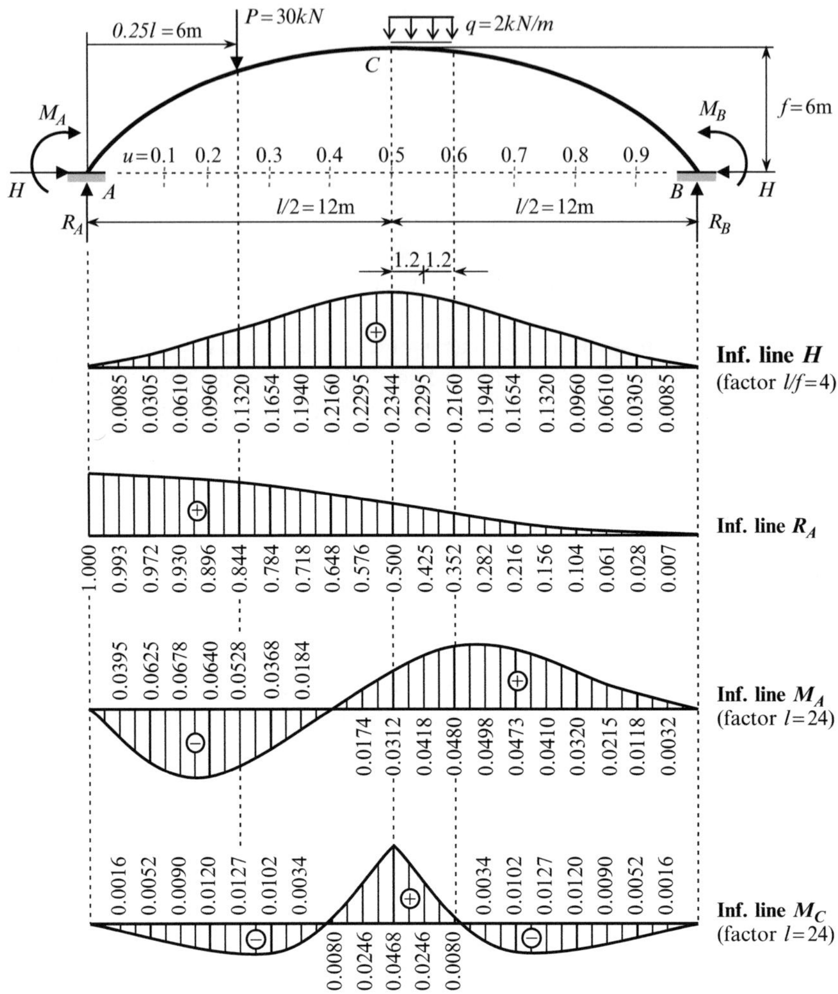

Fig. 3.35 Parabolic nonuniform arch. Design diagram and influence lines

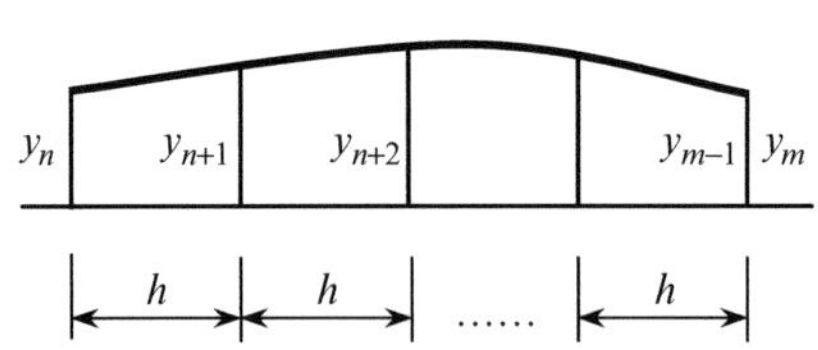

Fig. 3.36 Approximation for calculation of area of curvilinear influence line

$$M_C = R_A \frac{l}{2} - Hf + M_A - P\left(\frac{l}{2} - 0.25l\right)$$
$$= 27.36 \times 12 - 20.20 \times 6 - 33.32 - 30(12 - 6) = -6.2\ \text{kNm}.$$

Thus, we use the fixed and moving load approaches in parallel [Kar10].

The bending moment at crown C using the influence line is

$$M_C = l\left[-P \times 0.0127 + q\left(\frac{0.0468}{2} + 0.0246 + \frac{0.008}{2}\right)1.2\right] = -6.15\ \text{kNm}.$$

Relative error is $(6.2 - 6.15/6.175)100\% = 0.8\%$. This error is due to the approximate calculation of the area of influence lines.

Shear force at crown C is obtained by projecting all forces, located to the left of this section, onto the vertical: $Q_C = R_A - P = 27.36 - 30 = -2.64$ kN.

We can see that influence lines for primary unknowns have a fundamental meaning – they allow us to easily calculate reactions of a statically indeterminate arch subjected to any fixed load. After that, calculating of internal forces at any section of the arch may be performed as for a statically determinate structure.

3.10 Arch Subjected to Radial Pressure

This section presents the internal forces for circular arch subjected to in-plane uniform radial load. Two approaches are considered. They are analysis of the arch on the basis of integration of differential equations and by the Force method.

Behavior of a curvilinear rod of arbitrary radius of curvature ρ is described by the differential equations (1.21–1.23)

$$\frac{\mathrm{d}N}{\mathrm{d}s} - \frac{Q}{\rho} + p = 0, \quad \frac{\mathrm{d}Q}{\mathrm{d}s} + \frac{N}{\rho} + q = 0, \quad \frac{\mathrm{d}M}{\mathrm{d}s} - Q + m = 0.$$

Here p, q, and m present the intensity of the tangential, normal, and moment loads distributed along the rod (Fig. 1.19). Let us apply these formulas for the analysis of a circular arch of radius R with central angle 2α. The arch is subjected to radial uniform pressure $q_0 = q_\varphi = q$. In this case, the system of equations (1.21–1.23) is simplified. In polar coordinates $\mathrm{d}s = R\mathrm{d}\varphi$, the system becomes

$$\frac{\mathrm{d}N}{\mathrm{d}\varphi} = Q, \tag{3.41a}$$

$$\frac{\mathrm{d}Q}{\mathrm{d}\varphi} = -N - qR, \tag{3.41b}$$

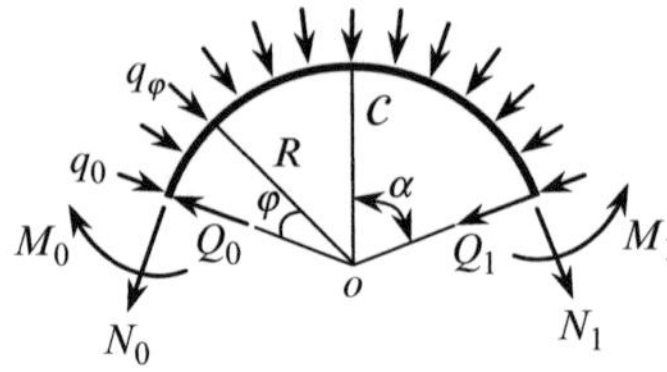

Fig. 3.37 Free-body diagram of circular rod subjected to uniform radial load

$$\frac{\mathrm{d}M}{\mathrm{d}\varphi} = QR. \tag{3.41c}$$

Assume that the forces and moment act on the ends of the arch, as shown in Fig. 3.37. Here subscript 0 is related to initial section of the arch ($\varphi = 0$) and subscript 1 is related to the final section of the arch ($\varphi = 2\alpha$).

3.10.1 Internal Forces Taking into Account and Neglecting Shrinkage

Expressions (3.41a)–(3.41c) do not take into account shrinkage of material of the arch. On the basis of these equations, we can easily derive expressions for internal forces in arbitrary section, which is characterized by the angle φ.

Differentiate (3.41b) with respect to φ and take into account (3.41a). We obtain the following differential equation $\mathrm{d}^2Q/\mathrm{d}\varphi^2 = -Q$. Solution of this equation is $Q = C_1 \cos\varphi + C_2 \sin\varphi$. Constant of integration is determined from the boundary conditions:

1. At $\varphi = 0$, the shear $Q = Q_0$. This condition leads to $C_1 = Q_0$.
2. Since $\mathrm{d}Q/\mathrm{d}\varphi = -C_1 \sin\varphi + C_2 \cos\varphi$, then according to (1.22), we get $\mathrm{d}Q_0/\mathrm{d}\varphi = Q_0' = -N_0 - q_0R$. At $\varphi = 0$, $-N_0 - q_0R = -C_1 \times 0 + C_2 \times 1$. This condition leads to $C_2 = -N_0 - q_0R$. Thus, for shear in any section, we get

$$Q = Q_0 \cos\varphi - (N_0 + q_0R)\sin\varphi. \tag{3.42}$$

Equation (3.41a) becomes $\mathrm{d}N/\mathrm{d}\varphi = Q_0 \cos\varphi - (N_0 + q_0R)\sin\varphi$. Solution of this equation is

$$N = Q_0 \sin\varphi + N_0 \cos\varphi + (\cos\varphi - 1)\, q_0R. \tag{3.43}$$

According to (3.41c), $\mathrm{d}M/\mathrm{d}\varphi = QR = Q_0R\cos\varphi - (N_0 + q_0R)R\sin\varphi$. Solution of this equation is

$$M = M_0 + Q_0R\sin\varphi - \left(N_0R + q_0R^2\right)(1 - \cos\varphi). \tag{3.44}$$

Expressions (3.42)–(3.44) give a complete picture of the distribution of internal forces in the circular rod subjected to uniform distributed radial load.

Apply derived formulas for the analysis of uniform hingeless arch with central angle 2α. Equations (3.42)–(3.44) contain unknown initial parameters Q_0, N_0, and M_0. Since the structure is symmetrical, then antisymmetrical unknown on the axis of symmetry $(\varphi = \alpha)$ is $Q_C = 0$, thus

$$Q_0 \cos \alpha - (N_0 + q_0 R) \sin \alpha = 0.$$

On the right end of the arch $(\varphi = 2\alpha)$, we have $Q_1 = Q_0$, thus

$$Q_0 \cos 2\alpha - (N_0 + q_0 R) \sin 2\alpha = Q_0.$$

Solving the last two equations, we get the axial and shear forces, which arise in the initial section

$$N_0 = -qR, \quad Q_0 = 0.$$

Since $\mathrm{d}M/\mathrm{d}\varphi = QR = 0$, then $M = M_0$, i.e., the bending moments along the arch are constant. It is known that in the case of a closed uniform ring, subjected to the in-plane uniform radial pressure, bending stresses do not arises [Bir68, vol. 1]. Thus, in such condition, there is a hingeless and two-hinged uniform circular arches with *arbitrary* central angle. Therefore, $M_0 = M(\varphi) = 0$. The tabulated data for arches with nonsymmetrical boundary conditions may be found in ref. [Roa75], [You89].

Now let us determine the internal forces taking into account the shrinkage of material of an arch. For hingeless uniform circular arch with arbitrary central angle subjected to action of radial uniform load, the final results are presented below. These results are obtained on the basis of the Bussinesk's equation (1.43) [Sni66].

The shear force and bending moment at the left end are

$$\begin{aligned} Q_0 &= -(q_0 R + N_0) \tan \alpha, \\ M_0 &= -\frac{(q_0 R + N_0)}{\alpha} R(\tan \alpha - \alpha). \end{aligned} \tag{3.45}$$

The axial force on the left end is $N_0 = -q_0 R f(\alpha, \lambda)$, where function $f(\alpha, \lambda)$ takes into account shrinkage of the material:

$$\begin{aligned} f(\alpha, \lambda) &= \frac{n - \lambda(\alpha + 1)}{n - \lambda}, \\ n &= (\alpha - 1) \tan \alpha - 1 + (\alpha + 2 \tan \alpha) \cot \alpha. \end{aligned}$$

where $\lambda = r^2/R^2$, while $r = \sqrt{I/A}$ is the radius of gyration of the cross section of the arch. The function $f(\alpha, \lambda)$, in terms of half-central angle α and dimensionless parameter λ, is presented in Table 3.13

It can be seen that shrinkage of the material leads to a slight decrease in axial force N_0 due to the factor $f(\alpha, \lambda)$.

Table 3.13 Function $f(\alpha, \lambda)$

λ	$\alpha = \pi/6$	$\pi/4$	$\pi/3$	$\pi/2$
0.001	0.9997	0.9995	0.9994	1.0
0.005	0.9984	0.9975	0.9969	1.0
0.01	0.9968	0.9950	0.9938	1.0
0.05	0.9834	0.9742	0.9680	1.0

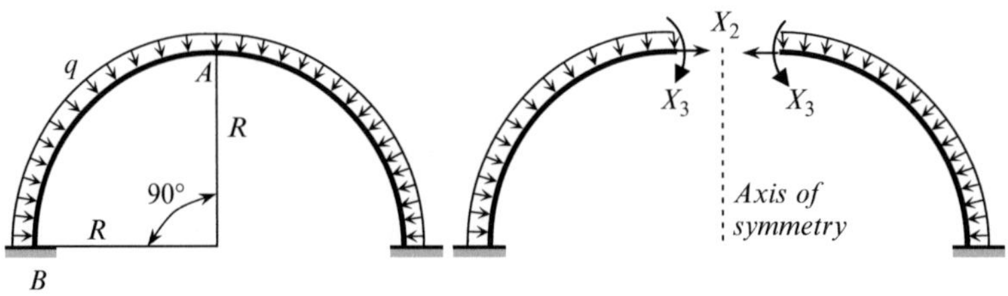

Fig. 3.38 Design diagram of hingeless arch and primary system

3.10.2 Complex Loading of Circular Arch

First of all let us consider the solution of the classic problem. A hingeless semicircular uniform arch is subjected to a uniform radial load within the total arc. For analysis of this problem, we use the Force method together with the tabulated data. Design diagram of the arch and primary system are shown in Fig. 3.38.

Primary unknowns are axial force X_2 and bending moment X_3 at crown A. Numeration of unknowns is adopted according to Table A.7. Since the loading is symmetrical, the antisymmetrical unknown shear force $X_1 = 0$. Canonical equations of the Force method are

$$\delta_{22}(90^\circ)X_2 + \delta_{23}(90^\circ)X_3 + \Delta_{2P} = 0,$$
$$\delta_{32}(90^\circ)X_2 + \delta_{33}(90^\circ)X_3 + \Delta_{3P} = 0.$$

Free terms of canonical equations, i.e., the axial displacement and slope at point A, caused by given the load q, according to Table A.6 are

$$\Delta_{2P} = \xi_A = qR^3 a_y\left(\frac{3\gamma}{2} - 2\sin\gamma + \frac{\sin 2\lambda}{4}\right) = \frac{qR^4}{EI}\left(\frac{3}{2}\frac{\pi}{2} - 2\sin 90^\circ + \frac{\sin 180^\circ}{4}\right)$$
$$= \frac{qR^4}{EI}\left(\frac{3\pi}{4} - 2\right),$$

$$\Delta_{3P} = \psi_A = qR^2 a_y(\gamma - \sin\gamma) = \frac{qR^3}{EI}\left(\frac{\pi}{2} - \sin 90^\circ\right) = \frac{qR^3}{EI}\left(\frac{\pi}{2} - 1\right).$$

Unit displacements at the free end of the half-arch, according to Table A.7, are

$$\delta_{22}(90^\circ) = 1 \times R^2 a_y \left(\frac{3}{2}\gamma + \frac{\sin 2\gamma}{4} - 2\sin\gamma\right) = \frac{R^3}{EI}\left(\frac{3}{2}\frac{\pi}{2} + \frac{\sin 180^\circ}{4} - 2\sin 90^\circ\right)$$
$$= \frac{R^3}{EI}\left(\frac{3\pi}{4} - 2\right),$$
$$\delta_{23}(90^\circ) = \delta_{32}(90^\circ) = 1 \times R a_y(\gamma - \sin\gamma) = \frac{R^2}{EI}\left(\frac{\pi}{2} - \sin 90^\circ\right) = \frac{R^2}{EI}\left(\frac{\pi}{2} - 1\right),$$
$$\delta_{33}(90^\circ) = 1 \times a_y \gamma = \frac{R}{EI}\frac{\pi}{2}.$$

Canonical equations become

$$R^3\left(\frac{3\pi}{4} - 2\right)X_2 + R^2\left(\frac{\pi}{2} - 1\right)X_3 = -qR^4\left(\frac{3\pi}{4} - 2\right),$$
$$R^2\left(\frac{\pi}{2} - 1\right)X_2 + R\frac{\pi}{2}X_3 = -qR^3\left(\frac{\pi}{2} - 1\right).$$

Solutions of these equations are $X_2 = -qR, \quad X_3 = 0$.
Reactions at the clamped support are

$$N_B^{\text{axial}} = N_B(q) + N_B(X_2) + N_B(X_3) = -qR(1 - \cos 90^\circ) + (-qR)\cos 90^\circ = -qR,$$

$$M_B = M_B(q) + M_B(X_2) + M_B(X_3) = qR^2(1 - \cos 90^\circ) + (-qR)R(1 - \cos 90^\circ) = 0.$$

Thus, in the case of a uniform circular hingeless arch subjected to uniform distributed radial load q, the axial force at any section of arch is $-qR$, while the bending moment is zero.

Now let us consider a complex loading of the symmetrical arched structure, partially subjected to a radial load with two concentrated forces as shown in Fig. 3.39a. Uniform hingeless circular arch of radius $R = 24$ m has a central angle $\gamma = 150^\circ$. Two cantilevered rods CD and $C'D'$ of length 6 m have rigid connections with the arch at points C and C'. Forces $F = 10$ kN act at the end point of the cantilevers. The portion of the arch C–A–C' is subjected to uniform distributed radial load $q = 2$ kN/m.

For analysis of this structure, let us modify the left part, BA, of a structure so we can use data that has already been tabulated, and is given in Tables A.6 and A.7. Transfer the force F into the rigid joint C on the arch and in doing so, add a couple $M = 60$ kN/m. Resolve the force F into the radial and axial components

$$F_{\text{rad}} = F\sin(15 + \beta) = 5 \text{ kN}, \quad F_\tau = -F\cos(15 + \beta) = -8.66 \text{ kN}.$$

Load the total portion AB by a uniformly distributed radial load q, and in doing so add a compensating load q of opposite direction within the portion BC. The same procedure should be done over the right part AB' of the structure. Thus, the initial design diagram of the arched structure is presented in the equivalent form: hingeless arch BAB' subjected to radial load q across the entire arch, uniform load $-q$ within

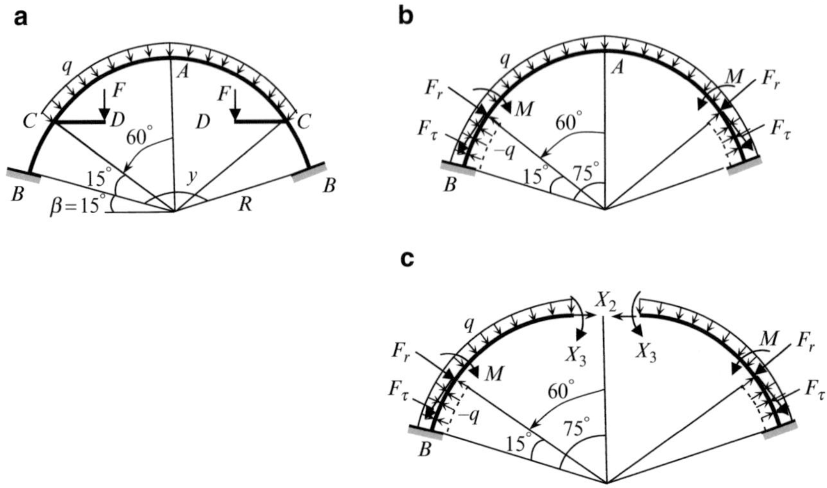

Fig. 3.39 (**a**) Design diagram of arch structure; (**b**) modified design diagram; (**c**) primary system

the portions BC and $B'C'$, as well as the axial force F_τ, radial force F_r, and couple M at the points C and C' (Fig. 3.39b).

Primary unknowns are the axial force X_2 and bending moment X_3 at crown A (Fig. 3.39c).

Antisymmetrical unknown (shear force), according to the properties of symmetrical structures is $X_1 = 0$.

Canonical equations of the Force method are

$$\delta_{22}(75^\circ)X_2 + \delta_{23}(75^\circ)X_3 = -\Delta_{2P} = -\xi_A = -\frac{109,504}{EI},$$

$$\delta_{32}(75^\circ)X_2 + \delta_{33}(75^\circ)X_3 = -\Delta_{3P} = -\psi_A = -\frac{9,863}{EI}.$$

Free terms of canonical equations ξ_A and ψ_A had been calculated in Chap. 1, Sect. 1.9, Example 1.8.

Displacements of the free end of the half-arch, according to Tables A.6 and A.7 are

$$\delta_{22}(75^\circ) = 1 \times R^2 a_y\left(\frac{3}{2}\gamma + \frac{\sin 2\gamma}{4} - 2\sin\gamma\right) = \frac{24^3}{EI}\left(\frac{3}{2}\frac{5\pi}{12} + \frac{\sin 150^\circ}{4} - 2\sin 75^\circ\right)$$
$$= \frac{2,165.5}{EI},$$

$$\delta_{23}(75^\circ) = 1 \times R a_y(\gamma - \sin\gamma) = \frac{24^2}{EI}\left(\frac{5\pi}{12} - \sin 75^\circ\right) = \frac{197.6}{EI},$$

$$\delta_{32}(75^\circ) = 1 \times R a_y(\gamma - \sin\gamma) = \frac{197.6}{EI},$$

$$\delta_{33}(75^\circ) = 1 \times a_y\gamma = \frac{R}{EI} \times \frac{5\pi}{12} = \frac{31.4}{EI}.$$

Canonical equations become

$$
\begin{aligned}
2,165.5X_2 + 197.6X_3 &= -109,504, \\
197.6X_3 + 31.4X_3 &= -9,863.
\end{aligned}
$$

The primary unknowns are

$$
\begin{aligned}
X_2 &= -50.81 \text{ kN}, \\
X_3 &= 5.68 \text{ kN m}.
\end{aligned}
$$

Reactions at Support

Now these primary unknowns should be considered as initial parameters and for computation of reactions at support, we can use the modified expressions (1.25) and (1.26). For this, it is necessary to add corresponding terms due to all loads (q along arc AB, $-q$ along arc BC, couple M and forces F_{rad} and F_{axial} at section C). These terms are denoted as $[N(\gamma)]$ and $[M(\gamma)]$; they should be calculated according to Tables A.6 and A.7

$$
\begin{aligned}
N &= N_0 \cos\varphi + Q_0 \sin\varphi + [N(\gamma)], \\
M &= M_0 + Q_0 R \sin\varphi - N_0 R(1 - \cos\varphi) + [M(\gamma)].
\end{aligned}
\tag{3.46}
$$

For the structure in Fig 3.39a, we have

$$
N_0 = X_2, \quad M_0 = X_3, \quad Q_0 = 0.
$$

$$
N(\gamma) = \underbrace{-qR(1 - \cos 75)}_{q \text{ along } AB} + \underbrace{qR(1 - \cos 15)}_{-q \text{ along } BC} - \underbrace{F_{\text{rad}} \sin 15}_{F_{\text{rad}} \text{at } C} - \underbrace{F_{\tau} \cos 15}_{F_{\tau} \text{at } C} + \underbrace{0}_{M \text{ at } C}.
$$

The expression for $M(\gamma)$ may be constructed in a similar manner.

3.11 Deflections of the Arches

This section is devoted to computation of displacements of redundant arches. Two approaches are considered. They are displacement of the specified points along the given direction (linear, angular, mutual) and displacements of the arches in continuous form. The first approach is based on the Maxwell–Mohr integral and multiplication of two bending moment diagrams, while the second approach is based on the integration of Boussinesq's equation of the arch (this approach is considered in Sect. 1.9).

3.11.1 Deflections at the Discrete Points of Redundant Arches

As in the case of a statically determinate arch, computation of displacement of some special section of two-hinged and hingedless arches may be performed using the Mohr integral. For this, we need to construct bending moment diagram M in the entire state, form a unit state, construct bending moment diagram $\overline{M}$ in the unit state, and multiply both diagrams. In the general case, the flexural stiffness EI of the arch is not constant and basis line is curvilinear. Therefore, the arch should be substituted by a set of chords, followed by the application of Vereshchagin rule (or Simpson or trapezoid rule) within each straight portion.

The construction of bending moment diagram M_P in the entire state is discussed above using the Force method. Now the following principal question arises: *how to construct the unit state*? It is obvious that unit load must correspond to the required deflection. But which structure must carry this unit load? It is obvious that unit load may be applied to the given statically indeterminate structure; for construction of bending moment diagram $\overline{M}$ and for statically indeterminate structure, the additional analysis is required. Therefore, computation of deflections for redundant structure becomes cumbersome. However, solution of this problem can be significantly simplified, taking into account a following fundamental concept.

Bending moment diagram M_P of any statically indeterminate structure can be considered as a result of application of two types of loads to a *statically determinate structure*. They are the given external loads and primary unknowns. It means that a given statically indeterminate structure may be replaced by *any statically determinate structure* subjected to a given load and primary unknowns, which are treated now as *external* forces. It does not matter which primary system has been used for final construction of bending moment diagram, since on the basis of *any* primary system the *final* bending moment diagram will be the same. Therefore, the unit load (force, moment, etc.), which corresponds to required displacement (linear, angular, etc.) should be applied in *any statically determinate (!) structure*, obtained from a given structure by the elimination of *any* redundant constraints.

Application of this fundamental concept for an arbitrary redundant structure is presented in [Kar10]. For example, for the calculation of vertical displacement Δ_C of the crown C for a two-hinged arch, the unit state may be adopted according to any version shown in Fig. 3.40. In the schemes (a) and (b), the primary unknown is horizontal reaction of one of the supports, therefore the primary system presents a simply supported curvilinear rod. In the version (c), the primary unknown is the bending moment at an arbitrary section k. Therefore, the primary system presents a three-hinged nonsymmetrical arch.

Thus, the following procedure may be applied:

1. Construct the bending moment diagram for the *entire* redundant structure M_P.
2. Show *any* statically determinate structure, apply unit load which corresponds to the required displacement and construct bending moment diagram $\overline{M}$.
3. Apply the procedure discussed in Sect. 2.8.2.

This procedure is very effective for the computation of displacement of nonsymmetrical arches.

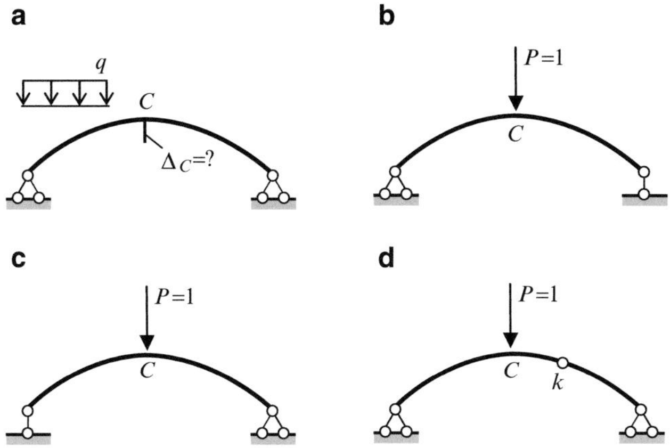

Fig. 3.40 Design diagram of two-hinged arch and versions of unit states for calculation of displacement Δ_{1C}

3.11.2 *Effect of Axial Forces*

Let us evaluate the influence of the axial forces in the arch on the unit and loaded displacements of canonical equations. These displacements should be determined by the formulas

$$\delta_{11} = \delta_{11}^{(M)} + \delta_{11}^{(N)} = \int_{(s)} \frac{\overline{M}_1^2}{EI} \mathrm{d}s + \int_{(s)} \frac{\overline{N}_1^2}{EA} \mathrm{d}s, \tag{3.47}$$

$$\Delta_{1\mathrm{P}} = \Delta_{1\mathrm{P}}^{(M)} + \Delta_{1\mathrm{P}}^{(N)} = \int_{(s)} \frac{\overline{M}_1 \times M_\mathrm{P}^0}{EI} \mathrm{d}s + \int_{(s)} \frac{\overline{N}_1 \times N_\mathrm{P}^0}{EA} \mathrm{d}s. \tag{3.48}$$

Here, $\overline{M}_1, \overline{N}_1$ are the bending moment and axial force, respectively, in the primary system caused by the primary unknown (e.g., thrust) $X = 1$; $M_\mathrm{P}^0, N_\mathrm{P}^0$ are, respectively, the bending moment and axial force in the primary system caused by the external load; I and A are moment of inertia and area of a cross section of the arch.

If a cross section of the arch is constant, then (3.47)–(3.48) may be rewritten as follows [Rzh82]

$$\delta_{11} = \frac{1}{EI} \int_{(s)} \overline{M}_1^2 \mathrm{d}s + \frac{1}{EA} \int_{(s)} \overline{N}_1^2 \mathrm{d}s = \frac{1}{EI} \int_{(s)} \left(\overline{M}_1^2 + r^2 \overline{N}_1^2 \right) \mathrm{d}s, \tag{3.47a}$$

$$\Delta_{1\mathrm{P}} = \frac{1}{EI} \int_{(s)} (\overline{M}_1 \times M_\mathrm{P}^0 + r^2 \overline{N}_1 \times N_\mathrm{P}^0) \mathrm{d}s, \tag{3.48a}$$

where $r^2 = I/A$ is a radius of gyration of the cross section of the arch.

Table 3.14 Computation of additional terms $\delta_{11}^{(N)}$ and $\Delta_{1P}^{(N)}$

Portion	$\frac{l_i}{6}$	Unit state $\overline{N}_1$ a_1	c_1	b_1	$\overline{N}_1^2 l_i$	Loaded state N_P^o a_2	c_2	b_2	$\overline{N}_1 N_P^o l_i$
0–1	0.6644	−0.7070	−0.7535	−0.8000	2.2661	−16.968	−13.884	−10.800	41.5123
1–2	0.5896	−0.8000	−0.8472	−0.8944	2.5418	−10.800	−7.9890	−5.1780	23.7874
2–3	0.5340	−0.8944	−0.9322	−0.9701	2.7861	−5.1780	−3.3160	−1.4550	9.8309
3–4	0.5039	−0.9701	−0.9850	−1.000	2.9338	−1.4550	−0.7275	0.00	2.1556
					10.5278				**77.2862**

Returning to the example in Sect. 3.3.1, the two-hinged uniform arch of span 24 м and rise 6 м is loaded by uniformly distributed load $q = 2$ kN/m. Analysis of the arch with such parameters, neglecting axial forces leads to terms of displacement (3.47) and (3.48), which according to Table 3.5 are

$$\delta_{11}^{(M)} = \frac{479.4484}{EI} \qquad \Delta_{1P}^{(M)} = -\frac{11,506.74}{EI}.$$

For calculation of additional terms, we will use modified design diagram (parameters of this diagram are presented in Tables 3.1 and 3.2) and multiplication the normal forces diagram in the form of the Simpson rule. All computations are presented in Table 3.14.

Because symmetrical structure obtained values must be doubled, we obtain

$$\delta_{11}^{(N)} = \frac{21.0555}{EA}, \qquad \Delta_{1P}^{(N)} = \frac{154.57}{EA}.$$

Note the precise value of the second term in (3.47) is [Rzh82]

$$\delta_{11}^{(N)} = \int_{(s)} \overline{N}_1^2 \mathrm{d}s = \int_{-l/2}^{l/2} \frac{\overline{N}_1^2 \mathrm{d}x}{\cos\varphi} = \int_{-l/2}^{l/2} \cos\varphi \, \mathrm{d}x$$
$$= \int_{-l/2}^{l/2} \frac{\mathrm{d}x}{\sqrt{1+4x^2/l^2}} 0.8814 \times 24.0 = 21.1536.$$

The unit displacement and loaded term of the canonical equation becomes

$$\delta_{11} = \frac{1}{EI}\left(479.4484 + 21.0555\frac{I}{A}\right),$$

$$\Delta_{1P} = \frac{1}{EI}\left(-11,506.74 + 154.57\frac{I}{A}\right).$$

Assuming a height of the rectangular cross section of the arch is 1.2м, then $I/A = 0.12$ and the primary unknown becomes $X_1 = -\Delta_{1P}/\delta_{11} = 23.83$ kN. If the

Table 3.15 Computation of the bending moments

Points	$\overline{M}_1$	$\overline{M}_1 X_1$	M_P^0	M(kNm)
0	0.0	0.0	0.0	0.0
1	−2.625	−62.685	63	0.315
2	−4.500	−107.46	108	0.55
3	−5.625	−134.32	135	0.675
4	−6.000	−143.28	144	0.72

axial compressed forces are neglected, then the thrust $H = 24$ kN. The change of thrust becomes nonsignificant. Computation of the bending moments is performed by the formula $M = \overline{M}_1 X_1 + M_P^0$ and presented in Table 3.15.

It is evident that the axial compressed forces lead to insignificant change of the thrust and the occurrence of minor bending moments. However, the uniqueness of this example lies in the following. If axial forces are taken into account, then the behavior of arch change qualitatively; in addition to axial forces which arise in the arch, bending moments are present as well. Neglecting the bending moments may leads to the collapse of the arch.

For two-hinged arch with rise/span ratio of $f < l/3$ and thickness/span ratio of $h < l/10$, the shear may be neglected for the calculation of δ_{11}, and shear and axial forces may be neglected for the calculation of Δ_{1P} [Kle80].

Let us present the results of analysis of a two-hinged parabolic symmetrical uniform arch subjected to single force P at the crown. If axial forces are neglected, then thrust $X = 0.785P$. If the axial compressed forces are taking into account, and dimensionless parameter $\lambda = (l/i) = 20, \quad i = \sqrt{I/A}$, then for thrust we get $X = 0.725P$[Rzh82].

A significantly greater effect causes horizontal compliance of the arch. Assume the two-hinged arch has a thin tie on the elevation of support. With this, a loaded term Δ_{1P} remains unchanged, while the unit displacement should be calculated by thc formula

$$\delta_{11} = \delta_{11}^{(M)} + \delta_{11}^{(N)} + \delta_{11}^{(N_{\text{tie}})} = \int_{(s)} \frac{\overline{M}_1^?}{EI} \mathrm{d}s + \int_{(s)} \frac{\overline{N}_1^?}{EA} \mathrm{d}s + \frac{1^2 \times l}{E_t A_t},$$

where l is a span of the arch. Therefore, the thrust of the arch (force in a tie) becomes

$$X = -\frac{\Delta_{1P}}{\delta_{11}} = -\frac{\Delta_{1P}}{\delta_{11}^0 + \frac{l}{E_t A_t}},$$

where δ_{11}^0 is determined by (3.47).

Numerous reference data for hingeless nonuniform parabolic arches under different types of loads (reactions and bending moments taking into account axial forces) is presented in Table A.31.

3.12 Arch Loaded Orthogonally to the Plane of Curvature

This paragraph contains some information about the analysis of hingeless arch subjected to a load which acts perpendicular to the plane of the arch.

Let us assume the following coordinate system: the arch lies in the plane x–z, the y-axis is perpendicular to the plane of the arch, x-axis coincides with tangential direction, the z-axis is directed to the principal normal of the axis of the arch. Let the y- and z-axis be the principal axis of the cross section of the arch. In the cross-section of a structure, in the general case of loading, the following internal forces arise: axial force N_x, shear forces Q_x, Q_z, bending moments M_y, M_z, as well as torque or twisting moment T.

The bending moment M_y acts in the plane x–z, bending moment M_z acts in the plane x–y, twisting moment T acts in the plane y–z; the vector of moments are shown by double-arrow lines (Fig. 3.41a).

We define a planar system to be a system that satisfies the following conditions:

1. Longitudinal axis of all the rods and one of the principal axis of all the cross-sections lie in the same plane P.
2. The supports are positioned in such a way that a load acting in the plane P leads to all the deformations and forces arise in a plane that is parallel to P.

For this structure, according to mutual displacement and reaction theorem $\left(r_{jk} = -\delta'_{kj}\right)$, one can show that if a load acts orthogonally to plane P, then axial and shear forces, as well as the bending moment in plane P are all equal to zero. Thus, if a load F is perpendicular to plane P (Fig. 3.41a, b), then $N_x = 0$, $Q_z = 0$, and $M_y = 0$.

Let us consider a uniform circular arch of radius R and central angle $2\varphi_0$ [Rab54a]. The arch is subjected to uniformly distributed load q which acts perpendicular to the plane of the arch (into the page) (Fig. 3.42a).

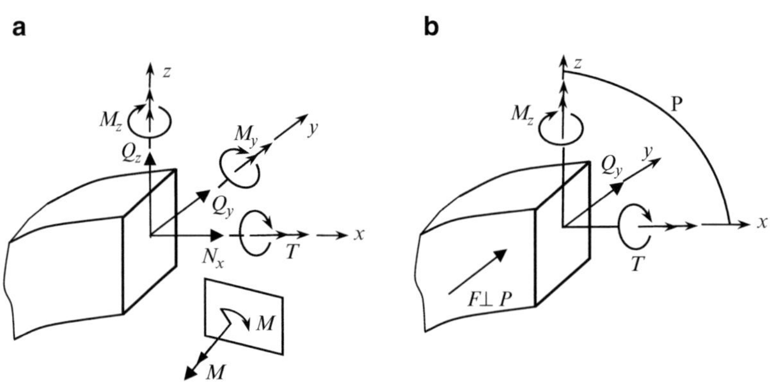

Fig. 3.41 (**a**) General notation, (**b**) planar system, load F acts normally to plane P

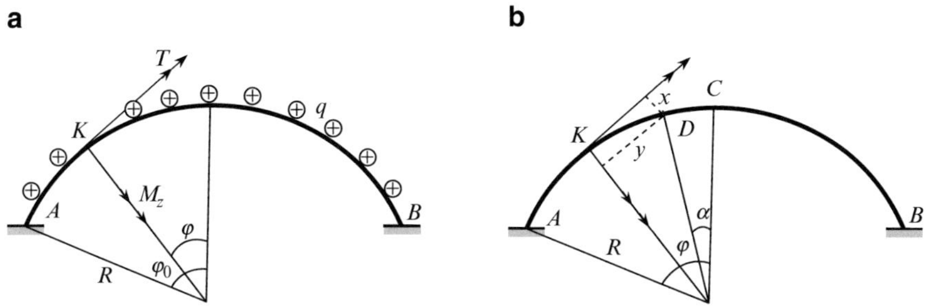

Fig. 3.42 (**a**) Design diagram of the arch, (**b**) coordinates of the any point *D*

Fig. 3.43 Primary system. Internal forces at the arbitrary section K are *T* and *Mz*. Vector X_1 at the crown C is resolved into two components, parallel to vectors *T* and *Mz*

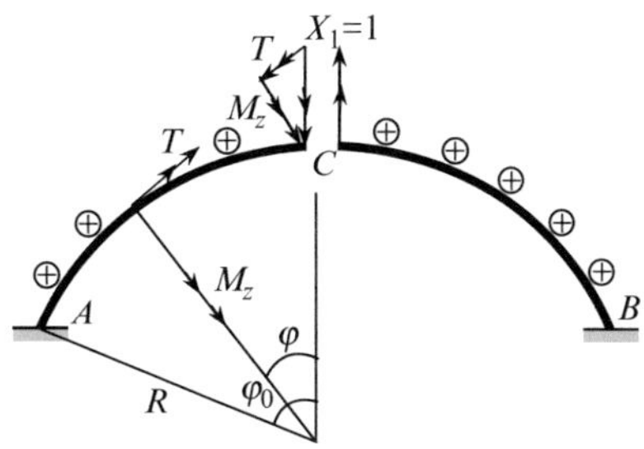

The feature of this design diagram is that thrust of the arch for the given loading does not arise. It means that the arch behaves as a clamped–clamped curvilinear beam. The second feature of this loading is the appearance of torque.

Due to symmetry, the structure has one unknown of the Force method. The primary unknown X_1 is the bending moment M_z at the middle section *C* of the span, $X_1 = M_z$. The primary system is shown in Fig. 3.43.

The bending moment and torque at any section caused by unit primary unknown are

$$\overline{M}_z = 1 \times \cos\varphi, \quad T = -1 \times \sin\varphi.$$

Canonical equation of the Force method is $\delta_{11}X_1 + \Delta_{1P} = 0$. The unit displacement δ_{11} is a mutual angular displacement in the plane *x*–*y* of two sections at point *C*.

Figure 3.42b presents coordinates *x* and *y* of an arbitrary point *D* with respect to tangent and radial axis with an origin at the point *K*

$$x = R[1 - \cos(\varphi - \alpha)], \quad y = R\sin(\varphi - \alpha).$$

Moments of external load q with respect to K are

$$M_z^P = -R\int_0^{\varphi} qy\mathrm{d}\alpha = -qR^2\int_0^{\varphi} \sin(\varphi-\alpha)\mathrm{d}\alpha = -qR^2(1-\cos\varphi),$$
$$T^P = R\int_0^{\varphi} qx\mathrm{d}\alpha = qR^2\int_0^{\varphi} [1-\cos(\varphi-\alpha)]\mathrm{d}\alpha = qR^2(\varphi-\sin\varphi).$$

The unit displacement is

$$\delta_{11} = \delta_{11}^M + \delta_{11}^T = 2R\int_0^{\varphi_0}\left(\frac{\cos^2\varphi}{EI}+\frac{\sin^2\varphi}{GI_T}\right)\mathrm{d}\varphi.$$

After integration, we get

$$EI\delta_{11} = \frac{R}{2}\left[2\left(\frac{EI}{GI_T}+1\right)\varphi_0 - \left(\frac{EI}{GI_T}-1\right)\sin 2\varphi_0\right].$$

The combined quantity GI_T is referred to as the torsional rigidity. For a square cross-section $T = 0.1426a^2$, for a rectangular cross-section, $a > b$, $I_T = (b^3/3)(a-0.63b)$. For steel structures $E/G \cong 2.57$, for concrete $E/G \cong 2.33$.

The free term of canonical equation is

$$EI\Delta_{1P} = EI\left(\Delta_{1P}^M + \Delta_{1P}^T\right) = 2R\int_0^{\varphi_0} M_z^P\overline{M}_z\mathrm{d}\varphi + 2R\int_0^{\varphi_0} T^P\overline{T}\frac{EI}{GI_T}\mathrm{d}\varphi,$$
$$= -\frac{qR^3}{2}\left[\left(\frac{EI}{GI_T}+1\right)(4\sin\varphi_0-2\varphi_0) - 4\frac{EI}{GI_T}\varphi_0\cos\varphi_0 + \left(\frac{EI}{GI_T}-1\right)\sin 2\varphi_0\right].$$

Primary unknown is $X_1 = -(\Delta_{1P}/\delta_{11})$.

The bending moment and torque, in terms of the angle φ, become

$$M_z(\varphi) = M_z^P + \overline{M}_z X_1 = -qR^2(1-\cos\varphi) + X_1\cos\varphi,$$
$$T(\varphi) = T^P + \overline{T}X_1 = qR^2(\varphi-\sin\varphi) - X_1\sin\varphi.$$

For circular arch subjected to out-of-plane load, the following relationship holds: $M_z = -(\mathrm{d}T/\mathrm{d}\varphi)$.

Numerous formulas for arches loaded orthogonally to the plane of curvature are presented in refs. [Bir68], [Roa75], and [You89].

Part II
Stability Analysis

Chapter 4
Elastic Stability of Arches

Theory of structural stability is a special branch of structural analysis. This theory explores very important phenomenon that is observed in the behavior of the structures subjected to compressed loads. This phenomenon lies in the abrupt change of initial form of equilibrium. Such phenomenon is called a loss of stability. As a rule, the loss of stability of a structure leads to it collapse. Engineering practice knows a lot of examples when ignoring this feature of a structure led to its failure.

This chapter is devoted to stability analysis of arches and arched structures subjected to compressed loads. Different types of arches and their loadings are considered. Analytical methods of determining the critical loads on arches are considered on the basis of integration of differential equations of the arch.

4.1 General

The mathematical basis of stability theory of arches was first *implemented* by Kirchhoff (1824–1887) [Kir76]. The first systematic analysis of equilibrium stability was performed by Bryan [Bry88]. Southwell [Sou13] continued further studies of the general stability theory of equilibrium of elastic bodies. Fundamental research on the stability of rings and circular arches was investigated by Nikolaii [Nik18]. Significant contributions solving the problem of stability of arches added by Timoshenko [Tim61], [Tim72], Federhofer [Fed34], Lokshin [Lok34], Shtaerman [Sht35], Morgaevsky [Mor39], [Mor40], [Mor61], Dinnik [Din46], Pflüger [Pfl50], Chudnovsky [Chu52]. Numerical methods of stability analysis of arches are largely the works of Smirnov [Smi84], [Pi02]. Experimental work on the subject was contributed by Gaber [Gab34], Dinnik and Morgaevsky [Din46], Pavlenko [Pav66].

I.A. Karnovsky, *Theory of Arched Structures: Strength, Stability, Vibration*,
DOI 10.1007/978-1-4614-0469-9_4, © Springer Science+Business Media, LLC 2012

4.1.1 Fundamental Concepts

Stable equilibrium state means that if the structure, under compressed load is disturbed from an initial equilibrium state and after all disturbing factors are removed, then the structure returns to the initial equilibrium state. This is concerning to the elastic structures. If a structure consists of plastic or elasto-plastic elements, then a complete returning to the initial state is impossible. However, equilibrium state is assumed to be stable, if a structure even tends to return to the initial equilibrium state.

In case of absolutely rigid bodies, we are talking about stable *position* of a structure, while in case of deformable elements we are talking about stable equilibrium *form* of a deformable state. In all these cases, we assume that the acting compressed load is less than the critical one.

The critical force Pcr is the maximum force at which the structure holds its initial equilibrium form (the structure is still stable), or minimum force, at which the structure no longer returns to the initial state (the structure is already unstable) if all disturbing factors are removed.

Unstable equilibrium state means that if a structure under compressed load is disturbed from an initial equilibrium state and after all disturbing factors are removed, then the structure does not return to the initial equilibrium state. In this case, we say that the acting compressed load is larger than the critical one.

Change of configuration of a structure under the action of compressed load is called a loss of stability of the initial form of equilibrium or a buckling. If the compressed load is a static one, then this case is referred as the static loss of stability. In this chapter, we consider elastic arches subjected to static loads only. If a structure switches to other state (as a result of loss of stability) and remains in this state in equilibrium, then this new equilibrium state is called the adjacent form of equilibrium.

The main types of loads for in-plane stability analysis of the arch, as follows:

(a) Tracking load. For this load, an angle δ between the load and deformable axis of an arch remains constant (Fig. 4.1a); (b) hydrostatic load, which is directed perpendicular to the deformable axis of an arch. This load is a special case of the tracking load at $\delta = \pi/2$ (Fig. 4.1b); (c) polar (radial for circular arch) load directed to a fixed center (Fig. 4.1c); (d) gravity load. Direction of this load does not depend on the deflections of the arch (Fig. 4.1d).

The state of a structure that corresponds to a critical load is called the critical state. The switching of a structure into a new state occurs suddenly and as a rule leads to the collapse of a structure. The theory of static stability of structures is devoted to methods of calculation of critical loads.

4.1.2 Forms of the Loss of Stability of the Arches

The loss of stability can manifest itself in several ways. Note the most important of them: (1) the emergence of qualitatively new adjacent forms of equilibrium; (2) the

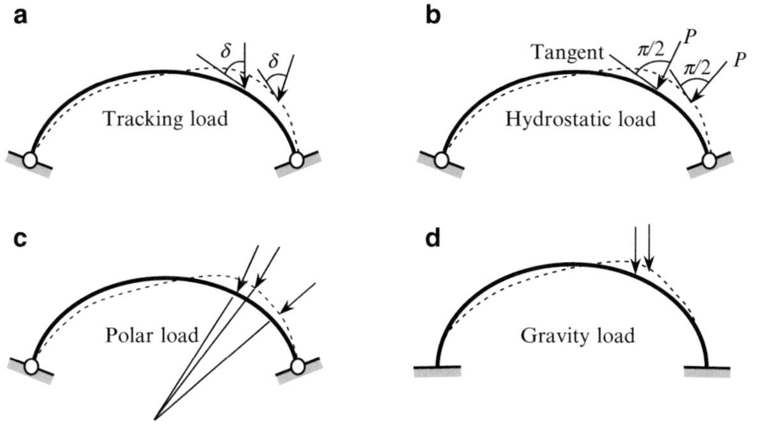

Fig. 4.1 Types of loads on the arches

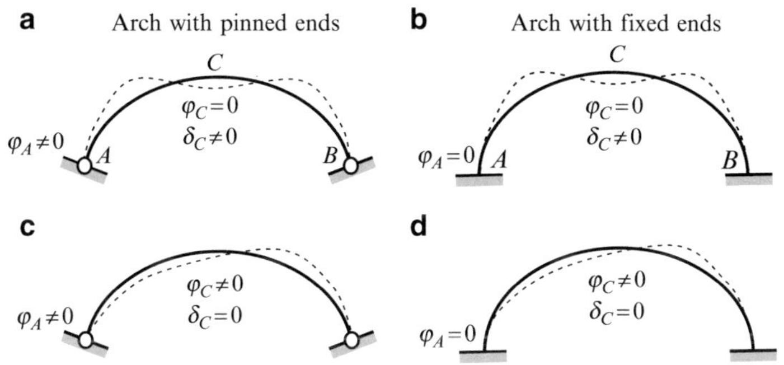

Fig. 4.2 Arches with different boundary conditions. Symmetrical and antisymmetrical forms of stability loss

emergence of nonadjacent forms of equilibrium; (3) complete disappearance of any forms of equilibrium.

Arches with pinned and fixed supports are shown in Fig. 4.2a, b. If an arch is subjected to symmetric loading, then the loss of stability may occur in two simplest forms. They are symmetrical form (a), when elastic curve is symmetrical with respect to vertical axis of symmetry and, otherwise, is antisymmetrical (b). Qualitatively new adjacent forms of equilibrium correspond to the loss of stability, in the Euler sense [Din46]. The buckling of arches in the Eulerian sense are considered in this chapter.

In the case of a very shallow arch, its compression should be taken into account. Due to this, the loss of stability of two-hinged and hingeless arches occurs differently, as shown in Fig. 4.2. Specifically, the lowest critical load corresponds to the symmetric form of the loss of stability along one sinusoidal half-wave (Fig. 4.3).

With this two cases are possible. (1) The sign of the curvature remains the same (line 1); (2) sign of the curvature changes (line 2). Structures that abruptly transform

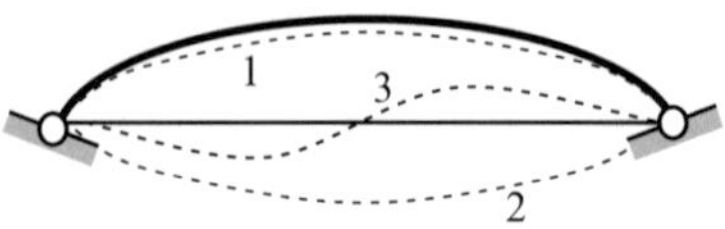

Fig. 4.3 Shallow arch. The forms of loss of stability

to state 2 (or 3) are called structures with a jump. Note that antisymmetrical form for loss of stability with one nodal point at the middle span (3) is possible.

Note that the general sensitivity theory in problems of stability of elastic structures is considered in detail in [God00].

4.1.3 Differential Equations of Stability of Curvilinear Rod

Let us present a set of differential equations of stability for planar curvilinear rod for some special cases. Assume that distributed radial load q is tracking.

The stability equation of a uniform rod ($EI = \text{const}$) of a variable radius of curvature ρ may be presented as follows (derivative, with respect to the curvilinear coordinate s, is denoted by the prime symbol):

$$EI\left\{\left[\rho(\rho u')^{\mathrm{IV}}\right]' + \left[\rho\left(\frac{u}{\rho}\right)'''\right]' + \frac{1}{\rho}(\rho u')''' + \frac{1}{\rho}\left(\frac{u}{\rho}\right)''\right\} + \left\{\rho^2 q\left[(\rho u')'' + \left(\frac{u}{\rho}\right)'\right]\right\}' = 0.$$

This differential equation of sixth order, with respect to tangential displacement u, takes into account the radial distributed load q [Rzh55].

In case of circular rod of radius $\rho = R = \text{const}$ and variable flexural stiffness EI we get

$$R^4(EIu''')''' + R^2(EIu')''' + R^2(EIu''')' + (EIu')' + R^5(qu''')' + R^3(qu')' = 0.$$

For circular uniform rod ($\rho = R = \text{const}$, $EI = \text{const}$) subjected to uniform distributed radial load $q = \text{const}$ we have

$$R^6 u^{\mathrm{VI}} + 2R^4 u^{\mathrm{IV}} + R^2 u'' + \frac{qR^3}{EI}\left(R^4 u^{\mathrm{IV}} + R^2 u''\right) = 0.$$

In the polar coordinate $\varphi = s/R$ the last equation may be rewritten as Lamb's equation [Lam1888] (see Sect. 1.7.3)

$$\frac{\mathrm{d}^6 u}{\mathrm{d}\varphi^6} + 2\frac{\mathrm{d}^4 u}{\mathrm{d}\varphi^4} + \frac{\mathrm{d}^2 u}{\mathrm{d}\varphi^2} + \frac{qR^3}{EI}\left(\frac{\mathrm{d}^4 u}{\mathrm{d}\varphi^4} + \frac{\mathrm{d}^2 u}{\mathrm{d}\varphi^2}\right) = 0.$$

Note that the stability equation of a planar curvilinear rod in the general case (nonuniform cross-section EI, variable radius of curvature (ρ) may be found in [Mor39], [Rzh55]. This book also contains a set of equations for case of non-tracking load.

4.1.4 Methods of Analysis

The problem of determining critical load on the arches may be obtained in the precise analytical form only in simplest cases. Such cases include uniform circular arches, which are subjected to uniform pressure normal to the axis of the arch.

Analytical solution to the issue of stability of arches of various shapes is based on the integration of differential equations of the arch and generally leads to a transcendental equation of stability. The roots of these equations are the parameters of the critical loads. Each parameter of a critical load corresponds to the specific form of loss of stability. For practical purposes, only the first form of loss of stability and the corresponding smallest critical load is of interest.

Approximate method of solving stability problems for arches consists of approximating the arch by a framed structure, and proceeding with analysis by the three-moment equations [Sni66] and the displacement method [Kar10]. In general cases of arbitrary equations of the axial line of the arch, Smirnoff's matrix method [Smi84] is most effective. Peculiarities and limitations of this method are considered in Chap. 5.

In the case of a nonuniform arch of arbitrary shape, strictly speaking, it is possible to derive the corresponding differential equation, however, its order may be high. In the case of nonuniform arch, the coefficients are variable; it becomes impossible to present the solution in analytical form. Therefore, in most cases the solution may be obtained using only approximate methods, in particular, variational ones [Vol67]. Many important solutions related to stability of arches have been obtained by Dinnik [Din46]. In case of shallow arches, the solution of differential equation of stability may be obtained in closed form. However, the solution becomes very cumbersome. For the analysis of very shallow arches, it is preferable to utilize the Bubnov–Galerkin method [Rzh55].

Unless stated otherwise, we consider stability of arches under the following assumptions:

1. The arch material is linearly elastic (Hooke's law applies)
2. The center line of an arch is incompressible.
3. Loads on arches are conservative.

4.2 Circular Arches Subjected to Radial Load

This section is devoted to analysis of stability of the uniform circular arch with different boundary conditions. In all cases, arches are subjected to uniform distributed radial load. Analysis of stability is based on the Boussinesq's and Lamb's equations. Analytical solutions for critical load are presented.

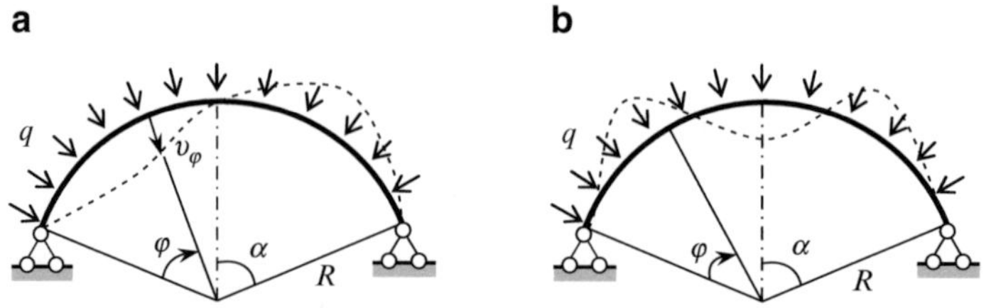

Fig. 4.4 Design diagram of two-hinged arch and the forms of the lost of stability

4.2.1 Solution Based on the Boussinesq's Equation

Behavior of the uniform circular arch of radius R and flexural rigidity EI is described by differential equation of the second order with respect to radial displacement υ of any point on the axis of the arch

$$\frac{d^2\upsilon}{d\varphi^2} + \upsilon = -\frac{MR^2}{EI}, \tag{4.1}$$

where υ is a displacement point of the arch in radial direction, and M is bending moment which is produced in the cross sections of the arch when it loss a stability [Pro48], [Bou1883].

This equation allows us to determine in the close form a critical load on the arch with different boundary conditions. The following procedure may be proposed:

1. Find analytical expression for bending moment M
2. Integrate the differential equation (4.1)
3. Use the boundary conditions to compute the constants of integration
4. Nontrivial solution presents the stability equation

4.2.1.1 Two-Hinged Arch

Circular two-hinged arch of radius R and constant cross-section is subjected to uniform radial pressure q. The central angle of the arch is 2α and flexural rigidity is EI. Antisymmetric and symmetric forms of the loss of stability of the arch are shown in Fig. 4.4 by dotted lines.

Antisymmetric buckling mode gives smaller values of critical load than symmetrical form. Therefore, we limit ourselves to the antisymmetric form.

In case of uniform radial load at any section of the arch only axial force $N = qR$ arises. Bending moment at any section is $M = N\upsilon = qR\upsilon$. Boussinesq's equation (4.1) becomes

$$\frac{d^2\upsilon}{d\varphi^2} + \upsilon = -\frac{q\upsilon R^3}{EI} \quad \text{or} \quad \frac{d^2\upsilon}{d\varphi^2} + n^2\upsilon = 0, \tag{4.2}$$

where

$$n^2 = 1 + \frac{qR^3}{EI}. \tag{4.3}$$

A homogeneous second-order linear differential equation (4.3) has the following general solution:

$$\upsilon = A \cos n\varphi + B \sin n\varphi,$$

where A and B are constants of integration. These constants should be evaluated using boundary conditions.

1. When $\varphi = 0$ (left support) the radial displacement $\upsilon = 0$. This condition leads to $A = 0$.
2. At $\varphi = \alpha$ (section at the axis of symmetry) $\upsilon = 0$. This condition leads to equation $B \sin n\alpha = 0$. Solution $B = 0$ is trivial and should be rejected. Thus, the stability equation becomes $\sin n\alpha = 0$. The roots of this equation are $n\alpha = \pi, 2\pi, \ldots, k\pi$.

The minimum critical parameter $n = \pi/a$ corresponds to smallest critical load. According to (4.3)

$$q_{\mathrm{cr}}^{\min} = \left(n^2 - 1\right)\frac{EI}{R^3} = \left(\frac{\pi^2}{\alpha^2} - 1\right)\frac{EI}{R^3}. \tag{4.4}$$

Axial critical force for two-hinged uniform arch becomes [Kis80], [Rzh55]

$$N_{\mathrm{cr}}^{\min} = q_{\mathrm{cr}}^{\min} R = \left(\frac{\pi^2}{\alpha^2} - 1\right)\frac{EI}{R^2}.$$

Special Cases

1. In case of semicircular arch the expression (4.4) leads to Levy' formula [Lev84] for critical load

$$q_{\mathrm{cr}}^{\min} = \frac{3EI}{R^3}.$$

Problems of this nature arise during analysis of a cylindrical shells enforced by circular ribs. Central angle $2\alpha = \pi$ corresponds to an arc located between two points of inflection of elastic curve of the ring. Circular arches (ribs) function as receivers of radial forces from the connected cylindrical shell.

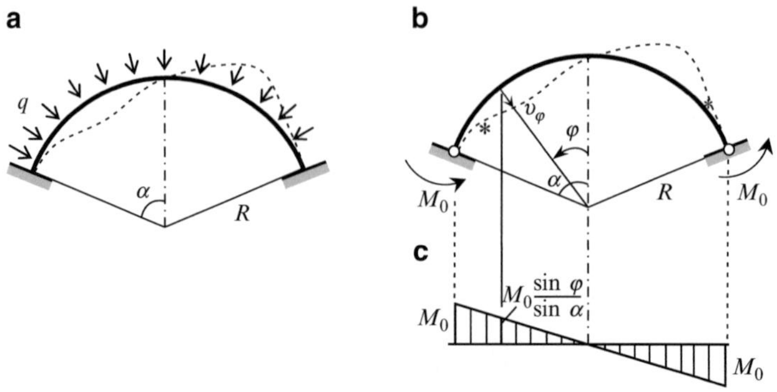

Fig. 4.5 Design diagram of hingedless arch, form of the stability lost, and bending moment diagram

2. For shallow arches with $f/l \le 0.2$–0.3 the angle $\alpha \ll \pi$, therefore $\pi^2/\alpha^2 \gg 1$ and the unity can be neglected, then $N_{cr}^{min} \cong \pi^2 EI/S_0^2$ where $S_0 = R\alpha$. Thus, the critical axial force for the gentle arch is approximately equal to that of the force found by Euler's formula for column which has a buckling length of half the length of the axis of the arch.

Note, that stability analysis of a two hinged uniform arch may be effectively performed using (1.27). In case of a uniformly distributed radial load, the two last terms of (1.27) should be omitted [Rzh82].

4.2.1.2 Hingeless Arch

As in case of two-hinged arch the most probable loss of stability realization will occur in an antisymmetric form (Fig. 4.5a) [Kis80].

In contrast to the two-hinged arch, the bending moments M_0 arise at the support sections. They lead to the appearance of inflection points. These points are specified by sign $^{(*)}$ (Fig. 4.5b). Bending moment diagram caused by support moments M_0 is shown in Fig. 4.5c. Additional bending moment at the arbitrary section is $M_0(\sin\varphi/\sin\alpha)$, where the angle φ is measured from the axis of symmetry.

The total bending moment at any section, which is characterized by displacement υ, is

$$M = qR\upsilon - M_0\frac{\sin\varphi}{\sin\alpha}.$$

Boussinesq's equation becomes

$$\frac{d^2\upsilon}{d\varphi^2} + n^2\upsilon = \frac{M_0R^2}{EI\,\sin\alpha}\sin\varphi = C\sin\varphi, \quad C = \frac{M_0R^2}{EI\,\sin\alpha}. \tag{4.5}$$

where n is still calculated by (4.3). Pay attention that C is unknown, since moment M_0 is unknown.

Solution of nonhomogeneous differential equation (4.5) is $\upsilon = \upsilon_1 + \upsilon_2$. The partial solution υ_2 should be presented in the form of the right part of (4.5), mainly $\upsilon_2 = k \sin \varphi$. Substituting this expression into (4.5) leads to formula

$$k = \frac{1}{n^2 - 1} \times \frac{M_0 R^2}{EI \sin \alpha} = \frac{C}{n^2 - 1}.$$

Thus the total solution of (4.5) becomes

$$\upsilon = A \cos n\varphi + B \sin n\varphi + \frac{1}{n^2 - 1} \times \frac{M_0 R^2}{EI \sin \alpha} \sin \varphi.$$

Boundary condition:

1. For point of the arch on the axis of symmetry ($\varphi = 0$) the radial displacement $\upsilon = 0$ (because the antisymmetric form of the loss of stability). This condition leads to $A = 0$.

   Given this, equations for radial displacement and slope are

$$\upsilon = B \sin n\varphi + \frac{C}{n^2 - 1} \sin \varphi,$$

$$\frac{d\upsilon}{d\varphi} = Bn \cos n\varphi + \frac{C}{n^2 - 1} \cos \varphi. \tag{4.6}$$

   Two unknowns B and C may be obtained from the following boundary conditions:

2. At $\varphi = \alpha$ (support point) the radial displacement $\upsilon = 0$.
3. At $\varphi = \alpha$ $d\upsilon/d\varphi = 0$.

Taking into account these conditions, the system of equations (4.6) becomes

$$B \sin n\alpha + \frac{C}{n^2 - 1} \sin \alpha = 0,$$
$$Bn \cos n\alpha + \frac{C}{n^2 - 1} \cos \alpha = 0.$$

Since these equations are algebraic and homogeneous, then nontrivial solution are possible if determinant of the matrix consisting of coefficients of the unknowns is equal to zero. After simplification we get the equation of critical loads

$$\begin{vmatrix} \sin n\alpha & \dfrac{1}{n^2 - 1} \sin \alpha \\ n \cos n\alpha & \dfrac{1}{n^2 - 1} \cos \alpha \end{vmatrix} = 0.$$

Table 4.1 Minimum root n of transcendental equation (4.7)

α (°)	30	45	60	90
n	8.621	5.782	4.375	3.000

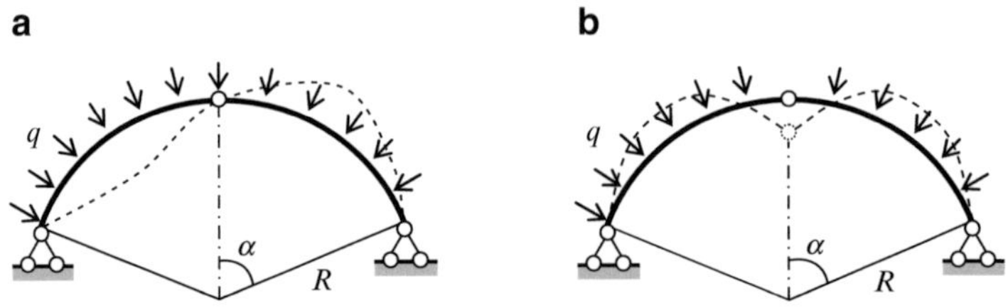

Fig. 4.6 Design diagram of three-hinged arch and the loss of stability forms

This equation may be presented in different forms, in particular

$$\frac{\alpha}{\tan\alpha} = \frac{n\alpha}{\tan n\alpha}. \tag{4.7}$$

Transcendental equation (4.7) allows us to calculate parameter n of critical load; they are for different angles α, which are presented in Table 4.1.

According to (4.3) the critical load is

$$q_{\mathrm{cr}} = \left(n^2 - 1\right)\frac{EI}{R^3}. \tag{4.8}$$

Corresponding compressed axial force is $N_{\mathrm{cr}} = q_{\mathrm{cr}}R = (EI/R^2)(n^2 - 1)$.

Special Case

If $\alpha = \pi/2$ (semicircular arch) then (4.7) can be presented in form $\cot(n\pi/2) = 0$, so $(n\pi/2) = (\pi/2), (3\pi/2), \ldots$. Solution $n = 1$ is trivial because this solution, accordingly (4.3), corresponds to $q = 0$. Thus, smallest root is $n = 3$. Thus, for semicircular arch with clamped supports the radial critical load equals $q_{\mathrm{cr\,min}} = 8(EI/R^3)$ [Rzh55].

Three-Hinged Arch

If loss of stability occurs by antisymmetric form (Fig. 4.6a) then a critical load will be same as for two-hinged arch. This happens because in case of two-hinged arch by antisymmetric loss of stability form (Fig. 4.4a) in the crown C the bending moment equals to zero. So this point may be treated as the hinge of three-hinged arch [Kis80].

Table 4.2 Circular three-hinged arch. Critical parameter η for symmetrical form of the loss of stability

α (°)	30	45	60	90
η	1.3872	1.4172	1.4584	$\pi/2 = 1.5708$

Table 4.3 Parameter K for critical radial load of circular uniform arches with different boundary conditions [Din46], [Kle72], [Sni66]

Types of arch		$\alpha = 15°$	30°	45°	60°	75°	90°
Hingeless	First form	294	73.3	32.4	18.1	11.6	8
	Second form	484	120	53.2	29.7	18.8	12.9
Two-hinged	First form	143	35	15	8	4.76	3
	Second form	320	79.2	34.7	19.1	11.9	8.0
Three-hinged (symmetrical form)		108	27.6	12	6.75	4.32	3
One-hinged arch		162	40.2	17.4	10.2	–	4.61

Symmetric loss of stability is shown in Fig. 4.6b. In this case, the hinge at the crown has a vertical displacement. Calculations and experiments showed that a smallest critical load corresponds to symmetrical loss of stability [Pav66], [Mor40]. Critical load is determined by the formula

$$q_{\mathrm{cr}} = \left[\left(\frac{2\eta}{\alpha}\right)^2 - 1\right]\frac{EI}{R^3}, \tag{4.9}$$

where a critical parameter η is a root of transcendental equation [Din46], [Kis80]

$$\frac{4(\tan\alpha - \alpha)}{\alpha^3} = \frac{\tan\eta - \eta}{\eta^3}.$$

The roots of this equation for different angle α are presented in Table 4.2.

Expression for radial critical loads (4.4), (4.8), and (4.9) for different types of arches may be combined using general formula $q_{\mathrm{cr}} = K(EI/R^3)$. Coefficients K are presented in Table 4.3.

Axial compressed force in all cases equals

$$N_{\mathrm{cr}} = q_{\mathrm{cr}}R = K\frac{EI}{R^2}.$$

The critical stress is $\sigma_{\mathrm{cr}} = N_{\mathrm{cr}}/A$ where A is area of the cross section of the arch.

All calculations above are legitimate if the critical stresses are less than the yield stress [Kle72].

4.2.2 Solution Based on the Lamb's Equation

This equation allows us to find the critical radial load for circular arches with different boundary conditions.

Behavior of the arch of radius R, with central angle 2α and a constant cross section subjected to uniform radial load q is described by Lamb's equation with respect to tangential displacement u [Lam1888]

$$\frac{\mathrm{d}^6 u}{\mathrm{d}\varphi^6}+2\frac{\mathrm{d}^4 u}{\mathrm{d}\varphi^4}+\frac{\mathrm{d}^2 u}{\mathrm{d}\varphi^2}+\frac{qR^3}{EI}\left(\frac{\mathrm{d}^4 u}{\mathrm{d}\varphi^4}+\frac{\mathrm{d}^2 u}{\mathrm{d}\varphi^2}\right)=0. \tag{4.10}$$

where EI is a flexural rigidity of the arch.

Two approaches of finding the critical load are presented below.

Exact Solution

Characteristic equation and corresponding roots are

$$\lambda^6+2\lambda^4+\lambda^2+k\left(\lambda^4+\lambda^2\right)=0, \quad k=\frac{qR^3}{\mathrm{EI}},$$

$$\lambda_{1,2}=0, \quad \lambda_{3,4}=\pm i, \quad \lambda_{5,6}=\pm i\beta, \quad \beta=\sqrt{1+k}.$$

General solution is

$$u=C_1+C_2\varphi+C_3\sin\varphi+C_4\cos\varphi+C_5\sin\beta\varphi+C_6\cos\beta\varphi.$$

Since the axis of the arch is incompressible, then $\varepsilon=(\mathrm{d}u/R\,\mathrm{d}\varphi)-(\upsilon/R)=0$, therefore

$$\upsilon=\frac{\mathrm{d}u}{\mathrm{d}\varphi}=C_2+C_3\cos\varphi-C_4\sin\varphi+C_5\beta\cos\beta\varphi-C_6\beta\sin\beta\varphi.$$

4.2.2.1 Two-Hinged Arch

In this case, the boundary conditions at $\varphi=\pm\alpha$ are

$$u=0, \quad u'=\upsilon=0, \quad u'''=0.$$

The last condition means that bending moment $M=0$. Now we consider two forms of the loss of stability separately.

Symmetric Form of the Loss of Stability

In this case, the radial displacement υ is even function, so the tangential displacement u should be presented in terms of odd functions

$$u=C_2\varphi+C_3\sin\varphi+C_5\sin\beta\varphi.$$

We subject this solution to the boundary conditions. This yields the set of homogeneous algebraic equations

$$\begin{aligned} C_2\alpha + C_3 \sin\alpha + C_5 \sin\beta\alpha &= 0, \\ C_2 + C_3 \cos\alpha + C_5\beta \cos\beta\alpha &= 0, \\ -C_3 \cos\alpha - C_5\beta^3 \cos\beta\alpha &= 0. \end{aligned}$$

Equation for critical load becomes

$$\begin{vmatrix} \alpha & \sin\alpha & \sin\beta\alpha \\ 1 & \cos\alpha & \beta\cos\beta\alpha \\ 0 & -\cos\alpha & -\beta^3\cos\beta\alpha \end{vmatrix} = 0.$$

In expanded form we get

$$\alpha\left(\beta - \beta^3\right) + \beta^3 \tan\alpha = \tan\beta\alpha.$$

Antisymmetric Form of the Loss of Stability

In this case, tangential displacement u should be presented in terms of even functions

$$u = C_1 + C_4 \cos\varphi + C_6 \cos\beta\varphi.$$

According to boundary conditions

$$\begin{aligned} C_1 + C_4 \cos\alpha + C_6 \cos\beta\alpha &= 0, \\ -C_4 \sin\alpha - C_6\beta \sin\beta\alpha &= 0, \\ C_4 \sin\alpha + C_6\beta^3 \sin\beta\alpha &= 0. \end{aligned}$$

Equation for critical load becomes

$$\begin{vmatrix} 1 & \cos\alpha & \cos\beta\alpha \\ 0 & -\sin\alpha & -\beta\sin\beta\alpha \\ 0 & \sin\alpha & \beta^3\sin\beta\alpha \end{vmatrix} = 0.$$

In expanded form we get $\left(\beta - \beta^3\right)\sin\alpha \sin\beta\alpha = 0$.
Two cases are possible.

1. Assume $\sin\alpha = 0$. In this case $\alpha = n\pi$. Since α cannot be more than π then only $n = 1$ may be considered. However, in this case we get a closed ring with one hinged support, i.e., a geometrically changeable system.

2. Let $\sin \beta\alpha = 0$. In this case $\beta\alpha = m\pi$, $\beta^2 = m^2\pi^2/\alpha^2$, critical parameter $k = \beta^2 - 1$, so critical load

$$q_{cr} = \left(\frac{m^2\pi^2}{\alpha^2} - 1\right)\frac{EI}{R^3}.$$

For semicircular arch $(2\alpha = \pi)$ we get $q_{cr\ min} = 3(EI/R^3)$.

4.2.2.2 Hingeless Arch

In this case, the boundary conditions at $\varphi = \pm\alpha$ are

$$u = 0, \quad u' = \upsilon = 0, \quad u'' = 0.$$

Antisymmetric Form of the Loss of Stability

In this case

$$u = C_1 + C_4 \cos \varphi + C_6 \cos \beta\varphi.$$

Proceeding as before, we subject this expression to the boundary conditions. Equation for critical parameter β becomes $\beta \tan \alpha = \tan \beta\alpha$. Critical load is $q_{cr} = (\beta^2 - 1)(EI/R^3)$. In case $\alpha = \pi/2$ stability equation can be presented in the form $\cot(\beta\pi/2) = 0$, so $\beta\pi = \pi, \quad 3\pi, \ldots$. Solution $\beta = 1$ is trivial because this solution corresponds to $q = 0$. Thus, the minimum root is $\beta = 3$, so for a semicircular arch with clamped supports the critical load equals $q_{cr\ min} = 8(EI/R^3)$.

Variational Approach

Let us consider two-hinged arch. Limiting ourselves to the approximate solution, set

$$u(\varphi) = -f\frac{\alpha}{\pi}\left(1 + \cos\frac{\pi\varphi}{\alpha}\right).$$

This expression satisfies all the boundary conditions. Namely, the tangential and radial displacements and bending moment at the supports are equal to zero.

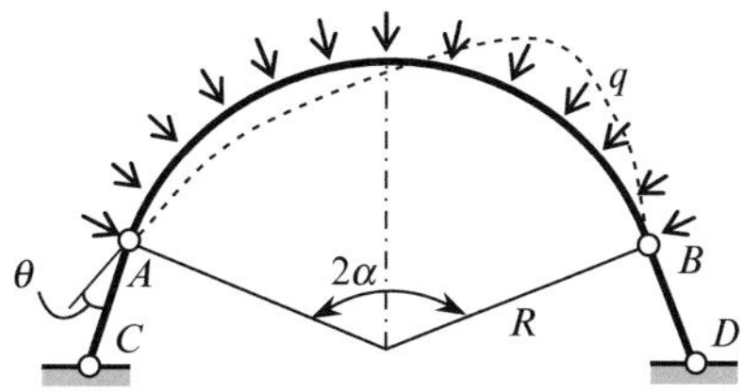

Fig. 4.7 Circular arch with hinged ends support rods

Bubnov–Galerkin procedure [Vol67]

$$\int_{-\alpha}^{\alpha}\left[\frac{\mathrm{d}^6u}{\mathrm{d}\varphi^6}+2\frac{\mathrm{d}^4u}{\mathrm{d}\varphi^4}+\frac{\mathrm{d}^2u}{\mathrm{d}\varphi^2}+\frac{qR^3}{EI}\left(\frac{\mathrm{d}^4u}{\mathrm{d}\varphi^4}+\frac{\mathrm{d}^2u}{\mathrm{d}\varphi^2}\right)\right]\left(1+\cos\frac{\pi\varphi}{\alpha}\right)\mathrm{d}\varphi=0.$$

As a result, we get the following relationships

$$-\left(\xi^6-2\xi^4+\xi^2\right)+\frac{qR^3}{EI}\left(\xi^4-\xi^2\right),\quad \xi=\frac{\pi}{\alpha}.$$

From here we immediately obtain expression for smallest critical load

$$q_{\mathrm{cr\,min}}=C\frac{EI}{R^3},\quad C=\frac{\left(\xi^3-\xi\right)^2}{\xi^4-\xi^2}. \tag{4.11}$$

For semicircular arch $(2\alpha=\pi)$ we get $q_{\mathrm{cr\ min}}=3EI/R^3$, which coincides with the exact solution.

If a central angle of the arch is $2\alpha=\pi/3$, then $C=35$.

4.2.3 Arch with Specific Boundary Conditions

Circular uniform arch of radius R and central angle 2α is subjected to uniform radial load q. Support constraints are straight rods AB and BD with hinges at the ends. The reaction which arises at the constraint may be resolved into the vertical and horizontal (thrust) components at the points A and B. Thus, this structure in fact represents a specific two-hinged arch. Indeed, inclination of the support constraint defines the direction of the resultant of the vertical reaction and the thrust for any type of load. Assume these rods are directed along the tangent to the arch at the support points (Fig. 4.7).

Therefore, before the loss of stability of the arch, reactions at the points A and B coincide with this tangent. At the instant the loss of stability occurs, the axial compressed force $N=qR$, at the points A and B will be directed at any angle θ to the deformable axis of the arch.

Table 4.4 Critical parameter β in terms of half of central angle α

α (rad)	0.2	0.4	0.6	0.8	1.0	1.2	1.4	$\pi/2$
β	1.58	1.61	1.66	1.73	1.81	1.90	2.01	2.11

Exact solution of the stability problem is obtained on the basis of (4.10). In the case of symmetrical loss of stability at $\varphi = \pm\alpha$, the boundary conditions (tangential displacement u, bending moment and shear) are $u = 0;\quad M = 0;\quad Q = qR\theta$. These conditions, in terms of displacement u may be presented as follows:

$$u = 0; \qquad u''' + u' = 0; \qquad u^{\mathrm{IV}} + u'' + \frac{qR^3}{EI}(u'' + u) = 0.$$

If we denote $(qR^3/EI) = \beta^2 - 1$, then for stability equation and corresponding critical load we get [Rzh55]

$$\frac{\tan \beta\alpha}{\beta\alpha} = 1 - \beta^2,$$

$$q_{\mathrm{cr}} = \left(\beta^2 - 1\right)\frac{EI}{R^3}.$$

Critical parameter β, in terms of half of central angle is presented in Table 4.4.

For arch with these specific supports a smallest critical load corresponds to symmetrical form of loss of stability. In case of antisymmetrical form of buckling the critical load coincides with critical load for two-hinged arch, i.e., $q_{\mathrm{cr}} = ((\pi^2/\alpha^2) - 1)(EI/R^3)$.

4.3 Circular Arches with Elastic Supports

This section is devoted to stability analysis of circular arches with elastic supports. The case of uniform radial load is considered. Stability equation is derived and classical boundary conditions are considered.

Stability analysis of a complex arched structure (supports of the arch are deformable frames) is presented.

4.3.1 *General Solution and Special Cases*

Assume that symmetrical circular uniform arch of radius R has the elastic-fixed supports; their rotational stiffness coefficient is k [kN m/rad]. The central angle of the arch is 2α; the flexural rigidity is EI, and the intensity of the radial uniformly distributed load is q (Fig. 4.8a).

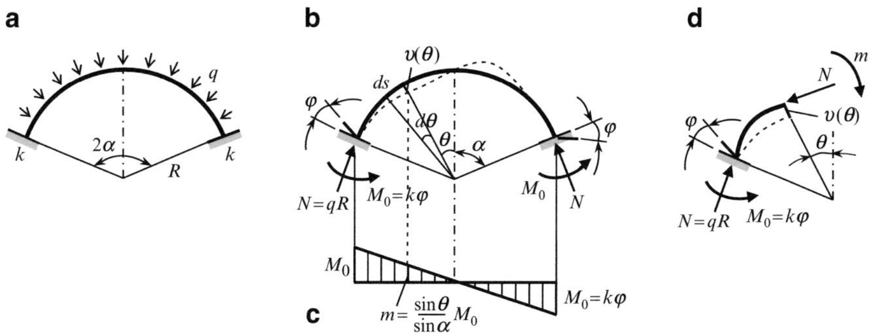

Fig. 4.8 (**a**) Design diagram of the circular arch with elastic supports; (**b**) reactions and antisymmetrical buckling form; (**c**) distribution of bending moments caused by two reactive moments M_0; (**d**) free-body diagram for portion of the arch (load q is not shown)

In case of arch with elastic supports, symmetrical and antisymmetrical forms mean that both supports rotate in the opposite directions or in the same directions, respectively. As it is shown by analytical analysis and experiments, the smallest critical load for hingeless and two-hinged arch corresponds to antisymmetrical form of the loss of stability.

For stability of the arch with elastic supports, we use the Boussinesq's equation

$$\frac{d^2\upsilon}{d\varphi^2} + \upsilon = -\frac{MR^2}{EI}, \tag{4.12}$$

where υ is a displacement point of the arch in radial direction (Fig. 4.8b).

It is easy to show that the horizontal and vertical components of reaction N are $H = qR\cos\alpha$ and $V = qR\sin\alpha$, so the axial compressive force of the arch caused by uniformly distributed hydrostatic load q is $N = qR$ [Kar10]. The slope φ at the elastic support and corresponding reactive moment M_0 are related as $M_0 = k\varphi$. Distribution of bending moments caused by two antisymmetrical angular displacements φ of elastic supports (or reactive moments M_0) is presented in Fig. 4.8c. Bending moment at section with central angle θ caused by only reactive moments M_0 equals m. The total bending moment at any section, which is characterized by displacement υ free-body diagram is shown in Fig. 4.8d, equals

$$M = qR\upsilon - \frac{\sin\theta}{\sin\alpha}k\varphi.$$

The second term takes into account additional moment due to elastic supports.

Thus, differential equation (4.12) becomes

$$\frac{d^2\upsilon}{d\theta^2} + \left(1 + \frac{qR^3}{EI}\right)\upsilon = \frac{k\varphi}{EI}\frac{R^2}{\sin\alpha}\sin\theta. \quad (4.13)$$

Denote

$$n^2 = 1 + \frac{qR^3}{EI} \quad (4.13a)$$

$$C = \frac{k\varphi R^2}{EI\sin\alpha} \quad (4.13b)$$

Pay attention that C is unknown, since the angles of rotation φ of the supports are unknown. Differential equation (4.13) may be rewritten as follows:

$$\frac{d^2\upsilon}{d\theta^2} + n^2\upsilon = C\sin\theta. \quad (4.14)$$

Solution of this equation is

$$\upsilon = A\cos n\theta + B\sin n\theta + \upsilon^*. \quad (4.15)$$

The partial solution υ^* should be presented in the form of the right part of (4.14), mainly $\upsilon^* = C_0\sin\theta$, where C_0 is a new unknown coefficient. Substituting of this expression into (4.14) leads to equation

$$-C_0\sin\theta + n^2C_0\sin\theta = C\sin\theta,$$

so $C_0 = C/(n^2 - 1)$. Thus the solution of (4.14) becomes

$$\upsilon = A\cos n\theta + B\sin n\theta + \frac{C}{n^2 - 1}\sin\theta. \quad (4.16)$$

Unknown coefficients A, B, and C may be obtained from the following boundary conditions:

1. For point of the arch on the axis of symmetry ($\theta = 0$), the radial displacement $\upsilon = 0$ (because the antisymmetrical form of the loss of stability); this condition leads to $A = 0$.
2. For point of the arch at the support ($\theta = \alpha$), the radial displacement is $\upsilon = 0$, so

$$B\sin n\alpha + \frac{C}{n^2 - 1}\sin\alpha = 0. \quad (4.17)$$

Table 4.5 Critical parameter n for circular arch with elastic clamped supports, $2\alpha = 60°$

k_0	0.0	1.0	10	100	1,000	10^5
n	6.0000	6.2955	7.5294	8.4628	8.6051	8.621

3. Using (4.17), the slope is

$$\frac{dv}{d\theta} = Bn \cos n\theta + \frac{C \cos \theta}{n^2 - 1}. \tag{4.18}$$

The slope at the support is $dv/ds = -\varphi$, the negative sign means the reactive moment M_0 and angle φ have the opposite directions. On the other hand $dv/ds = dv/R\,d\theta$, so $dv/d\theta = -R\varphi$. According to (4.13b) we get

$$\varphi = C\frac{EI \sin \alpha}{kR^2}, \quad \text{so} \quad \frac{dv}{d\theta} = -C\frac{EI \sin \alpha}{kR}$$

If $\theta = \alpha$ then the expression (4.18) becomes

$$Bn \cos n\alpha + C\left(\frac{\cos \alpha}{n^2 - 1} + \frac{EI \sin \alpha}{kR}\right) = 0. \tag{4.19}$$

Equations (4.17) and (4.19) are homogeneous linear algebraic equations with respect to unknown parameters B and C. The trivial solution $B = C = 0$ corresponds to state of the arch before the loss of stability. Nontrivial solution occurs if the following determinant is zero:

$$D = \begin{vmatrix} \sin n\alpha & \dfrac{\sin \alpha}{n^2 - 1} \\ n \cos n\alpha; & \dfrac{\cos \alpha}{n^2 - 1} + \dfrac{EI \sin \alpha}{kR} \end{vmatrix} = 0. \tag{4.20}$$

This stability equation may be presented as follows [Kle72]:

$$\tan n\alpha = \frac{nk_0}{k_0 \cot \alpha + (n^2 - 1)}, \quad k_0 = \frac{kR}{EI}. \tag{4.21}$$

Solution of this transcendental equation for given α and dimensionless parameter k_0 is the critical parameter n. According to (4.13a) the critical load

$$q_{cr} = \left(n^2 - 1\right)\frac{EI}{R^3}.$$

If the central angle $2\alpha = 60°$, then the roots of (4.21) for different k_0 are presented in Table 4.5.

Parameter $k_0 = 10^5$ in fact corresponds to the fixed support.

For an arch with elastic supports, parameter k satisfy to condition $k_1 \le k \le k_2$, where $k_1 = 0$ and $k_2 = \infty$ related for two-hinged and hingeless arches, respectively.

Limiting Cases

1. *Two-hinged arch.* In this case, the stiffness of support $k = 0$ and stability equation (4.21) is $\tan n\alpha = 0$. The minimum roots of this equation is $n\alpha = \pi$, so $n = \pi/\alpha$ and corresponding critical load equals

$$q_{\text{cr min}} = \left(\frac{\pi^2}{\alpha^2} - 1\right)\frac{EI}{R^3}.$$

   This formula coincides with (4.4). Critical load for $\alpha = \pi/2$ (semicircular arch) equals $q_{\text{cr min}} = 3EI/R^3$.
2. *Arch with fixed supports.* In this case, the stiffness $k = \infty$ and stability equation (4.21) becomes

$$\tan n\alpha = n \tan \alpha$$

   This equation coincides with (4.7). The minimum roots in terms of α are presented in Table 4.1.

Two Cases of the Ring with Braces (Circular Tunnel Lining)

1. The ring of radius R with one brace in the diameter is subjected to uniform radial load q. Each half of the ring may be considered as semicircular arch with elastic fixed ends. If flexural rigidity of the arch and the cross-bar are I_0 and I_1, respectively, then the stability equation becomes

$$\cot\frac{n\pi}{2} = \frac{2(n^2-1)}{3n}\frac{I_0}{I_1}, \quad n = \sqrt{1 + \frac{qR^3}{EI_0}}.$$

   If $I_1 = 4I_0$, then $q_{\text{cr}} = 6.6(EI_0/R^3)$ [Pro48].
2. The ring of radius R with two braces in the diameters is subjected to the same loading. Braces are perpendicular to each other and connected in the center of the ring by a fixed joint or by a multiple hinge. We assume that $I_1 = 2I_0$. In both cases the stability equation is

$$\cot\frac{n\pi}{4} = \frac{n^2+2}{3n} \quad \text{and} \quad q_{\text{cr}} = 20.9\frac{EI_0}{R^3}.$$

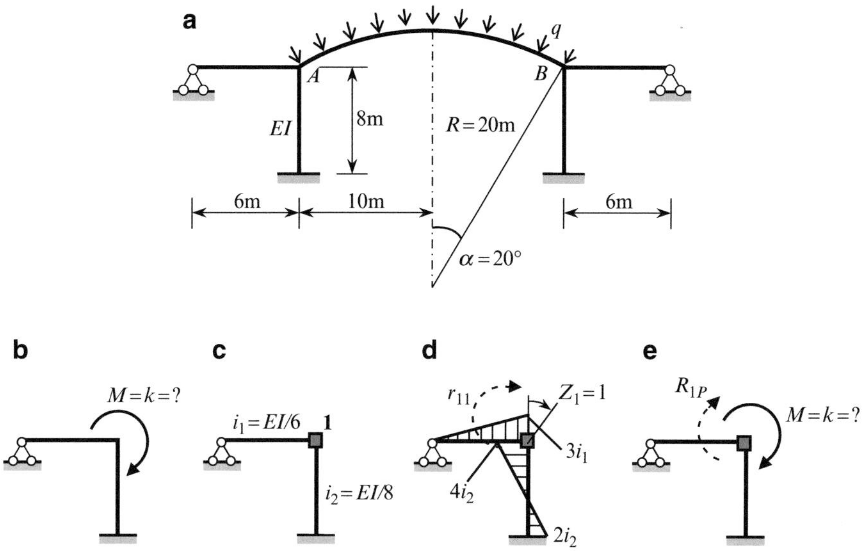

Fig. 4.9 (**a**) Design diagram of the structure. (**b**–**e**) Calculation of the stiffness k of the elastic supports of the arch

4.3.2 Complex Arched Structure

In practical engineering, the stiffness coefficient k of the elastic supports is not given as a clear value. However, in special cases, it can be determined from an analysis of adjacent parts of the arch [Kle72].

Let us calculate the critical load for a structure shown in Fig. 4.9a. The central part of the structure presents the circular arch; supports of the arch are rigid joints A and B of the frames. The arch is subjected to uniformly distributed radial load q. Assume that $R = 20$ m and the central angle $2\alpha = 60^\circ$. The stiffness of all members of the structure is EI.

Since the left and right frames are deformable structures, then the each joint A and B has some angle of rotation, so the arch AB should be considered as arch with elastic supports with rotational stiffness k. For this case of circular arch with given type of load, the stability equation according to (4.21) becomes:

$$\tan n\alpha = \frac{n}{\cot \alpha + \dfrac{(n^2 - 1)EI}{kR}}. \tag{4.23}$$

Rotational stiffness coefficient k is a couple M, which arises at elastic support of the arch if this support rotates through the angle $\varphi = 1$. Since the joints A and B are rigid, so the angle of rotation for frame and arch are same. Therefore, for calculation of the stiffness k we have to calculate the couple M, which should be applied at the rigid joint A of the frame in order to rotate this joint by angle $\varphi = 1$.

The frame subjected to unknown moment $M = k$ is shown in Fig. 4.9b. For solving of this problem we can use the displacement method [Kar10].

The primary system of the displacement method is obtained by introducing additional constraint 1 (Fig. 4.9c). The primary unknown Z_1 is angular displacement of introduced constraint. Canonical equation is

$$r_{11}Z_1 + R_{1P} = 0.$$

Displacement $Z_1 = 1$ and corresponding bending moment diagram is shown in Fig. 4.8d. The unit reaction

$$r_{11} = 3i_1 + 4i_2 = 0.5EI + 0.5EI = 1EI.$$

The primary system subjected to external unknown couple M is presented in Fig. 4.9e, so $R_{1P} = -M$. The canonical equation becomes $1EI \times Z_1 - M = 0$. If the angle of rotation $Z_1 = 1$, then $M = k = 1EI$. For given parameters R and α, the stability equation (4.23) of the structure becomes

$$\tan n\frac{\pi}{6} = \frac{n}{\cot\frac{\pi}{6} + \frac{(n^2-1)}{20}} \quad \text{or} \quad \tan(0.5236n) = \frac{20n}{33.64 + n^2}.$$

The root of this equation $n = 7.955$. The critical load is

$$q_{cr} = \left(7.955^2 - 1\right)\frac{EI}{R^3} = 62.28\frac{EI}{R^3}.$$

According to Table 4.3, the critical load for arch with fixed supports and for two-hinged arch (the central angle in both cases is $2\alpha = 60°$) are $q_{cr} = 73.3(EI/R^3)$ and $q_{cr} = 35(EI/R^3)$, respectively. Above calculated critical load is located between two limiting cases.

4.4 Gentle Circular Arch Subjected to Radial Load

Earlier we considered arches with an incompressible axis. As a result, in case of two-hinged and hingeless arches the antisymmetrical form was the most unstable. In a gently circular arch ($f/l \leq 0.3$), we must consider the change in length of the arch's axis due to compression. This leads to a characteristically new form for loss of stability. In Fig. 4.10, the dotted line 1 corresponds to the loss of stability by symmetric form with one half-wave. Given this, the curvatures in the deformable and initial states have the same sign. Dotted line 2 also corresponds to the symmetric buckling with one half-wave. However, the curvatures in the deformable and initial states have the different signs. Antisymmetrical form of the loss of stability

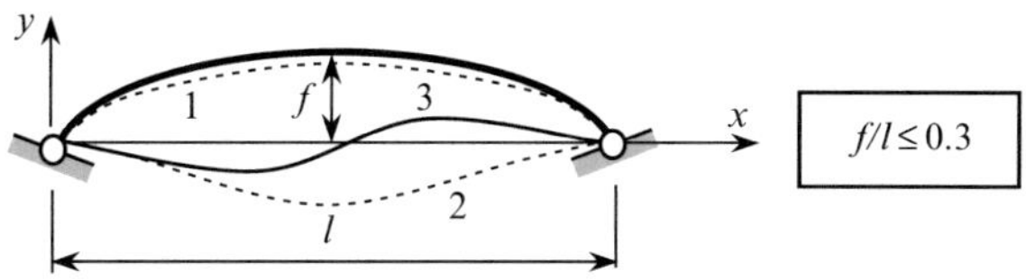

Fig. 4.10 Gentle arch and loss of stability forms

(3) with two half-waves is also possible. Structures that abruptly transform to state 2 (or 3) are called structures with a jump; this phenomenon is called *snap transformation.*

4.4.1 Mathematical Model and Bubnov–Galerkin Procedure

Let us consider a shallow circular uniform arch of the span l and radius R. Arch is subjected to radial uniformly distributed load q. Differential equation of stability is [Rzh55]

$$EI\frac{d^4\upsilon}{dx^4}+P\frac{d^2\upsilon}{dx^2}=q-\frac{P}{R}. \tag{4.24}$$

Here P is compressed load which arise in the cross-sections of the arch. A rigorous solution of this equation can be found in [Rzh55], however, such solution is very cumbersome. It is easier to solve this problem by Bubnov–Galerkin's method [Vol67].

The procedure for solving the problem is as follows:

1. Assume the radial displacement of the arch is $\upsilon(x)=\upsilon_0 f(x)$ which satisfies the boundary conditions. The assumed functions $\upsilon(x)$ for different boundary conditions are covered adequately in the book [Kar01].
2. According to Bubnov–Galerkin procedure,

$$\int_0^l\left(EI\frac{d^4\upsilon}{dx^4}+P\frac{d^2\upsilon}{dx^2}+\frac{P}{R}-q\right)f(x)\,dx=0.$$

   This procedure leads to the relationships which includes the unknown critical load q, radial displacement at the crown υ_0, and axial compressive load P.
3. The expression for difference in distances between the support points of the arch

$$\Delta l=-\frac{1}{R}\int_0^l\upsilon dx+\frac{1}{2}\int_0^l(\upsilon')^2dx+\frac{Pl}{EA},$$

   where A is the cross-sectional area of the arch.
4. Condition of intractability of supports $\Delta l=0$ yields to additional relationships between P and υ_0.

5. Eliminating P we obtain a relationships between q and υ_0.
6. The critical load q is found through the condition $\mathrm{d}q/\mathrm{d}\upsilon_0 = 0$.
   We show applications of this procedure for stability analysis of the different types of arches.

4.4.2 Two-Hinged Arch

Assume that the symmetrical form of the loss of stability may be described by formula $\upsilon(x) = \upsilon_0 \sin(\pi x/l)$ [Rzh55]. This expression satisfies to boundary conditions $\upsilon = 0$ and $\upsilon'' = 0$ at $x = 0$ and $x = l$.

Bubnov–Galerkin's procedure

$$\int_0^l \left(EI \frac{\mathrm{d}^4 \upsilon}{\mathrm{d}x^4} + P \frac{\mathrm{d}^2 \upsilon}{\mathrm{d}x^2} + \frac{P}{R} - q \right) \sin \frac{\pi x}{l} \, \mathrm{d}x = 0 \tag{4.25}$$

leads to relationships

$$q = \frac{\pi^3 \upsilon_0}{4l^4} \left(\pi^2 EI - Pl^2 \right) + \frac{P}{R}. \tag{4.26}$$

This equation includes unknowns critical load q, radial displacement at the crown υ_0, and axial compressive load P.

The difference in distances between the support points of the arch is

$$\Delta l = -\frac{\upsilon_0}{R} \frac{2l}{\pi} + \frac{\upsilon_0^2 \pi^2}{4l} + \frac{Pl}{EA}.$$

Equation of intractability of supports $\Delta l = 0$ yields

$$P = -\frac{\pi^2 \upsilon_0^2 \gamma}{4l} + \frac{2l \upsilon_0 \gamma}{\pi R}, \quad \gamma = \frac{EA}{l}.$$

We notice that the relationships between the displacement υ_0 and compressive force P in arch are nonlinear.

Substituting expression for P into (4.26) leads to

$$q = \frac{\pi^3 \upsilon_0}{4l^4} \left(\pi^2 EI + \frac{\pi^2 \upsilon_0^2 \gamma l}{4} - \frac{2 \upsilon_0 \gamma l^3}{\pi R} \right) - \frac{\pi^2 \upsilon_0^2 \gamma}{4lR} + \frac{2l \upsilon_0 \gamma}{\pi R^2}. \tag{4.27}$$

This expression relates the required critical load q and parameter υ_0 of deformable arch. The critical q is found from the condition $\mathrm{d}q/\mathrm{d}\upsilon_0 = 0$. From this

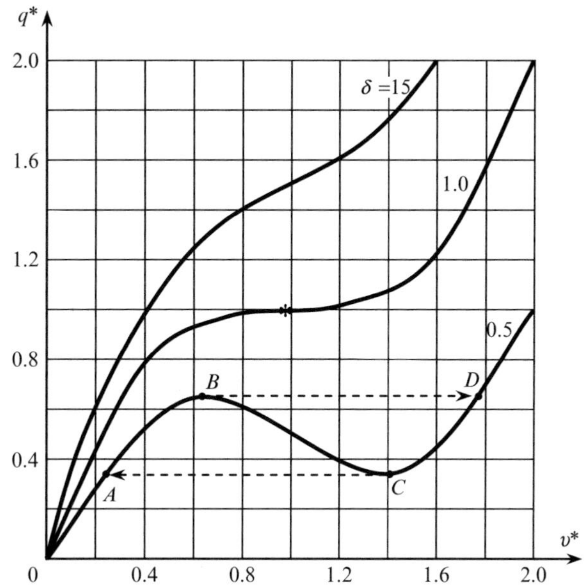

Fig. 4.11 Curves of equilibrium states for two-hinged arch

condition we get $\upsilon^* = 1 \pm \sqrt{(1-\delta)/3}$. Substituting of this expression into (4.27) leads to the final expression for the critical load

$$q_{\mathrm{cr}} = \frac{4l^2 EA}{\pi^4 R^3} q^*,$$

$$q^* = \delta \pm 2\left(\frac{1-\delta}{3}\right)^{3/2}; \quad \delta = \frac{\pi^6 i^2 R^2}{4l^4} = \frac{\pi^6}{4} \frac{IR^2}{Al^4}.$$

The dimensionless parameter δ in terms of central angle 2α and slenderness ratio of the arch $\lambda = l/i$ ($i = \sqrt{I/A}$ is the radius of gyration of the cross section of the arch) becomes $\delta = 240.4/(2\alpha)^2\lambda^2$.

4.4.3 Graphical Interpretation of Results

For two-hinged arch, the relationship between load $q^* = q(\pi^4 R^3/4l^2 EA)$ and radial displacement $\upsilon^* = \upsilon_0(\pi^3 R/4l^2)$ for special parameters δ is shown in Fig. 4.11. All parameters q^*, υ^*, δ are dimensionless.

All curves characterize the state of equilibrium. Curve $\delta = 1$ divides the state diagram into two regions. In the upper region $\delta > 1.0$ arch does not lose stability, but deforms smoothly when acted upon by a load. In the bottom region $\delta < 1.0$ loss of stability is possible. In this region state curve has two extreme points. The upper extreme point B corresponds to the critical state of the arch under a monotonal increase in the load. For this point in the expression for q^* we assign a positive sign.

After moving past this critical state the deformation of the arch increases by a jump under constant load. With this, the sign of initial curvature of the arch changes.

This state is shown by dotted arrow BD. The lower critical point C corresponds to the critical conditions which are achieved with a monotonal decrease of the load. Reaching the load which corresponds to point C a jump in the system occurs in the opposite direction. The dotted arrow CA corresponds to this jump. With this, the line BC corresponds to unstable equilibrium. This form of equilibrium may be realized only with additional constraints.

For $\delta = 1$ both critical loads are coincides. The inflection point (*) has coordinates $\upsilon^* = 1.0, \quad q^* = 1.0$.

Thus at $\delta > 1$ for two-hinged arch the jump does not occur, and the arch smoothly deforms under the load [Rzh55].

Notes

1. Since arch is gentle, then its shape and type of load (tracking or fixed direction) does not significantly influence the value of critical load [Uma72-73].
2. In case of very shallow arch, we should take into account the change in length of the arch caused by the compressive forces. Two-hinged sinusoidal arch $y = f \times (\sin(\pi x)/l)$ have been considered by Timoshenko [Kle72], and Dinnik [Din46] under the above assumptions. Parameter u of critical load, under which the jump occurs is $u = 1 + \sqrt{4(1-m)^3/27m^2}$, where $m = 4I/Af^2 < 1$, A-area of cross-section of an arch. In case of vertical uniformly distributed load q within all span l parameter $u = (5ql^4/384\mathrm{EI}) \times (1/f)$. In case of vertical force P at the crown we have $u = (Pl^3/48\mathrm{EI}) \times (1/f)$. If $m \geq 1$, then exists only unity stable form of equilibrium.

   A jump of shallow two-hinged arch of the arbitrary shape by antisymmetric form of the loss of stability with nodal point in the mid-span [Uma72-73] is possible. In this case, the critical load is $q_{\mathrm{cr}} = 32\pi^2(f/l)(EI/l^3)$. Detailed analysis of very gentle parabolic two-hinged arch is presented in [Din46].
3. The jump phenomena characterizes the shallow arch as nonlinear structure [Kaz09]. This nonlinearities turn out to be static type [Kar01].
4. In case of the arch of the variable cross-section instead of (4.24) should be considered equation

$$\frac{\mathrm{d}^2}{\mathrm{d}x^2}\left(EI\frac{\mathrm{d}^2\upsilon}{\mathrm{d}x^2}\right) + P\frac{\mathrm{d}^2\upsilon}{\mathrm{d}x^2} = q - \frac{P}{R}.$$

   Stability analysis of the such arch of any shape and arbitrary boundary conditions can be carried out by numerical methods.

4.4.4 Hingeless Arch

An approximation of the radial displacement of the arch has the form [Rzh55]

$$\upsilon = \frac{\upsilon_0}{2}\left(1 - \cos\frac{2\pi x}{l}\right).$$

This expression satisfies the boundary conditions at $\upsilon = 0$ and $\upsilon' = 0$ at $x = 0$ and $x = l$. Bubnov–Galerkin procedure

$$\int_0^l \left(EI\frac{d^4\upsilon}{dx^4} + P\frac{d^2\upsilon}{dx^2} + \frac{P}{R} - q\right)\left(1 - \cos\frac{2\pi x}{l}\right)dx = 0.$$

Next we apply the above-described algorithm, the result of which gives us the critical load [Rzh55]

$$q_{\mathrm{cr}} = \frac{l^2 EA}{64\pi^2 R^3} q^{**},$$

$$q^{**} = 3\delta_1 - 30 \pm \sqrt{\frac{19 - \delta_1}{3}\frac{2\delta_1 - 38}{3}}; \quad \delta_1 = \frac{64\pi^4 i^2 R^2}{l^4} = \frac{64\pi^4 IR^2}{Al^4}.$$

In terms of the central angle 2α and the slenderness of the arch $\lambda = l/i$, the dimensionless parameter δ_1 becomes $\delta_1 = 6,234/(2\alpha)^2\lambda^2$. At $\delta_1 > 19$ for hingeless arch the jump does not occur, and the arch smoothly deforms under the load [Rzh55].

The cases of shallow two-hinged and hingeless arches with elastic supports are considered in [Rzh55].

4.5 Parabolic Arch

This section is devoted to analysis of stability of the uniform parabolic arches with different boundary conditions. In all cases arches are subjected to vertical loads only. Analysis of stability for uniformly distributed load is based on the Dinnik's equation. Some important results of numerical solutions of critical load for typical arches are presented.

4.5.1 Dinnik's Equation

The parabolic uniform symmetrical arch of span l and rise f is subjected to vertical uniform distributed load q within span of an arch. Let the radius of curvature at the crown be R_0 and arbitrary section of the arch is defined by angle θ which is measured from the vertical. The radius for some section of the arch according to (A.2) is $\rho = R_0/\cos^3\theta$. Assume that during loading the load is tracking. Under these assumptions the stability of such arch with in-plane bending describes by Dinnik's equation [Din46]

$$\left(\frac{d^2}{d\theta^2}+1\right)\left(\cos^3\theta\frac{dM}{d\theta}\right)+\bar{q}\frac{d}{d\theta}\left(M\sec^4\theta\right)=0. \tag{4.28}$$

This is a homogeneous differential equation of third order with variable coefficients with respect to bending moment M which arises in the arch when the loss of stability occurs. Parameter of critical load $\bar{q} = qR_0^3/EI$. Dinnik's equation may be derived by two different ways: the first way is based on the Kirchhoff equations [Kir76], [Din46]. The second way is based on Lokshin's equation [Lok34], [Kle72], which is the general equation of the stability for plane curvilinear rod.

Analytical solutions of stability equation for parabolic arch are generally more complicated then the corresponding equation for circular arches. Stability problem of parabolic arches under the given assumptions reduces to the determining the lowest parameter $\bar{q}$, for which a nonzero solution of (4.28) for bending moment M may exist and satisfy the boundary conditions for M. Numerical procedure is described in [Din46]. For two-hinged arch the bending moment at support $M(\alpha) = 0$; we need to find the parameter qR_0^2/EI, under which the moment at the axis of symmetry is $M(0) = 0$. For hingeless arch we have the same condition with additional boundary conditions, which means that the angle of rotations of terminal cross-sections is zero.

Let us show some numerical results. Just as with radial load, for two-hinges and hingeless arch a smallest critical load corresponds antisymmetric form the loss of stability. Critical load may be calculated by formula

$$q_{\mathrm{cr}} = K\frac{EI}{l^3}. \tag{4.29}$$

Parameter K for parabolic uniform different types of arches is presented in Table 4.6. In case of three-hinged arch the coefficient K are presented for symmetric and antisymmetric form of the loss of stability.

Coefficient K for conservative load is shown in parenthesis [Uma72-73]. Dashes represent unavailable data.

For two-hinged and three-hinged arches at antisymmetric forms of the loss of stability coefficients K coincides. This fact was discussed earlier. Note that the coefficients K for the cases of nonuniform arches $(I(\varphi) = I_C/\cos^3\varphi$ and $I(\varphi) = I_C/\cos\varphi)$ subjected to tracking load, as well as for uniform arches subjected to gravity load are presented in [Mor73].

For three-hinged arch relationships stability coefficient K for symmetrical (1) and antisymmetrical (2) form of the loss of stability vs. dimensionless parameter f/l is presented in Fig. 4.12.

For very gentle arches ($f/l<0.3$) the symmetrical form of the loss of stability is realized.

Table 4.6 Parameter K for parabolic uniform arch [Kle72], [Mor73], [Sni66]

$\frac{f}{l}$	Hingeless arch antisymmetric form	Two-hinged arch antisymmetric form	One-hinged arch	Three-hinged arch: Antisymmetric form	Three-hinged arch: Symmetric form
0.1	60.7	28.5 (28.8)	33.8	See two-hinged arch (antisymmetric form)	22.5 (22.7)
0.2	101.0	45.4 (46.1)	59.0		39.6 (40.2)
0.3	115.0	46.5 (48.4)	84.0		47.3 (49.8)
0.4	111.0	43.9 (45.0)	96.0		49.2 (54.5)
0.5	97.4	38.4 (–)	87.0		43 (–)
0.6	83.8	30.5 (31.7)	80.0		38.0 (–)
0.8	59.1	20.0 (–)	63.0		28.8 (–)
1.0	43.7	14.7 (15.4)	48.0		22.1 (–)

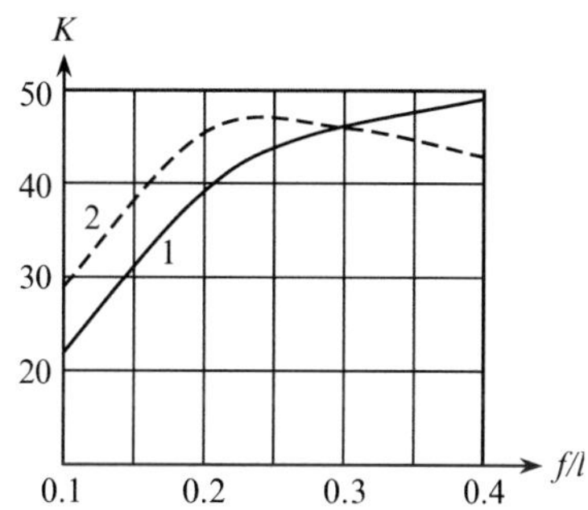

Fig. 4.12 Coefficient of stability K for three-hinged arch vs. parameter f/l

4.5.2 *Nonuniform Arches*

Let us consider arches with variable cross-section loaded by uniformly distributed load within the entire span. Assume $A = A_0/\cos\varphi$, where A_0 is area of the cross section of the arch at the crown. Two cases are considered:

1. *Rectangular cross-section of the arch has constant width.* In this case $h(\varphi) = h_0/\cos\theta$ and $I - I_0/\cos^3\theta$ where h_0 and I_0 are height and inertia moment of the cross section at the crown. Critical load may be calculated by formula $q_{\mathrm{cr}} = K_1(EI_0/l^3)$ where coefficient K_1 is presented in Table 4.7.
2. *Rectangular cross section of the arch has constant height.* In this case $b(\varphi) = b_0/\cos\theta$ and $I = I_0/\cos\theta$, where b_0 is a width of the cross section at the crown. Critical load may be calculated by formula $q_{\mathrm{cr}} = K_2(EI_0/l^3)$ where coefficient K_2 is presented in Table 4.8.

The book [Din46] has a list of differential equations for in-plane and out-of-plane stability problems for different shapes of the arches with variable cross section, different loading and corresponding critical loads.

Table 4.7 Coefficient K_1 $(I = I_0/\cos^3\theta)$ [Din46], [Kis80]

$\frac{f}{l}$	Type of the arch			
	Hingeless	One-hinged	Two-hinged	Three-hinged
0.1	65.5	36.5	30.7	24
0.2	134	75.8	59.8	51.2
0.3	204	–	81.1	81.1
0.4	277	187	101	118
0.6	444	332	142	–
0.8	587	497	170	–
1.0	700	697	193	–

Table 4.8 Coefficient K_2; $I = I_0/\cos\theta$ [Din46], [Kis80]

$\frac{f}{l}$	Type of the arch			
	Hingeless	One-hinged	Two-hinged	Three-hinged
0.1	62.3	34.3	29.5	23.2
0.2	112	70	49	43.6
0.3	–	94	–	59
0.4	154	115	57	68
0.6	152	139	52	70
0.8	133	140	44	–
1.0	118	133	37	–

4.5.3 Partial Loading

We present approximate formulas for critical loads for parabolic uniform arch loaded by partial uniformly distributed load and concentrated force at the crown. To derive these formulas, the elastic loads method was applied [Pro48].

1. Two-hinged arch of span l and rise f is subjected to simple load P at the crown (Fig. 4.13a). For such loading we have symmetrical bending but antisymmetrical form of the loss of stability. Let $m = f/l \le 0.2$. In this case the critical load [Pro48]

$$P_{\text{cr}} = \frac{76.8m\mu}{(1 + 6m^2\mu)\,\overline{S}}\frac{EI}{l^2}, \quad \mu = \frac{H_3}{H_2} \cong 1.25, \quad \overline{S} = \frac{S}{l}$$

   where S is a length of an axial line of a half-arch; H_3 and H_2 are trust of the three-hinged and two-hinged arches for given loading.

2. Partial loading of the arch by uniform distributed load q within portion $2\underline{z}$ (Fig. 4.13b). Let $m = f/l \le 0.2$
   Critical thrust is

$$H_{\text{cr}} = \frac{19.2(2-\eta)}{(2-\eta+12m^2\mu)\,\overline{S}}\frac{EI}{l^2} \qquad \text{if} \quad z < \frac{l}{4}$$

$$H_{\text{cr}} = \frac{19.2(2\eta-\eta^2)}{(2\eta-\eta^2+6m^2\mu)\,\overline{S}}\frac{EI}{l^2} \qquad \text{if} \quad z > \frac{l}{4}$$

   Here $\mu = H_3/H_2$ for given loading.

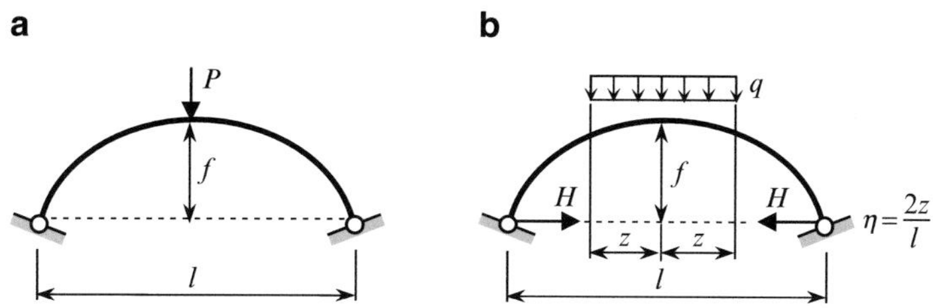

Fig. 4.13 Special loading of two-hinged arch

Limiting Cases

1. If we set $\eta = 0$, then the formula for critical thrust for arch subjected to single force at the crown becomes

$$H_{\text{cr}(P)} = \frac{Pl}{4f\mu} = \frac{19.2}{(1+6m^2\mu)\overline{S}} \frac{EI}{l^2}$$

2. If we set $\eta = 1$, then the formula for critical thrust for arch subjected to uniformly distributed load within all span of the arch. In this case $\mu = H_3/H_2 = 1$ (Tables A.10, A.16) and for critical thrust we get

$$H_{\text{cr}(q)} = \frac{ql^2}{8f} = \frac{19.2}{(1+6m^2)\overline{S}} \frac{EI}{l^2}$$

4.6 Parabolic Arch with Tie

Let us consider uniform two-hinged symmetric parabolic uniform arch with tie at the level of supports. The arch is loaded by the vertical uniformly distributed load q within all span l (Fig. 4.14). A feature of this problem is that even before loss of stability the arch is compressed-bent [Kle72], [Smi84]. The tie may be connected with arch itself by hinges at the ends (flexible tie) or by rigid ends (rigid tie).

Differential equation of the bending for parabolic arch with tie at the level of supports is described by Shtaerman's equation [Sht35], [Kle72]:

$$\left(\frac{\text{d}^2}{\text{d}\theta^2}+1\right)\left(\cos^2\theta\frac{\text{d}M_1}{\text{d}\theta}\right)+\bar{q}\left(\eta-1+\sec^2\theta\right)\sec^2\theta\frac{\text{d}M_1}{\text{d}\theta}+$$
$$\bar{q}\left(\eta-1+4\sec^2\theta\right)\sec^2\theta\times\tan\theta\times M_1=0 \qquad (4.30)$$

where $\bar{q} = (qR_0^3/EI)$; $\eta = (H_{\text{tie}}/H)$, R_0 is a radius of curvature of the arch at the crown, H and H_{tie} are the thrust of two-hinged parabolic arch without tie and the

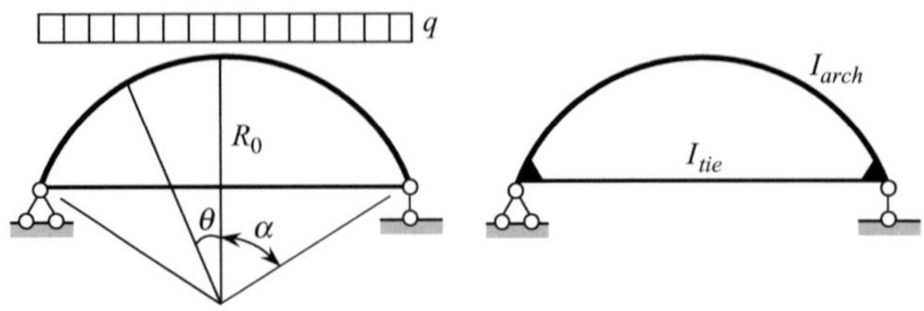

Fig. 4.14 Design diagram of two-hinged arch with flexible and rigid tie

Table 4.9 Critical parameter K for two-hinged arch with tie

$\eta = \dfrac{H_{\text{tie}}}{H}$	1.0	0.9	0.8
K	45.4	50.5	57

such arch with tie, respectively, M_1 is additional bending moment which arises at the buckling.

This is a differential equation of third order with variable coefficients. Integration of this equation with respect to M_1 may done by numerical methods. Shtaerman's equation allows us to consider the flexible or rigid tie as shown above.

Antisymmetric form of the loss of stability is characterized by two half-waves. In case of flexible tie the boundary conditions are $M(0) = 0; \quad M''(0) = 0$. Thus, the boundary conditions at the crown and at the support are

$$M_1(0) = 0; \quad \left.\frac{\mathrm{d}^2 M_1}{\mathrm{d}\theta^2}\right|_{\theta=0} = 0$$

$M_1(\alpha) = 0$, respectively.

Numerical integration of (4.30) leads to the following result for critical load of the arch with flexible tie $q_{\text{cr}} = K(EI/l^3)$, where coefficient K for arch with ratio $f/l = 0.2$ are presented in Table 4.9 [Kle72].

One should pay attention to the unusual fact; decreasing of the stiffness EA of the tie at the fixed stiffness of the arch EI leads to decreasing of the axial force which arise in a tie. It means that the ratio $\eta = (H_{\text{tie}}/H)$ is also decreasing. However, from Table 4.9 it is evident that coefficient K is increasing. This can be explained by the fact that decreasing the axial load in a tie leads to decreasing axial compressed force in the arch [Kle72], [Pav51].

In case of rigid tie we need to take into account an additional boundary condition

$$\int_0^{\alpha} \tan\varphi \sec^3\varphi M \,\mathrm{d}\varphi = -\frac{1}{6\alpha}\frac{I_{\text{arch}}}{I_{\text{tie}}} \tan\alpha\, M(\alpha),$$

where 2α is a central angle, $M(\alpha)$ is a moment at the rigid joint. This condition expresses continuity of the angular deflection of the arch and tie [Kle72]. Critical load is

Table 4.10 Critical parameter K for different ratio $I_{\text{arch}}/I_{\text{tie}}$ [Kle72]

	Type of arches				
			Arch with tie, parameter $\frac{I_{\text{arch}}}{I_{\text{tie}}}$		
	Two hinged (Table 4.6)	Hingeless (Table 4.6)	0.28	1.6	4.45
K	45.4	101	82	61.5	53

$$q_{\text{cr}} = K\frac{EI}{l^3}.$$

Some results for arch with parameter $f/l = 0.2$ and $\eta = H_{\text{tie}}/H = 1$ are presented in Table 4.10.

4.7 Out-of-Plane Loss of Stability of a Single Arch

In the case of loaded and bent in-plane arches the height of the cross section usually more than its width. If the moments of inertia of cross-sections significantly differ from each other, then a gradual increase in load on the arch leads to the flat shape of the bending of the arch to become unstable. The phenomenon of the loss of stability is characterized by the fact that the arch starts to bend in other planes and a new form of equilibrium becomes the spatial form [Mor39]. In Fig 4.15 a new form of equilibrium is shown by dotted line.

The spatial stability of a single arch does not depend on the type of supports in the plane of the arch, but on the type of out-of-plane supports of the arch) [Uma72–73].

Below we consider a circular arch of radius R with central angle 2α and a parabolic arch. Assume that load is applied in the plane of the arch, material of the arch obeys to Hook's law, the axis of the arch is uncompressed and out-of-plane rotation of the support sections are not possible (fixed in out-of-plane directions).

4.7.1 *Circular Arch Subjected to Couples on the Ends*

Critical couple M is determined by formula [Smi84]

$$M_{\text{cr}} = \frac{EI_y + GI_{\text{d}}}{2R} \pm \sqrt{\frac{\left(EI_y - GI_{\text{d}}\right)^2}{4R^2} + \frac{\pi^2}{4\alpha^2}\frac{EI_yGI_{\text{d}}}{R^2}}. \tag{4.31}$$

Here, EI_y is a flexural rigidity of the arch in direction which is perpendicular to the plane of the arch; GI_{d} is a torsional rigidity of the arch. If for rectangle cross

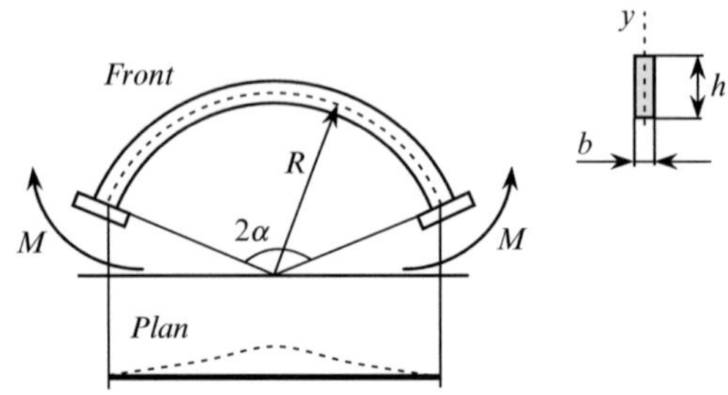

Fig. 4.15 Out-of-plane buckling of the single arch

section $h \gg b$ (Fig. 4.15), then I_d may be calculated by Timoshenko's formula [Smi84]

$$I_d = \frac{hb^3}{3}\left(1 - 0.630\frac{b}{h}\right).$$

If support moment M decreases the curvature of the arch, as shown in Fig. 4.15, then in (4.31) we must put a negative sign in front of the square root.

4.7.2 Circular Arch Subjected to Uniform Radial Load

If a load remains parallel to the initial plane of the arch then critical load is determined by formula

$$q_{cr} = K\frac{EI_y}{R^3}$$

Coefficient K may be determined from transcendental equation [Mor73]

$$\frac{\tan r_1\alpha}{\tan r_2\alpha} = \frac{r_1}{r_2}\left(\frac{1 - r_1^2}{1 - r_2^2}\right)^2,$$

$$r_{1,2} = 1 + \frac{K}{2} \pm \sqrt{\left(1 + \frac{K}{2}\right)^2 + K\frac{EI_y}{GI_d} - 1} \tag{4.32}$$

Table 4.11 gives the coefficient K in terms of the central angle 2α of the arch and ratio EI_y/GI_d between flexural and torsional rigidity.

Table 4.11 Coefficient K in terms of α and parameter EI_y/GI_d [Mor73]

	Parameter $\frac{EI_y}{GI_d}$					
2α (°)	0.7	2.0	5.0	10	20	30
90°	13.8	13.3	12.1	10.86	9.25	7.9
120°	7.05	6.7	5.88	5.06	3.94	3.3

Table 4.12 Coefficient K_1 in terms of f/l and EI_y/GI_d

	Parameter $\frac{EI_y}{GI_d}$		
$\frac{f}{l}$	0.7	1.0	2.0
0.1	28.5	28.5 (31.7)	28.0
0.2	41.5	41.0 (65.0)	40.0
0.3	40.0	38.5 (137)	36.5

4.7.3 Parabolic Arch Subjected to Uniform Vertical Load

Let a load preserve its initial direction and be distributed within the entire span. Then the critical load may be determined by formula $q_{cr} = K_1(EI_y/l^3)$, where coefficient K_1 depending on parameter f/l and ratio EI_y/GI_d between the flexural and torsion rigidity (Table 4.12) [Mor73].

In the case of a tracking load, the coefficient K_1 is shown in parenthesis [Mor73].

Chapter 5
Matrix and Displacement Methods

This chapter is devoted to numerical analysis of stability for different types of arches. Among them are two approaches – Smirnov's matrix method and classical methods of structural analysis. A procedure for constructing the stability equation for different approaches is discussed.

5.1 General

Determination of critical loads on the arches can be achieved by the precise or approximate methods. By "precise methods" we will assume that this is done by integration of differential equation of the arch, or by some other method under which the initial design diagram of the arch does not have to be modified to accommodate a numerical procedure. Precise methods for finding critical loads have been discussed in Chap. 4.

The only way to determine critical loads of an arch with variable stiffness is by approximate methods. The terms "Approximate methods" means replacing the arch by set of the members, followed by precise methods of analyzing the modified design diagram. Here the engineer is faced with two important issues. The first is how to approximate the arch, and the second is which method of analysis to choose for analysis of the modified design diagram. In the general case, one must choose such an approximation of the design diagram, and such a method for analysis that will simplify the numerical procedures without compromising the numerical accuracy. There exists a variety of approaches and their variations. Among them are Smirnov's matrix method, classical methods of structural analysis, and of course, finite element method.

Smirnov's matrix method [Smi47], [Smi84] considers the arch as a series of *curvilinear* segments, each of which *coincides* with the corresponding portion of the arch.

DOI 10.1007/978-1-4614-0469-9_5,

No modification of the design diagram is performed under this method. Therefore, this method should be treated as an exact method in matrix form. Smirnov's method is based on the classic concepts of the fictitious (conjugate) beam, elastic loads, and utilizes the tools of the matrix algebra.

Classical methods of structural analysis (the Force method, Displacement method and Mixed method in canonical form) [Kar10] are precise, but an approximate result is the consequence of a change in the design diagram of the arch by its approximate (modified) diagram.

Finite element method is implemented in modern computer software and allows the user to obtain the value of critical loads with a high accuracy. Nowadays this is an effective method for stability analysis of arches with peculiarities (piecewise-linear stiffness, nonlinearities, the need to account for secondary effects, etc.).

5.2 Smirnov Matrix Method

This method allows us to numerically determine the critical loads on the arches. The shape of the arch and the law under which the moment of inertia of the cross section of the arch changes along the axis line are unspecified. At the heart of the method lies a discretization of the system in association with elastic load method (see Sect. 1.6).

In the case of a circular arch, we can use two fundamental differential equations. The first equation is constructed with respect to radial displacements (Boussinesq's equation)

$$\frac{\mathrm{d}^2 \upsilon}{\mathrm{d}s^2} + \frac{\upsilon}{R^2} = -\frac{M}{EI}.$$

The second equation is constructed with respect to bending moments in the curvilinear bar

$$\frac{\mathrm{d}^2 M}{\mathrm{d}s^2} + \frac{M}{R^2} = -q.$$

These equations are similar, so computation of radial displacements υ may be replaced by the computation of fictitious bending moments caused by the load $q = M/EI$; this load is replaced by elastic loads W [Smi84].

5.2.1 *Matrix Form for Elastic Loads*

Ignoring the axial forces, elastic loads, according to (1.20), become

$$W_n = \frac{s_n}{6EI_n}(M_{n-1} + 2M_n) + \frac{s_{n+1}}{6EI_{n+1}}(2M_n + M_{n+1}), \tag{5.1}$$

where s_n is a length of an structural element between joints $(n - 1)$ and n (Fig. 1.15).

In the stability problems (as opposed to the strength problems), ordinates of the bending moment diagram M and even the general profile of design diagram are not known in advance. Therefore, in the stability problems the bending moment diagram is approximated either by the straight line segments or a set of quadratic polynomials each of which passes through the three neighboring points (Sect. 1.6, Fig. 1.15, dotted line). Each of these approximations allows us to represent the elastic load at joint n in the form [Smi47], [Smi84]

$$W_n = \frac{S_0}{6EI_0}\left(\beta_{n(n-1)}M_{n-1} + \beta_{nn}M_n + \beta_{n(n+1)}M_{n+1}\right). \tag{5.2}$$

In the case of approximating the bending moment diagram by parabolas, the coefficients β may be presented in terms of length and moment of inertia of each element (s, I) as well as length and moment of inertia in terms of some basic element (S_0, I_0)

$$\begin{aligned}
\beta_{n(n-1)} &= \frac{s_n + 2s_{n+1}}{2(s_n + s_{n+1})}\rho_n - \frac{s_{n+1}^2}{2s_n(s_n + s_{n+1})}\rho_{n+1},\\
\beta_{nn} &= \left(2 + \frac{s_n}{2s_{n+1}}\right)\rho_n + \left(2 + \frac{s_{n+1}}{2s_n}\right)\rho_{n+1},\\
\beta_{n(n+1)} &= -\frac{s_n^2}{2s_{n+1}(s_n + s_{n+1})}\rho_n + \frac{s_{n+1} + 2s_n}{2(s_n + s_{n+1})}\rho_{n+1}, \quad \rho_n = \frac{s_n I_0}{I_n S_0}.
\end{aligned} \tag{5.3}$$

The vector of elastic loads is expressed by a matrix of elastic loads B_W and the vector of moments at the nodal points

$$\vec{W} = \frac{S_0}{6EI_0}B_W\vec{M}. \tag{5.4}$$

Vector of bending moments of any broken rod during buckling is presented as

$$\vec{M} = \lfloor M_1 \quad M_2 \quad \ldots \quad M_n \rfloor^{\mathrm{T}}.$$

The matrix of elastic loads reduces to a Jakobi-like tri-diagonal matrix

$$B_W = \frac{S_0}{6EI}\begin{bmatrix} \beta_{11} & \beta_{12} & 0 & 0 & \ldots & \ldots & 0\\ \beta_{21} & \beta_{22} & \beta_{23} & 0 & \ldots & \ldots & 0\\ 0 & \beta_{32} & \beta_{33} & \beta_{34} & \ldots & \ldots & 0\\ \vdots & \vdots & \vdots & \vdots & \vdots & \vdots & \vdots\\ 0 & 0 & 0 & 0 & \ldots & \beta_{n(n-1)} & \beta_{nn} \end{bmatrix}. \tag{5.5}$$

Special Case

Assume that the rod is divided into elements of same length ($s_n = s_{n+1} = S_0 = \text{const}$). In this case, expressions for parameters β, according to formulas (5.3) may be simplified

$$\begin{aligned} \beta_{n(n-1)} &= \frac{3}{4}\rho_n - \frac{1}{4}\rho_{n+1}; \\ \beta_{nn} &= \frac{5}{2}\left(\rho_n + \rho_{n+1}\right); \\ \beta_{n(n+1)} &= -\frac{1}{4}\rho_n + \frac{3}{4}\rho_{n+1}. \end{aligned} \tag{5.6}$$

Let us make an additional assumption $I_n = I_{n+1} = I_0$. In this case $\rho_n = 1$ and for parameters β we get

$$\begin{aligned} &\beta_{21} = \beta_{32} = \ldots = 0.5, \\ &\beta_{11} = \frac{5}{2}(\rho_1 + \rho_2) = 5; \quad \beta_{22} = \frac{5}{2}(\rho_2 + \rho_3) = 5, \ldots \\ &\beta_{12} = \beta_{23} = \ldots = 0.5, \end{aligned} \tag{5.7}$$

Matrix of elastic loads becomes

$$B_W = \frac{S_0}{6EI_0}\begin{bmatrix} 5 & 0.5 & 0 & 0 & \ldots & \ldots & 0 \\ 0.5 & 5 & 0.5 & 0 & \ldots & \ldots & 0 \\ 0 & 0.5 & 5 & 0.5 & \ldots & \ldots & 0 \\ \vdots & \vdots & \vdots & \vdots & \vdots & \vdots & \vdots \\ 0 & 0 & 0 & 0 & \ldots & 0.5 & 5 \end{bmatrix}. \tag{5.8}$$

If curvilinear bending moment diagram M_P within portion s_n and s_{n+1} are replaced by straight elements, then

$$\beta_{n(n-1)} = \rho_n; \qquad \beta_{nn} = 2\left(\rho_n + \rho_{n+1}\right); \qquad \beta_{n(n+1)} = \rho_{n+1}. \tag{5.9}$$

In a special case, if the length of all the elements is $s_n = s_{n+1} = S_0 = \text{const}$ and $I_n = I_{n+1} = I$ then we get following matrix of elastic loads:

$$B_W = \frac{S_0}{6EI}\begin{bmatrix} 4 & 1 & 0 & 0 & \ldots & \ldots & 0 \\ 1 & 4 & 1 & 0 & \ldots & \ldots & 0 \\ 0 & 1 & 4 & 1 & \ldots & \ldots & 0 \\ \vdots & \vdots & \vdots & \vdots & \vdots & \vdots & \vdots \\ 0 & 0 & 0 & 0 & \ldots & 1 & 4 \end{bmatrix}. \tag{5.9a}$$

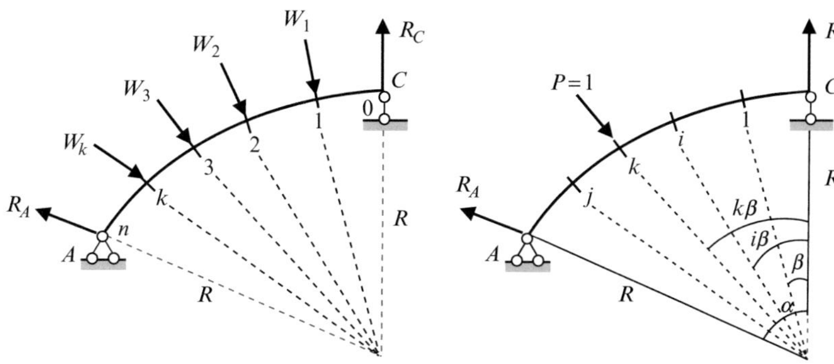

Fig. 5.1 Equivalent half-arch of entire symmetric two-hinged circular arch, elastic loads at joints 1, . . ., k and the angles notation

The matrix B_W is a square matrix of $(n - 1)$th order, where n is the number of elements of the arch.

Matrices (5.8) and (5.9a) are as presented assuming that the moments at the ends points are equal to zero [Smi47], [Smi84].

5.2.2 *Moment Influence Matrix*

Let us consider two-hinged circular symmetrical arch of radius R and central angle 2α. The arch is loaded by uniform radial load q. In case of antisymmetric form of loss of stability the initial arch is replaced by its equivalent half-arch with rolled support at the axis of symmetry (Fig. 5.1) [Kar10]. Moment influence matrix L is constructed for a fictitious (conjugate) structure. For design diagram in Fig. 5.1 fictitious structure coincides with real structure. Divide the axis of the half-arch into n equals segments and number of all nodal points from 0 to n. The central angle for each segment is β, therefore the length of each portion is $S_0 = R\beta$.

Each column of the moment influence matrix L contains the moments at the nodal points 1, 2, . . . caused by unit load which act at these arch points.

Let a unit radial force $P = 1$ be applied at arbitrary nodal point k; this point is defined by the angle $k\beta$ (Fig. 5.1). The vertical reaction R_C and radial reaction R_A are

$$R_A = \frac{\sin k\beta}{\sin \alpha}, \quad R_C = \frac{\sin(\alpha - k\beta)}{\sin \alpha}.$$

Bending moments at points i, k, j are

$$m_{ik} = R\frac{\sin i\beta \sin(\alpha - k\beta)}{\sin \alpha} \quad (i<k)$$
$$m_{kk} = R\frac{\sin k\beta \sin(\alpha - k\beta)}{\sin \alpha}$$
$$m_{jk} = R\frac{\sin k\beta \sin(\alpha - j\beta)}{\sin \alpha} \quad (i>k). \tag{5.10}$$

If half-arch is divided into two equal parts, then $k = 1$ and matrix L_{1m} according to the second formula (5.10) becomes

$$L_{1m} = R\frac{\sin^2 \beta}{\sin \alpha}[1]. \tag{5.10a}$$

If half-arch is divided into three equal parts, then for matrix L_{2m} we get

$$L_{2m} = \begin{bmatrix} l_{11} & l_{12} \\ l_{21} & l_{22} \end{bmatrix} = R\frac{\sin^2 \beta}{\sin \alpha}\begin{bmatrix} 2\cos \beta & 1 \\ 1 & 2\cos \beta \end{bmatrix}. \tag{5.10b}$$

If half-arch is divided into four equal parts, then for matrix L_{3m} we get

$$L_{3m} = R\frac{\sin^2 \beta}{\sin \alpha}\begin{bmatrix} 4\cos^2 \beta - 1 & 2\cos \beta & 1 \\ 2\cos \beta & 4\cos^2 \beta & 2\cos \beta \\ 1 & 2\cos \beta & 4\cos^2 \beta - 1 \end{bmatrix}. \tag{5.10c}$$

The matrix L_m is symmetric and square order $(n - 1)$.

5.2.3 Stability Equation in Matrix Form

Bending moment at any section is $M = qR\upsilon$. Vector of elastic loads is

$$\vec{\upsilon} = L_m\vec{W} = qRL_mB_W\vec{\upsilon}.$$

Thus, the stability equation becomes

$$\det(C - \lambda E) = 0,$$

where E is identity matrix, L_m is moment influence matrix,

$$C = L_mB_W, \quad \lambda = 1/(qR).$$

Matrix Procedures

In the case of a two-hinged circular arch subjected to uniform radial load q the following procedure may be applied [Smi47], [Smi84]:

1. Divide the arch into n equal curvilinear segments. Larger values of n ensure greater numerical accuracy.
2. Construct the moment influence matrix L_m and matrix B_W of elastic loads [Smi84].
3. Compute the matrix product $C = L_m B_W$ and forms the stability equation $\det(C - \lambda E) = 0$, where $\lambda = 1/(qR)$ is eigenvalue of the stability problem.
4. Find the greatest eigenvalue λ and calculate the smallest critical load $q_{\min} = 1/(\lambda R)$.

Application of this procedure for stability analysis of symmetrical arches of two different shapes is discussed in the following sections.

5.3 Two-Hinged Symmetrical Arches

Two types of two-hinged arches are considered. They are circular and parabolic symmetrical arches. Circular uniform and nonuniform arches are loaded by uniform radial load, while the parabolic arch is loaded by uniform vertical load and simple force at crown. Assume the antisymmetrical form of the loss of stability occurs. In this case, equivalent half-arch has a rolled support on the axis of symmetry and half-arch itself is statically determinate structure.

5.3.1 Circular Uniform Arch

This arch is loaded by uniform radial load. A half-arch is divided into two equal portions with one nodal point (1). In this case, the moment influence matrix according to (5.10a) has a single entry [Smi84]

$$L_{1m} = R\frac{\sin^2\beta}{\sin\alpha}[1].$$

Elastic load is applied at point 1 and have radial direction. Matrix of elastic load according to (5.8) is

$$B_W = \frac{5S_0}{6EI}[1].$$

The matrix product

$$C = L_m B_W = R\frac{\sin^2\beta}{\sin\,\alpha}\frac{5S_0}{6EI}[1].$$

Stability equation is $\det(C - \lambda E) = 0$.
We get the following expression for the required eigenvalue

$$\lambda = \frac{5S_0 R}{6EI}\frac{\sin^2\beta}{\sin\,\alpha}.$$

Since $S_0 = R\beta$, then expression for λ may be rewritten as

$$\lambda = \frac{5R^2\beta}{6EI}\frac{\sin^2\beta}{\sin\,\alpha}.$$

The critical load becomes

$$q_{\rm cr} = \frac{1}{\lambda R} = \frac{12}{5\beta\,\tan\,\beta}\frac{EI}{R^3}.$$

If $\alpha = \pi/2, \quad \beta = \pi/4$, then

$$q_{\rm cr} = 3.055\frac{EI}{R^3}.$$

Exact result according to Levy's formula is $q_{\rm cr} = 3.0(EI/R^3)$ (see Sect. 4.2.1). Relative error is 1.83%.

If half-arch is divided into three equal parts the critical load becomes [Smi47]

$$q_{\rm cr} = \frac{12(2\,\cos\,\beta - 1)}{11\beta\,\sin\,\beta}\frac{EI}{R^3}.$$

If $\alpha = \pi/2, \quad \beta = \pi/6$, then $q_{\rm cr} = 3.0504(EI/R^3)$. Thus, approximating of the uniform circular arch by only two segments leads to very good results.

5.3.2 *Circular Nonuniform Arch*

Design diagram of symmetrical two-hinged circular arch ACB is shown in Fig. 5.2a. The radius of the arch is R and central angle $2\alpha = 120^{\circ}$. The relative moments of inertia for special portions are underlined, i.e., $I_{C-1} = 1.0I_0$, $I_{1-2} = 0.8I_0$, $I_{2-3} = 0.6I_0$, $I_{3-A} = 0.5I_0$. The arch is subjected to uniform radial load q; this

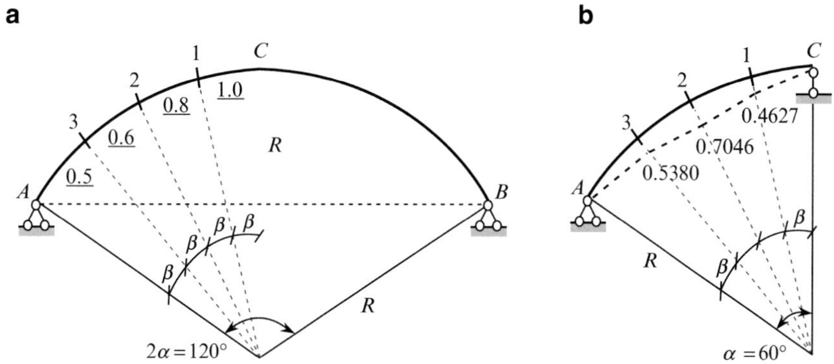

Fig. 5.2 Design diagram of symmetrical two-hinged circular arch and its equivalent design diagram for antisymmetrical buckling

load is not shown. In case of antisymmetrical form of buckling the equivalent design diagram of the half-arch is shown in Fig. 5.2b [Smi47], [Smi84].

The half-arch is divided into four equal portions with central angle $\beta = \alpha/4 = \pi/12$ for each portion. For given presentation of the arch we get

$$\sin\beta = 0.2588, \quad \cos\beta = 0.9659,$$
$$\sin^2\beta = 0.0670, \quad \cos^2\beta = 0.9330, \quad \sin^2\beta/\sin\alpha = 0.07737.$$

Matrix of moments according to (5.10c) becomes

$$L_m = 0.07737R\begin{bmatrix} 2.732 & 1.9318 & 1.0 \\ 1.9318 & 3.732 & 1.9318 \\ 1.0 & 1.9318 & 2.732 \end{bmatrix}$$
$$= R\begin{bmatrix} 0.2114 & 0.1495 & 0.07737 \\ 0.1495 & 0.2887 & 0.1495 \\ 0.07737 & 0.1495 & 0.2114 \end{bmatrix}.$$

Since the arch is divided into equal segments, one must use (5.6) for construction of the elastic loads matrix B_W. The entries b_{kk} of matrix B_W are as follows:

$$b_{11} = \frac{5}{2}(\rho_1 + \rho_2) = \frac{5}{2}\left(1 \times \frac{1}{1} + \frac{1}{0.8}\right) = 5.625;$$
$$b_{22} = \frac{5}{2}\left(1 \times \frac{1}{0.8} + \frac{1}{0.6}\right) = 7.2917; \quad b_{33} = \frac{5}{2}\left(1 \times \frac{1}{0.6} + \frac{1}{0.5}\right) = 9.1667.$$

For entries $b_{k(k+1)}$ of matrix B_W we get:

$$b_{12} = -\frac{1}{4}\rho_1 + \frac{3}{4}\rho_2 = -\frac{1}{4}\times\frac{1}{1} + \frac{3}{4}\times\frac{1}{0.8} = 0.6875;$$
$$b_{13} = 0; \quad b_{23} = -\frac{1}{4}\times\frac{1}{0.8} + \frac{3}{4}\times\frac{1}{0.6} = 0.9375.$$

The entries $b_{k(k-1)}$ of matrix B_W are

$$b_{21} = \frac{3}{4}\rho_2 - \frac{1}{4}\rho_3 = \frac{3}{4}\times\frac{1}{0.8} - \frac{1}{4}\times\frac{1}{0.6} = 0.5208;$$
$$b_{31} = 0; \quad b_{32} = \frac{3}{4}\times\frac{1}{0.6} - \frac{1}{4}\times\frac{1}{0.5} = 0.7500.$$

Elastic loads matrix B_W becomes

$$B_W = \frac{S_0}{6EI_0}\begin{bmatrix} 5.6250 & 0.6875 & 0.0 \\ 0.5208 & 7.2917 & 0.9375 \\ 0.0 & 0.7500 & 9.1667 \end{bmatrix}.$$

For matrix C we get

$$C = L_m B_W = \frac{RS_0}{6EI_0}\begin{bmatrix} 1.2669 & 1.2935 & 0.8494 \\ 0.9913 & 2.3200 & 1.6418 \\ 0.5130 & 1.3018 & 2.0780 \end{bmatrix}.$$

Maximum eigenvalue of the stability equation

$$\det(L_m B_W - \lambda E) = 0$$

is $\lambda = 4.2241(RS_0/6EI_0)$. Corresponding eigenvector is $[-0.4627\ -0.7046\ -0.5380]$. The form of the loss of stability is shown in Fig. 5.2b by a dotted line.

Since $\lambda = (1/qR) = 4.2244(RS_0/6EI_0)$ and $S_0 = \beta R = (\pi R/12)$ then for radial critical load we get $q_{\mathrm{cr}} = 5.4251EI/R^3$.

5.3.3 Parabolic Uniform Arch

This section is devoted to stability analysis of two-hinged parabolic arch subjected to two types of loading: vertical uniformly distributed load along the span of the arch and concentrated force applied at the crown of the arch. In both cases, arches are symmetric. In case of antisymmetric form of the loss of stability the initial two-hinged arch is replaced by its equivalent half-arch. This scheme contains rolled

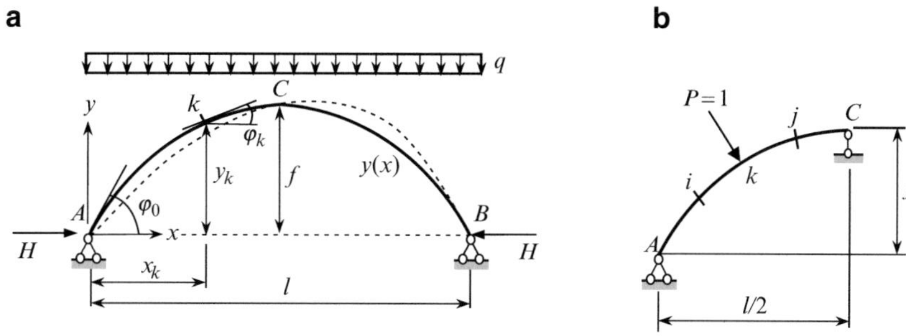

Fig. 5.3 Design diagram of symmetrical two-hinged parabolic arch and equivalent half-arch for antisymmetrical form of the loss of stability and its equivalent design diagram for antisymmetrical buckling

support at the axis of symmetry and half-arch itself presents a statically determinate structure. Therefore, computation of elastic loads is easily evaluated using only the equilibrium conditions.

5.3.3.1 Uniformly Distributed Load

The arch of span l and rise f is loaded by a uniform vertical load distributed within the entire span (Fig. 5.3a). The equivalent half-arch, assuming the antisymmetric form of the loss of stability, is shown in Fig. 5.3b. The half-arch is a statically determinate structure.

The axis of the arch is divided into equal curvilinear portions and the nodal points are denoted as 1, 2, ...

The nodal point i has coordinates $x_i = \xi_i l$ and $y_i = \eta_i l$.

Geometrical Relationships

Equation of the axial line of the arch is $y = 4fx(l-x)(1/l^2)$.

The general expressions for slope and slope at the support A are

$$\tan\varphi = 4\frac{f}{l^2}(l-2x), \quad \tan\varphi_0 = 4m, \quad m = \frac{f}{l}.$$

Length of half-arch is determined as

$$S = \frac{l}{4}\left[\sec\varphi_0 + \frac{1}{4m}\ln(4m + \sec\varphi_0)\right]. \tag{5.11}$$

The length S_k of axis of the arch from origin (point A) to arbitrary point $\mathcal{K}$ with coordinates $x_k = \xi_k l$, $y_k = \eta_k l$ is

$$S_k = S - \frac{l}{16m}\left(\frac{\tan \varphi_k}{\cos \varphi_k} + \ln\frac{1+\sin \varphi_k}{\cos \varphi_k}\right). \tag{5.12}$$

Stability equation is $\det(C - \lambda E) = 0$, where $C = L_m B_W G$. The eigenvalue and thrust are related by the formula $\lambda = 1/H$.

For constructing the influence matrix L_m it is necessary to apply the unit force at point k and to determine the bending moment at the neighboring points (Fig. 5.3b); expressions for the entries of this matrix can be found in [Smi84]. The matrix B_W is constructed in the same manner as for circular arch. The diagonal matrix G contains the entries $\sec \varphi_k$.

Let us show this procedure and evaluate numerical results if the length of half-arch is divided only into two equal portions; the nodal point 1 in Fig. 5.3 is not shown. Assume $m = f/l = 0.5$.

Geometrical Parameters

Slope at the support A is $\tan \varphi_0 = 4m = 2.0$ and $\sec \varphi_0 = 2.2361$.

The length of half-arch with adopted parameter m is

$$S = \frac{l}{4}\left[2.2361 + \frac{1}{4 \cdot 0.5}\ln(4 \cdot 0.5 + 2.2361)\right] = 0.740l. \tag{5.12a}$$

The length of each curvilinear segment is $S_0 = S/2 = 0.370l$.

The length S_0 of the segment $A - 1$ and slope at point 1 are related by

$$0.370l = \frac{l}{16m}\left(\frac{\tan \varphi_1}{\cos \varphi_1} + \ln\frac{1+\sin \varphi_1}{\cos \varphi_1}\right). \tag{5.12b}$$

Solution of this equation determines the slope at the point 1

$$\varphi_1 = 50.7^\circ, \quad \tan \varphi_1 = 1.2217, \quad \sin \varphi_1 = 0.7738, \quad \cos \varphi_1 = 0.6334.$$

Coordinates of joint 1 Dimensionless coordinates of point 1 are $\xi_1 = x_1/l$, $\eta_1 = y_1/l$. For this point 1, we have $\tan \varphi_1 = 4m(1 - 2\xi_1)$ or $1.2217 = 4 \cdot 0.5(1 - 2\xi_1)$. Solution of this equation is $\xi_1 = 0.1945$, so $x_1 = 0.1945l$. For ordinate of point 1 we get $y_1 = 0.3133l$.

Stability equation Since the arch is divided into two portions, then each matrix from $C = L_m B_W G$ will be a scalar.

Entry of matrix L_m is

$$\begin{aligned}[L_m] &= l(\xi_1 \cos \varphi_1 + \eta_1 \sin \varphi_1) \times (1 - 2\xi_1)\\ &= l(0.1945 \times 0.6334 + 0.3133 \times 0.7738) \times (1 - 2 \times 0.1945) = 0.2234l[1].\end{aligned}$$

Table 5.1 Parameter K for critical load [Smi84], [Din46]

$m = \frac{f}{l}$	x_1 (factor l)	S_0 (factor l)	K
0.1	0.24524	0.25652	28.2 (28.5)
0.2	0.23333	0.27456	44.5 (45.4)
0.3	0.21896	0.30109	47.6 (46.5)
0.4	0.20562	0.33343	43.3 (43.9)
0.5	0.19463	0.36974	36.8 (38.4[a])
0.6	0.18601	0.40882	30.6 (30.5)
0.7	0.17934	0.44992	25.3 (–)
0.8	0.17416	0.49251	21.1 (20)
0.9	0.17009	0.53623	17.7 (–)
1.0	0.16685	0.58085	15.0 (14.1)

[a]See comment in [Smi84, p. 333]

Entries of matrices B_W and G are

$$[B_W] = \frac{5S_0}{6EI}[1] = \frac{5 \cdot 0.370l}{6EI}[1] = \frac{0.3083l}{EI}[1],$$

$$[G] = \sec \varphi_1[1] = 1.5788[1].$$

For matrix C we get $C = L_m B_W G = 0.2234l \times (0.3083l/EI) \times 1.5788[1] = (0.1087l^2/EI)$.

Stability equation $\det(C - \lambda E) = 0$ leads to the following eigenvalue $\lambda = 0.1087l^2/EI$.

Since $\lambda = 1/H$, while thrust $H = ql^2/(8f)$, then for critical load we get $q_{cr} = 36.8EI/l^3$.

For different parameters m the critical load may be presented as $q_G = K(EI)/l^3$. Parameter K is presented in Table 5.1; coefficients K, obtained by Dinnik [Din46] are shown in parentheses. This table also contains the abscissa x_1 for point 1 and length S_0.

Analysis of Results

With the half-arch divided only into two segments, Smirnov's method leads to acceptable results. Even for arches with parameters ($m = 0.8$–1.0) the relative error is not more than 6%. If the axis of the arch is divided into arbitrary number of the segments, the analytical expressions for entries of C can be found in [Smi84].

5.3.3.2 Concentrated Load

The arch AB of span l and rise f is loaded by the single force P at a crown C. The form of the loss of stability is shown in Fig. 5.4 by dotted line.

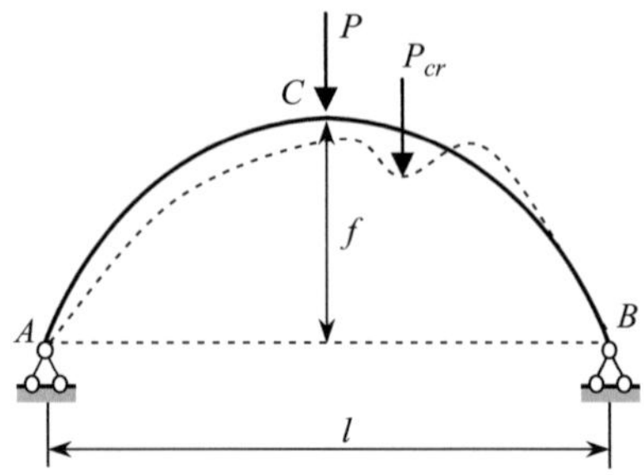

Fig. 5.4 Design diagram of symmetrical two-hinged parabolic arch subjected to load P and form of the loss of stability and its equivalent design diagram for antisymmetrical buckling

Table 5.2 Parameter K_P for critical load [Smi84]

m	0.1	0.2	0.3	0.4	0.5	0.6	0.7	0.8	0.9	1.0
K_P	15	23.8	25.9	24.3	21.4	18.4	15.6	13.3	11.4	9.8

If half-arch is divided into two segments, then Smirnov's method leads to the following critical load

$$P_{\text{cr}} = \frac{K_P EI}{l^2}.$$

Parameter K_P in terms of $m = f/l$ is presented in Table 5.2.

5.4 Hingeless Symmetrical Arches

In case of symmetrical hingeless arch we can replace it by equivalent half-arch. However, in this case, in contrast to two-hinged arch, the half-arch now presents a redundant structure. This fact adds additional features in Smirnov's procedure: the matrix C in the stability equation may be constructed after solution of corresponding canonical equation [Smi47].

5.4.1 Duality of Bending Moment Diagram and Influence Line

For stability analysis of hingeless symmetrical arch we use the following theorem:

> Bending moment diagram in the real structure caused by unit load (force or couple) which acts at point k coincides with influence line of corresponding fictitious load factor at the same point k for fictitious (conjugate) structure [Smi47].

This correspondence for some structures is shown in Table 5.3

We explain the concept of "corresponding load." In the fictitious structure, the bending moment in the point k due to the fictitious load is a linear displacement at

Table 5.3 Correspondence of bending moment diagram for real structure and influence line for fictitious structure

Bending moment diagram for real structure	Influence line of corresponding factor for fictitious structure

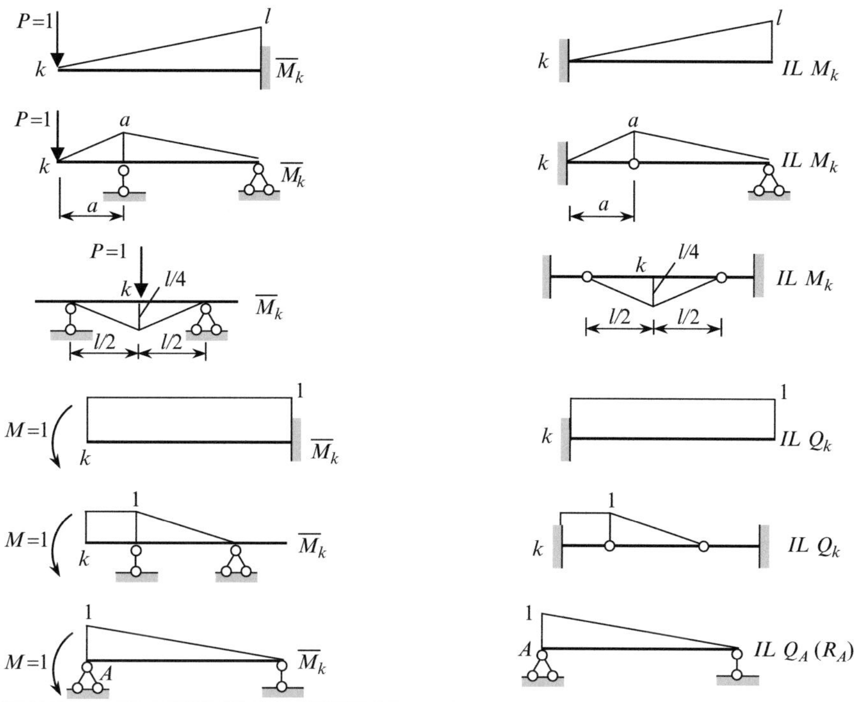

the same point of the real structure; this displacement and unit force P in the real structure correspond to each other, so $P = 1$ corresponds to M_{fict}. Similarly, $M = 1$ corresponds to fictitious shear force Q_{fict}.

This theorem allows us to determine the displacements of the real structure by applying elastic loads to the bending moment diagram and treating them as the influence line [Smi47], [Kle80]. This theorem will also be used for stability analysis of the hingeless arch.

Let us demonstrate the application of this theorem for computation of displacement at the free end of the cantilever uniform beam of the span l (Fig. 5.5a).

Subdivide the beam into two equal parts (0–1 and 1–2). The specified points are labeled as 0, 1, and 2. The bending moment diagram for actual beam is shown in Fig. 5.5b. Fictitious beam and elastic loads W_0 and W_1 are shown in Fig. 5.5c.

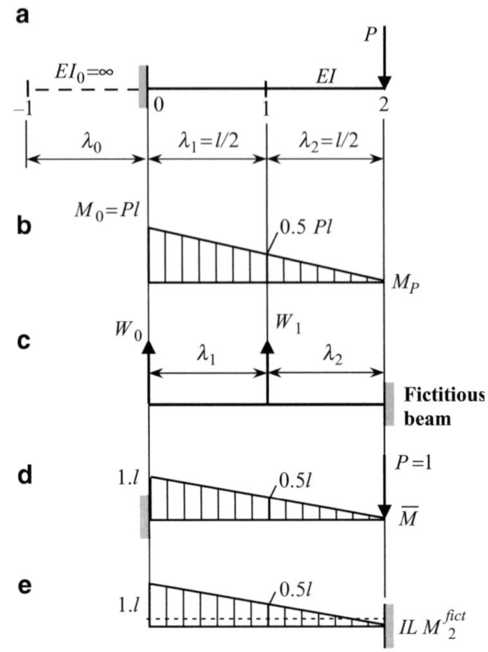

Fig. 5.5 (**a**) Design diagram of the beam; (**b**) bending moment diagram of the real beam; (**c**) fictitious beam; (**d**) unit state and corresponding bending moment diagram; (**e**) influence line for bending moment at the clamped support of the fictitious beam

For calculation of W_0 we need to know bending moments at three consecutive points; dotted line shows additional portion of the beam with end points -1 and 0; the length and stiffness of this portion are λ_0 and $EI_0 = \infty$. The elastic loads are

$$W_0 = \frac{\lambda_0}{6EI_0}(M_{-1} + 2M_0) + \frac{\lambda_1}{6EI_1}(2M_0 + M_1)$$
$$= \frac{\lambda_0}{\infty}(M_{-1} + 2M_0) + \frac{l}{12EI}\left(2Pl + \frac{Pl}{2}\right) = \frac{5Pl^2}{24EI},$$

$$W_1 = \frac{\lambda_1}{6EI_1}(M_0 + 2M_1) + \frac{\lambda_2}{6EI_2}(2M_1 + M_2)$$
$$= \frac{l}{12EI}\left(Pl + 2\frac{Pl}{2}\right) + \frac{l}{12EI}\left(2\frac{Pl}{2} + 0\right) = \frac{Pl^2}{4EI}.$$

Now these elastic loads should be applied to fictitious beam. Since the bending moment diagram is traced on the tensile fibers and ordinates of M diagram are located above the axis, then the elastic loads should be directed upward. Unit state which corresponds to the required displacement and corresponding bending moment diagram is shown in Fig. 5.5d. Influence line for bending moment at the clamped support of the fictitious beam is shown in Fig. 5.5e. To determine vertical

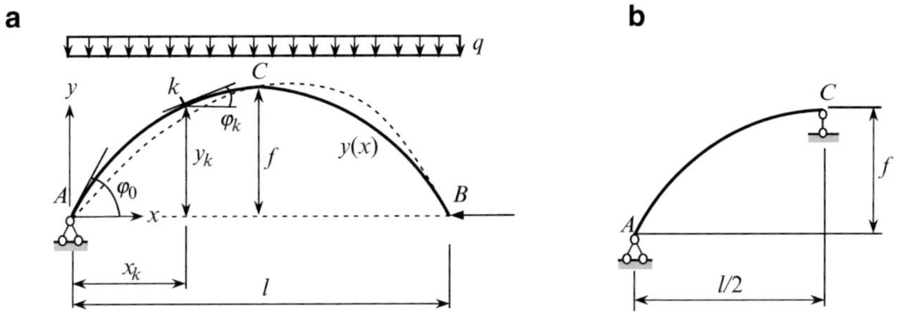

Fig. 5.6 Design diagram of symmetrical hingeless parabolic arch and equivalent half-arch for antisymmetrical buckling and its equivalent design diagram for antisymmetrical buckling

displacement at point 2 of the real structure we need to load the influence line M_2^{fict} by the elastic loads

$$y_2 = W_0 \times 1 \times l + W_1 \times 0.5l = \frac{5Pl^2}{24EI} \times l + \frac{Pl^2}{4EI} \times \frac{l}{2} = \frac{Pl^3}{3EI}.$$

5.4.2 *Parabolic Uniform Arch*

Design diagram of hingeless parabolic uniform arch is shown in Fig. 5.6a. Let us show the application of the theorem considered in the previous section for the stability analysis by the Smirnov's method. If the loss of stability occurs according to the antisymmetrical form, then the equivalent half-arch is shown in Fig. 5.6b.

Stability equation, as before, is $\det(C - \lambda E) = 0$, where E is identity matrix, eigenvalue $\lambda = 1/H$, and thrust $H = ql^2/(8f)$. However, since the half-arch is a redundant structure, then matrix C should be determined by a different method [Smi84]. The structure in Fig. 5.6b has one redundant constraint. Canonical equation of the Force method and primary unknown are

$$\delta_{11}X_1 + \Delta_{1P} = 0 \rightarrow X_1 = -\frac{\Delta_{1P}}{\delta_{11}}.$$

Let the primary unknown X_1 be the moment at the fixed support. The axis of the arch is divided into n equal curvilinear parts. The nodal points are denoted by 1, 2, ..., $n-1$, the coordinates of which need to be calculated. After that we show the bending moment diagram for unit state (Fig. 5.7a). The matrix moments in the unit state and its transposed matrix are

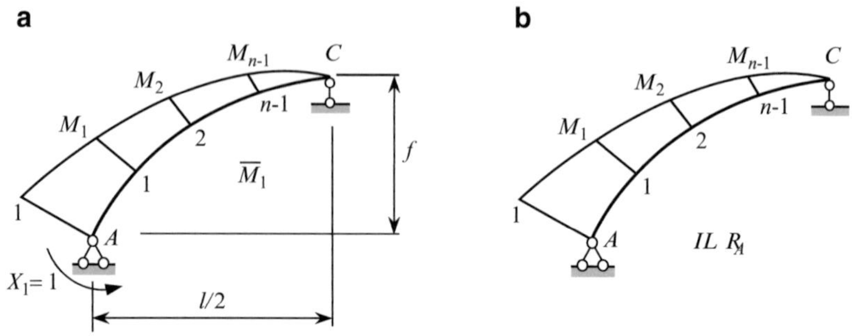

Fig. 5.7 Primary system, bending moment diagram due $X_1 = 1$ and influence line R_A

$$\overline{M}_1 = \begin{bmatrix} M_{1A} \\ M_{11} \\ \vdots \\ M_{1(n-1)} \end{bmatrix}, \overline{M}_1^{\mathrm{T}} = \begin{bmatrix} M_{1A} & M_{11} & \cdots & M_{1(n-1)} \end{bmatrix},$$

where first subscript 1 represents the primary unknown X_1, while the second subscript represents the index of the nodal point.

It can be shown [Smi84] that the unit displacement is

$$\delta_{11} = \frac{1}{128m^2EI}\left(\sec^3 \varphi_0 - 2\overline{S}\right),$$

where

$$m = \frac{f}{l}; \quad \varphi_0 = \arctan 4m, \quad \overline{S} = \frac{S}{l} = \frac{1}{4}\left[\sec \varphi_0 + \frac{1}{4m} \ln(4m + \sec \varphi_0)\right].$$

Free term Δ_{1P} of canonical equation presents an angular displacement at support A. This displacement may be presented in terms of shear at point A of a fictitious structure. For given supports of the primary system (Fig. 5.7a) the fictitious structure and real half-arch coincides. Fictitious shear at support A equals to the reaction at A. Influence line of this reaction due to load $P = 1$, which is directed normally to the axis of the arch, coincides with bending moment diagram $\overline{M}_1$ (see Table 5.3).

The final results are as follows [Smi84]. Matrix C of stability equation is determined by formula

$$C = L_m B_W \left(E - \frac{1}{\delta_{11}} \overline{M}_1 \overline{M}_1^{\mathrm{T}} B_W^* \right) G. \tag{5.13}$$

Moment influence matrix L_m will be the same as the one for a two-hinged parabolic arch.

The matrices of elastic load for primary system and given structure are

$$B_W^* = \frac{S_0}{6EI}\begin{bmatrix} 2 & 1 & 0 & . & & & \\ 1 & 4 & 1 & . & & & \\ . & . & . & . & . & . & . \\ & & & . & 1 & 4 & 1 \\ & & & . & 0 & 1 & 4 \end{bmatrix},$$

$$B_W = \frac{5S_0}{6EI}\begin{bmatrix} 0.5 & 0.1 & 0 & . & & & \\ 0.1 & 1.0 & 0.1 & . & & & \\ . & . & . & . & . & . & . \\ & & & . & 0.1 & 1.0 & 0.1 \\ & & & . & 0 & 0.1 & 1.0 \end{bmatrix} \tag{5.14}$$

At point A only one portion $(A-1)$ exist. Therefore, the first entry of the matrices B_W and B_W^* are twice as small as the rest of the diagonal entries. Last diagonal entry of these matrices remains unchanged since at the last point C the bending moment is zero. Diagonal matrix

$$G = \text{diag}\left[\sec\varphi_0 \quad \sec\varphi_1 \quad \ldots \quad \sec\varphi_{(n-1)}\right].$$

Detailed stability analysis of uniform parabolic arch with ratio $f/l = 1/6$ is presented in [Smi84]. If half-arch is divided into four equal parts then the critical load becomes

$$q_{\text{cr}} = (89.5EI)/l^3.$$

The critical uniformly distributed load may be determined by the formula (4.29) $q_{\text{cr}} = K(EI/l^3)$. Parameter K for different types of parabolic uniform arches in terms of f/l is presented in Table 4.6 [Din46], [Mor73].

5.5 Arch with Complex Tie

A tie of the arch may be represented as the single member at the level of supports (or the elevated tie), as well as a complex tie. One example of such complex tie is shown in Fig. 5.8a. If a load is applied to the tie, then internal forces in the hangers create the effect of elastic foundation for the arch. If load acts on the arch itself, then with the aid of hangers the load is transferred onto the tie.

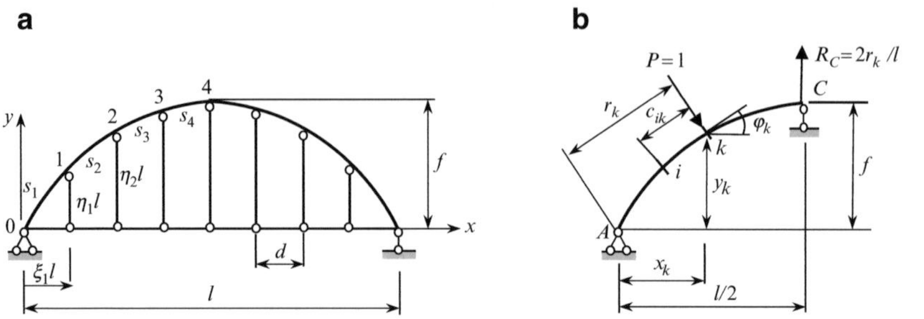

Fig. 5.8 (**a**) Design diagram of two-hinged arch with complex tie; (**b**) half-arch and geometrical notation

Table 5.4 Geometrical parameters of the parabolic arch [Smi84]

Point	$x_k = \xi_k l$	$y_k = \eta_k l$	tan $s\varphi_k$	sin φ_k	cos φ_k	s_k
0	0.0	0.0	1.6	0.8480	0.5300	–
1	0.125	0.175	1.2	0.76822	0.64018	$s_1 = 0.21506$
2	0.250	0.300	0.8	0.62470	0.78087	$s_2 = 0.17678$
3	0.375	0.375	0.4	0.37138	0.92848	$s_3 = 0.14577$
4	0.500	0.400	0.0	0.0	1.00	$s_4 = 0.12748$
Factor	l	l				l

The stability problem for arch with complex tie may be effectively solved by Smirnov's matrix method as shown below.

Design diagram of the uniform symmetric parabolic arch of span l and rise f is shown in Fig. 5.8. The complex tie includes a tie at the level of supports and vertical hangers. Connections of the hangers with arch itself are realized by means of simple hinges, while the connections with horizontal part of the complex tie are realized by means of multiple hinges. The tie of the arch is subjected to uniformly distributed load (this load is not shown). Distances d between all hangers are equal. The span of the arch is divided into $2n$ equal portions, so the length s_k of each curvilinear portion of the arch are different; in our case $n = 4$.

Assume that parameter $f/l = 0.4$; coordinates x and y for each joint (points $k = 0, 1, 2, 3, 4$), the trigonometric functions for these joints, and the length of curvilinear portions of the arch are shown in Table 5.4.

As before, the stability equation is $\det(C - \lambda E) = 0$. However, matrix C is calculated by taking into account the features of design diagram.

Numerical procedure is based on the following additional structural analysis [Smi84]:

1. Calculation of the vertical displacements of joints of a tie at the moment loss of stability occurs and corresponding change of internal force in each hanger.
2. Calculation of additional forces which prevent the loss of stability and corresponding additional bending moments in the arch.

This analysis can be presented in matrix form [Smi84]. For antisymmetrical loss of stability the first stage of numerical procedure for given arched structure includes the construction of the following matrices:

Initial Matrices

(The total number of panels for half-arch is $n = 4$)

1. Matrix of elastic loads according to (5.9) is

$$B_{WV} = \frac{l}{6EI}\begin{bmatrix} 2(\rho_1+\rho_2) & \rho_2 & 0 \\ \rho_2 & 2(\rho_2+\rho_3) & \rho_3 \\ 0 & \rho_3 & 2(\rho_3+\rho_4) \end{bmatrix} = \frac{l}{6EI}\{b_{ik}\},$$

where $\rho_k = s_k I/I_k l$ $(k = 1,\ldots,4)$, so the relative flexibility ρ_k of a member k of an arch equals to its relative length, i.e., $\rho_1 = s_1/l = 0.21506$, $\rho_2 = 0.17678,\ldots$. For entry b_{11} of matrix B_{WV} we get

$$b_{11} = 2(0.21506 + 0.17678) = 0.78368.$$

2. For primary system the matrix (3 by 3) of elastic loads

$$B_W^0 = \frac{l}{6EI}\{b_{ik}^0\}$$

is determined according to (5.3). For entry b_{11}^0 of matrix B_W^0 we get

$$\begin{aligned} b_{11}^0 &= \left(2+\frac{s_1}{2s_2}\right)\rho_1 + \left(2+\frac{s_2}{2s_1}\right)\rho_2 \\ &= \left(2+\frac{0.21506}{2\cdot 0.17678}\right)0.21506 + \left(2+\frac{0.17678}{2\cdot 0.21506}\right)0.17678 = 0.98715. \end{aligned}$$

3. The moment influence matrix L_m is determined by *radial* forces

$$L_m = \frac{1}{4}\begin{bmatrix} 3r_1 & 3r_2-4c_{12} & 3r_3-4c_{13} \\ 2r_1 & 2r_2 & 2r_3-4c_{23} \\ r_1 & r_2 & r_3 \end{bmatrix} = \frac{1}{4}\{l_{ik}\},$$

where segments, according to Fig. 5.8b, are

$$\begin{aligned} r_1 &= \bar{d}\cos\varphi_1 + \eta_1\sin\varphi_1, & c_{12} &= \bar{d}\cos\varphi_2 + (\eta_2-\eta_1)\sin\varphi_2, \\ r_2 &= 2\bar{d}\cos\varphi_2 + \eta_2\sin\varphi_2, & c_{13} &= 2\bar{d}\cos\varphi_3 + (\eta_3-\eta_1)\sin\varphi_3, \\ r_3 &= 3\bar{d}\cos\varphi_3 + \eta_3\sin\varphi_3, & c_{23} &= \bar{d}\cos\varphi_3 + (\eta_3-\eta_2)\sin\varphi_3. \end{aligned}$$

For entry $l_{12} = 3r_2 - 4c_{12}$ of matrix L_m we get

$$\begin{aligned} r_2 &= 2\bar{d}\cos\varphi_2 + \eta_2 \sin\varphi_2 = 2 \times 0.125l \times 0.78087 + 0.300l \times 0.62470 \\ &= 0.38263l, \\ c_{12} &= \bar{d}\cos\varphi_2 + (\eta_2 - \eta_1)\sin\varphi_2 = 0.125l \times 0.78087 + (0.30 - 0.175)l \\ &\quad \times 0.62470 = 0.17570l, \\ l_{12} &= 3r_2 - 4c_{12} = 3 \times 0.38263l - 4 \times 0.17570l = 0.44509l. \end{aligned}$$

In our case we get [Smi84]

$$L_m = \frac{l}{4}\begin{bmatrix} 0.64338 & 0.44509 & 0.23675 \\ 0.42892 & 0.76526 & 0.39926 \\ 0.21446 & 0.38263 & 0.48745 \end{bmatrix}.$$

4. The moment influence matrix L_{mV} is determined by *vertical* forces for simply supported beam of length $l/2$.

$$L_{mV} = \frac{l}{32}\begin{bmatrix} 3 & 2 & 1 \\ 2 & 4 & 2 \\ 1 & 2 & 3 \end{bmatrix}.$$

5. Diagonal matrices G and G_V are

$$G = \begin{bmatrix} \sec\varphi_1 & 0 & 0 \\ 0 & \sec\varphi_2 & 0 \\ 0 & 0 & \sec\varphi_3 \end{bmatrix},$$
$$G_V = \begin{bmatrix} \cos\varphi_1 & 0 & 0 \\ 0 & \cos\varphi_2 & 0 \\ 0 & 0 & \cos\varphi_3 \end{bmatrix}, \qquad (5.15)$$

where φ_k is the angle between tangent at point k and horizontal line.

Matrix Procedures

1. Matrix C_0 relates to the two-hinged arch without tie $C_0 = L_m B_W^0 G$.
2. Matrix C_1 takes into account effect of tie $C_1 = L_m B_{WV} L_{mV} G_V$.
3. Stability equation of the arched structure is $\det(C - \lambda E) = 0$. Stability matrix $C = C_0 - C_1 = (l^2/24EI)\{c_{ik}\}$.

4. Maximum stability parameter $\lambda_{\max}$ should be calculated numerically. For given arch $\lambda_{\max} = 0.895l^2/24EI$. Corresponding thrust is $H_{\min} = 1/\lambda_{\max} = 24EI/0.895l^2$ and smallest critical load is

$$q_{\min} = \frac{8H_{\min}f}{l} = 85.81\frac{EI}{l^3}.$$

According to the formula $C = C_0 - C_1$, the critical load for arch with tie is more than the critical load for arch without tie. For arch with tie in Fig. 5.8 the critical load is almost twice as large as the critical load for the same arch without tie.

More stability problems of the arches (with elastic supports and with overarched structure, out-of-plane stability, etc.) are presented in [Smi47], [Mor61], [Smi84].

5.6 Displacement Method

If an arch has certain features (e.g., nonsymmetrical or skew arch) then easiest approach for stability analysis is based on substituting the arch by frame. The frame is constructed from chords of the arch. Stability analysis of this substitute structure may be performed by any classical method, in particular, by Displacement method in canonical [Kar10], [Uma72-73] or expanded form [Sni66]. The Displacement method is exact for stability analysis of frames. However, it leads to approximate results when analyzing arches because the initial design diagram of arch is replaced by its modified scheme. This method has also some disadvantages – the modified design diagram contains the inclined members and diagram itself has sidesway. Increasing the number of substitute straight members leads to an increase in the overall computational complexity.

5.6.1 General

The distributed load which acts on the arch may be replaced by the set of concentrated forces P. Position of the nodal points and their number depend on the position of the flexural rigidity changes of the arch and required computational numerical accuracy. Increasing the number of substitute straight members leads to a increase in the numerical accuracy. Thus, this approximation may be performed by several different ways.

Different variations of approximating the arch by straight members in the vicinity of the crown C are shown in Fig. 5.9.

In all cases in Fig. 5.9 we obtained a framed structure and for its stability analysis we can apply the Displacement method. In case of a symmetrical structure we can replace the substitute frame by its equivalent half-frame.

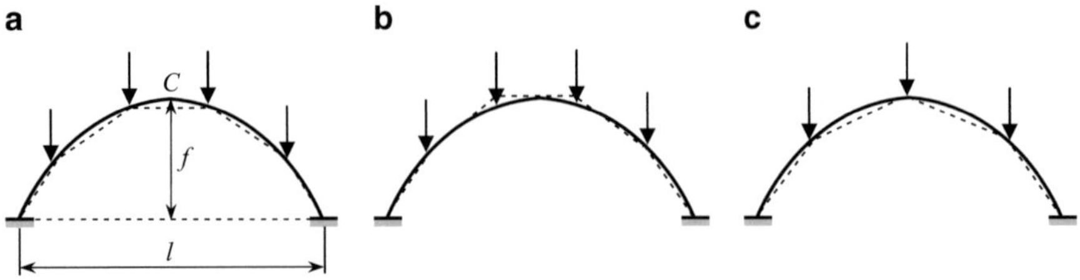

Fig. 5.9 Approximation of initial design diagram of the arch by substitute frame

Canonical equations of the Displacement method for structure with n unknowns Z_j ($j = 1,2, \ldots, n$) are

$$\begin{aligned} r_{11}Z_1 + r_{12}Z_2 + \cdots + r_{1n}Z_n &= 0 \\ r_{21}Z_1 + r_{22}Z_2 + \cdots + r_{2n}Z_n &= 0 \\ &\vdots \\ r_{n1}Z_1 + r_{n2}Z_2 + \cdots + r_{nn}Z_n &= 0. \end{aligned} \tag{5.16}$$

Features of (5.16)

1. Since the forces P_i are applied only at the joints, then the canonical equations are homogeneous ones.
2. Bending moment diagrams, caused by unit displacements of introduced constrains, within the *compressed* members are *curvilinear*. Reactions of constraints depend on axial forces in the members of the frame, i.e., contain parameter υ of critical load. If a frame is subjected to different forces P_i, then critical parameters should be formulated for each compressed member $\upsilon_i^2 = P_i l_i^2/(EI)_i$ and after that all of these parameters should be expressed in terms of parameter υ for specified basic member. Thus, the unit reactions are functions of parameter, i.e., $r_{ik}(\upsilon)$.

The trivial solution ($Z_i = 0$) of (5.16) corresponds to initial nondeformable design diagram. Nontrivial solution ($Z_i \neq 0$) corresponds to the new form of equilibrium. This occurs if the determinant, which is consisting of coefficients of unknowns, equals zero, i.e.,

$$\det \begin{bmatrix} r_{11}(\upsilon) & r_{12}(\upsilon) & \cdots & r_{1n}(\upsilon) \\ r_{21}(\upsilon) & r_{22}(\upsilon) & \cdots & r_{2n}(\upsilon) \\ \vdots & \vdots & \vdots & \vdots \\ r_{n1}(\upsilon) & r_{n2}(\upsilon) & \cdots & r_{nn}(\upsilon) \end{bmatrix} = 0. \tag{5.17}$$

Condition (5.17) is called the stability equation of a structure in a form of the Displacement method. For practical engineering, it is necessary to calculate the

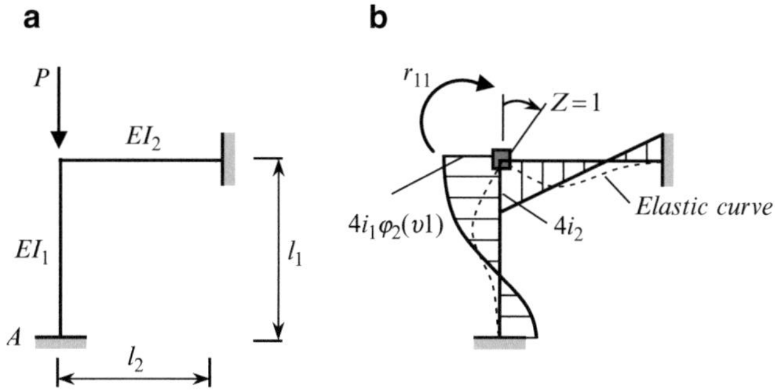

Fig. 5.10 (**a**) Design diagram; (**b**) primary system of the Displacement method and unit bending moment diagram

smallest root of the above equation. This root defines the smallest parameter υ of critical force, or smallest critical force. Canonical equation of the Displacement method in the form (5.17) for stability analysis of the frames was developed by Leites (1937) [Uma72-73, Chapters 17.1–17.10].

It is obvious that condition (5.17) leads to a transcendental equation with respect to parameter υ. The functions $\varphi(\upsilon)$ and $\eta(\upsilon)$ are tabulated (Table A.33). Since the determinant is very sensitive with respect to parameter υ, it is recommended to solve (5.17) using a computer. The functions $\varphi(\upsilon)$ and $\eta(\upsilon)$ may be presented in approximate form [Bol64]; in this case (5.17) leads to an algebraic equation.

The only limitation for applying this method is that the flexural rigidity and axial compressed force should be constant within the each member.

Let us derive the stability equation and determine the critical load for frame shown in Fig. 5.10a. This frame has one unknown of the Displacement method. The primary unknown is the angle of rotation of rigid joint. Figure 5.10b shows the primary system, elastic curve, and bending moment diagram caused by unit rotation of introduced constrain. The bending moment diagram for compressed vertical member of the frame is curvilinear. The ordinate for this member is taken from Table A.32, row 1.

The bending moment diagram yields $r_{11} = 4i_1\varphi_2(\upsilon_1) + 4i_2$, where $i_1 = EI_1/l_1$, $i_2 = EI_2/l_2$ and parameter of critical load $\upsilon_1 - l_1\sqrt{P/EI_1}$. Note that subscript 2 at function φ is related to the clamped–clamped member subjected to angular displacement of one support (Table A.32), while the subscript 1 at the parameter υ is related to the compressed-bent member 1. Canonical equation of the Displacement method is $r_{11}(\upsilon_1)Z = 0$.

Nontrivial solution of this equation leads to equation of stability $r_{11} = 0$ or in expanded form

$$r_{11} = \frac{4EI_1}{l_1}\varphi_2(\upsilon_1) + \frac{4EI_2}{l_2} = 0.$$

Note, that Displacement method allows us to take into account some changes in the design diagram; for example, if a fixed support A is replaces by a pinned support, then stability equation becomes

$$r_{11} = 3i_1\varphi_1(\upsilon_1) + 4i_2 = 0.$$

Special Cases

1. Assume that $l_2 \to 0$. In this case the second term $(4EI_2/l_2) \to \infty$, rigid joint is transformed to clamped support and the initial frame is transformed into the vertical clamped–clamped column. Stability equation becomes $\varphi_2(\upsilon_1) = -\infty$. The smallest root of this equation is $\upsilon_1 = 2\pi$ and critical force becomes

$$P_{\text{cr}} = \frac{\upsilon_1^2 EI}{l_1^{12}} = \frac{4\pi^2 EI}{l_1^2} = \frac{\pi^2 EI}{(0.5l_1)^2},$$

   where $\mu = 0.5$ is effective-length factor for clamped–clamped column.
2. Assume $EI_2 \to 0$. In this case, the rigid joint is transformed to hinge and the initial frame is transformed into the vertical clamped–pinned column. Stability equation becomes $\varphi_2(\upsilon_1) = 0$. The smallest root of this equation is $\upsilon_1 = 4.488$ and for critical force we get

$$P_{\text{cr}} = \frac{\upsilon_1^2 EI}{l_1^2} = \frac{4.488^2 EI}{l_1^2} = \frac{\pi^2 EI}{(0.7l_1)^2}, \quad \mu = 0.7.$$

3. If $l_1 = l_{2,}\quad EI_1 = EI_2$, then stability equation becomes $\varphi_2(\upsilon_1) + 1 = 0$. The smallest root of this equation is $\upsilon_1 = 5.3269$ and critical load equals $P_{\text{cr}} = (\upsilon_1^2 EI/l_1^2) = (28.397EI/l_1^2)$.

Modified Approach of the Displacement Method

In general case, the Displacement method requires introducing constraints, which prevent angular displacement of rigid joints and independent linear displacements of joints. However, in stability problems of a frame with sidesway, it is possible for some modification of the classical Displacement method. Using modified approach, we can introduce a new type of constraint, mainly the constraint, which prevents angular displacement, but *simultaneously* has a linear displacement (Table A.32, row 3). Modified approach of the Displacement method is presented in [Kar10].

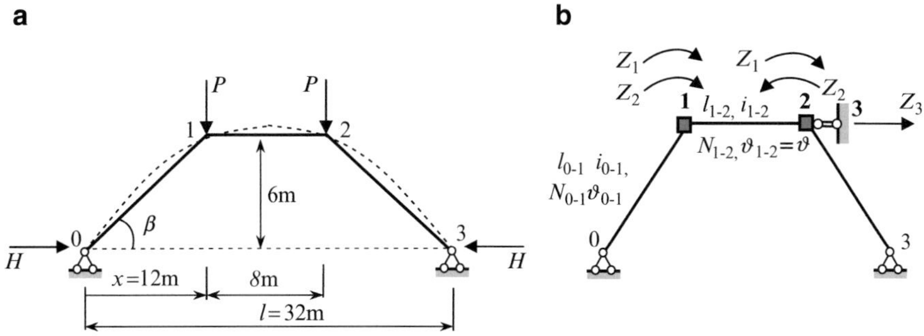

Fig. 5.11 (**a**) Design diagram of symmetrical two-hinged uniform arch and its substituted frame; (**b**) primary system of Displacement method and group unknowns

5.6.2 *Two-Hinged Arch*

Let us demonstrate application of the Displacement method in canonical form for stability analysis of two-hinged uniform symmetric arch loaded by two forces P (Fig. 5.11a). Simplest version of the substituted frame is shown by solid lines; it is constructed in such way so that forces are applied at the rigid joints of the frame. The unavoidable disadvantage of this system is that the substituted frame is with sidesway.

To obtain the primary system of Displacement method we need to introduce two rigid joints 1 and 2 and support 3 into the design diagram; they are shown in bold in Fig. 5.11b. Constraints 1 and 2 prevent angular displacements of joint 1 and 2 and constraint 3 prevent linear displacement of the cross bars 1–2.

For stability analysis of this symmetrical frame it is very effective to adopt the group unknowns of the Displacement method as shown in Fig. 5.11b; Z_1 represents the *simultaneous* angular displacement rotation of introduced constraints 1 and 2 in *one* direction, while Z_2 represents the *simultaneous* angular displacement of same introduced constraints in *opposite* directions (Bresse's method, 1854) [Bre54, 59]; Z_3 represents a linear displacement of cross bars 1–2.

By using the group unknowns we can separate the full system of canonical equations into two separate independent subsystems. First subsystem allows us to determine the critical load for symmetrical form of loss of stability and second for antisymmetrical form of loss of stability.

Stability analysis consists of the following steps:

1. Determine some parameters of the frame

$$l_{0-1} = 13.4164m, \quad \tan\beta = 0.5, \quad \sin\beta = 0.44721, \quad \cos\beta = 0.89443.$$

2. For a given frame we find the thrust H by any analytical approach, for example, by the Force method or using tabulated data. Axial loads in members 0–1 and 1–2 in terms of H are $N_{0-1} = H/\cos\beta$; $N_{1-2} = H$.

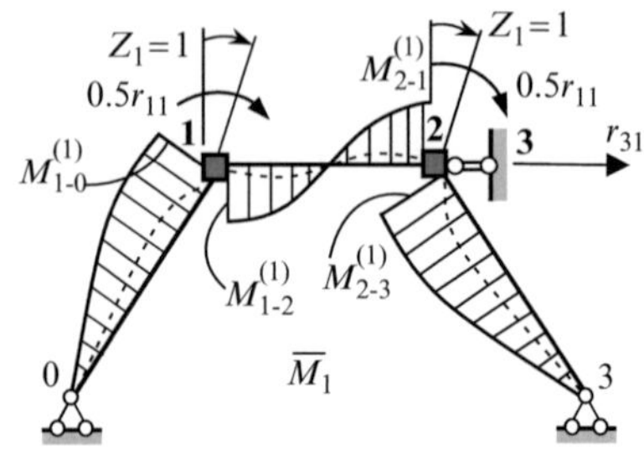

Fig. 5.12 Bending moment diagram in the first unit state

3. For each member of length l_i of the frame, calculate the parameter $\vartheta_i = l_i\sqrt{N_i/EI}$ of critical force. For one specific member of the frame the parameter of a critical load is adopted as the base parameter and all remaining parameters are expressed in terms of the base parameter.
   Parameter of critical loads for members 0–1 and 1–2 are

$$\vartheta_{0-1} = l_{0-1}\sqrt{\frac{N_{0-1}}{EI}}, \quad \vartheta_{1-2} = l_{1-2}\sqrt{\frac{N_{1-2}}{EI}} = 8\sqrt{\frac{H}{EI}}.$$

   Assume the parameter $\vartheta_{1-2} = \vartheta$ is a base parameter. Then for ϑ_{0-1} we get

$$\vartheta_{0-1} = 13.4164\sqrt{\frac{H}{0.89443EI}} = 1.7733 \times 8\sqrt{\frac{H}{EI}} = 1.7733\vartheta.$$

   Figure 5.11b contains the length l, stiffness per unit length i, axial forces N, and critical parameter ϑ for each member of the substituted frame.
4. Now we need to construct bending moment diagrams caused by unit primary unknowns of Displacement method and calculated unit reactions.

First unit state ($Z_1 = 1$). Bending moment diagram caused by the unit rotation of introduced constraints $Z_1 = 1$ (antisymmetrical unknown) is shown in Fig. 5.12. This diagram is antisymmetrical. Elastic curve is shown by dotted line.

$$\begin{aligned}
M^{(1)}_{1-0} &= 3i_{0-1}\varphi_1(\vartheta_{0-1}) \\
M^{(1)}_{2-3} &= M^{(1)}_{1-0} \\
M^{(1)}_{1-2} &= 4i_{1-2}\varphi_2(\vartheta) + 2i_{1-2}\varphi_3(\vartheta) \\
M^{(1)}_{2-1} &= M^{(1)}_{1-2}
\end{aligned}$$

Top subscript (1) at M denotes the first state. Subscript 1 at function φ_1 relates to pinned–fixed beam in case of angular displacement of fixed support while

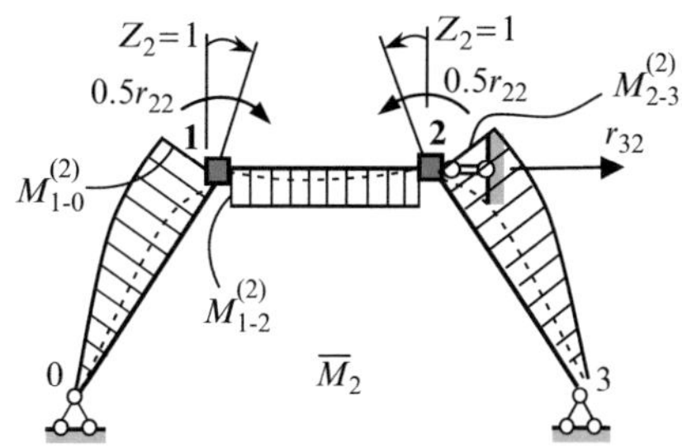

Fig. 5.13 Bending moment diagram in the second unit state

subscripts 2 and 3 relates to fixed–fixed beam (member 1–2) in case of angular displacement of fixed support (Table A.32).

Unit reaction

$$r_{11} = 2\left(M^{(1)}_{1-0} + M^{(1)}_{1-2}\right) = 2[3i_{0-1}\varphi_1(\vartheta_{0-1}) + 4i_{1-2}\varphi_2(\vartheta) + 2i_{1-2}\varphi_3(\vartheta)].$$

After substituting the corresponding quantities we get

$$r_{11} = 2[0.2236\varphi_1(1.7733\vartheta) + 0.5\varphi_2(\vartheta) + 0.25\varphi_3(\vartheta)]EI.$$

Secondary reactions will be calculated later.

Second unit state ($Z_2 = 1$). Bending moment diagram caused by the unit rotation of introduced constraints $Z_2 = 1$ (symmetrical unknown) is shown in Fig. 5.13. This diagram is symmetrical.

$$M^{(2)}_{1-0} = 3i_{0-1}\varphi_1(\vartheta_{0-1})$$
$$M^{(2)}_{2-3} = M^{(2)}_{1-0}$$
$$M^{(2)}_{1-2} = 4i_{1-2}\varphi_2(\vartheta) - 2i_{1-2}\varphi_3(\vartheta)$$

Unit reaction

$$r_{22} = 2\left(M^{(2)}_{1-0} + M^{(2)}_{1-2}\right) = 2[3i_{0-1}\varphi_1(\vartheta_{0-1}) + 4i_{1-2}\varphi_2(\vartheta) - 2i_{1-2}\varphi_3(\vartheta)].$$

After substituting the corresponding quantities we get

$$r_{22} = 2[0.2236\varphi_1(1.7733\vartheta) + 0.5\varphi_2(\vartheta) - 0.25\varphi_3(\vartheta)]EI.$$

Since $\overline{M}_1$ and $\overline{M}_2$ diagrams are antisymmetrical and symmetrical, respectively, then $r_{12} = r_{21} = 0$. This is an important result due to application of group unknowns. Other secondary reactions are discussed below.

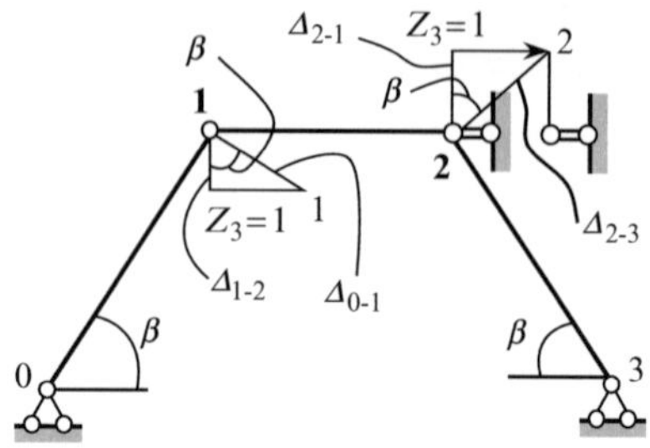

Fig. 5.14 Hinged scheme of the frame and displacement of the joints in the third unit state

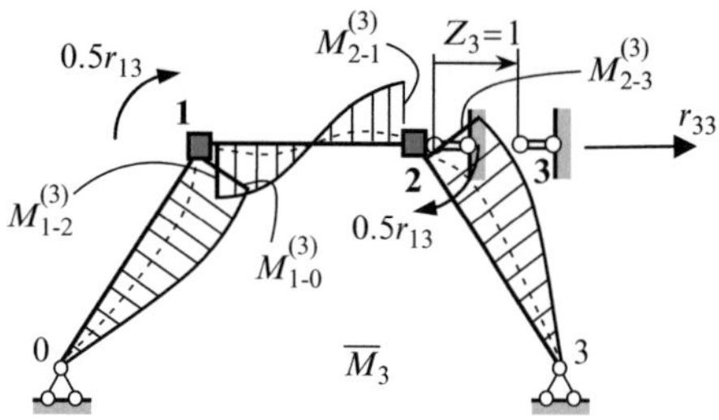

Fig. 5.15 Bending moment diagram in the third unit state

Third unit state ($Z_3 = 1$). Hinged scheme of the frame and its position caused by $Z_3 = 1$ is shown in Fig. 5.14. The displacement of point 1 is directed perpendicularly to member 0–1 and its new position is denoted as 1′. Point 2 is moved perpendicularly to member 2–3 and its new position is denoted as 2′.

$$\begin{aligned}\Delta_{0-1} &= \csc\beta,\\ \Delta_{1-2} &= \cot\beta,\\ \Delta_{2-3} &= \csc\beta,\\ \Delta_{2-1} &= \cot\beta.\end{aligned}$$

For member 1–2 relative vertical displacement of the ends is

$$\Delta = \Delta_{1-2} + \Delta_{2-1} = 2\cot\beta.$$

Bending moment diagram in the third unit state is shown in Fig. 5.15.

$$\begin{aligned}M_{1-0}^{(3)} &= 3\frac{i_{0-1}}{l_{0-1}}\Delta_{0-1} \times \varphi_1(\vartheta_{0-1})\\ M_{2-3}^{(3)} &= M_{1-0}^{(3)}\\ M_{1-2}^{(3)} &= 6\frac{i_{1-2}}{l_{1-2}}\Delta \times \varphi_4(\vartheta)\\ M_{2-1}^{(3)} &= M_{1-2}^{(3)}\end{aligned}$$

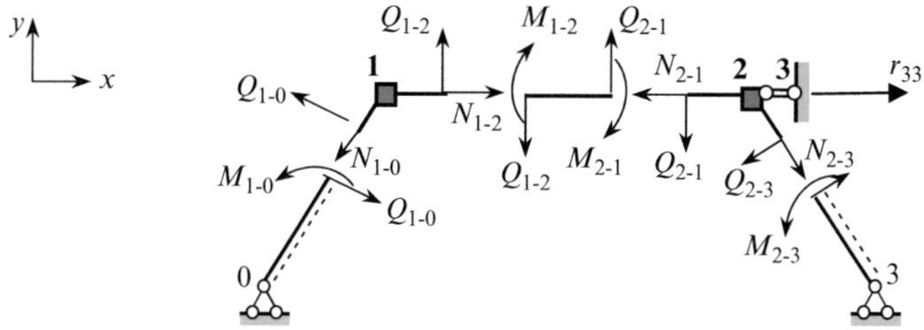

Fig. 5.16 Calculation of r_{33}

Design diagram for calculation of $r_{33.}$ is shown in Fig. 5.16. The section passing through member 0–1 is infinitely close to joint 1. Bending moment $M_{1\text{–}0}$ is directed according to position of extended fibers, which are shown by dotted line; the top subscript "3" is omitted. This moment is equilibrated by shear force $Q_{1-0} = (3EI/l_{0-1}^3)\eta_1(\vartheta_{0-1})\Delta_{0-1}$. Then this force is transferred on the part which is adjacent to joint 1. Also we need to show the axial force $N_{1\text{–}0.}$

Similar procedure is applied to portions 1–2 and 2–3. Shear force for fixed–fixed member 1–2 subjected to relative vertical displacement Δ of the ends equals $Q_{1-2} = Q_{2-1} = (EI/l_{0-1}^3)\eta_2(\vartheta)\Delta$. Equilibrium equations $\sum X = 0$ and $\sum Y = 0$ for joints 1 and 2 allow us to calculate the normal forces. Relationship $N_{1-2} = N_{2-1}$ can be used for verification of computed results. Moreover, $N_{0-1} = -N_{2-3}$. Therefore, equation $\sum X = 0$ for cross bar in whole leads to the following result

$$r_{33} = 2Q_{1-0}^{(3)} \sin\beta = 2 \times \frac{3EI}{l_{0-1}^3}\Delta_{0-1}\eta_1(\vartheta_{0-1}) \sin\beta = 2 \times \frac{3EI}{l_{0-1}^3}\eta_1(\vartheta_{0-1}).$$

Performing a similar procedure over the bending moment diagram $\overline{M}_1$ we get

$$r_{31} = -2\left[\frac{3EI}{l_{0-1}^2 \sin\beta}\varphi_1(\vartheta_{0-1}) - \frac{12EI}{l_{1-2}^2 \tan\beta}\varphi_4(\vartheta)\right].$$

This result may be obtained by considering equilibrium of joints 1 and 2 from diagram $\overline{M}_3$. In this case, moment is

$$r_{31} = \underbrace{-M_{1-0}^{(3)} + M_{1-2}^{(3)}}_{\text{Joint 1}} + \underbrace{M_{2-1}^{(3)} - M_{2-3}^{(3)}}_{\text{Joint 2}}.$$

Substituting the numerical data, we get

$$\begin{aligned}
r_{11} &= 2[0.2236\varphi_1(1.7733\vartheta) + 0.5\varphi_2(\vartheta) + 0.25\varphi_3(\vartheta)]EI,\\
r_{22} &= 2[0.2236\varphi_1(1.7733\vartheta) + 0.5\varphi_2(\vartheta) - 0.25\varphi_3(\vartheta)]EI,\\
r_{33} &= 2 \times 0.001242\eta_1(1.7733\vartheta)EI,\\
r_{13} &= r_{31} = -2[0.037476\varphi_1(1.7733\vartheta) - 0.375\varphi_4(\vartheta)]EI.
\end{aligned}$$

Since the bending moment diagrams $\overline{M}_1$ and $\overline{M}_3$ are antisymmetrical while $\overline{M}_2$ is symmetrical, then

$$r_{12} = r_{21} = 0 \quad \text{and} \quad r_{32} = r_{23} = 0.$$

Now we can form the stability equation, calculate the critical parameter, and determine the critical load.

In our case, the set of canonical equations of stability

$$\begin{aligned} r_{11}Z_1 + r_{12}Z_2 + r_{13}Z_3 &= 0 \\ r_{21}Z_1 + r_{22}Z_2 + r_{23}Z_3 &= 0 \\ r_{31}Z_1 + r_{32}Z_2 + r_{33}Z_3 &= 0. \end{aligned}$$

are separated into two independent systems

$$\begin{aligned} r_{11}Z_1 + r_{13}Z_3 &= 0 \\ r_{31}Z_1 + r_{33}Z_3 &= 0 \end{aligned} \quad \text{and} \quad r_{22}Z_2 = 0.$$

Stability equations become

$$\begin{vmatrix} r_{11} & r_{13} \\ r_{31} & r_{33} \end{vmatrix} = 0 \quad \text{and} \quad r_{22} = 0.$$

The first subsystem describes the antisymmetrical loss of stability, while second equation describes symmetrical loss of stability. For solution of these equations we need to take into account expressions for functions φ_i, η according to Table A.33. For stability parameters we get

$$\vartheta_{\text{ant}} = 2.0344 \quad \text{and} \quad \vartheta_{\text{sym}} = 2.6007.$$

Critical thrust becomes

$$H_{\text{asym}} = 2.0344^2 \frac{EI}{8^2}, H_{\text{sym}} = 2.6007^2 \frac{EI}{8^2}.$$

After that we can calculate the critical load P. Note, these results corresponds to a crude model approximating entire arch.

Numerical results for three-hinged and two-hinged arches subjected to uniformly distributed vertical load within the whole span are presented in [Sni66]. If entire arch is represented as five chords with equal horizontal projections then relative error of the Displacement method is no greater than 2%.

5.7 Comparison of the Smirnov's and Displacement Methods

Advantages and Disadvantages of Smirnov's Method

In Smirnov' method the axis of the arch is replaced by a set of curvilinear segments which coincides with arch itself. Therefore, errors of the method arise from numerically inaccurate calculation of nodal point coordinates obtained from the solution of transcendental equation. It means that the results which are obtained from Smirnov' method should be treated as ground truth.

According to Smirnov's method, the arch should be presented as a set of curvilinear segments of equal length and constant stiffness within each segment. Meeting these requirements is not always possible in case of arches with variable stiffness (of course, with a reasonable number of the finite portions). The other disadvantage is related to the difficulties of analyzing nonsymmetric arches.

Any change in the design diagram of the arch (addition of the overarched structure, combined tie, elevated tie, fixed supports, etc.) leads to a procedure that is not easily generalizable. Even if the governing general stability equation $\det(C - \lambda E) = 0$ holds true, the matrix C depends on the features of the arched structure. These peculiarities limit the scope of the method.

We note that the *span* of a parabolic arch can be divided into a set of equal lengths portions. In this case, we can use the concept of the parabolic chain [Rab54b], [Rab58, Vol. II]. Its properties and important relationships are considered for free vibration analysis in Chap. 6.

Advantages and Disadvantages of the Classical Methods in Canonical Form

Stability problems of the arches may be solved using the classical Force and Displacement methods in canonical form. Approximate stability analysis by both of these methods requires the construction of a substitute frame, and thus these methods lead to the approximate results. A poor choice for the primary system of the Force method can lead to significant difficulties of computational nature [Smi47]. Displacement method does not have similar disadvantages, because the primary system is constructed according to strong rules. These rules allow considering nonsymmetrical arches, take into account the elastic supports as well as *any* type of loading. It is very convenient that the Displacement method in canonical form is easily generalizable. As in case of Smirnov's method, computational procedure becomes more cumbersome with increasing the number of elements which approximate an arch.

Displacement method, generally, allows constructing the stability equation for the full spectra of critical forces. Given this, it is unnecessary to assume the initial form of loss of stability. The group unknowns allow simplifying numerical procedure and separate stability analysis for symmetrical and antisymmetrical forms of the loss of stability.

Part III
Vibration Analysis

Chapter 6
Free Vibration of Arches

The theory of vibration is a special branch of structural analysis. This theory allows us to evaluate internal forces and displacements in the structure caused by dynamical loads of different nature. Often it is the case that these forces and displacements are significantly greater than the forces and displacement for the case of static loading. Engineering practice has seen a lot of cases when underestimation of this feature of dynamical loads leads to the collapse of structure.

This chapter is devoted to free vibration analysis of arches. Different types of arches are considered. Among them are circular and parabolic arches with different boundary conditions. Analytical methods of analysis are demonstrated.

6.1 Fundamental Concepts

The mathematical basis for the theory vibration of the arches was laid by Kirhhoff (1824–1887) [Kir76]. Fundamental investigation of vibration of deformable structures was performed by Lord Rayleigh (1842–1912) [Ray77], Love (1863–1940) [Lov20], Timoshenko (1878–1972) [Wea90]. Duhamel (1797–1872) gave a general method for analyzing the forced vibration of elastic bodies [Duh43]. A significant contribution to the problem of vibration of arches was done by Morgaevsky [Mor40], Demidovich [Dem49], Rabinovich [Rab51], [Rab58], Chudnovsky [Chu52], Pratusevich [Pra52], Bolotin [Bol78], [Bol84]. Prokofiev and Smirnov [Pro48], Laura [Lau87a,b], [Lau88a–c], to name a few, devoted a significant amount of work to developing numerical methods of dynamical analysis of arches.

6.1.1 General

We begin with considering some of the key concepts.

I.A. Karnovsky, *Theory of Arched Structures: Strength, Stability, Vibration*,
DOI 10.1007/978-1-4614-0469-9_6,

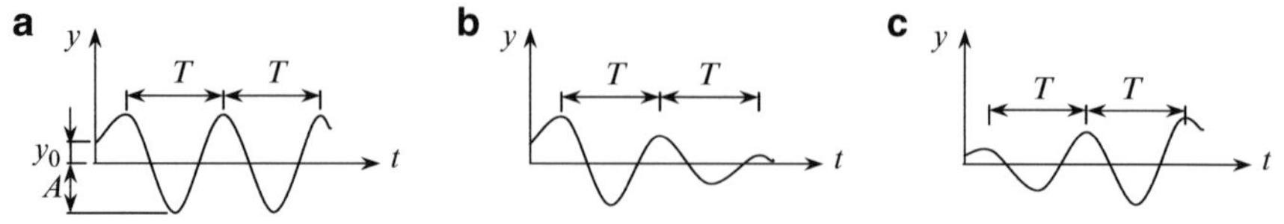

Fig. 6.1 Types of oscillatory motions

Kinematics of Vibrating Processes

The simplest periodic motion can be written as

$$y(t) = A\sin(\omega t + \varphi_0),$$

where A is the amplitude of vibration; φ_0 is the initial phase of vibration; t is the time. This case is presented in Fig. 6.1a. The initial displacement $y_0 = A\sin\varphi_0$ is measured from the static equilibrium position. The number of cycles of oscillation during 2π seconds is referred to as circular (angular or natural) frequency of vibration $\omega = 2\pi/T$(radians per second or s^{-1}), T(sec) is the period of vibration. Figure 6.1b, c presents the damped and increased vibration with constant period.

Vibration Forces

During vibration, a structure is subjected to different forces. These forces are different in nature and exert a different influence on the vibrating process. All forces may be divided into the following groups: disturbing forces, positional (restoring) forces, resisting forces, and forces of the mixed character.

1. *Disturbing forces* are functions of time. These forces are usually subdivided into the following types: immovable periodical loads, impact (impulsive) loads, moving loads, seismic loads.
2. *Restoring forces* depend on the displacement of the structure, arise due to deviation of system from a static equilibrium position and tend to return the system to its initial position. Restoring properties of a system are described by its elastic characteristic $P = P(y)$, where P is a static force which is applied to the structure. Characteristic $P - y$ may be linear or nonlinear. Some types of characteristics $P - y$ are presented in Table 6.1; in all cases y is the displacement at the point of P.
3. *Resisting forces*. The forces of inelastic resistance (friction or damping forces) depends on the velocity υ of motion, $R = R(\upsilon)$, and always act in the opposite direction of velocity. These forces are the result of internal friction in the material of a structure and/or in the connections of a system.

Table 6.1 Types of elastic members and their characteristics

Design diagram	Characteristic *P*-*y*	Design diagram	Characteristic *P*-*y*
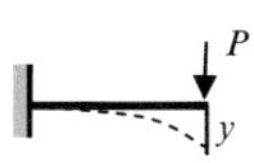	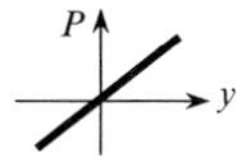	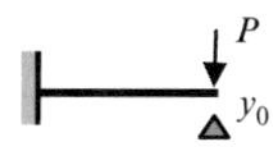	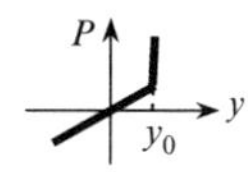
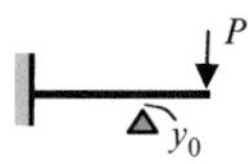	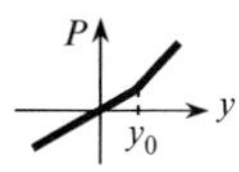		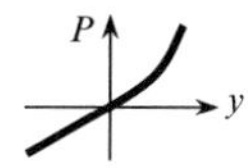

Degrees of Freedom

Fundamental concept in the structural dynamics is the degrees of freedom of a structure. They are independent geometrical parameters that describe positions of a structure at any moment in time. The difficulties and features of dynamical analysis of structures depends first of all on the number of degrees of freedom.

All structures may be divided into two principal classes according to their number of degrees of freedom. They are structures with concentrated and distributed parameters. Members with concentrated parameters assume that the distributed mass of the member itself may be neglected in comparison with the lumped mass, which is located on the member. The structure with distributed parameters is characterized by uniform or nonuniform distribution of mass within its parts. From a mathematical point of view, the difference between the two types of systems is the following: systems of the first class are described by ordinary differential equations, while systems of the second class are described by partial differential equations.

The fundamental difference of the concept of "degrees of freedom" in static and structural dynamics and computation of degrees of freedom for different deformable structures is discussed in [Kar10].

Free and Forced Vibrations

Different types of forces acting on a structure lead to different types of vibrations. Among them are two general classes – free and forced vibration.

Vibrations of a system in which disturbing forces are absent are called free vibrations. In the case of free vibrations, the system is subjected to forces inherent to the system itself, i.e., the restoring and resisting forces. In order to impose free vibrations, nonzero initial conditions should be created, which means that the system is subjected to some initial displacement and/or initial velocity. Free vibration may be linear or nonlinear depending on the characteristics of restoring and resisting forces. Absence of resisting forces leads to the *free undamped* vibrations; in this case, the system is subjected only to a restoring force.

The free vibration occurs at a certain frequency inherent to the system itself. Therefore, this frequency is called eigenfrequency or just frequency of free vibration. If a system has two or more degrees of freedom, and the system vibrates at a single frequency, then the ratio of displacements of two arbitrary points of the system remains constant. The shape of vibration is the profile of the system at the specified frequency of the system and is therefore known as the eigenfunction or just mode shapes of vibration. Due to the fact that dampening forces are always present in any structure, free vibrations will always decrease in magnitude as a function of time. However, knowing the frequencies and shapes of free vibrations is necessary for further analysis of the system and determining these frequencies and shapes is the primary problem of free vibration.

Vibration of a system caused by any disturbing forces is called a forced vibration. Neglecting the resisting forces in the system leads to *forced undamped* vibration. Just as in the case of free vibrations, forced vibrations may be linear or nonlinear. The ratio of any dynamical quantity (displacement, reaction, etc.) to a static quantity due to the maximal disturbing force is called dynamic coefficient.

6.1.2 Discrete Models of the Arches

In the general case, transversal vibrations of the arch are described by a complicated partial differential equation; special modification of the equation are discussed in [Chu52], [Rek73]. Generally, it is extremely difficult to obtain analytical solutions for vibration of an arch as a system with distributed parameters. For approximate vibration analysis the arch will be presented in discrete form as a system with finite number of degrees of freedom. Discretization of an arch may be performed by several different ways. Perhaps the most common one is to replace the mass distributed along the axis of an arch by a series of point masses. Is it possible to approximate the arch by a set of absolutely rigid chords with elastic connections between them [Ter54], [Kis80].

Vibration of arches has some features. In the arch, unlike beam, the different masses are displaced in nonparallel directions [Rab58]. The second feature of dynamical behavior of the arch is that during vibrations horizontal inertial forces are of the same order as the vertical ones, and in the case of a tall arch the magnitude of horizontal forces can exceed that of vertical forces.

Figure 6.2a presents an arch with distributed mass, which is replaced by three point masses; the axis of the arch remains curvilinear. In this case, we have a system with six degrees of freedom.

It is possible to replace the curvilinear axis of the arch by a set of straight members; in fact, the arch is replaced by some frame. Different types of possible approximating frames are shown in Fig. 6.2b–d.

In Fig. 6.2b, the arch is replaced by four straight rods and masses are concentrated in the joint of the frame. This structure has two degrees of freedom.

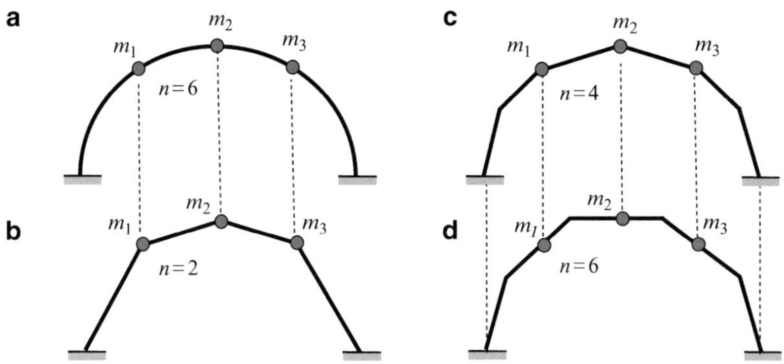

Fig. 6.2 Possible discretization of the arch and degrees of freedom

In Fig. 6.2c, the arch is replaced by six rods and as before by three lumped masses, which are still located at the joints. This structure has four degrees of freedom.

In Fig. 6.2d, the arch is replaced by five rods and as before by three lumped masses, which differs from the one in Fig. 6.2b by the fact that the masses are located at the center of straight member. This structure has six degrees of freedom.

We can see that the simplest scheme is the one under which the mass of each straight member is replaced by two point masses, applied at the ends of this member. It is evident that any type of shown discretization types may be applied to arches with variable cross-section, arbitrary shape, and various boundary conditions.

The difference in each arch-approximating scheme is that for each of scheme in Fig. 6.2 the number of the frequencies, correspondingly, mode shape of vibrations, is different. However, the lowest frequencies of vibration for various approximation schemes of the arch are close. The increase in the number of elements that approximate the frame leads to a rapid pursuit of the true frequency.

In determining the fundamental frequency of vibration, one should choose a scheme of approximation which leads to a simple and obvious computational procedure that yields accurate results. It must be remembered that the replacement of the arch by a frame leads to a decrease in fundamental frequency in comparison with the frequency for the arch [Pro48].

Approximate Solutions of Dynamic Analysis of the Discrete Models of the Arches

If a frame with sidesway, which is an approximate model of an arch, has two or more point masses then theirs displacements (and the number of degrees of freedom and the choice of generalized coordinates) are not always obvious.

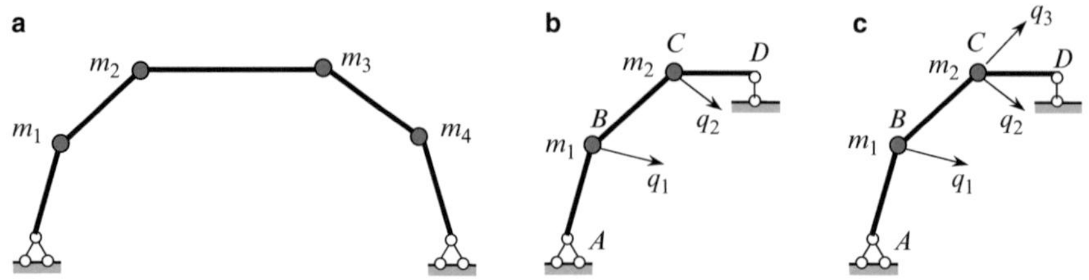

Fig. 6.3 Discrete arch models and generalized coordinates

Figure 6.3a presents an arched structure after it has been replaced by a set of straight members with point masses at the rigid joints. A structure contains four point masses. In the case of antisymmetric vibration, the equivalent half-structure is shown in Fig. 6.3b. It is evident that virtual displacement q_1 of mass m_1 is directed perpendicular to rod AB. The support D does not prevent horizontal displacement. Therefore, the displacement of the joint C is not obvious. Assume that this structure has two degrees of freedom. For convenience, let us show q_2 perpendicular to the rod BC. It is clear that if rod BC was an extension of the rod AB, then this representation of q_2 would be exact. In our case, q_2 does not account for displacement m_2 along the rod BC. Therefore, given such a choice of the generalized coordinates, the corresponding analysis of structure with two degrees of freedom is only approximate. Since the frame approximating the arch is piecewise-linear, it is difficult to show the exact direction of q_2 [Pro48].

More exact solution can be obtained by considering the structure as a system with three degrees of freedom. For this, we must show a virtual displacement of joint C, consisting of two components: one of them is q_2, which is directed perpendicular to the rod BC and other component, q_3, is directed along the rod BC (Fig. 6.3c). Both of schemes in Fig. 6.3b, c will be analyzed later.

Dynamic analysis of arch is approximate not only because we replace initial design diagram of an arch by its discrete model, but also because different terms of the Maxwell–Mohr integral may be taken into account while calculating unit and load displacements. Neglecting of shear and/or axial forces also leads to approximate analysis of the arch.

Unless stated otherwise, we will consider vibration of arches under the following assumptions:

1. The arch material is linearly elastic (Hooke's law applies).
2. The center line of an arch is incompressible.
3. Resisting forces in the material of a structure and at connections are ignored.
4. Effects of shear and rotary inertia are neglected.

6.2 Eigenvalues and Eigenfunctions of Arches with Finite Number Degrees of Freedom

Behavior of structures with finite number degrees of freedom may be described by two types of differential equations. They are equations in displacement (i.e., in the form of the Force method) and equations in reactions (i.e., in the form of the Displacement method). Below we consider only undamped vibration in form of the Force method. Analysis of free vibration of the framed structures in form of the Displacement method is presented in refs. [Kle72], [Kar01], [Kar04], and [Kar10].

6.2.1 *Differential Equations of Vibration*

Let us consider a structure with concentrated masses (Fig. 6.4). Express the forces of inertia as function of *unit displacements*.

In case of free vibration, each mass is subjected to forces of inertia only. Displacement of each mass may be presented as

$$\begin{aligned} y_1 &= \delta_{11}F_1^{\text{in}} + \delta_{12}F_2^{\text{in}} + \cdots + \delta_{1n}F_n^{\text{in}}, \\ y_2 &= \delta_{21}F_1^{\text{in}} + \delta_{22}F_2^{\text{in}} + \cdots + \delta_{2n}F_n^{\text{in}}, \\ &\cdots\cdots\cdots\cdots \\ y_n &= \delta_{n1}F_1^{\text{in}} + \delta_{n2}F_2^{\text{in}} + \cdots + \delta_{nn}F_n^{\text{in}}, \end{aligned} \tag{6.1}$$

where δ_{ik} is displacement in ith direction caused by unit force acting in the kth direction. Since the force of inertia of mass m_i is $F_i^{\text{in}} = -m_i\ddot{y}_i$, then the differential equations become

$$\begin{aligned} &\delta_{11}m_1\ddot{y}_1 + \delta_{12}m_2\ddot{y}_2 + \ldots + \delta_{1n}m_n\ddot{y}_n + y_1 = 0, \\ &\cdots\cdots\cdots\cdots \\ &\delta_{n1}m_1\ddot{y}_1 + \delta_{n2}m_2\ddot{y}_2 + \ldots + \delta_{nn}m_n\ddot{y}_n + y_n = 0. \end{aligned} \tag{6.1a}$$

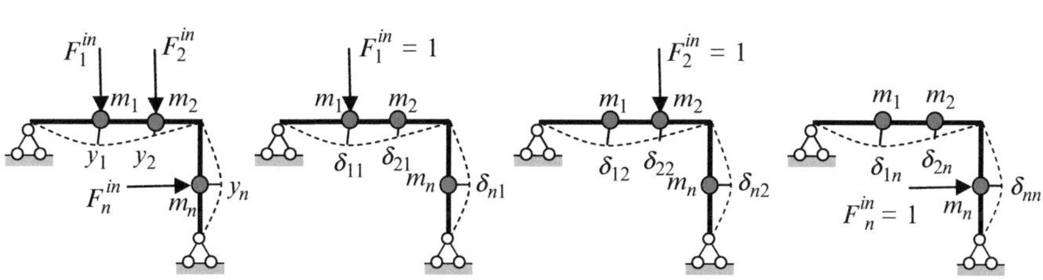

Fig. 6.4 Design diagram and unit conditions

In matrix form, this system may be presented as

$$\mathbf{FM\ddot{Y}} + \mathbf{Y} = \mathbf{0}, \tag{6.1b}$$

where **F** is the flexibility matrix (or matrix of unit displacements), **M** is the diagonal mass matrix, and **Y** represents the vector displacements

$$\mathbf{F} = \begin{bmatrix} \delta_{11} & \delta_{12} & \dots & \delta_{1n} \\ \delta_{21} & \delta_{22} & \dots & \delta_{2n} \\ \dots & \dots & \dots & \dots \\ \delta_{n1} & \delta_{n2} & \dots & \delta_{nn} \end{bmatrix}, \ \mathbf{M} = \begin{bmatrix} m_1 & 0 & \dots & 0 \\ 0 & m_2 & \dots & 0 \\ \dots & \dots & \dots & \dots \\ 0 & 0 & \dots & m_n \end{bmatrix}, \mathbf{Y} = \begin{bmatrix} y_1 \\ y_2 \\ \dots \\ y_n \end{bmatrix}.$$

The term $\delta_{ik}m_k\ddot{y}_k$ represents the displacement in the ith direction caused by the force of inertia $-m_k\ddot{y}_k$, which acts in the kth direction. Each equation in (6.1a) presents the compatibility condition. The differential equations of motion are coupled dynamically because the second derivative of all coordinates appears in each equation.

6.2.2 Frequency Equation

Solution of system of differential equations (6.1) is

$$y_1 = A_1 \sin(\omega t + \varphi_0), \quad y_2 = A_2 \sin(\omega t + \varphi_0), \quad y_n = A_n \sin(\omega t + \varphi_0), \tag{6.2}$$

where A_i are the amplitudes of the corresponding masses m_i and φ_0 is the initial phase of vibration. The second derivatives of these displacements over time are

$$\begin{aligned} \ddot{y}_1 &= -A_1\omega^2 \sin(\omega t + \varphi_0), \quad \ddot{y}_2 = -A_2\omega^2 \sin(\omega t + \varphi_0), \cdots \\ \ddot{y}_n &= -A_n\omega^2 \sin(\omega t + \varphi_0). \end{aligned} \tag{6.2a}$$

By substituting (6.2) and (6.2a) into (6.1a) and reducing by $\omega^2 \sin(\omega t + \varphi_0)$, we get

$$\begin{aligned} &\left(m_1\delta_{11}\omega^2 - 1\right)A_1 + m_2\delta_{12}\omega^2 A_2 + \cdots + m_n\delta_{1n}\omega^2 A_n = 0, \\ &m_1\delta_{21}\omega^2 A_1 + \left(m_2\delta_{22}\omega^2 - 1\right)A_2 + \cdots + m_n\delta_{2n}\omega^2 A_n = 0, \\ &\cdot \quad \cdot \quad \cdot \quad \cdot \quad \cdot \quad \cdot \quad \cdot \quad \cdot \quad \cdot \quad \cdot \quad \cdot \quad \cdot \quad \cdot \quad \cdot \quad \cdot \\ &m_1\delta_{n1}\omega^2 A_1 + m_2\delta_{n2}\omega^2 A_2 + \cdots + \left(m_n\delta_{nn}\omega^2 - 1\right)A_n = 0. \end{aligned} \tag{6.3}$$

Equations (6.3) are homogeneous algebraic equations with respect to unknown amplitudes A. Trivial solution $A_i = 0$ corresponds to the system at rest. Nontrivial solution (nonzero amplitudes A_i) is possible, if the determinant of the coefficients of amplitude is zero

$$D = \begin{bmatrix} m_1\delta_{11}\omega^2 - 1 & m_2\delta_{12}\omega^2 & \ldots & m_n\delta_{1n}\omega^2 \\ m_1\delta_{21}\omega^2 & m_2\delta_{22}\omega^2 - 1 & \ldots & m_n\delta_{2n}\omega^2 \\ \ldots & \ldots & \ldots & \ldots \\ m_1\delta_{n1}\omega^2 & m_2\delta_{n2}\omega^2 & \ldots & m_n\delta_{nn}\omega^2 - 1 \end{bmatrix} = 0. \qquad (6.4)$$

This equation is called the frequency equation in terms of displacement. Solution of this equation ω_1, ω_2, ..., ω_n presents the natural frequencies of a structure. The number of the frequencies of free vibration equals to the number of degrees of freedom.

6.2.3 Mode Shape of Vibration

The set of equations (6.3) are homogeneous algebraic equations with respect to unknown amplitudes A_i. This system does not allow us to find unknown amplitudes. However, we can find ratios between different amplitudes. If a structure has two degrees of freedom, then system (6.3) becomes

$$\begin{aligned} &\left(m_1\delta_{11}\omega^2 - 1\right)A_1 + m_2\delta_{12}\omega^2 A_2 = 0, \\ &m_1\delta_{21}\omega^2 A_1 + \left(m_2\delta_{22}\omega^2 - 1\right)A_2 = 0. \end{aligned}$$

From these equations, we can find the following ratios

$$\frac{A_1}{A_2} = -\frac{m_2\delta_{12}\omega^2}{m_1\delta_{11}\omega^2 - 1}$$

or

$$\frac{A_1}{A_2} = -\frac{m_2\delta_{22}\omega^2 - 1}{m_1\delta_{21}\omega^2}. \qquad (6.5)$$

If we substitute the first frequency of vibration ω_1 into any of the two equations (6.5), then we can find $(A_1/A_2)_{\omega_1}$. Then we can assume that $A_2 = 1$ and calculate the corresponding A_1 (or vice versa). The numbers $A_2 = 1$ and A_1 defines the distribution of amplitudes at the first frequency of vibration ω_1; such distribution is referred as the first mode shape of vibration. This distribution is presented in form of vector-column φ_1, whose elements are $A_2 = 1$ and the calculated A_1; this column vector is called a first eigenvector φ_1. Thus, the set of equations (6.3) for ω_1 define the first eigenvector to within an arbitrary constant.

Second mode shape of vibration or second eigenvector, which corresponds to the second frequency vibration ω_2, can be found in a similar manner. After that we can construct a modal matrix $\Phi = \lfloor \varphi_1 \quad \varphi_2 \rfloor$.

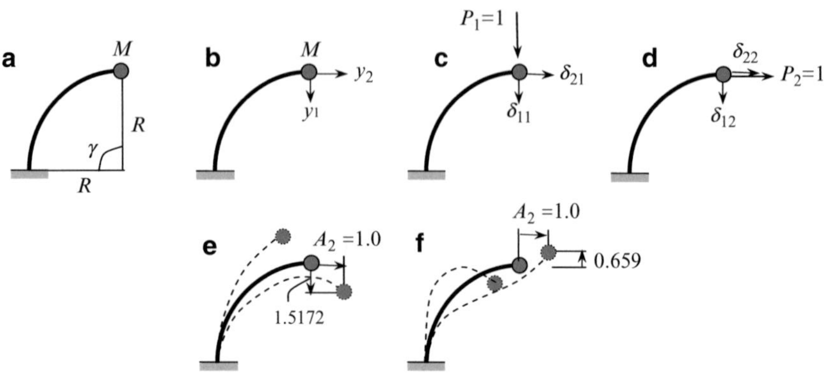

Fig. 6.5 (**a**) Design diagram; (**b**) generalized coordinates; (**c**, **d**) unit states; (**e**, **f**) mode shapes vibrations

If a structure has n degrees of freedom, then the modal matrix becomes [Tho81], [Wea90], [Kar10].

$$\Phi = \lfloor \varphi_1 \quad \varphi_2 \quad \cdots \quad \varphi_n \rfloor.$$

6.3 Examples

Example 6.1. Circular clumped-free uniform rod of radius R and central angle $\gamma = 90°$ has lumped mass M at the free end (Fig. 6.5a). Flexural rigidity of the arch is EI, distributed mass of the arch itself is neglected. Calculate the frequencies of the free vibrations.

Solution. The structure has two degrees of freedom. The first and second generalized coordinates y_1 and y_2 are directed along the radial and tangential lines (Fig. 6.5b), respectively; corresponding unit states and unit displacements are shown in Fig. 6.5c, d.

According to Table A.7, the unit displacements are

$$\delta_{11} = R^2 a_y\left(\frac{\gamma}{2} - \frac{\sin 2\gamma}{4}\right), \quad \delta_{12} = \delta_{21} = R^2 a_y \frac{(1-\cos\gamma)^2}{2},$$
$$\delta_{22} = R^2 a_y\left(\frac{3\gamma}{2} + \frac{\sin 2\gamma}{4} - 2\sin\gamma\right), \quad a_y = \frac{R}{EI}.$$

For given central angle $\gamma = 90°$ we get

$$\delta_{11} = \frac{0.7854R^3}{EI}, \quad \delta_{12} = \delta_{21} = \frac{0.5R^3}{EI}, \quad \delta_{22} = \frac{R^3}{EI}\left(\frac{3\pi}{4} - 2\right) = 0.3562\frac{R^3}{EI}.$$

Since mass M moves in the directions y_1 and y_2, then (6.3) become

$$\begin{aligned}&\left(M\delta_{11}\omega^2-1\right)A_1+M\delta_{12}\omega^2A_2=0,\\&M\delta_{21}\omega^2A_1+\left(M\delta_{22}\omega^2-1\right)A_2=0.\end{aligned}$$

Let denote eigenvalue $\lambda = EI/MR^3\omega^2$. The frequency equation becomes

$$\begin{vmatrix}0.7854-\lambda & 0.5\\ 0.5 & 0.3562-\lambda\end{vmatrix}=0.$$

Roots in descending order are $\lambda_1 = 1.1150$ and $\lambda_2 = 0.02655$. Frequencies of free vibrations in increasing order are

$$\omega_1^2=\frac{1}{\lambda_1}\frac{EI}{MR^3}=0.8969\frac{EI}{MR^3}\left(\mathrm{s}^{-2}\right),\quad \omega_2^2=\frac{1}{\lambda_2}\frac{EI}{MR^3}=37.665\frac{EI}{MR^3}\left(\mathrm{s}^{-2}\right).$$

Mode shape vibration may be determined on the basis of (6.5). For the first mode ($\lambda_1 = 1.1150$), ratio of amplitudes is

$$\left.\frac{A_1}{A_2}\right|_{\omega_1}=-\left.\frac{M\delta_{12}\omega^2}{M\delta_{11}\omega^2-1}\right|_{\omega_1}=-\frac{M\dfrac{0.5R^3}{EI}0.8969\dfrac{EI}{MR^3}}{M\dfrac{0.7854R^3}{EI}0.8969\dfrac{EI}{MR^3}-1}=1.5172$$

or

$$\left.\frac{A_1}{A_2}\right|_{\omega_1}=-\left.\frac{M\delta_{22}\omega^2-1}{M\delta_{21}\omega^2}\right|_{\omega_1}=-\frac{0.3562\times0.8969-1}{0.5\times0.8969}=1.5175.$$

Assume that $A_2 = 1$, so the first eigenvector becomes $\varphi = \lfloor \varphi_{11} \quad \varphi_{21} \rfloor^{\mathrm{T}} = \lfloor 1.5175 \quad 1.0 \rfloor^{\mathrm{T}}$.

For the second mode ($\lambda_2 = 0.02655$), ratio of amplitudes is

$$\left.\frac{A_1}{A_2}\right|_{\omega_2}=-\left.\frac{M\delta_{12}\omega^2}{M\delta_{11}\omega^2-1}\right|_{\omega_2}=-\frac{0.5\times37.665}{0.7854\times37.665-1}=-0.6589$$

or

$$\left.\frac{A_1}{A_2}\right|_{\omega_2}=-\left.\frac{M\delta_{22}\omega^2-1}{M\delta_{21}\omega^2}\right|_{\omega_2}=-\frac{0.3562\times37.665-1}{0.5\times37.665}=-0.6593.$$

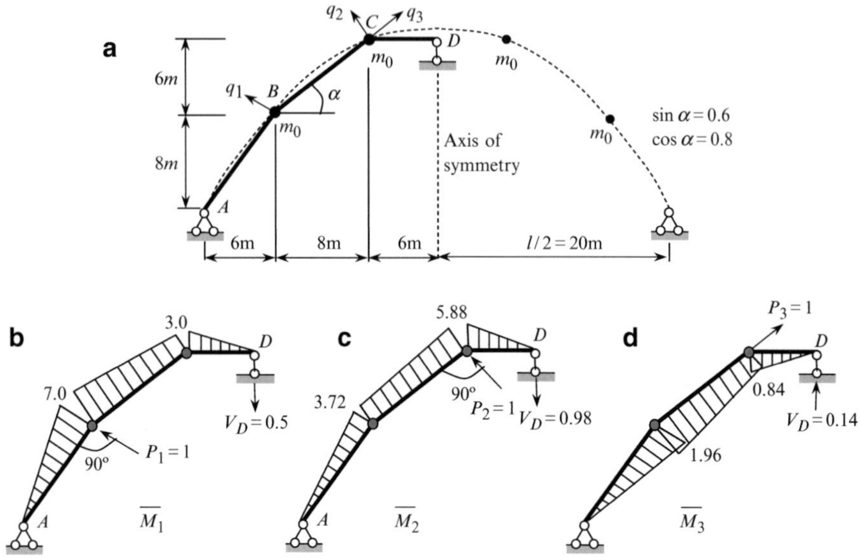

Fig. 6.6 (**a**) Approximate design diagram of half-arch; (**b**) unit states and corresponding bending moment diagrams

The modal matrix is defined as

$$\Phi = \begin{bmatrix} 1.5175 & -0.6590 \\ 1.0 & 1.0 \end{bmatrix}.$$

Corresponding mode shapes of vibration are shown in Fig. 6.5e, f.

Orthogonality condition: $1.5175 \times (-0.6590)M + 1.0 \times 1.0M = 0.0$.

Example 6.2. Two-hinged symmetrical uniform arch with four lumped masses m_0 is shown by a dotted line in Fig. 6.6a. The flexural rigidity of arch is EI; the span and rise of arch is $l = 40$ m, and $f = 14$ m, respectively. Compute the frequencies of antisymmetrical vibration.

Solution. Approximate model of the half-arch is a frame with two lumped masses m_0, as shown in Fig. 6.6a by solid lines. The rolled support D on the axis of symmetry corresponds to the antisymmetrical vibration. Generalized coordinates q_1 and q_2 are directed perpendicular to elements AB and BC, respectively, while coordinate q_3 is directed along the member BC. The unit states due to the generalized forces and corresponding bending moment diagrams are shown in Fig. 6.6b–d.

The unit displacements are $\delta_{ik} = \sum \int (\overline{M}_i \overline{M}_k / EI) \mathrm{d}s$. In our case, we get

$$EI\delta_{11} = \frac{1}{2} \times 10 \times 7 \times \frac{2}{3} \times 7 + \frac{10}{6}(2 \times 7 \times 7 + 2 \times 3 \times 3 + 7 \times 3 + 3 \times 7)$$
$$+ \frac{1}{2} \times 6 \times 3 \times \frac{2}{3} \times 3 = 444.66\,(\mathrm{m}^3),$$
$$EI\delta_{22} = 349.56; \quad EI\delta_{33} = 34.86; \quad EI\delta_{12} = 354.88; \quad EI\delta_{13} = -124.50;$$
$$EI\delta_{23} = -99.37.$$

Now we will consider two cases. In the first case, we take into account only *two* generalized coordinates q_1 and q_2, while in the second case *all* coordinates q_1, q_2, and q_3 are considered. Note that in both cases we are dealing with the *same* design diagram.

1. Approximate design diagram of half-arch has *two* degrees of freedom. In this case, the equation for frequency of vibration (6.4) becomes

$$\begin{vmatrix} 444.67 - \theta & 354.88 \\ 354.88 & 349.56 - \theta \end{vmatrix} = 0, \quad \text{where } \theta = \frac{EI}{m_0 \omega^2}\,(\mathrm{m}^3).$$

Roots of this equation are called eigenvalues and they are $\theta_1 = 755.206$, $\theta_2 = 39.114$.

Frequency vibrations are

$$\omega = \frac{1}{\sqrt{\theta}} \sqrt{\frac{EI}{m_0}}\,(\mathrm{s}^{-1}),$$

so

$$\omega_1 = 0.0364\sqrt{\frac{EI}{m_0}}, \quad \omega_2 = 0.160\sqrt{\frac{EI}{m_0}}.$$

2. The same design diagram of half-arch is treated as a structure with *three* degrees of freedom. In this case, the equation for frequency of vibration (6.4) becomes

$$\begin{vmatrix} 444.67 - \theta & 354.88 & -124.51 \\ 354.88 & 349.56 - \theta & -99.37 \\ -124.51 & -99.37 & 34.86 - \theta \end{vmatrix} = 0.$$

Roots of this equation are $\theta_1 = 788.56, \quad \theta_2 = 40.331, \quad \theta_3 \cong 0$.

The last root $\theta_3 \cong 0$ means that system in fact has **two** degrees of freedom. Indeed, we can see that a ratio between elements of the first and third rows is constant

$$\frac{444.67}{-124.51} = \frac{354.88}{-99.37} = \frac{-124.51}{34.86} = -3.571.$$

i.e., displacements q_1 and q_3 are dependent. This is expected, since diagrams $\overline{M}_1$ and $\overline{M}_3$ are dependent ($7.0/1.96 = 3.0/0.84 = 0.5/0.14 = 3.571$).

Frequencies of vibrations are

$$\omega_1 = 0.0356\sqrt{\frac{EI}{m_0}}, \quad \omega_2 = 0.157\sqrt{\frac{EI}{m_0}}.$$

Presentation of q_2 as a perpendicular to BC is approximate; however, this leads to simple calculations with a desired accuracy for the first and second frequencies. Possible options for consideration of the forces of inertia of mass at joint C are discussed in [Pro48].

Let us find the shape of vibration (eigenfunction) considering the structure with two degrees of freedom (case 1).

For the first frequency of vibration $\omega_1 = 0.0364\sqrt{EI/m_0}$; according to (6.5) we have

$$\frac{A_1}{A_2} = -\frac{m_2\delta_{12}\omega^2}{m_1\delta_{11}\omega^2 - 1} = -\frac{m_0\dfrac{354.88}{EI}0.0364^2\dfrac{EI}{m_0}}{m_0\dfrac{444.66}{EI}0.0364^2\dfrac{EI}{m_0} - 1} = 1.144,$$

or

$$\frac{A_1}{A_2} = -\frac{m_2\delta_{22}\omega^2 - 1}{m_1\delta_{21}\omega^2} = -\frac{349.56 \times 0.0364^2 - 1}{354.88 \times 0.0364^2} = 1.142.$$

The relative error is 0.23%. Thus, if the displacement of joint B is equals to 1.0, then the displacement of joint C is equal to 1.142. The positive sign means that displacements occur in the same directions as the forces P_1 and P_2, as shown in Fig. 6.6b, c (or in the opposite direction).

The second eigenfunction may be constructed similarly. Approximate presentation of q_2 as a perpendicular to BC leads to a larger error for shape A_1/A_2, according to the two expressions (6.5).

Example 6.3. Parabolic three-hinged symmetric arch with three equal lumped masses is shown in Fig. 6.7a. Parameters of the arch are $l = 16$ m, $f = 0.25l = 4$m. Lumped masses are $m_1 = m_2 = m_3 = m_0$ [Pro48]. Flexural stiffness of the arch is $EI = \text{const}$. Calculate the frequencies of the free vibrations.

Solution. Equation of the axis of the arch is $y(x) = (4f/l^2)x(l - x)$, therefore $y_1 = 3$ m. Axis of the arch is replaced by straight members. All dimensions of the arch and the straight segments (in meters) are shown in Fig. 6.7b.

This structure has two degrees of freedom; however, presentation of generalized coordinates, as in the previous example, is not entirely clear. So the Bresse group unknowns will be applied [Bre54c], [Ber57]. For symmetric vibration, we have

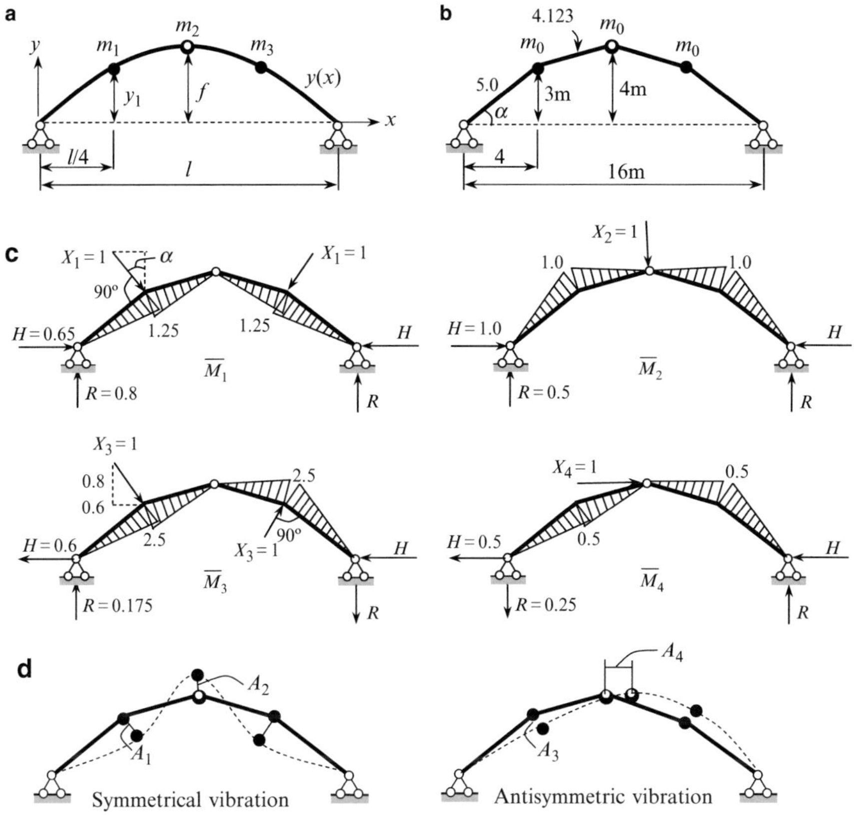

Fig. 6.7 (**a**) Design diagram of three-hinged arch with lumped masses; (**b**) modified design diagram; (**c**) Unit states and corresponding bending moment diagrams; (**d**) Mode shape of vibrations

group (or paired) unknowns X_1 and simple unknown X_2. For antisymmetric vibration, we have group unknowns X_3 and simple unknown X_4. Unit states, corresponding reactions, and bending moment diagrams are shown in Fig. 6.7c.

Unit displacements are

$$EI\delta_{11} = \overline{M}_1 \times \overline{M}_1 = 2\left(\frac{1}{2} \times 5 \times 1.25 \times \frac{2}{3} \times 1.25 + \frac{1}{2} \times 4.123 \times 1.25 \times \frac{2}{3} \times 1.25\right) = 9.5\ (\text{m}^3),$$

$$EI\delta_{22} = \overline{M}_2 \times \overline{M}_2 = 6.08, \quad EI\delta_{33} = 38.01, \quad EI\delta_{44} = 1.52,$$
$$EI\delta_{12} = EI\delta_{21} = \overline{M}_1 \times \overline{M}_2 = -7.6, \quad EI\delta_{34} = EI\delta_{43} = 7.6,$$
$$\delta_{13} = \delta_{31} = 0, \quad \delta_{14} = \delta_{41} = 0, \quad \delta_{23} = \delta_{32} = 0, \quad \delta_{24} = \delta_{42} = 0.$$

Equations (6.3) is separated into subsystems; since all masses are equal then

$$\begin{aligned}&\left(m_0\delta_{11}\omega^2-1\right)A_1+m_0\delta_{12}\omega^2A_2=0,\\&m_0\delta_{21}\omega^2A_1+\left(m_0\delta_{22}\omega^2-1\right)A_2=0,\end{aligned}\quad \text{and} \quad \begin{aligned}&\left(m_0\delta_{33}\omega^2-1\right)A_3+m_0\delta_{34}\omega^2A_4=0,\\&m_0\delta_{34}\omega^2A_3+\left(m_0\delta_{44}\omega^2-1\right)A_4=0.\end{aligned}$$

Let us denote the eigenvalue by

$$\lambda=\frac{EI}{m_0\omega^2}\ (\mathrm{m}^3)\rightarrow\omega^2=\frac{EI}{\lambda m_0}\ (\mathrm{s}^{-2}).$$

Since we use the paired unknown, the frequency equations should be modified, introducing coefficient 0.5 as shown below:

Symmetrical Vibration

The frequency equation becomes [Pro48]

$$D_{\text{sym}}=\begin{vmatrix}\frac{1}{2}\delta_{11}-\lambda & \delta_{12}\\ \frac{1}{2}\delta_{21} & \delta_{22}-\lambda\end{vmatrix}=0\quad\text{or}\quad D_{\text{sym}}=\begin{vmatrix}0.5\times\frac{9.5}{EI}-\lambda & -\frac{7.60}{EI}\\ -0.5\times\frac{7.60}{EI} & \frac{6.08}{EI}-\lambda\end{vmatrix}=0.$$

Roots of this equation are $\lambda_1=0$, $\lambda_2=10.83$, therefore $\omega_{\text{sym}}^2=(1/10.83)\times(EI/m_0)\rightarrow\omega_{\text{sym}}=0.304\sqrt{EI/m_0}$.

Antisymmetrical Vibration

Similarly, the frequency equation should be presented in the form

$$D_{\text{antisym}}=\begin{vmatrix}\frac{1}{2}\delta_{33}-\lambda & \delta_{34}\\ \frac{1}{2}\delta_{43} & \delta_{44}-\lambda\end{vmatrix}=0\quad\text{or}$$

$$D_{\text{antisym}}=\begin{vmatrix}0.5\times\frac{38.01}{EI}-\lambda & \frac{7.60}{EI}\\ 0.5\times\frac{7.60}{EI} & \frac{1.52}{EI}-\lambda\end{vmatrix}=0.$$

Roots of this equation are $\lambda_1 = 0, \quad \lambda_2 = 20.526$, therefore

$$\omega^2_{\text{antisym}} = \frac{1}{20.526} \times \frac{EI}{m_0} \to \omega_{\text{antisym}} = 0.221\sqrt{\frac{EI}{m_0}}.$$

Verification

The sum of the eigenvalues is $10.83 + 20.526 = 31.356$. On the other hand, the sum of traces of two matrices is $9.5/2 + 6.08 + 38.01/2 + 1.52 = 31.355$.

Mode Shape of Vibration

We can use expressions for mode shapes in terms of eigenvalues. For symmetrical form, we get

$$\frac{A_1}{A_2} = -\frac{\delta_{12}}{0.5\delta_{11} - \lambda} = -\frac{\delta_{22} - \lambda}{0.5\delta_{21}}.$$

For nonzero eigenvalue ($\lambda = 10.83$), we have

$$\frac{A_1}{A_2} = -\frac{(-7.6)}{0.5 \times 9.5 - 10.83} = -\frac{6.08 - 10.83}{0.5 \times (-7.6)} = -1.25.$$

For antisymmetrical form, we have

$$\frac{A_3}{A_4} = -\frac{\delta_{34}}{0.5\delta_{33} - \lambda} = -\frac{\delta_{44} - \lambda}{0.5\delta_{43}}.$$

For nonzero eigenvalue of antisymmetrical vibration ($\lambda = 20.526$), we have

$$\frac{A_3}{A_4} = -\frac{7.6}{0.5 \times 38.01 - 20.526} = -\frac{1.52 - 20.526}{0.5 \times 7.6} = 5.0.$$

Corresponding mode shapes of vibrations are shown in Fig. 6.7d.

Discussion

1. For symmetric and antisymmetric vibrations, we obtained two roots $\lambda = 0$.
   It means that two degrees of freedom (two zero roots) should be eliminated from the total number of degrees of freedom (which was assumed to be four).

Thus, in fact, the arch with three lumped masses shown in Fig. 6.7a, b has two degrees of freedom.

2. The lowest frequency of free vibration corresponds to the antisymmetric form of vibration.

6.4 Vibration of Circular Uniform Arches

This section is devoted to determination of frequencies of in-plane bending vibration of circular arches based on the Lamb's equation. Two approaches are considered. The first approach presents exact solution of the Lamb's differential equation, while the second one is based on the application of the variational Bubnov–Galerkin method [Vol67].

6.4.1 *Lamb's Differential Equation of In-Plane Bending Vibration*

Let us consider a symmetrical circular arch of constant cross-section with a central angle 2α and radius of curvature R. The moment of inertia of cross-section is I and mass of the arch per unit its length is m. We will consider two types of in-plane vibration. They are the bending and radial vibrations. General assumptions in Sect. 6.1.2 allow us to rewrite the equation of in-plane vibration of the arch in the form of a partial differential equation [Lam1888], [Rek73], [Chu52], [Dem49].

$$\frac{\partial^6 u}{\partial \varphi^6}+2\frac{\partial^4 u}{\partial \varphi^4}+\frac{\partial^2 u}{\partial \varphi^2}=\frac{mR^4}{EI}\frac{\partial^2}{\partial t^2}\left(u-\frac{\partial^2 u}{\partial \varphi^2}\right). \tag{6.6}$$

where φ defines the position of the point on the nondeformable axial line of the arch $-\alpha\leq\varphi\leq\alpha$, $u=u(\varphi,t)$ is tangential displacement of the circle in the direction in which φ increases (Fig. 6.8).

The radial and tangential displacements are related by $\upsilon=\mathrm{d}u/\mathrm{d}\varphi$.

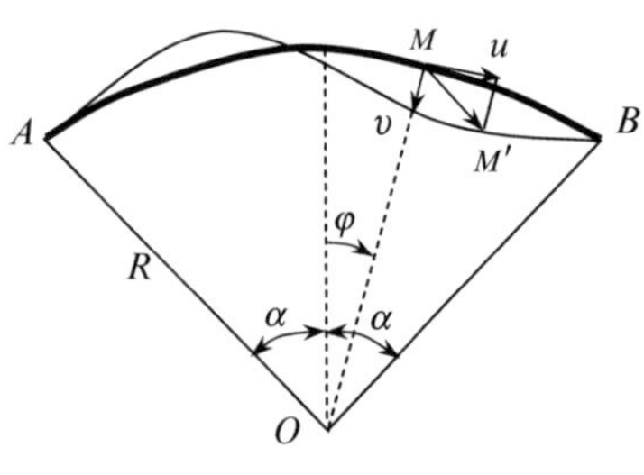

Fig. 6.8 Circular arch and notation of displacements

6.4.2 Frequency Equation of Bending Vibration. Demidovich's Solution

Let the displacement be $u = U(\varphi)\cos\omega t$, then for U we have the Lamb's ordinary differential equation [Lam1888]

$$\frac{\mathrm{d}^6 U}{\mathrm{d}\varphi^6} + 2\frac{\mathrm{d}^4 U}{\mathrm{d}\varphi^4} + \frac{\mathrm{d}^2 U}{\mathrm{d}\varphi^2} - \frac{mR^4\omega^2}{EI}\left(\frac{\mathrm{d}^2 U}{\mathrm{d}\varphi^2} - U\right) = 0. \tag{6.7}$$

Assume the expression for U has the form [Lov20]

$$U = \sum_{i=0}^{2}\left(A_i \cos n_i\varphi + B_i \sin n_i\varphi\right). \tag{6.8}$$

where A_i and B_i are constant. Substitution of (6.8) into (6.7) leads to the equation

$$n^2\left(n^2 - 1\right)^2 = \left(n^2 + 1\right)\frac{m\omega^2 R^4}{EI}. \tag{6.9}$$

It can be shown that if $\omega^2 \geq 0$, (6.9) has at least one pair of the real roots $\pm n_0$ [Dem49]. This root is uniquely and will be referred to as the basic root. Equation (6.9) may be rewritten as

$$n_0^2\left(n_0^2 - 1\right)^2 = \left(n_0^2 + 1\right)\frac{m\omega^2 R^4}{EI}. \tag{6.10}$$

Hence, it follows that

$$\frac{m\omega^2 R^4}{EI} = \frac{n_0^2\left(n_0^2 - 1\right)^2}{\left(n_0^2 + 1\right)}. \tag{6.11}$$

So the circular frequency of vibration may be expressed in terms of the basic root

$$\omega = \frac{C}{R^2}\sqrt{\frac{EI}{m}}\ \left(\mathrm{s}^{-1}\right), \quad C = \frac{n_0\left(n_0^2 - 1\right)}{\sqrt{n_0^2 + 1}}. \tag{6.12}$$

However, the basic root is still unknown. Substitution of (6.11) in (6.9) leads to the governing equation

$$n^2\left(n^2 - 1\right)^2 - \frac{n_0^2\left(n_0^2 - 1\right)^2}{\left(n_0^2 + 1\right)}\left(n^2 + 1\right) = 0. \tag{6.13}$$

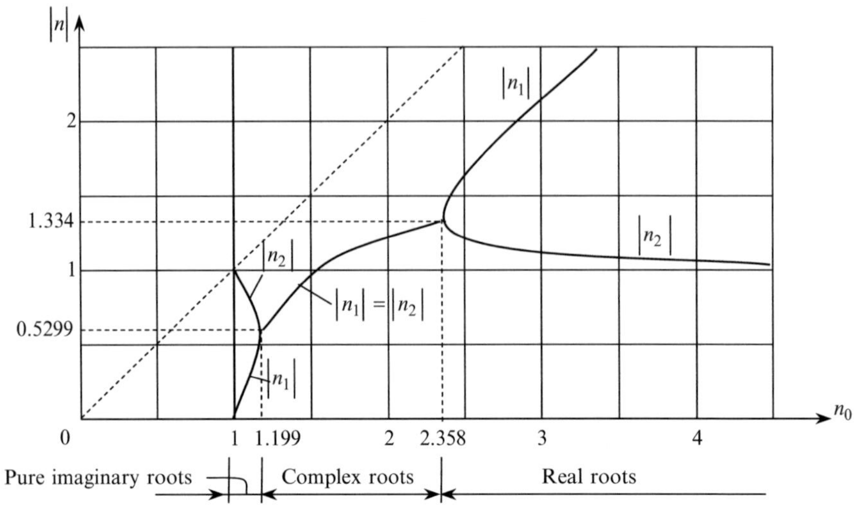

Fig. 6.9 Absolute values of roots n_1 and n_2 in terms of base root n_0

Detailed analysis of (6.13) [Dem49] shows that the other two pairs of roots of (6.9) are purely imaginary, namely, $\pm n_1 i$ and $\pm n_2 i$, $\left(i = \sqrt{-1}\right)$. They may be expressed in terms of basic root n_0 as follows

$$n_{1,2} = \sqrt{\frac{1}{2}\left(n_0^2 - 2\right) \pm \frac{1}{2} n_0 \sqrt{\frac{n_0^4 - 7n_0^2 + 8}{n_0^2 + 1}}}. \tag{6.14}$$

Analysis of the roots of the characteristic equation was done by Waltking [Wal34] and Demidovich [Dem49]. The roots are multiple only in the three special cases:

1. $n_0 = 1, \quad n_1 = 0, \quad n_2 = i,$
2. $n_0 = \sqrt{\frac{1}{2}\left(7 - \sqrt{17}\right)} = 1.199, \quad n_1 = n_2 = \frac{1}{2} i \sqrt{\sqrt{17} - 3} = 0.5299i,$
3. $n_0 = \sqrt{\frac{1}{2}\left(7 + \sqrt{17}\right)} = 2.358, \quad n_1 = n_2 = \sqrt{\sqrt{17} + 3} = 1.334.$

The graphs of $|n_1|$ and $|n_2|$ as a function of the base root n_0 is shown in Fig. 6.9 [Dem49]

Solution (6.7) must be conformed to the boundary conditions. If both ends of the arch are fixed then the tangential and radial displacements, as well the slopes at the supports are zero, i.e., $U = U' = U'' = 0$.

If both ends of the arch are hinged then $U = U' = U''' = 0$.

Since the roots of (6.9) are not multiple, general solution (6.7) should be adopted in the modified form, which involves the hyperbolic functions

$$\begin{aligned} U = {} & A_0 \cos n_0\varphi + A_1 \cosh n_1\varphi + A_2 \cosh n_2\varphi \\ & + B_0 \sin n_0\varphi + B_1 \sinh n_1\varphi + B_2 \sinh n_2\varphi. \end{aligned}$$

Let us consider the antisymmetrical and symmetrical vibration separately.

Antisymmetrical Vibration of Two-Hinged and Hingeless Arches

In this case, the function U should be assumed in the form

$$U = A_0 \cos n_0\varphi + A_1 \cosh n_1\varphi + A_2 \cosh n_2\varphi.$$

Hingeless Arch

Taking into account the boundary conditions, we get the following expressions (at $\varphi = \pm\alpha$)

$$\begin{aligned} U &= A_0 \cos n_0\varphi + A_1 \cosh n_1\varphi + A_2 \cosh n_2\varphi = 0, \\ U' &= -A_0 n_0 \sin n_0\varphi + A_1 n_1 \sinh n_1\varphi + A_2 n_2 \sinh n_2\varphi = 0, \\ U'' &= -A_0 n_0^2 \cos n_0\varphi + A_1 n_1^2 \cosh n_1\varphi + A_2 n_2^2 \cosh n_2\varphi = 0. \end{aligned}$$

Nontrivial solution occurs if

$$\begin{bmatrix} \cos n_0\varphi & \cosh n_1\varphi & \cosh n_2\varphi \\ -n_0 \sin n_0\varphi & n_1 \sinh n_1\varphi & n_2 \sinh n_2\varphi \\ -n_0^2 \cos n_0\varphi & n_1^2 \cosh n_1\varphi & n_2^2 \cosh n_2\varphi \end{bmatrix}_{\varphi=\pm\alpha} = 0. \tag{6.15}$$

After rearrangements, the frequency equation (6.15) may be presented in the form

$$\tan n_0\alpha = K_1 \tanh n_1\alpha + K_2 \tanh n_2\alpha,$$

where

$$K_1 = \frac{n_1\left(n_0^2 + n_2^2\right)}{n_0\left(n_1^2 - n_2^2\right)}, \quad K_2 = -\frac{n_2\left(n_0^2 + n_1^2\right)}{n_0\left(n_1^2 - n_2^2\right)}.$$

Two-Hinged Arch

In this case, the boundary conditions are $U = U' = U''' = 0$. The frequency equation may be derived in a similar manner:

$$\cot n_0\alpha = -F_1 \coth n_1\alpha - F_2 \coth n_2\alpha,$$

where

$$F_1 = -\frac{n_0\left(n_0^2 + n_2^2\right)}{n_1\left(n_1^2 - n_2^2\right)}, \quad F_2 = \frac{n_0\left(n_0^2 + n_1^2\right)}{n_2\left(n_1^2 - n_2^2\right)}.$$

Symmetrical Vibration

In this case, the function U should be assumed in the form

$$U = B_0 \sin n_0\varphi + B_1 \sinh n_1\varphi + B_2 \sinh n_2\varphi.$$

The final frequency equation for arches with specific boundary conditions is shown below.

Hingeless arch: $\cot n_0\alpha = -K_1 \coth n_1\alpha - K_2 \coth n_2\alpha$.

Two-hinged arch: $\tan n_0\alpha = F_1 \tanh n_1\alpha + F_2 \tanh n_2\alpha$.

In these formulas, coefficients $K_{1,2}$ and $F_{1,2}$ are assumed as in the case of antisymmetric vibration.

According to (6.14), coefficients K_1, K_2, F_1, and F_2 are functions of one unknown basic root n_0. The frequency equations are transcendental and theirs solutions can be obtained by numerical methods. The first and second roots n_0 of the frequency equation and the corresponding parameter C, according to (6.12), are presented in Table 6.2.

Finally, the circular frequency of bending in-plane vibration of an arch is determined by (6.12). The lowest frequency of the free vibration for circular arches, as for parabolic ones, corresponds to antisymmetric mode of vibration [Dem49], [Rab51], [Rab54b], [Rab58].

6.4.3 *Variational Approach*

The bending in-plane vibration of the arch is described by (6.6); as before, the central angle of the arch is 2α. Angle φ is measured from a vertical axis of

Table 6.2 Basic roots n_0 and parameters C_k of circular frequency of vibrations

Central angle	Antisymmetric vibration				Symmetrical vibration			
	Arch with fixed ends		Two-hinged arch		Arch with fixed ends		Two-hinged arch	
2α	n_0	C_k	n_0	C_k	n_0	C_k	n_0	C_k
$\pi/6$	$n_{01} = 14.962$	$C_1 = 222.36$	11.959	141.52	20.114	403.07	17.526	305.63
	$n_{02} = 26.973$	$C_2 = 726.04$	23.980	573.54	32.488	1053.9	29.732	882.49
$\pi/3$	7.431	53.735	5.925	34.033	10.047	93.45	8.750	75.07
	13.448	179.35	11.962	141.59	16.235	262.08	14.858	219.26
$\pi/2$	4.908	22.623	3.900	13.764	6.689	43.262	5.820	32.397
	8.929	78.238	7.947	61.668	10.813	115.43	9.896	96.439
$2\pi/3$	3.645	11.848	2.886	6.925	5.008	23.614	4.353	17.492
	6.665	42.941	5.937	33.772	8.100	64.123	7.413	53.468
$5\pi/6$	2.892	6.959	2.283	3.858	4.000	14.552	3.472	10.623
	5.306	26.683	4.731	20.920	6.471	40.394	5.923	33.606
π	2.398	4.384	1.888	2.266	3.328	9.649	2.8857	6.919
	4.402	17.921	3.9237	13.944	5.384	27.516	4.9279	22.81

symmetry, so $-\alpha \le \varphi \le \alpha$. For the first form of antisymmetric vibration of two-hinged arch, we assume

$$u(t) = f(t)\cos\frac{\pi\varphi}{\alpha}. \tag{6.16}$$

According to Bunnov–Galerkin procedure [Vol67], we take the derivatives

$$\frac{\partial^2 u}{\partial\varphi^2} = -f(t)\left(\frac{\pi}{\alpha}\right)^2\cos\frac{\pi\varphi}{\alpha};\quad \frac{\partial^4 u}{\partial\varphi^4} = f(t)\left(\frac{\pi}{\alpha}\right)^4\cos\frac{\pi\varphi}{\alpha};\quad \frac{\partial^6 u}{\partial\varphi^6} = -f(t)\left(\frac{\pi}{\alpha}\right)^6\cos\frac{\pi\varphi}{\alpha};$$

$$\frac{\partial^2 u}{\partial t^2} = \ddot{f}(t)\cos\frac{\pi\varphi}{\alpha};\quad \frac{\partial^4 u}{\partial t^2\partial\varphi^2} = -\ddot{f}(t)\left(\frac{\pi}{\alpha}\right)^2\cos\frac{\pi\varphi}{\alpha},$$

substitute them into (6.6), multiply by $\cos(\pi\varphi/\alpha)$, integrate and equate to zero

$$\int_{-\alpha}^{\alpha}\left[\frac{\partial^6 u}{\partial\varphi^6} + 2\frac{\partial^4 u}{\partial\varphi^4} + \frac{\partial^2 u}{\partial\varphi^2} - \frac{mR^4}{EI}\frac{\partial^2}{\partial t^2}\left(u - \frac{\partial^2 u}{\partial\varphi^2}\right)\right]\cos\frac{\pi\varphi}{\alpha}\mathrm{d}\varphi = 0.$$

As a result, we get the following ordinary second-order differential equation with respect to the unknown time-dependent function $f(t)$

$$\frac{mR^4}{EI}\left(\xi^2 + 1\right)\ddot{f}(t) + \xi^2\left(\xi^2 - 1\right)^2 f(t) = 0,\quad \xi = \frac{\pi}{\alpha}\quad \text{or}\quad \ddot{f}(t) + \omega_1^2 f(t) = 0,$$

For the first frequency of antisymmetric vibration, we get

$$\omega_1^2 = \frac{\xi^2\left(\xi^2 - 1\right)^2}{\xi^2 + 1} \frac{1}{R^4} \frac{EI}{m}. \tag{6.16a}$$

Approximate formula for frequencies of free bending in-plane vibration of uniform two-hinged and hingeless circular arches (α_0 is a central angle) [Uma72–73] is

$$\omega_i = \frac{k_i}{R^2\alpha_0^2}\sqrt{\frac{EI}{m}}\ (\mathrm{s}^{-1}). \tag{6.16b}$$

Coefficients k_i for first and second symmetric and antisymmetric forms are presented in Tables A.43 and A.44; these coefficients are presented in terms of a central angle α_0.

In the case of two-hinged uniform circular arch with lumped mass M at the crown, the lowest frequency of free antisymmetric vibration may be obtained by Bolotin's approximate formula [Bol64]:

$$\omega = \frac{\pi}{R^2\alpha}\left(\frac{\pi^2}{\alpha^2} - 1\right)\sqrt{\frac{EI}{\frac{4M}{R\alpha} + m\left(\frac{\pi^2}{\alpha^2} + 3\right)}}, \tag{6.16c}$$

where 2α is the central angle.

Comparison of the lowest frequencies of antisymmetric vibration of the circular arch according to the different approaches is presented in Table 6.3.

Computation of frequencies of free vibration of circular arches taking into account additional effects (damping, shear forces, etc,) is presented in refs. [Hen81], [Tuf98].

6.4.4 Radial Vibration

Such vibrations occur due to the change in length of the axis of the arch. The frequency of the radial vibration of the circular uniform arch of radius R, with central angle α_0 and distributed mass m is [Uma72-73]

$$\omega_i = \frac{1}{R}\sqrt{1 + \frac{\lambda_i^4}{\alpha_0^4}\frac{I}{R^2 A}}\sqrt{\frac{EA}{m}}\ (\mathrm{s}^{-1}),$$

Table 6.3 Parameter K for circular frequency of vibration, $\omega_i = (K/R^2)\sqrt{EI/m}$ (s^{-1})

Type of circular arch	Central angle	Exact result (Table 6.2)	Galerkin procedure (6.16a)	Approximate approach (Table A.43)	Bolotin formula (6.16c)
Hingeless	$\pi/3$	53.735	–	53.69	–
	$\pi/2$	22.623	–	22.62	–
	π	4.384	–	4.38	–
Two-hinged	$\pi/3$	34.033	34.52	33.61	33.62
	$\pi/2$	13.764	14.55	12.79	13.76
	π	2.266	2.68	2.26	2.27

where A is an area of the cross-section of an arch, I is the moment of inertia of a cross-section. Coefficients λ_i for the first and second frequencies of free vibrations of two-hinged arch are $\lambda_1 = 3.1416, \quad \lambda_2 = 6.2832$; for arch with fixed supports $\lambda_1 = 4.7300, \quad \lambda_2 = 7.8532$. It should be noted that the frequencies of radial vibration are significantly higher than the frequencies of the bending vibration; the frequencies of radial vibration would probably be difficult to excite [Lov20].

6.5 Rabinovich's Method for Parabolic Arch

At the present moment in time, there are several approaches for approximate vibration analysis of arches. The most natural approach consists of replacing the curvilincar axis of the arch by a discrete set of straight elements while the distributed mass is approximated by a set of lumped masses. Given these approximations, the following different modifications are possible.

1. One can replace the arch by a frame with straight, absolutely rigid members and lumped masses at the midpoints of these members. Given this, it is assumed that the members are connected by means of the elastic constraints.
2. Terenin's method [Ter54] takes into account the *finite* stiffness of the rods, their rigid connections, and lumped masses at the rigid joints. The method is based on the following assumptions: Curvilinear axis of the arch is replaced by *six straight members* of *equal length;* the members are not compressed and nonextended, but able to exert bending moments. The mixed method [Dar89] was applied as a possible solution to this problem.
3. Rabinovich' method [Rab51, 58] also takes into account the finite stiffness of the rods, their rigid connections, and lumped masses at the rigid joints. Contrary to Terenin's method, Rabinovich' method is based on the concept of a hinged chain; this approach is more effective than Terenin's method. Concept of kinematical chain allows establishing the simple geometrical relationships between displacements of the joints. Corresponding vibration model of the parabolic arch is called Rabinovich's model.

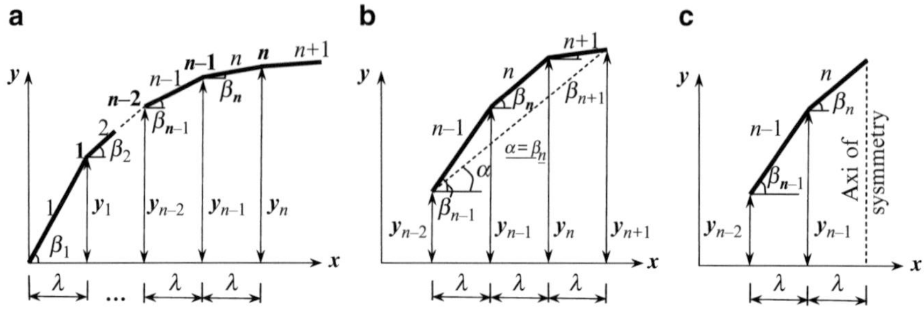

Fig. 6.10 Geometry of parabolic polygon

6.5.1 Geometry of Parabolic Polygon

Equation of axis of a parabolic arch is given by $y = (4f/l^2)x(l - x)$, where l and f are span and rise of the arch, respectively. The arch is replaced by an inscribed polygon; the horizontal projections of all sides of polygon are equal to $\lambda = l/k$, where k is the number of sides of a polygon. Such polygon is called parabolic. Numeration of the rods, joints (bold), theirs coordinates, and angles of inclination from the x-axis are shown in Fig. 6.10a.

Parameters of parabolic polygon are

$$x_n = n\lambda; \quad y_n = \frac{4f}{l^2}x_n(l - x_n) = \frac{4f}{k^2}n(k - n). \tag{6.17}$$

The length of the side n of a parabolic polygon is

$$s_n = \frac{\lambda}{\cos\beta_n} = \lambda\sqrt{1 + \tan^2\beta_n}. \tag{6.17a}$$

The slope of the side n is

$$\tan\beta_n = \frac{y_n - y_{n-1}}{\lambda} = \frac{4f}{k^2\lambda}[n(k - n) - (n - 1)(k - n + 1)].$$

Last equation may be presented in terms of λ, or l as follows

$$\tan\beta_n = \frac{4f}{k^2\lambda}(k - 2n + 1) = \frac{4f}{kl}(k - 2n + 1). \tag{6.17b}$$

The angles of inclination of the parabolic polygon have the following properties:

1. The slopes form the arithmetic progression

$$\tan\beta_n - \tan\beta_{n+1} = \frac{8f}{k^2\lambda} = \frac{8f}{kl} = \text{const.}$$

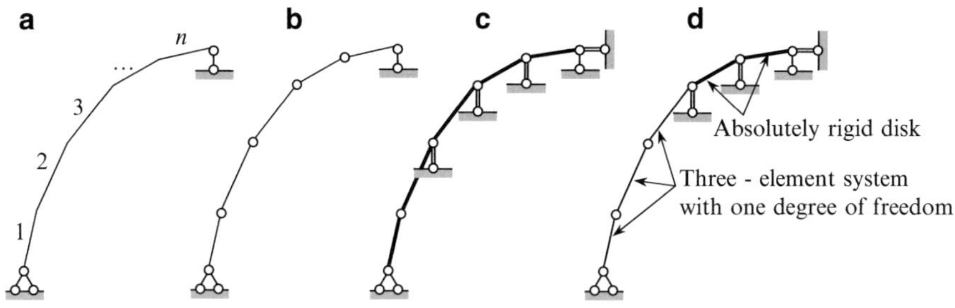

Fig. 6.11 Kinematical analysis of parabolic polygon

2. If one connects the ends of the chain consisting of three consecutive members $n-1, n$, and $n+1$, then this chord will be parallel to the element n, i.e., $\alpha = \beta_n$ (Fig. 6.10b).
3. Assume that number $\mathcal{K}$ is even. For the two last elements, which belong to the left half-arch and are adjacent to the axis of the symmetry, the ratio $\tan\beta_{n-1}/\tan\beta_n = 3$ (Fig. 6.10c). Indeed, the number of sides of polygon is $n = k/2$. Two last members are denoted by $n-1$ and n. According to (6.17b), the following properties hold true

$$\tan\beta_n = \frac{4f}{kl}(k - 2n + 1) = \frac{4f}{kl}\left(k - 2\frac{k}{2} + 1\right) = \frac{4f}{kl},$$

$$\tan\beta_{n-1} = \frac{4f}{kl}\left[k - 2\left(\frac{k}{2} - 1\right) + 1\right] = 3\tan\beta_n.$$

Such an approximation of a symmetrical parabolic arch is called Rabinovich's model [Rab56]. This model also includes some additional information. They are supports of half-arch, lumped masses, kinematical properties of the kinematical chain, etc. These will be considered later.

6.5.2 Kinematics of Parabolic Polygon

Figure 6.11a presents a frame (parabolic polygon), consisting of n rigid members. The structure has three support constraints. This structure has $n-1$ independent joint displacements.

Indeed, if one were to construct a hinged scheme of the frame (Fig. 6.11b), then introducing $n-1$ arbitrary-oriented additional constraints (shown by double lines) leads to an absolutely rigid structure (Fig. 6.11c). It is clear that if one were to introduce only $n-2$ additional constraints, then the structure would be divided into two distinct parts. They are three-element system and absolutely rigid disk (shown by a solid line) (Fig. 6.11d). Three-element system has only

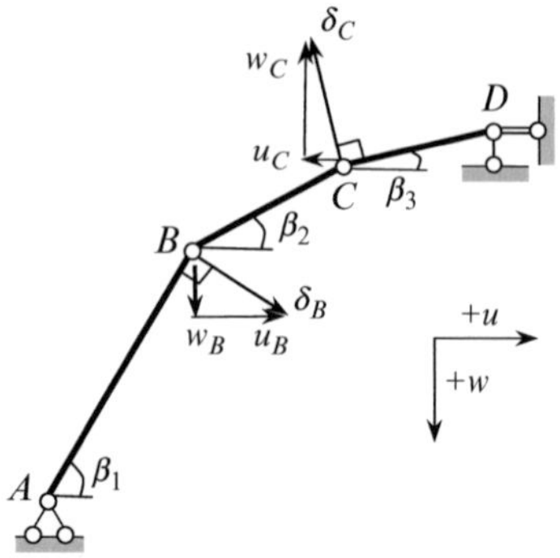

Fig. 6.12 Kinematics of the hinged three-element chain

one degree of freedom. Thus, the displacement of the entire chain depends on the sum of displacements of the simplest three-element systems. We now focus on the analysis of the three-element system.

6.5.2.1 Kinematics of the Three-Element Chain

Three-element hinged chain is shown in Fig. 6.12. The support constraint at point D prevents vertical displacements, so this scheme has two degrees of freedom. If one were to introduce additional constraint at point D (shown by double line), then new structure will have only one degree of freedom, as a bottom part in Fig. 6.11d. We establish the relationships between displacements of the joints B and C.

The displacement of point B, δ_B, is directed perpendicular to AB. Displacement of point C, δ_C, is perpendicular to CD. Projections of these displacements on the vertical and horizontal axis are denoted by w_B and u_B for point B, and w_C and u_C for point C. Let the generalized coordinate for three-member system be the vertical displacement w_B. It is clear that

$$u_B = w_B \tan \beta_1. \tag{6.18}$$

We express w_C and u_C in terms of independent variable w_B. During displacements of joint B, the member BC executes planar motion and therefore, from the theorem of the projections of displacements δ_B and δ_C on the line BC, we get

$$\delta_B \cos[90 - (\beta_1 - \beta_2)] = \delta_C \cos[90 - (\beta_2 - \beta_3)].$$

This leads to the formula

$$\delta_C = \delta_B \frac{\sin(\beta_1 - \beta_2)}{\sin(\beta_2 - \beta_3)} = \frac{w_B}{\cos \beta_1} \times \frac{\sin(\beta_1 - \beta_2)}{\sin(\beta_2 - \beta_3)}.$$

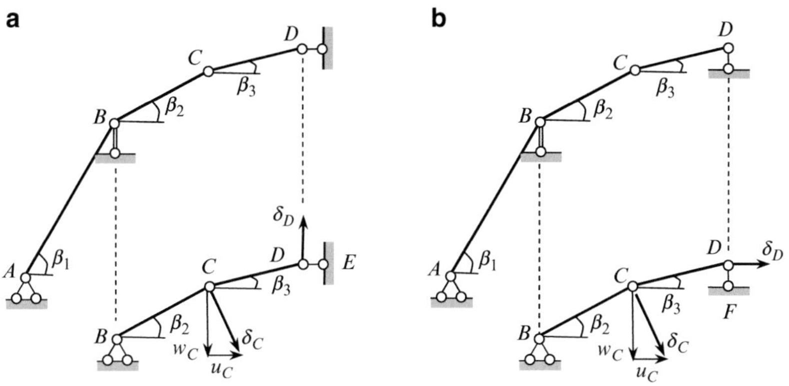

Fig. 6.13 Kinematics of the two element chain

From displacement triangle at joint C, we have

$$w_C = \delta_C \cos\beta_3 = \frac{w_B}{\cos\beta_1} \times \frac{\sin(\beta_1 - \beta_2)}{\sin(\beta_2 - \beta_3)} \times \cos\beta_3.$$

After some elementary simplifications, we get

$$w_C = w_B \frac{\tan\beta_1 - \tan\beta_2}{\tan\beta_2 - \tan\beta_3}. \tag{6.19}$$

Taking into account the first property of the parabolic polygon, we get $w_C = -w_B$. The negative sign is injected on the basis of the rule of signs (Fig. 6.12). Projection δ_C onto the horizontal axis give us the relationship

$$u_C = w_C \tan\beta_3 = -w_B \tan\beta_3. \tag{6.20}$$

6.5.2.2 Kinematics of a Two-Element Chain

In the case of symmetrical arch, it is worthwhile to investigate the symmetric and antisymmetric vibration of the arch separately. For this, we must move to the equivalent half-arch. In case of symmetric vibration of the three-hinged arch on the axis of symmetry, it is necessary to introduce a constraint which prevents horizontal displacements (Fig. 6.13a). In case of antisymmetrical vibration on the axis of symmetry, one should also include a constraint which prevents vertical displacements (Fig. 6.13b).

We proceed to construct a hinged chain and extract from it the three-element part $ABCD$ which is adjacent to the axis of symmetry. This part of structure has two degrees of freedom. If we introduce a constraint which prevents displacements of the joint B, then we get a subsystem BCD with one degree of freedom (Fig. 6.13). For such structures, we establish the relationships between the virtual displacements of the joints C and D.

6.5.2.3 Symmetrical Vibration (Fig. 6.13a)

Virtual displacement δ_C is directed perpendicularly to BC, and virtual displacement δ_D is directed perpendicularly to support constraint DE. It is obvious that

$$w_C = \delta_C \cos \beta_2; \quad u_C = \delta_C \sin \beta_2 \quad \text{and} \quad u_D = 0. \tag{6.21}$$

Projections of δ_C and δ_D on line CD are equal, so

$$\delta_C \cos(90 - \beta_2 + \beta_3) = \delta_D \cos(90 - \beta_3).$$

If follows that

$$\delta_D = \delta_C \frac{\sin(\beta_3 - \beta_2)}{\sin \beta_3} = \frac{w_C}{\cos \beta_2} \frac{\sin(\beta_3 - \beta_2)}{\sin \beta_3}.$$

After some elementary simplifications, we get

$$\delta_D = w_C \left(1 - \frac{\tan \beta_2}{\tan \beta_3} \right).$$

According to property 3 of parabolic polygon, $\tan \beta_2 / \tan \beta_3 = 3$, and finally we get

$$\delta_D = -2w_C. \tag{6.22}$$

6.5.2.4 Antisymmetric Vibration (Fig. 6.13b)

Virtual displacement δ_C is directed perpendicularly to BC, a virtual displacement δ_D is directed perpendicularly to support constraint DF. It is clear that $w_D = 0$. Projections of δ_C and δ_D onto the line CD are equal, thus

$$\delta_C \cos(90 - \beta_2 + \beta_3) = \delta_D \cos \beta_3.$$

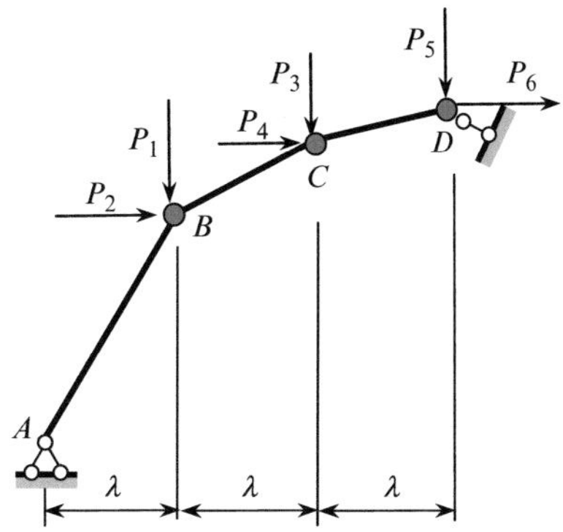

Fig. 6.14 Parabolic polygon for half-arch. Design diagram is loaded by the inertial forces $P_1 - P_6$

Thus,

$$\delta_D = \delta_C \frac{\sin(\beta_2 - \beta_3)}{\cos\beta_3} = \frac{w_C}{\cos\beta_2}\frac{\sin(\beta_2 - \beta_3)}{\cos\beta_3}.$$

After some elementary simplifications, we get

$$\delta_D = u_D = w_C(\tan\beta_2 - \tan\beta_3). \tag{6.23}$$

Kinematical relationships will be used later on.

6.5.3 *Inertial Forces*

The general method for determining the frequencies and mode shapes of vibration of elastic structures with lumped masses is discussed in Sect. 6.2. In order to form the approximate model of the free vibration of parabolic arch, we need to construct parabolic polygon as shown in Sect. 6.4. For practical purposes for symmetrical and antisymmetrical vibrations, it is sufficient to calculate two frequencies. For this purpose, each half-arch should contain three straight members with lumped masses at the ends (Fig. 6.14).

If vibrations of the structure occur with a frequency ω, then each mass is subjected to the following inertial forces

$$\begin{aligned} P_1 &= \omega^2 m_1 w_B; \quad P_2 = \omega^2 m_1 u_B; \\ P_3 &= \omega^2 m_2 w_C; \quad P_4 = \omega^2 m_2 u_C; \\ P_5 &= \omega^2 m_3 w_D; \quad P_6 = \omega^2 m_3 u_D. \end{aligned}$$

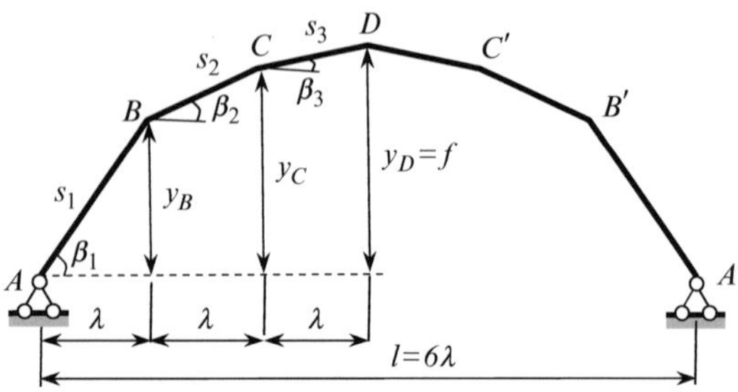

Fig. 6.15 Rabinovich model of symmetric parabolic arch for the case $k = 6$

Rabinovich's model of parabolic arch allows us to take into account arbitrary distribution of mass along the axis of the arch, so computation of lumped mass does not cause problems. It is then necessary to construct the bending moment diagrams due to all inertial forces, to choose generalized coordinates, to create unit states, which corresponds to the generalized coordinates and to calculate the unit displacements. For the last stage, it is convenient to apply Mohr method in the form of Vereshchagin or Simpson rule.

The concept of a kinematical chain allows us to significantly simplify the numerical procedure: instead of computing six displacements we can consider two partial systems, for each system with one degree of freedom calculate only one displacement, and for computation of all other displacements apply the kinematical relationships for the chain.

Additional Parameters for Parabolic Polygon ($k = 6$) for Different f/l

Rabinovich model for symmetrical parabolic arch of span l and rise f for $k = 6$ is presented in Fig. 6.15; lumped masses at the joints are not shown. The arch may be two-hinged, three-hinged, or hingeless.

Coordinates of the joint points are defined by (6.17), i.e.,

$$y_n = \frac{4f}{k^2} n(k - n).$$

In our case,

$$y_B = \frac{4f}{6^2} \times 1 \times (6 - 1) = \frac{5}{9} f; \quad y_C = \frac{4f}{6^2} \times 2 \times (6 - 2) = \frac{8}{9} f;$$
$$y_D = \frac{4f}{6^2} \times 3 \times (6 - 3) = f.$$

Slopes for each inclined member is defined by (6.17b), i.e.,

$$\tan \beta_n = \frac{4f}{kl} (k - 2n + 1).$$

In our case,

$$\tan \beta_1 = \frac{4f}{6l}(6 - 2 \times 1 + 1) = \frac{10f}{3l}; \quad \tan \beta_2 = \frac{4f}{6l}(6 - 2 \times 2 + 1) = 2\frac{f}{l};$$
$$\tan \beta_3 = \frac{4f}{6l}(6 - 2 \times 3 + 1) = \frac{2f}{3l}.$$

The length of each member is defined by (6.17a), i.e.,

$$s_n = \frac{\lambda}{\cos \beta_n} = \frac{l}{k}\sqrt{1 + \tan^2 \beta_n} = \frac{l}{6}\sqrt{1 + \tan^2 \beta_n}.$$

If an arch has a constant cross-section and the mass per unit length is μ, then the lumped masses at the joints are

$$m_A = \frac{s_1}{2}\mu, \quad m_B = \frac{s_1 + s_2}{2}\mu, \quad m_C = \frac{s_2 + s_3}{2}\mu, \quad m_D = \frac{s_3}{2}\mu.$$

Here m_D is the lumped mass at joint D considering portion DC. The same mass occurs at joint D considering portion DC', thus the total mass at joint D is equal to $2m_D$.

For arches with different ratios of f/l and $k = 6$, parameters y/l, $\tan \beta$, s/l, $m/\mu l$ of the Rabinovich model are presented in Table A.36.

Application of Rabinovich's method for symmetrical and antisymmetric vibrations of parabolic arch is shown in Sects. 6.6–6.8.

6.6 Symmetrical Vibrations of Three-Hinged Parabolic Arch

This section shows vibration analysis of a parabolic arch by Rabinovich's method, which required the following steps:

- Construction of equivalent design diagram, taking into account the type of vibration (symmetrical/antisymmetrical).
- Expressing nodal displacements in terms of generalized coordinates.
- Calculation of inertial forces based on expressions of nodal displacements.
- Finding displacements caused by inertial forces.
- Determining frequencies and shape modes of vibration from the frequency equation.
- Showing bending moment diagrams for internal forces due to free vibration.

These steps are considered below.

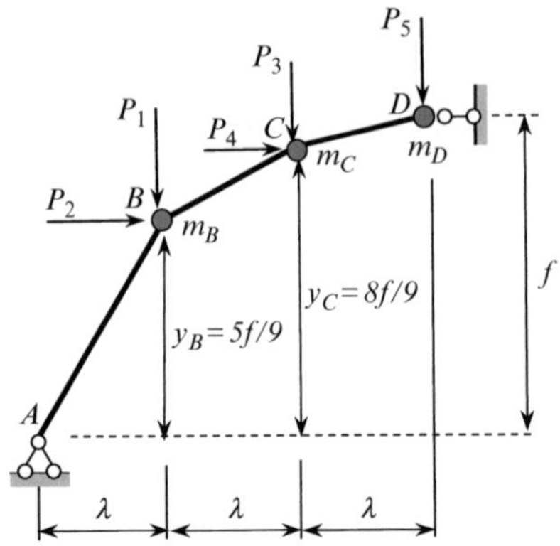

Fig. 6.16 Design diagram of half-arch for symmetrical vibration

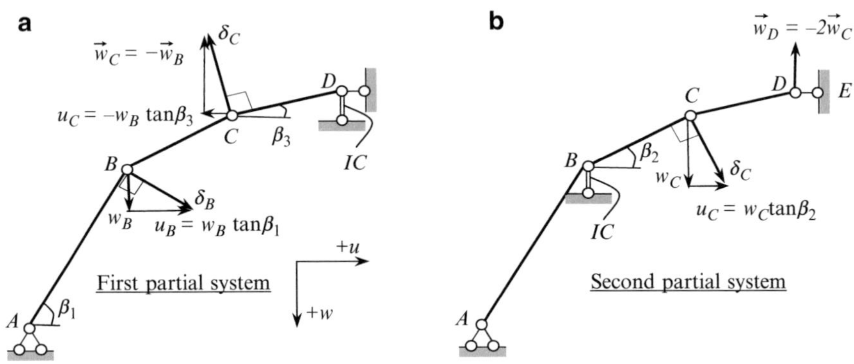

Fig. 6.17 Displacements in the partial systems, *IC* – introduced constraints

6.6.1 Equivalent Design Diagram. Displacements

In the case of symmetrical vibration, the section on the axis of symmetry is displaced vertically. Therefore, in the equivalent half-arch, it is necessary to introduce support constraint which prevents horizontal displacement and allows vertical displacements.

Design diagram of the equivalent half-arch for $k = 6$ is shown in Fig. 6.16. The mass m_A and horizontal force of inertia which acts on the mass m_D are not shown.

This structure has two degrees of freedom. Displacement of the structure may be considered as a sum of displacements which occur in two systems, each with one degree of freedom. Such systems are called partial systems [Kar01]. Figure 6.17 demonstrates two partial systems. In Fig. 6.17a, an additional constraint is introduced at joint *D* (presented by double line) and in Fig. 6.17b at joint *B*.

Let the generalized coordinate be the vertical displacements of joints *B* and *C*, i.e., w_B and w_C. Table 6.4 shows the kinematical relationships for joints *B*, *C*, and *D*. For example, let us consider total horizontal displacement of joint *C*. For the first

Table 6.4 Symmetrical vibrations: Total displacements of the joints *B*, *C*, and *D* in terms of displacements of the partial systems [Rab58]

Joint	Total vertical displacement w'	Total horizontal displacement u'	Notes
B	$w'_B = w_B$	$u'_B = w_B \tan \beta_1$	Geometrical representation of these formulas is shown in Figs. 6.12 and 6.17a, b
C	$w'_C = -w_B + w_C$	$u'_C = -w_B \tan \beta_3 + w_C \tan \beta_2$	
D	$w'_D = -2w_C$	$u'_D = 0$	

partial system (Fig. 6.17a) $u_C^{(1)} = -w_B \tan \beta_3$, while for the second partial system (Fig. 6.17b) $u_C^{(2)} = +w_C \tan \beta_2$

According to Table 6.4, the inertial forces which act on the half-arch are

$$\begin{aligned}
&P_1 = \omega^2 m_B w_B; \quad P_2 = \omega^2 m_B w_B \tan \beta_1; \\
&P_3 = \omega^2 m_C(-w_B + w_C); \quad P_4 = \omega^2 m_C(-w_B \tan \beta_3 + w_C \tan \beta_2); \\
&P_5 = \omega^2 m_D(-2w_C).
\end{aligned}$$

Table A.36 allows us to determine the values of the lumped masses in terms of rise–span ratio. In particularly, for a uniform arch with parameter $f/l = 0.5$[Rab58], we get

$$m_B = 0.280\mu\, l, \quad m_C = 0.206\mu\, l, \quad m_D = 0.0880\mu\, l.$$

Corresponding forces of inertia in terms of generalized coordinates becomes

$$\begin{aligned}
&P_1 = 0.280\omega^2\mu\, l w_B, \quad P_2 = 0.280 \cdot 1.667 w_B \times \omega^2\mu\, l = 0.4667\omega^2\mu\, l w_B, \\
&P_3 = 0.206\omega^2\mu\, l(-w_B + w_C), \quad P_4 = 0.206\omega^2\mu\, l(-0.3333 w_B + 1.00 \times w_C), \\
&P_5 = 0.0880\omega^2\mu\, l(-2w_C) = -0.176\omega^2\mu\, l w_C.
\end{aligned}$$

Now let us determine the displacements in the direction of the generalized coordinates w_B and w_C. For this, we need to construct the bending moment diagram caused by all inertial forces P_i.

The vertical and horizontal reactions at *A* (Fig. 6.18) caused by the forces of inertia are

$$\begin{aligned}
&V = P_1 + P_3 + P_5 = (0.074 w_B + 0.030 w_C)\omega^2\mu\, l, \\
&H = \frac{1}{f}(V \times 3\lambda - P_1 \times 2\lambda - P_3 \times \lambda) - \frac{1}{f}[P_2 \times (f - y_B) - P_4 \times (f - y_C)].
\end{aligned}$$

Since for points *B* and *C*, we have $y_B = 5f/9; \quad y_C = 8f/9$ (Fig. 6.16), then for horizontal reactions we get

$$H = (-0.2438 w_B - 0.062 w_C)\omega^2\mu\, l.$$

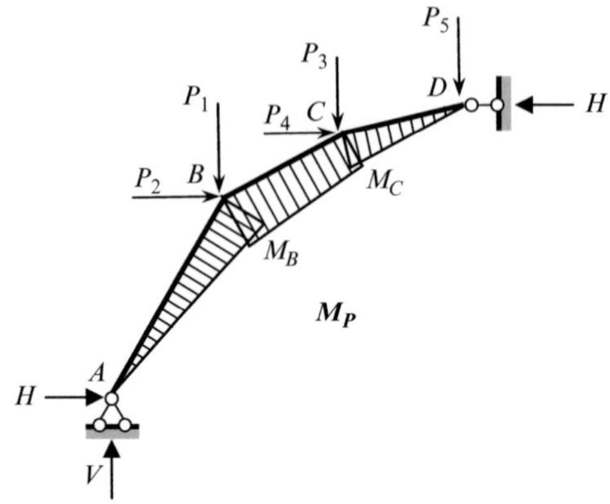

Fig. 6.18 Bending moment diagram caused by all inertial forces P_i

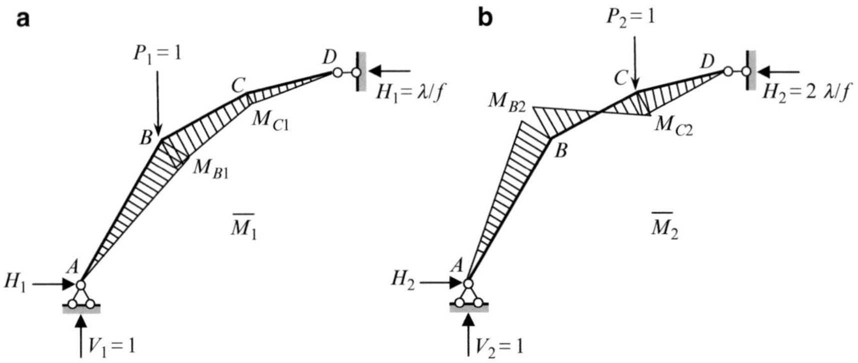

Fig. 6.19 Unit states and corresponding bending moment diagrams

The bending moments at joints B and C are

$$\begin{aligned} M_B &= V\lambda - Hy_B = (0.4803w_B + 0.1333w_C)\,\omega^2\lambda\,\mu\,l, \\ M_C &= V2\lambda - Hy_C - P_1\lambda - P_2(y_C - y_B) \\ &= (0.0514w_B + 0.2253w_C)\,\omega^2\lambda\,\mu\,l. \end{aligned} \tag{6.24}$$

Bending moment diagram caused by all inertial forces is shown in Fig. 6.18. Ordinates (6.24) of this diagram are functions of generalized coordinates w_B and w_C.

Unit states, which correspond to the generalized coordinates and the corresponding bending moment diagrams, are shown in Fig. 6.19.

Reactions and bending moments for both conditions are

State 1.

$$V_1 = 1; \quad H_1 = \frac{\lambda}{f}; \quad M_{B1} = V_1\lambda - H_1\frac{5}{9}f = \frac{4}{9}\lambda = \frac{2}{27}l;$$
$$M_{C1} = V_1 2\lambda - H_1\frac{8}{9}f = \frac{1}{54}l.$$

State 2.

$$V_2 = 1; \quad H_2 = \frac{2\lambda}{f}; \quad M_{B2} = -\frac{1}{54}l; \quad M_{C2} = \frac{1}{27}l.$$

Now we can calculate the displacement in direction of each generalized coordinate caused by all the forces of inertia. Vertical displacement of joint B is

$$w'_B = \sum \int \frac{M_P \overline{M}_1}{EI} \mathrm{d}s.$$

Multiplication of diagrams along s_1 and s_3 is done by formula "triangle by triangle" and along s_2 using the trapezoid formula, so

$$w'_B = \left[\frac{s_1}{3} M_B M_{B1} + \frac{s_2}{6}(2M_B M_{B1} + 2M_C M_{C1} + M_B M_{C1} + M_C M_{B1}) + \frac{s_3}{3} M_C M_{C1}\right] \frac{1}{EI}.$$

If we take into account the expressions for bending moments M_B and M_C caused by inertial forces (Fig. 6.18), then we get the following expression for total displacement in terms of these moments

$$\begin{aligned} w'_B &= \left[\left(\frac{4}{27}s_1 + \frac{1}{6}s_2\right) M_B + \left(\frac{1}{9}s_2 + \frac{1}{27}s_3\right) M_C\right] \frac{\lambda}{EI} \\ &= [(8s_1 + 9s_2)M_B + (6s_2 + 2s_3)M_C] \frac{\lambda}{54EI}. \end{aligned} \quad (6.25)$$

Vertical displacement of joint C caused by all inertial forces is

$$\begin{aligned} w'_C &= (-w_B + w_C) = \sum \int \frac{M_P \overline{M}_2}{EI} \mathrm{d}s \\ &= \left[\frac{s_1}{3} M_B M_{B2} + \frac{s_2}{6}(2M_B M_{B2} + 2M_C M_{C2} + M_B M_{C2} + M_C M_{B2}) + \frac{s_3}{3} M_C M_{C2}\right] \frac{1}{EI}. \end{aligned}$$

Upon rearrangement, this result can be rewritten as

$$\begin{aligned} w'_C &= \left[-\frac{1}{27}s_1 M_B + \left(\frac{1}{18}s_2 + \frac{2}{27}s_3\right) M_C\right] \frac{\lambda}{EI} \\ &= [-2s_1 M_B + (3s_2 + 4s_3)M_C] \frac{\lambda}{54EI}. \end{aligned} \quad (6.26)$$

All other displacements may be calculated using kinematical relationships shown in Table 6.4. Note that (6.25) and (6.26) are valid for symmetric arches with any ratio f/l.

6.6.2 Frequencies and Mode Shape of Vibrations

Expressions (6.24) for M_B and M_C should be substituted into (6.25), take into account the relationship $w'_B = w_B$ (Table 6.4) and the values for $s_1 - s_3$ according to Table A.36 for a given rise–span ratio. As a result, we obtain [Rab58]

$$w_B = (0.0436w_B + 0.019018w_C)\frac{\omega^2 \mu \lambda^2 l^2}{EI}.$$

If we introduce the dimensionless frequency parameter

$$\frac{EI}{\omega^2 \mu \lambda^2 l^2} = \theta \rightarrow \omega = \frac{1}{\sqrt{\theta}\lambda l}\sqrt{\frac{EI}{\mu}} = \frac{1}{\sqrt{\theta}}\frac{1}{\frac{l}{6}l}\sqrt{\frac{EI}{\mu}} = \frac{1}{\sqrt{\theta}}\frac{6}{l^2}\sqrt{\frac{EI}{\mu}},$$

then the previous equation may be rewritten as follows

$$(0.0436 - \theta)w_B + 0.019018w_C = 0. \tag{6.27}$$

Similarly, if (6.24) for M_B and M_C are substituted into (6.26), then take into account the relationship $w'_C = -w_B + w_C$ (Table 6.4) and the values for $s_1 - s_3$ according to Table A.36, we obtain

$$(-0.00442 + \theta)w_B + (0.00429 - \theta)w_C = 0. \tag{6.28}$$

Thus, for generalized coordinates w_B and w_C, we have obtained two homogeneous linear algebraic equations (6.27) and (6.28) with respect to w_B and w_C. Nontrivial solution of these equations presents the frequency equation

$$\det\begin{bmatrix} 0.0436 - \theta & 0.019018 \\ -0.00442 + \theta & 0.00429 - \theta \end{bmatrix} = 0.$$

The roots of this equation in decreasing order are

$$\theta_1 = 0.06257, \quad \theta_2 = 0.004333.$$

The frequencies of free vibration of an arch in increasing order become

$$\omega_1 = \frac{1}{\sqrt{\theta_1}}\frac{6}{l^2}\sqrt{\frac{EI}{\mu}} = \frac{1}{\sqrt{0.06257}}\frac{6}{l^2}\sqrt{\frac{EI}{\mu}} = \frac{23.98}{l^2}\sqrt{\frac{EI}{\mu}},$$

$$\omega_2 = \frac{1}{\sqrt{\theta_2}}\frac{6}{l^2}\sqrt{\frac{EI}{\mu}} = \frac{91.18}{l^2}\sqrt{\frac{EI}{\mu}}.$$

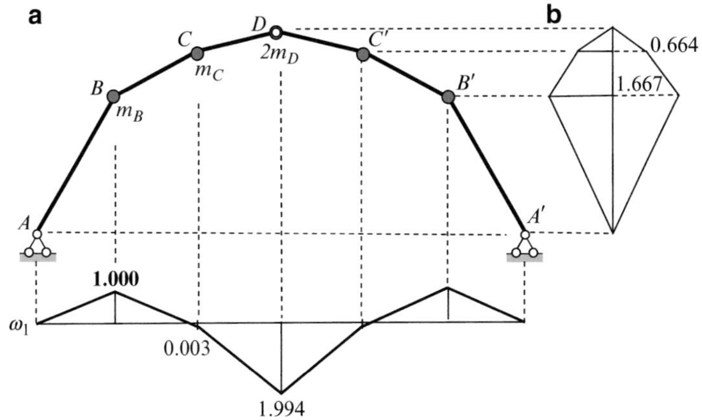

Fig. 6.20 First form of symmetrical vibration: (**a**) vertical displacements; (**b**) horizontal displacements

6.6.2.1 First Form of Symmetrical Vibrations

Equations (6.27) and (6.28) do not allow us to find w_B and w_C. However, we can find the ratios between generalized coordinates. Since the determinant is equal to zero, (6.27) and (6.28) are dependent, and the ratio may be calculated from *any* equation.

Vertical Displacements

From (6.27) and (6.28), respectively, we get

$$\frac{w_C}{w_B} = -\frac{0.0436 - \theta_1}{0.019018} = -\frac{0.0436 - 0.06257}{0.019018} = 0.997,$$

$$\frac{w_C}{w_B} = -\frac{-0.00442 + \theta_1}{0.00429 - \theta_1} = -\frac{-0.00442 + 0.06257}{0.00429 - 0.06257} = 0.997.$$

According to Table 6.4, the ratio of total vertical displacements of the joint points are

$$\frac{w'_C}{w_B} = \frac{-w_B + w_C}{w_B} = -1 + \frac{w_C}{w_B} = -1 + 0.997 = -0.003,$$
$$\frac{w'_D}{w_B} = \frac{-2w_C}{w_B} = -2 \cdot 0.997 = -1.994.$$

The vertical displacements, which correspond to the first mode of symmetrical vibration, are shown in Fig. 6.20a. If the mass B is displaced upward, then masses C and D are displaced downward, and vice versa.

Horizontal Displacements

Assume that $w_B = 1$ and determine the total horizontal displacements of the joints. According to Table 6.4, we have

$$u'_B = w_B \tan\beta_1 = 1 \times 1.667 = 1.667,$$
$$u'_C = -w_B \tan\beta_3 + w_C \tan\beta_2 = -1 \times 0.333 + 0.997 \times 1.000 = 0.664,$$
$$u'_D = 0.00.$$

Horizontal displacements, which correspond to the first form of symmetrical vibrations, are shown in Fig. 6.20b. It can be seen that masses which belongs to the left and right half-arches are displaced in the opposite direction, while masses which belong to a particular half-arch are all displaced in one direction. For a given ratio $f/l = 0.5$, even for symmetrical vibration, the horizontal and vertical displacements of the joints are of the same order. In particular, for joint B we get $u'_B = 1.667w_B$.

6.6.2.2 Second Form of Symmetrical Vibrations

Vertical Displacements

From (6.27) and (6.28), respectively, we have

$$\frac{w_C}{w_B} = -\frac{0.0436 - \theta_2}{0.019018} = -\frac{0.0436 - 0.004333}{0.019018} = -2.065,$$

$$\frac{w_C}{w_B} = -\frac{-0.00442 + \theta_2}{0.00429 - \theta_2} = -\frac{-0.00442 + 0.004333}{0.00429 - 0.004333} = -2.023.$$

The relative error is about 2.0%. Assume, that $w_C/w_B = -2.044$.

According to Table 6.4, the ratio of total vertical displacements of the joint points are

$$\frac{w'_C}{w_B} = \frac{-w_B + w_C}{w_B} = -1 + \frac{w_C}{w_B} = -1 - 2.044 = -3.044,$$
$$\frac{w'_D}{w_B} = \frac{-2w_C}{w_B} = -2 \cdot (-2.044) = 4.088.$$

The second form of vertical displacement is shown in Fig. 6.21a.

If masses B and D are displaced upwards, then mass C is displaced downwards, and vice versa.

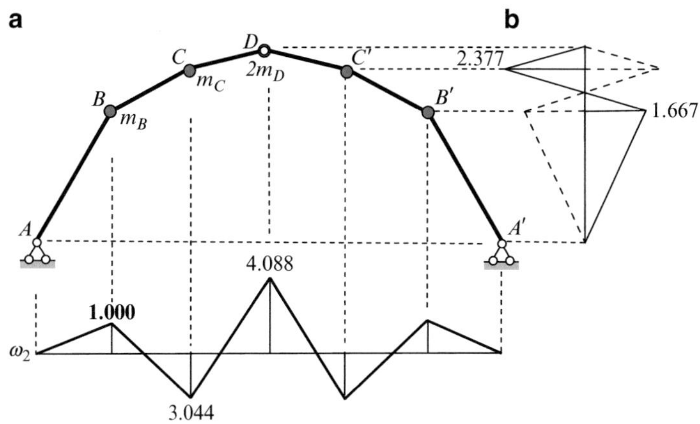

Fig. 6.21 Second form of symmetrical vibration: (**a**) vertical displacements; (**b**) horizontal displacements

Table 6.5 Symmetrical vibrations: the total vertical and horizontal displacements of the joints B, C, and D of the arch

Mode	Total vertical displacement of the joints w'		Total horizontal displacement of the joints u'
First	B	$w'_B = w_B = 1$	$u'_B = w_B \tan\beta_1 = 1 \times 1.667 = 1.667$
	C	$w'_C = -w_B + w_C$	$u'_C = -w_B \tan\beta_3 + w_C \tan\beta_2$
		$= -1 + 0.997 = -0.003$	$= -1 \times 0.333 + 0.997 \times 1.000 = 0.664$
	D	$w'_D = -2w_C = -1.994$	$u'_D = 0.00$
Second	B	$w'_B = w_B = 1$	$u'_B = w_B \tan\beta_1 = 1 \cdot 1.667 = 1.667$
	C	$w'_C = -w_B + w_C$	$u'_C = -w_B \tan\beta_3 + w_C \tan\beta_2$
		$= -1 - 2.044 = -3.044$	$= -1 \times 0.333 + (-2.044) \times 1.000 = -2.377$
	D	$w'_D = -2w_C = 4.088$	$u'_D = 0.00$

Horizontal Displacements

Assume that $w_B = 1$ and determine the total horizontal displacements of the joints. For this, we apply the formulas according to Table 6.4. The vertical and horizontal displacements of the joint points are presented in Table 6.5.

Horizontal displacements which arise at the second mode of symmetrical vibration are shown in Fig. 6.21b.

Orthogonality condition of the form of vibration can be verified by the formula

$$\sum m_i w'_1 w'_2 + \sum m_i u'_1 u'_2 = 0.$$

Here the subscripts 1 and 2 represent the number of the mode of vibration; summation is performed over the index i. In the expanded form, this expression becomes

$$\begin{aligned}&0.280 \times 1 \times 1 + 0.280 \times 1.667 \times 1.667\\ &\quad + 0.206 \times (-0.003) \times (-3.044) + 0.206 \times 0.664 \times (-2.377)\\ &\quad + 0.088 \times (-1.994) \times 4.088 = 1.05997 - 1.0424.\end{aligned}$$

The difference between positive and negative terms is 0.01757; the relative error is approximately 1.6%.

6.6.3 *Internal Forces for First and Second Modes of Vibration*

Now we can determine the bending moments which arise at the first and second mode of the free vibration. For this, we can use two approaches.

First Approach

Bending moments at joints B and C, caused by inertial forces, according to (6.24) are

$$\begin{aligned}M_B &= (0.4803w_B + 0.1333w_C)\,\omega^2\lambda\,\mu\,l,\\ M_C &= (0.0514w_B + 0.2253w_C)\,\omega^2\lambda\,\mu\,l.\end{aligned}$$

First Form of Vibration

$\theta_1 = 0.06257$. Let $w_B = 1$, then $w_C = 0.997$ and for bending moments we get

$$\begin{aligned}M_B &= (0.4803w_B + 0.1333w_C)\,\frac{1}{\theta_1}\,\frac{EI}{\mu l^2\lambda^2}\mu\,l\lambda,\\ &= (0.4803 \times 1.0 + 0.1333 \times 0.997)\,\frac{1}{0.06257}\,\frac{EI}{l\lambda} = 15.98\frac{6EI}{l^2} = 58.80\frac{EI}{l^2},\\ M_C &= (0.0514 \times 1.0 + 0.2253 \times 0.997)\,\frac{1}{0.06257}\,\frac{6EI}{l^2} = 26.47\frac{EI}{l^2}.\end{aligned}$$

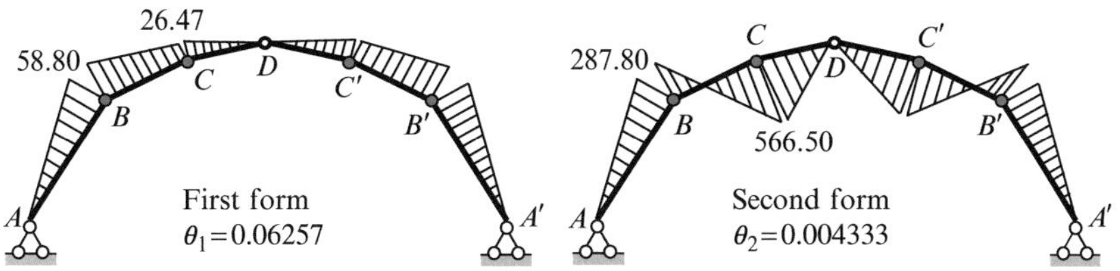

Fig. 6.22 Bending moment diagrams for the first and second forms of symmetrical vibration

<u>*Second Form of Vibration*</u>

$\theta_2 = 0.004333$. Similarly, let $w_B = 1$, then $w_C = -2.044$ and for bending moments, we get

$$M_B = (0.4803w_B + 0.1333w_C)\,\frac{1}{\theta_2}\,\frac{EI}{\mu l^2\lambda^2}\,\mu\, l\lambda,$$

$$= (0.4803 \cdot 1.0 - 0.1333 \cdot 2.044)\,\frac{1}{0.004333}\,\frac{6EI}{l^2} = 287.80\frac{EI}{l^2},$$

$$M_C = (0.0514 \cdot 1.0 - 0.2253 \cdot 2.044)\,\frac{1}{0.004333}\,\frac{6EI}{l^2} = -566.50\frac{EI}{l^2}.$$

Corresponding bending moment diagrams are shown in Fig. 6.22; (factor of EI/l^2).

Second Approach

Equations (6.25) and (6.26) may be solved with respect to the moments M_B and M_C

$$M_B = 36EI\,\frac{[(3s_2 + 4s_3)w'_B - 2(3s_2 + s_3)w'_C]}{lD_0},$$

$$M_C = 36EI\,\frac{[2s_1 w'_B + (8s_1 + 9s_2)w'_C]}{lD_0}, \quad D_0 = 4s_1(s_2 + s_3) + s_2(3s_2 + 4s_3).$$

If the values $s_1 = 0.324l$, $s_2 = 0.236l$, $s_3 = 0.176l$ (Table A.36) are substituted into expression for M_B, then we get

$$M_B = \frac{36EI}{l^2}(1.6283w'_B - 2.0388w'_C).$$

According to Table 6.5, for the first form of vibration, we get $w'_B = 1.0$, $w'_C = -0.003$. Therefore, the bending moment at point B becomes $M_B^{(1)} = 58.84EI/l^2$.

For the second form of vibration, $w'_B = 1.0, \quad w'_C = -3.037$. Therefore, $M_B^{(2)} = 281.5EI/l^2$.

Similarly, we can calculate the bending moments at point C for each form of vibration.

Bending moment diagrams $M^{(1)}$ and $M^{(2)}$ for first and second form of vibration satisfy the orthogonality condition

$$\sum \int_A^D \frac{M^{(1)}M^{(2)}}{EI} \mathrm{d}s = 0.$$

Multiplication of both diagrams should be performed via the Vereshchagin and Simpson formulas.

The frequencies, displacements, and bending moments for symmetric vibration of parabolic three-hinged uniform arches with different ratios f/l are presented in Tables A.37, A.38, A.39, and A.40 which serve for antisymmetrical vibrations.

6.7 Antisymmetrical Vibration of Three-Hinged Parabolic Arch

As in the previous paragraph, we will consider equivalent design diagram, find displacements caused by inertial forces, and determine frequencies and shape of antisymmetrical vibrations for three-hinged parabolic arch.

6.7.1 Equivalent Design Diagram. Displacements

In the case of *antisymmetrical vibration*, the section on the axis of symmetry is displaced in the horizontal direction. Therefore, in the equivalent half-arch, it is necessary to introduce a support constraint which prevents vertical displacement and allows horizontal displacement. Rabinovich's model for $k = 6$ is shown in Fig. 6.23; the mass m_A is not shown.

This structure has two degrees of freedom. As in the case of symmetrical vibration, displacement of structure may be considered as a sum of displacements, which occurs in two partial systems, each with one degree of freedom. Note that in the case of antisymmetrical vibration of a *two-hinged* arch, Rabinovich's model does not differ from design diagram in Fig. 6.23.

Figure 6.24a presents partial system with additional constraint at the joint D and the partial system in Fig. 6.24b contains additional constraint at joint B; all introduced constraints are shown by doubled lines. It can be seen that the first

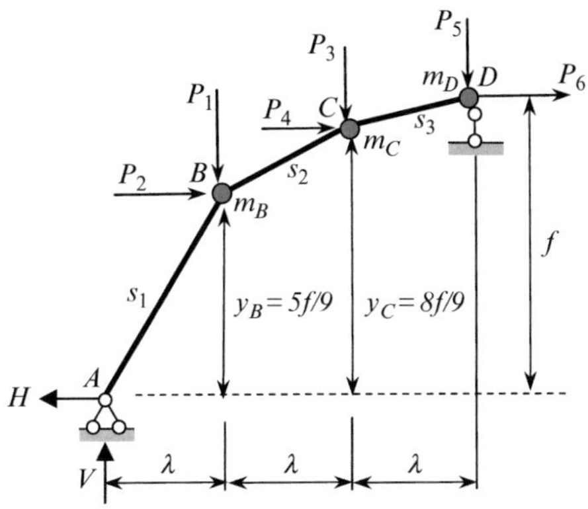

Fig. 6.23 Design diagram of the half-arch for antisymmetric vibration

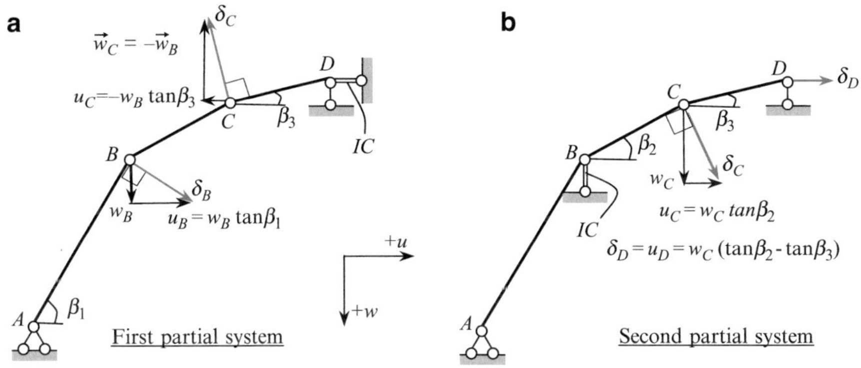

Fig. 6.24 Partial systems for the analysis of antisymmetrical vibration; *IC* – introduced constraints

partial system for symmetrical and antisymmetrical vibrations coincide with each other (Figs. 6.17a and 6.24a).

Let the generalized coordinates be the vertical displacements of the joints B and C, i.e., w_B and w_C. Table 6.6 presents the kinematical relationships for joints B, C, and D.

According to Table 6.6, the inertial forces which act on the half-arch are

$$P_1 = \omega^2 m_B w_B; \quad P_2 = \omega^2 m_B w_B \tan\beta_1;$$
$$P_3 = \omega^2 m_C(-w_B + w_C); \quad P_4 = \omega^2 m_C(-w_B \tan\beta_3 + w_C \tan\beta_2);$$
$$P_5 = 0; \quad P_6 = \omega^2 m_D w_C(\tan\beta_2 - \tan\beta_3).$$

Inertial forces in terms of generalized coordinates ($f = 0.5\ l$) become

Table 6.6 Antisymmetric vibrations: total displacements of the joints B, C, and D in terms of displacements of the partial systems

Joint	Total vertical displacement	Total horizontal displacement	Notes
B	$w'_B = w_B$	$u'_B = w_B \tan\beta_1$	See Fig. 6.12
C	$w'_C = -w_B + w_C$	$u'_C = -w_B \tan\beta_3 + w_C \tan\beta_2$	Geometrical representation is shown in Fig. 6.24a, b
D	$w'_D = 0$	$u'_D = w_C(\tan\beta_2 - \tan\beta_3)$	

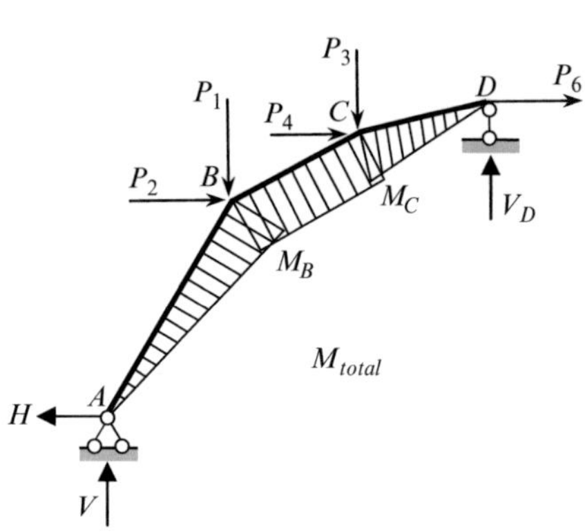

Fig. 6.25 Bending moment diagram due to all inertial forces

$$P_1 = 0.280\omega^2\mu\, l w_B; \quad P_2 = 0.280 \times 1.667\omega^2\mu\, l w_B = 0.4667\omega^2\mu\, l w_B,$$
$$P_3 = 0.206\omega^2\mu\, l(-w_B + w_C); \quad P_4 = 0.206\omega^2\mu\, l(-0.3333 w_B + 1.00 \times w_C),$$
$$P_5 = 0; \quad P_6 = 0.088\omega^2\mu\, l(1 - 0.333)w_C = 0.05867\omega^2\mu\, l w_C.$$

Now we can calculate the displacements caused by all inertial forces. The reactions of support A and bending moments at joints B and C (Fig. 6.25) are

$$H = P_2 + P_4 + P_6,$$
$$V = \frac{2}{3}P_1 + \frac{1}{3}P_3 - \frac{5}{27}\frac{f}{\lambda}P_2 - \frac{8}{27}\frac{f}{\lambda}P_4 - \frac{1}{3}\frac{f}{\lambda}P_6,$$

$$M_B = V\lambda + H\frac{5}{9}f = \left(V + H\frac{5}{9}\frac{f}{\lambda}\right)\lambda,$$
$$M_C = V \times 2\lambda + \frac{8}{9}fH - P_1\lambda - P_2\frac{f}{3} = \left(2V + \frac{16}{3}\frac{f}{l}H - P_1 - \frac{2f}{l}P_2\right)\lambda.$$

If we take into account the expressions for inertial forces, then bending moments at joints B and C may be presented in terms of generalized coordinates w_B and w_C, as follows

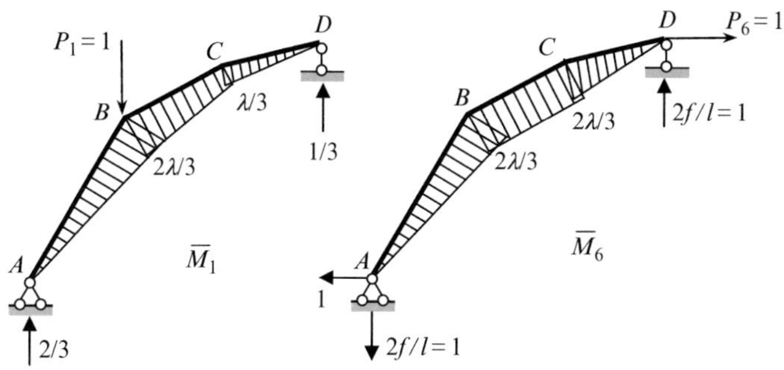

Fig. 6.26 Unit states and corresponding bending moment diagrams

$$M_B = \left(V + H\frac{5}{9}\frac{f}{\lambda}\right)\lambda = (0.58293w_B + 0.26801w_C)\,\omega^2\mu\, l\lambda,$$

$$M_C = \left(2V + \frac{16}{3}\frac{f}{l}H - P_1 - \frac{2f}{l}P_2\right)\lambda = (0.1538w_B + 0.35956w_C)\,\omega^2\mu\, l\lambda.$$

Let us determine the displacements in direction of P_1 and P_6. Corresponding unit states and bending moment diagrams are shown in Fig. 6.26.

The total vertical displacement w_B' in direction P_1 and horizontal displacement u_D' in direction P_6 is

$$EIw_B' = \sum \int M_{\text{total}}\overline{M}_1 \text{d}s = \frac{\lambda}{18}[(4s_1 + 5s_2)M_B + (4s_2 + 2s_3)M_C],$$

$$EIu_D' = \sum \int M_{\text{total}}\overline{M}_6 \text{d}s = \frac{\lambda}{9}[(2s_1 + 3s_2)M_B + (3s_2 + 2s_3)M_C].$$

These displacements are caused by all inertial forces. Substitution of s_i, M_B, and M_C into these formulas, allow us to present the displacements in terms of generalized coordinates w_B and w_C, as follows

$$\begin{aligned}EIw_B &= \frac{\omega^2\mu\, l^2\lambda^2}{18}[(4\times 0.324 + 5\times 0.236)\,(0.58293w_B + 0.26801w_C)\\ &\quad + (4\times 0.236 + 2\times 0.176)\,(0.1538w_B + 0.35956w_C)]\\ &= \omega^2\mu\, l^2\lambda^2(0.091254w_B + 0.06275w_C).\end{aligned}$$

Joint B is characterized by only one of the generalized coordinate (see Fig. 6.24 and Table 6.6) so $w_B' = w_B$, therefore, the prime superscript at the displacement in the left part of equation above is omitted.

Since

$$u_D' = w_C(\tan\beta_2 - \tan\beta_3) = w_C\left(\frac{f}{3\lambda} - \frac{f}{9\lambda}\right) = \frac{4f}{3l}w_C,$$

then

$$
\begin{aligned}
EIu'_D &= EI\frac{4f}{3l}w_C \\
&= \frac{\omega^2\mu l^2\lambda^2}{9}[(2\times 0.324 + 3\times 0.236)\,(0.58293w_B + 0.26801w_C) \\
&\quad + (3\times 0.236 + 2\times 0.176)\,(0.1538w_B + 0.35956w_C)] \\
&= \omega^2\mu l^2\lambda^2(0.10594w_B + 0.08273w_C).
\end{aligned}
$$

6.7.2 *Frequencies and Mode Shape of Vibrations*

If we denote $\theta = EI/\omega^2\mu l^2\lambda^2$, and take into account $\theta(4/3)(f/l)w_C = \theta(2/3)w_C$, then the two last equations may be rewritten as follows

$$
\begin{aligned}
&(0.09125 - \theta)w_B + 0.06275w_C = 0, \\
&0.10594w_B + \left(0.08273 - \frac{2}{3}\theta\right)w_C = 0.
\end{aligned}
\tag{6.29}
$$

Nontrivial solution of the set of homogeneous linear algebraic equations leads to the frequency equation

$$
\det\begin{vmatrix} 0.09125 - \theta & 0.06275 \\ 0.10594 & 0.08273 - \frac{2}{3}\theta \end{vmatrix} = 0.
$$

The roots of this equation, in decreasing order are $\theta_1 = 0.20887$; $\theta_2 = 0.006475$. The frequencies of free vibration in increasing order are

$$
\begin{aligned}
\omega_1 &= \frac{1}{\sqrt{\theta_1}\lambda l}\sqrt{\frac{EI}{\mu}} = \frac{1}{\sqrt{\theta_1}}\frac{6}{l^2}\sqrt{\frac{EI}{\mu}} = \frac{1}{\sqrt{0.20887}}\frac{6}{l^2}\sqrt{\frac{EI}{\mu}} = \frac{13\times 128}{l^2}\sqrt{\frac{EI}{\mu}}, \\
\omega_2 &= \frac{1}{\sqrt{\theta_2}}\frac{6}{l^2}\sqrt{\frac{EI}{\mu}} = \frac{74.56}{l^2}\sqrt{\frac{EI}{\mu}}.
\end{aligned}
$$

Mode Shapes of Antisymmetrical Vibration

First mode. For the first eigenvalue $\theta_1 = 0.20887$ from the first equation of (6.29), we have

Table 6.7 Antisymmetrical vibrations: total vertical and horizontal displacements of the joints B, C, and D of the arch

Mode	Joint	Total vertical displacement w'	Total horizontal displacement u'
First	B	$w'_B = 1$	$u'_B = w_B \tan\beta_1 = 1.667$
	C	$w'_C = w_B\left(-1 + \frac{w_C}{w_B}\right)$	$u'_C = w_B\left(-\tan\beta_3 + \frac{w_C}{w_B}\tan\beta_2\right)$
		$= -1 + 1.874 = 0.874$	$= -0.333 + 1.874 \times 1.0 = 1.541$
	D	$w'_D = 0$	$u'_D = \frac{w_C}{w_B}(\tan\beta_2 - \tan\beta_3)$
			$= 1.874 \times (1.0 - 0.333) = 1.249$
Second	B	$w'_B = 1$	$u'_B = w_B \tan\beta_1 = 1.667$
	C	$w'_C = w_B\left(-1 + \frac{w_C}{w_B}\right)$	$u'_C = w_B\left(-\tan\beta_3 + \frac{w_C}{w_B}\tan\beta_2\right)$
		$= -1 - 1.351 = -2.351$	$= -0.333 - 1.351 \times 1.0 = -1.684$
	D	$w'_D = 0$	$u'_D = \frac{w_C}{w_B}(\tan\beta_2 - \tan\beta_3)$
			$= -1.351 \cdot (1.0 - 0.333) = -0.901$

$$\frac{w_C}{w_B} = -\frac{0.09125 - \theta}{0.06275} = -\frac{0.09125 - 0.20887}{0.06275} = 1.874.$$

Second mode. For the second eigenvalue $\theta_2 = 0.006475$ from the first equation of (6.29), we have

$$\frac{w_C}{w_B} = -\frac{0.09125 - \theta}{0.06275} = -\frac{0.09125 - 0.006475}{0.06275} = -1.351.$$

The same results may be obtained from the second equation of the system (6.29).

Computation of the total displacements of the joints is presented in Table 6.7. We assume that $w_B = 1$.

The total displacements for the first and second forms of antisymmetric vibration are shown in Fig. 6.27.

Verification of orthogonality condition and computation of internal forces for each mode of vibration should be performed as in the case of symmetrical vibration (see Sects. 6.5.2–6.5.3).

The frequencies, eigenfunctions, and bending moments for antisymmetric vibration of parabolic three-hinged and two-hinged uniform arches with different ratios of f/l are presented in Tables A.39 and A.40.

Tables A.41 and A.42 serve for symmetric vibration of parabolic two-hinged uniform arches with different ratios of f/l.

Studies have shown [Rab58] that for two-hinged arches with the same f/l, the smallest frequency of vibration for circular and parabolic arches satisfy the condition $\omega_{circ} < \omega_{parab}$. This happens because the length of parabolic arch is less

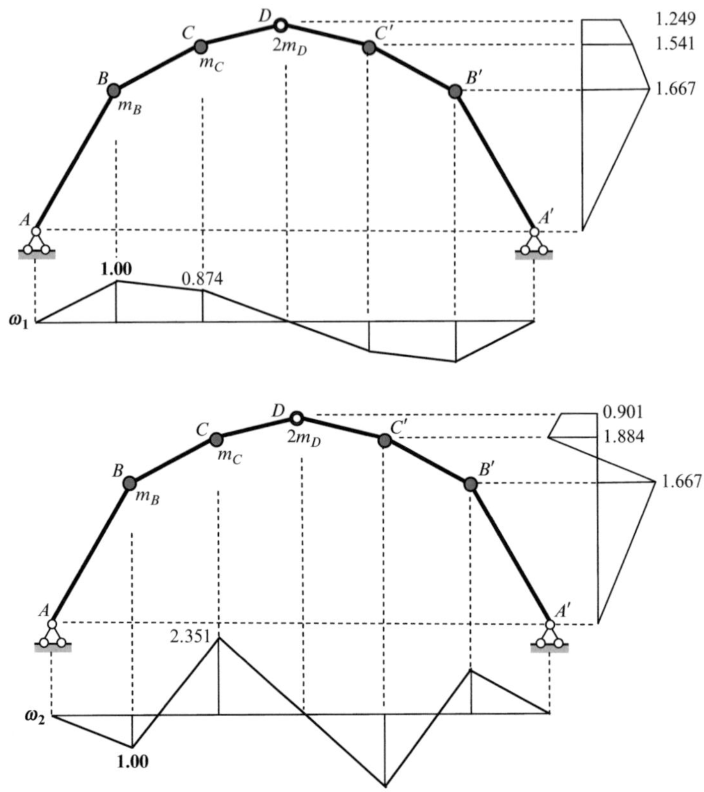

Fig. 6.27 First and second forms of antisymmetrical vibration

than the length of the circular arch. For small values of $f/l\ (\approx 0.1)$ we get $\omega_{\text{circ}} = (0.97 - 0.98)\omega_{\text{parab}}$. If $f/l = 0.5$, we get $\omega_{\text{circ}} = 0.69\omega_{\text{parab}}$. Indeed, for central angle $2\alpha = \pi$ of a circular arch, we have $\omega_{\text{antis}}^{\text{circ}} = 2.266/R^2\sqrt{EI/\mu} = (9.064/l^2)\sqrt{EI/\mu}$, while for parabolic arch $\omega_{\text{antis}}^{\text{parab}} = 13.13/l^2\sqrt{EI/\mu}$.

6.8 Parabolic Two-Hinged Uniform Arch

This section shows vibration analysis of symmetrical two-hinged arch based on Rabinovich's method.

Since the arch is symmetric, the symmetrical and antisymmetrical vibrations are considered separately.

A feature of this analysis is that for symmetrical vibrations, the equivalent half-arch is a redundant structure.

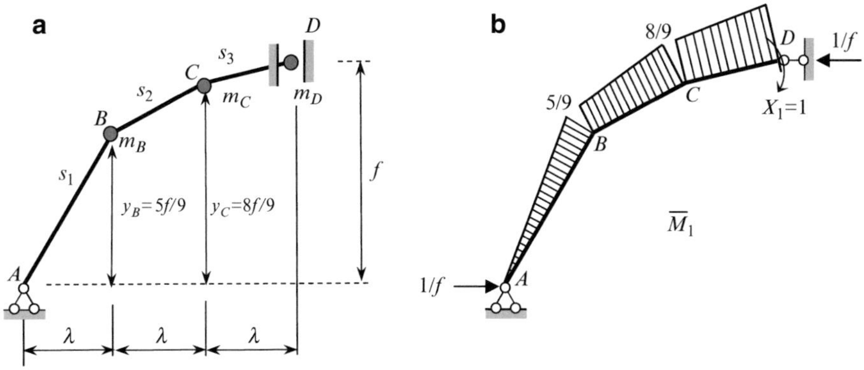

Fig. 6.28 (**a**) Design diagram of the half-arch for symmetrical vibration; (**b**) primary system and bending moment diagram due to $X_1 = 1$

6.8.1 Symmetrical Vibration

Design diagram of equivalent half-arch is shown in Fig. 6.28a. At point D on the axis of symmetry is a constraint which prevents horizontal and angular displacements. This structure is statically indeterminate of the first degree. Let the primary unknown be moment X_1 at support D. Bending moment diagram in primary system of the force method caused by $X_1 = 1$ is shown in Fig. 6.28b. This diagram does not depend on the ratio f/l.

Canonical equation of the force method is $\delta_{11}X_1 + \Delta_{1P} = 0$. Unit displacement is

$$\begin{aligned}EI\delta_{11} &= \sum \int \overline{M}_1 \times \overline{M}_1 \mathrm{d}s \\ &= \frac{s_1}{3}\left(\frac{5}{9}\right)^2 + \frac{s_2}{6}\left[2 \times \left(\frac{5}{9}\right)^2 + 2 \times \left(\frac{8}{9}\right)^2 + 2 \times \frac{5}{9} \times \frac{8}{9}\right] \\ &= \frac{25}{243}s_1 + \frac{129}{243}s_2 + \frac{434}{81}s_3.\end{aligned}$$

For an arch with a ratio of $f/l = 0.5$ [Rab58], we get

$$EI\delta_{11} = \left(\frac{25}{243} \times 0.324 + \frac{129}{243} \times 0.236 + \frac{434}{81} \times 0.178\right)l = 0.3157l.$$

For computation of the free term, the bending moment diagram in Fig. 6.18 should be used:

$$
\begin{aligned}
EI\Delta_{1P} &= \sum \int \overline{M}_1 \times M_P \mathrm{d}s, \\
&= \frac{s_1}{3} M_B \left(-\frac{5}{9}\right) + \frac{s_2}{6}\left[2M_B\left(-\frac{5}{9}\right) + 2M_C\left(-\frac{8}{9}\right) + M_B\left(-\frac{8}{9}\right) + M_C\left(-\frac{5}{9}\right)\right] \\
&\quad + \frac{s_3}{3}\left[2M_C\left(-\frac{8}{9}\right) + M_C(-1)\right] \\
&= -\frac{1}{54}[(10s_1 + 18s_2)M_B + (21s_2 + 25s_3)M_C].
\end{aligned}
$$

Expressions (6.24) for moments M_B and M_C caused by the all inertial forces allow us to present the free term of canonical equation in terms of w_B and w_C

$$
\begin{aligned}
EI\Delta_{1P} = &-\frac{\omega^2 \mu l^2 \lambda}{54}\Bigg[(10 \times 0.324 + 18 \times 0.236)\underbrace{(0.4803w_B + 0.1333w_C)}_{M_B,\ (6.24)} \\
&+ (21 \times 0.236 + 25 \cdot 0.176)\underbrace{(0.0514w_B + 0.2253w_C)}_{M_C,\ (6.24)}\Bigg] \\
&= -(0.07550w_B + 0.05750w_C)\,\omega^2 \mu l^2 \lambda.
\end{aligned}
$$

Primary unknown becomes

$$
X_1 = -\frac{\Delta_{1P}}{\delta_{11}} = (0.2391w_B + 0.1821w_C)\omega^2 \mu l \lambda.
$$

The bending moments for a redundant structure is determined by the formula $M = \overline{M}_1 X_1 + M_P$.

At the joints B, C, and D,the bending moments are

$$
\begin{aligned}
M'_B &= \left[\underbrace{(0.4803w_B + 0.1333w_C)}_{M_B} - \underbrace{\frac{5}{9}}_{\overline{M}_1}\underbrace{(0.2391w_B + 0.1821w_C)}_{X_1}\right]\omega^2 \mu l \lambda \\
&= (0.3475w_B + 0.0321w_C)\omega^2 \mu l \lambda, \\
M'_C &= \left[(0.0514w_B + 0.2253w_C) - \frac{8}{9}(0.2391w_B + 0.1821w_C)\right]\omega^2 \mu l \lambda \\
&= (-0.1611w_B + 0.1821w_C)\omega^2 \mu l \lambda, \\
M'_D &= -X_1 = -(0.2391w_B + 0.1821w_C)\omega^2 \mu l \lambda.
\end{aligned}
\tag{6.30}
$$

Fig. 6.29 Bending moment diagram due to $P_5 = 1$

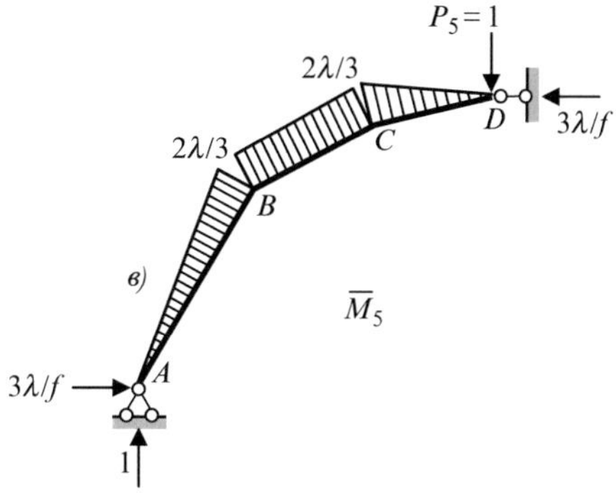

Frequency Vibration

The total vertical displacements of the joints B and D are

Joint B

$$EIw'_B = EIw_B = \sum \int M\overline{M}_1 \mathrm{d}s$$

$$= \frac{\lambda}{54}\left[(8s_1 + 9s_2)\underbrace{M_B}_{M'_B} + (6s_2 + 2s_3)M_C + 2s_3M_D\right],$$

where M is a just constructed bending moment diagram (6.30) of the redundant half-arch, and diagram $\overline{M}_1$ is shown in Fig. 6.19a.

Joint D

$$EIw'_D = \sum \int M\overline{M}_5 \mathrm{d}s = -\frac{\lambda}{9}[(2s_1 + 3s_2)M_B + (3s_2 + 2s_3)M_C + s_3M_D],$$

where bending moment diagram $\overline{M}_5$ is shown in Fig. 6.29.

The remainder of the procedures is the same as in the case of three-hinged arch. Substitution of s_i for a given span–rise ratio, the moments M_B, M_C, and M_D into formulas for displacements according to (6.30), and taking into account a kinematical relationship $w'_D = -2w_C$ allow us to construct the following expressions

$$EIw_B = (\alpha_{11}w_B + \alpha_{12}w_C)\omega^2\mu l^2\lambda^2,$$
$$2EIw_C = (\alpha_{21}w_B + \alpha_{22}w_C)\omega^2\mu l^2\lambda^2.$$

We then introduce the dimensionless parameter $\theta = EI/(\omega^2\mu l^2\lambda^2)$ for frequency of the free vibration. For determination of the generalized coordinates w_B and w_C, we get a system of linear homogeneous equations

$$(\alpha_{11} - \theta)w_B + \alpha_{12}w_C = 0,$$
$$\alpha_{21}w_B + (\alpha_{22} + 2\theta)w_C = 0.$$

where, according to Rabinovich et al. [Rab58], we have $a_{11} = 0.02351, a_{12} = 0.003692, a_{21} = 0.02871, a_{22} = -0.008745$.

Nontrivial solution presents the frequency equation. Construction of the mode of shape vibration, verification of the orthogonality condition, computation of the internal forces according to the first and second modes should be done as in the case of a three-hinged arch.

Tables A.41 and A.42 contain the frequencies of symmetric vibrations, eigenfunctions, and bending moments for two-hinged arches according to the first and second mode for different ratios *f*/*l*.

In order to form design diagram of the arch in the case of antisymmetrical vibration, we need to introduce a constraint which prevents the vertical displacement on the axis of entire arch. Such scheme had been considered previously for the analysis of antisymmetrical vibration of a three-hinged arch. Therefore, frequencies of free vibration, mode shape vibration, and internal forces for three-hinged and two-hinged arches coincide (see Tables A.39 and A.40).

Extensive numerical material for different shapes of arches (catenary, cycloid, spiral, parabola, etc.) are presented in [Rom72].

6.8.2 Advantages and Disadvantage of the Rabinovich' Method

1. Rabinovich' method provides effective dynamical analysis of arches. Effectiveness of the method is based on two fundamental propositions. The first proposition is associated with a very simple construction of an approximate model of the arch: the axis of the arch is replaced by a set of inscribed straight members with the equal horizontal projection. The second proposition is based on algebraic transformations over the fundamental relationships for displacements of the joints of a hinged chain; the structure and parameters of this chain are determined by the model of the arch.
2. Effectiveness of the method can be traced on the analysis of the approximate model of the arch, presented in Sect. 6.5. This model contains three masses and has two degrees of freedom. Its six displacements in the directions of the inertial forces can be represented as a linear combination of two independent kinematic parameters. Therefore, it suffices to calculate only two displacements, while for the calculation of others we can apply the kinematical relationships. This procedure may also be simplified by considering two partial systems, each with one degree of freedom.
3. Rabinovich's method allows us to perform analysis of nonsymmetrical arches with arbitrary distributed masses and arbitrary law of change of flexural stiffness

along the axis of the arch. The method leads to approximate results because the curvilinear axis of the arch is replaced by a set of chords.

4. Rabinovich's method admits different versions of the definition of the eigenvalues and eigenforms. Instead of two independent kinematic parameters (displacement joints of a kinematical chain), it is possible to take two ordinates of bending moments. Any version leads to the system of two linear homogeneous algebraic equations; they allow us to find the frequencies of free vibrations and corresponding shapes of vibrations. Next advantage of the Rabinovich's method is the following: the method allows us to easily investigate the transient vibration of the arch, to construct the dynamical bending moment diagram for any type of excitations, and to calculate the dynamical coefficients. This procedure is shown in the next chapter.
5. Construction of a Rabinovich's model and algorithms for determining the frequencies and mode shapes of vibrations are much easier, clearer, and more efficient than those which discussed in the method of Smirnov. However, limitation of the Rabinovich's method is associated with only the shape of the arch, which is parabolic, while Smirnoff's method does not impose limitation on the shape of the axis of an arch.
6. It should be mentioned that the Snitko method [Sni57] may be used for determining frequencies of free vibrations of arches. This method uses the concept of kinematical chain. However, Snitko method is not tied to a parabolic arch and the way of its approximation (as in the case of Rabinovich's model). The method consists of the following: constructing a kinematical chain of the arch and deriving its kinematical relations; in the case of the parabolic arch as a special case, we can use Rabinovich's kinematical relationships. For computation of the frequency of free vibration, it is necessary to construct expression for work done by the inertial forces and the joint moments along the virtual displacements. The concept of kinematical chain also may be applied for stability problems, as shown in [Kar10].

6.9 Parabolic Nonuniform Hingeless Arch

Symmetrical parabolic arch with span l and rise f is shown in Fig. 6.30. Assume that $I_x = I_C/\cos\varphi_x$, where φ_x is the angle between horizontal and the tangent to the axis of the arch at the section with coordinate x; I_C and I_x are moment of inertia at the crown and at section x away from crown, respectively.

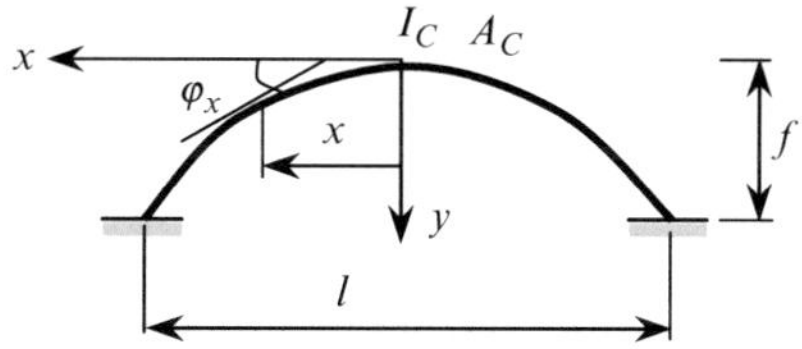

Fig. 6.30 Design diagram of nonuniform symmetric arch

Assume that the distributed mass m per unit length of the arch is constant. Area of the cross section and radius of inertia at the crown are A_C and r_C, respectively. The formulas for frequencies of free vibration are as follows [Bon52b], [Uma72–73].

Antisymmetrical Vibration

In this case,

$$\omega_i = \frac{k_i^2}{l^2}\sqrt{\frac{EI_C}{m}}\ (\text{s}^{-1}),$$

where $k_1 = 3.9266, k_2 = 7.0685, k_3 = 10.210, k_4 = 13.352,\quad k_5 = 16.494,$ ..., $k_n = (4n + 1/4)\pi.$

Symmetrical Vibration

In this case,

$$\omega_i = \frac{4k_i^2}{l^2}\sqrt{\frac{EI_C}{m}}\ (\text{s}^{-1}).$$

Eigenvalues k_i are the roots of the frequency equation

$$(\cosh k - \beta r)\sin k + (\cos k - \gamma r)\sinh k = 0,$$

where

$$\beta = \frac{1.33}{k^3}\left[-\left(1 + \frac{3}{k^2}\right)\sinh k + \frac{3}{k}\cosh k\right],$$
$$\gamma = \frac{1.33}{k^3}\left[\left(1 - \frac{3}{k^2}\right)\sin k + \frac{3}{k}\cos k\right],$$
$$r = \frac{f^2}{i_C^2},\quad i_C = \sqrt{\frac{I_C}{A_C}}.$$

The roots k_i of the frequency equation are presented in Table A.45.

6.10 Rayleigh–Ritz Method[1]

This section is devoted to numerical computation of frequency of vibration of the arch by the Rayleigh–Ritz method. According to this method, we need to aply the following procedure:

- Accept expressions for displacements, which satisfy the boundary conditions.
- Construct the expressions for strain and kinetic energies.
- By finding the extreme values of the two quantities, and equating them, we can calculate the corresponding frequency of vibration [Tho81].

This method is especially effective in the cases of the nonuniform arches and arches with nonuniform radius of curvature.

6.10.1 Circular Uniform Arch

Let us consider a circular uniform arch with arbitrary boundary conditions. Let R be the radius of the arch and 2α be the central angle. Assume that the axial strain ε is neglected. The strain and kinetic energy for bending vibration of arch may be written as [Mau90]

$$U = \frac{EI}{R^3}\int_0^{\alpha}\left(\frac{\partial u}{\partial\varphi}+\frac{\partial^2 \upsilon}{\partial\varphi^2}\right)^2 \mathrm{d}\varphi,$$

$$T = Rm\int_0^{\alpha}\left[\left(\frac{\partial u}{\partial t}\right)^2+\left(\frac{\partial \upsilon}{\partial t}\right)^2\right]\mathrm{d}\varphi. \tag{6.31}$$

where φ is the angle measured from the vertical axis of symmetry, E and m are Young's modulus and a mass of the arch per unit length, u and υ are tangential and radial displacement components, respectively.

For a two-hinged arch, the following coordinate functions satisfy the geometric conditions $u = \upsilon = 0$ at $\varphi = \pm\alpha$ [Den28], [Rom72]:

$$\upsilon = A\sin\left(\frac{\pi\varphi}{\alpha}\right)\sin\omega t,$$

$$u = -A\frac{\alpha}{\pi}\left(1+\cos\frac{\pi\varphi}{\alpha}\right)\sin\omega t. \tag{6.32}$$

Substitution of these expressions into (6.31) leads to the following expressions for U and T

$$U = \frac{EIA^2}{2R^3\alpha^3}\left(\alpha^2-\pi^2\right)^2\sin^2\omega\, t,$$

$$T = \frac{mRA^2}{2}\,\frac{\alpha}{\pi^2}\,\omega^2\left(3\alpha^2+\pi^2\right)\cos^2\omega\, t.$$

[1] This section was written in collaboration with Evgeniy Lebed.

The Rayleigh–Ritz method states that [Tho81]

$$U_{\max} = T_{\max}. \tag{6.32a}$$

From (6.32a), follows Bolotin's formula for smallest frequency vibration of the circular arch [Bol64]

$$\omega = \frac{\pi}{R^2\alpha}\left(\frac{\pi^2}{\alpha^2} - 1\right)\frac{1}{\sqrt{3 + \dfrac{\pi^2}{\alpha^2}}}\sqrt{\frac{EI}{m}}. \tag{6.33}$$

Since axial strain ε is neglected, then according to (1.32), we get $\upsilon = \partial u/\partial\varphi$ and (6.31) may be rewritten in terms of the tangential displacement only

$$\begin{aligned} U &= \frac{1}{EI}\int_0^{\alpha} M^2 \mathrm{d}\varphi = \frac{EI}{R^3}\int_0^{\alpha}\left(\frac{\partial u}{\partial\varphi} + \frac{\partial^3 u}{\partial\varphi^3}\right)^2 \mathrm{d}\varphi, \\ T &= mR\int_0^{\alpha}\left[\left(\frac{\partial u}{\partial t}\right)^2 + \left(\frac{\partial^2 u}{\partial\varphi\,\partial t}\right)^2\right]\mathrm{d}\varphi. \end{aligned} \tag{6.31a}$$

If we adopt expressions for energy in the form of (6.31a), then we only need to consider the expression for tangential displacement. Rayleigh–Ritz procedure remains the same: we need to determine the strain and the kinetic energy of an arch, calculate their extreme values and equate them according to (6.32a). It is then easy to check that this procedure leads to (6.33).

If an arch has peculiarities, then (6.31) for strain and kinetic energy should be modified. Such peculiarities include piecewise constant rigidity, variable radius of curvature, elastic supports, the presence of lumped masses on the arch, etc.

6.10.2 *Circular Arch with Piecewise Constant Rigidity*

In the case of a nonuniform arch, expressions for strain and kinetic energy for bending vibration of the arch may be written as [Mau90]

$$\begin{aligned} U &= \frac{E}{2R^3}\int_{-\alpha}^{\alpha} I(\varphi)\left(\frac{\partial u}{\partial\varphi} + \frac{\partial^2 \upsilon}{\partial\varphi^2}\right)^2 \mathrm{d}\varphi, \\ T &= \frac{R}{2}\int_{-\alpha}^{\alpha} m(\varphi)\left[\left(\frac{\partial u}{\partial t}\right)^2 + \left(\frac{\partial \upsilon}{\partial t}\right)^2\right]\mathrm{d}\varphi. \end{aligned} \tag{6.34}$$

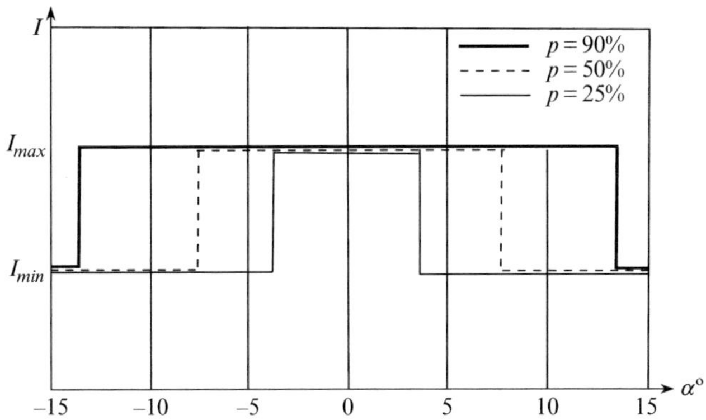

Fig. 6.31 Different shapes of the arch with a piecewise constant moment of inertia

Assume that an arch has a piecewise constant moment of inertia of the cross section $I(\varphi)$, which is symmetrical with respect to the axis of symmetry. The function $I(\varphi)$ describes the moment of inertia. We assume this function to be piecewise constant with two jump discontinuities. In this case, the function $I(\varphi)$ is defined as follows

$$I(\varphi,p) = \begin{cases} I_{max} & \text{if} \quad |\varphi| \leq \alpha p/2, \\ I_{min} & \text{if} \quad |\varphi| > \alpha p/2. \end{cases} \tag{6.35}$$

The variable p is fraction (in percent) of the portion of an axis of the arch with a moment of inertia of a cross section equal to $I_{\max}$. Correspondingly, the percentage of the length of the axis of the arch with a moment of inertia of cross section of $I_{\min}$ is $1 - p$.

Let the mass m per unit length for portions with $I_{\max}$ and $I_{\min}$ remain the same, therefore, as before, we can use the expression for kinetic energy (6.31). In case of antisymmetric vibration, we still use the assumed coordinate functions (6.32).

To evaluate the integral U, we use Romberg integration [Bur01]. Romberg integration uses the composite trapezoidal rule to give preliminary approximation to the integral and then applies Richardson extrapolation [Bur01] to improve the approximations thus giving a high-order numerical approximation to the integral. For the particular choice of $I(\varphi)$, Romberg integration produces approximations of forth order accuracy.

For a particular example, we take an arch with the following parameters: radius $R = 180$ (ft), central angle $2\alpha = 30°$, mass distribution $m = 30$ (lb s^2/ft^2), and $E = 400 \times 10^7$ (lb/ft^2). Let $I_{\min} = 1.4$ (ft^4) and $I_{\max} = 3.0$ (ft^4). Some particular functions $I(\varphi)$ for $p = 25, \; 50,$ and 90% ?are shown in Fig. 6.31.

The resulting smallest frequencies of vibration are shown in Fig. 6.32.

The extreme conditions ($p = 0$ and $p = 100\%$) can be verified analytically by Bolotin formula; $\omega(I_{\min}) = 14.18$ (1/sec) and $\omega(I_{\max}) = 20.76$ for the particular choices of $I_{\min}$ and $I_{\max}$. As one can see from Fig. 6.32, the function $\omega_{\min}(I(p))$ is a

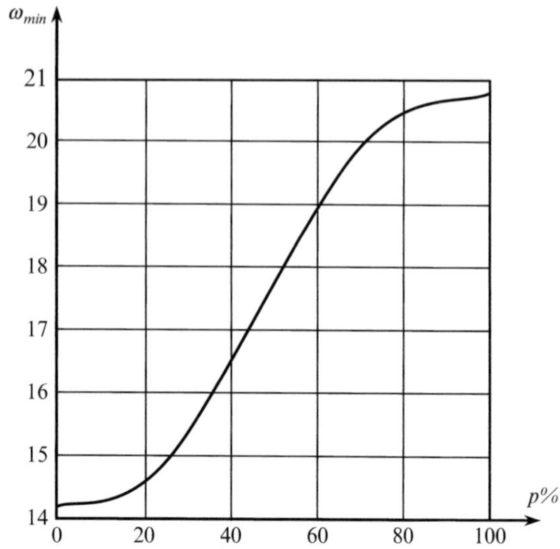

Fig. 6.32 Smallest frequencies of vibration as a function of the percentage p

monotonically increasing function. The values of this function at the end points can be verified analytically.

Note that evaluating the integrals in (6.34) numerically allows us to consider a whole new type of problems, namely, optimal design of arches. In this scope, two such problems may be formulated as follows [Gri79], [Olh77]:

1. The volume–frequency problem: find a configuration of the cross-sectional area $A(x)$ along the arch for the minimum (or maximum) frequency ω, if the volume of the arch V_0 is given.
2. The frequency–volume problem: Find a configuration of the cross-sectional area $A(x)$ along the arch for the minimum (or maximum) volume V of the arch is known, and the frequency $\omega = \omega_0$ of the arch is given.

We note that with the use of (6.31a), we can easily derive Lamb's equation for a uniform circular arch. Indeed, the Hamilton principle [Ham35] states that

$$\delta \int_{t_0}^{t} (U - T)\mathrm{d}t = 0.$$

For strain and kinetic energy given by (6.31a), we have

$$\delta \int_{t_0}^{t} \int_{-\alpha}^{\alpha} \left[\int \left[\frac{m}{2} \left(\frac{\partial u}{\partial t} \right)^2 + \frac{m}{2} \left(\frac{\partial^2 u}{\partial \varphi \, \partial t} \right)^2 - \frac{EI}{2R^4} \left(\frac{\partial u}{\partial \varphi} + \frac{\partial^3 u}{\partial \varphi^3} \right)^2 \right] R\mathrm{d}\varphi \right] \mathrm{d}t = 0.$$

The corresponding Euler–Lagrange equation exactly represents Lamb's equation (6.6) [Rek73].

6.11 Conclusion

This chapter presents the methods for the determination of the frequencies of free vibration of circular and parabolic arches with different boundary conditions. In the case of a circular uniform arch, analytical solution on the basis of the Lamb equation is presented. The Rayleigh–Ritz method becomes an effective tool for determining frequencies of free vibrations when the circular arch has a nonuniform moment of inertia. For parabolic uniform arches, Rabinovich method is utilized. Even if the exact analytical procedures are applied, the method leads to approximate result because the initial arch with distributed masses is replaced by a frame with lumped masses.

It is worth-while to also note that the well-known Dunkerley approximate formula [Bir68] [Kar10] and Bernshtein and Smirnov estimations [Bir68] allow us to calculate the smallest frequency of vibration.

We also note the importance of numerical methods, which allow us to determine the natural frequencies of arches with peculiarities (arches of different shapes, arches with continuously and discontinuously varying cross-section, non-symmetrical arches, arches with distributed and lumped masses, arches with elastic supports, etc.) as well as to take into account the secondary effects. Numerous and varied examples can be found in the works of Laura [Lau87], [Lau88], Romanelli [Rom72], Filipich [Fil88], [Fil90], Rossi [Ros89], Lee [Lee89], De Rosa [DeR91], [Cha69], [Den28], [Sak85], [Volter60], [Volter61a, b], [Wan72], [Wan73], [Wan75], [Was78], [Wol71], and others. Validation and comparison of results which are obtained using different numerical methods are presented by Gutierrez et al. [Gut89]. Out-of-plane vibration of arches is presented in the works of [Bon50], [Suz78], [Iri82]. A large set of the formulas for natural frequency and mode shapes of vibration of deformable structures can be found in [Ble79], [Kar04].

Chapter 7
Forced Vibrations of Arches

This chapter is devoted to forced vibration analysis of arches and arched structures subjected to disturbing loads. Different types of arches and theirs loading are considered. Analytical methods of analysis are applied.

7.1 General

7.1.1 Types of Disturbing Loads

Forced vibrations of engineering structures are brought about by dynamic (or disturbing) loads. These forces are functions of time. The nature of these loads is diverse. Generally, disturbing forces may be of the following types:

1. Immovable periodical loads produced by stationary units and mechanisms with moving parts. These loads have a periodical, but not necessary a harmonic (according to the law of sine or cosine) character and generally do not depend on the elastic properties of the structure.
2. Impulsive loads are produced by a blast, falling weights (pile drivers, hammers, etc.), or collision of bodies. Impulsive loads are characterized by a very short duration of their action and depend on the elastic properties of the structure which is subjected to such loads.
3. Moving loads act on structures by transference through wheels of a moving train or truck moving across the deck. The availability of the rail joints on a railway bridge or irregularities of the deck on a car bridge lead to appearance of inertial forces. This type of moving loads takes into account dynamical effects, and therefore, should be distinguished from moving loads, which has been

I.A. Karnovsky, *Theory of Arched Structures: Strength, Stability, Vibration*,
DOI 10.1007/978-1-4614-0469-9_7,

studied in the sections "Influence lines" where the term "moving load" implies only that position of the load is arbitrary, i.e., this is a static load, which may have different positions along the structure.

4. Seismic loads arise due to earthquakes. The reason for the presence of seismic load on a structure is acceleration of the supports caused by acceleration of the ground. This type of disturbance is called kinematical disturbance. The acceleration of supports leads to the acceleration of the individual parts of the structure, and as a result inertial forces act on these parts. Seismic forces which arise in the individual elements of the structure are dependent on the type and the amount of ground acceleration, distribution of the mass within the elements of the structure, as well as their elastic properties [Clo75].

7.1.2 Classification of Forced Vibration

Forced vibration may be classified according to:

1. The number of degrees of freedom of the structure (either lumped or distributed parameters). From methodological point of view, structures with lumped parameters are divided into those with one degree of freedom and structures with two or more degrees of freedom.
2. According to the type of relationship "*P*–*y*" (Table 6.1), the vibration is divided into linear and nonlinear vibrations.
3. Resisting forces which arise in the structure – these may be taken into account or neglected.
4. If forces which act on the structure are characterized by the preciseness of their parameters, then the corresponding vibrations are called determinate vibrations. There exists a series of dynamical loads which are characterized by the lack of preciseness of their parameters. These include loads created by wind, loads which arise on account of irregularities of the deck on car bridges, seismic loads, etc. In all these cases, it is impossible to set factual parameters to these loads. Such occurrences are called nondeterminate and the corresponding vibrations are random vibrations [Bol84].

We will consider determinate in-plane bending vibration of arches neglecting the secondary effects.

7.2 Structures with One Degree of Freedom

This section contains analysis of the forced vibration of the structure with one degree of freedom; the structure is subjected to typical disturbing forces.

7.2.1 Dugamel Integral

The forced undamped vibration of a system with a single degree of freedom, subjected to an arbitrary disturbing force $F(t)$, is described by the equation

$$m\ddot{y} + ky = F(t), \tag{7.1}$$

where y is the displacement of the mass m, the stiffness coefficients is denoted as k. General solution of this equation is

$$y(t) = y_0 \cos \omega t + \frac{\upsilon_0}{\omega} \sin \omega t + y^*(t). \tag{7.2}$$

The first two terms in (7.2) describes free vibration of a mass m with its initial position y_0 and initial velocity υ_0. These vibrations occur with the frequency of free vibration $\omega = \sqrt{k/m}$. Last term in (7.2), $y^*(t)$, describes the forced vibration which depends on the disturbing force $F(t)$. Equation (7.1) and its solution do not take into account the resisting forces; due to the inevitable presence of various damping forces in the structure, the free vibration rapidly disappears and only the purely forces component remains.

$$y^*(t) = \frac{1}{\omega m} \int_0^t F(\tau) \sin \omega(t - \tau)\, d\tau. \tag{7.3}$$

Expression (7.3) is called Duhamel's integral [Duh43] (convolution integral). This formula allow us to determine the response of the linear system with a single degree of freedom in the case of an arbitrary disturbing force $F(t)$. Formula (7.3) is derived on the basis of the superposition of the responses of the system to a sequence of impulses [Wea90]. The Handbooks [Har61], [Har88] contain forced component (7.3) for a system with a single degree of freedom subjected to numerous types of impact loads.

7.2.2 Application of the Duhamel Integral for a Bar Structure

Integral (7.3) may be presented in the form applicable for bar structure. Approximate analysis of a structure subjected to arbitrary loads (impact, harmonic, etc.) is based on the following assumptions:

1. Vibration of the bar structure occurs as in a system with a single degree of freedom.
2. The shape of vibration coincides with elastic line which corresponds to specific equivalent static load.

Assume that at moment $t = 0$, the structure is subjected to load [Rab54a]

$$q(x,t) = q(x)\,\varphi(t), \tag{7.4}$$

where $q(x)$ is any distributed function of x and $\varphi(t)$ is arbitrary function of t. Assume that the structure has one degree of freedom and the mode of shape vibration $X(x)$ is an elastic curve which corresponds to a static load $q(x)$. Since $1/(\omega m) = \omega\,\delta_{11}$, then

$$y^*(t) = \omega\,\delta_{11} \int_0^t F(\tau)\sin\,\omega(t-\tau)\,\mathrm{d}\tau. \tag{7.5}$$

This formula means that at any moment the displacement of the bar system is determined by the static equivalent load

$$F_{\text{eq}} = \omega \int_0^t F(\tau)\sin\,\omega(t-\tau)\,\mathrm{d}\tau. \tag{7.6}$$

Taking into account (7.4), we can say that at any moment t the displacement at any point of a structure will be such as if the structure would be subjected to a static load

$$q_{\text{stat}} = \omega\,q(x) \int_0^t \varphi(\tau)\sin\,\omega(t-\tau)\,\mathrm{d}\tau.$$

Now the displacements of the bar structure may be presented in terms of shape $X(x)$

$$y(x,t) = \omega\,X(x) \int_0^t \varphi(\tau)\sin\omega(t-\tau)\mathrm{d}\tau. \tag{7.7}$$

7.2.3 *Special Types of Disturbance Forces*

Let us show an application of (7.7) for different excitations [Rab54a], [Har88].

Constant-Force Excitation

A weight of mass m is attached to the end of a vertical spring of stiffness k. At moment $t = 0$, a force F_0 suddenly is applied to the mass and then remains constantly applied to the mass (Fig. 7.1a)

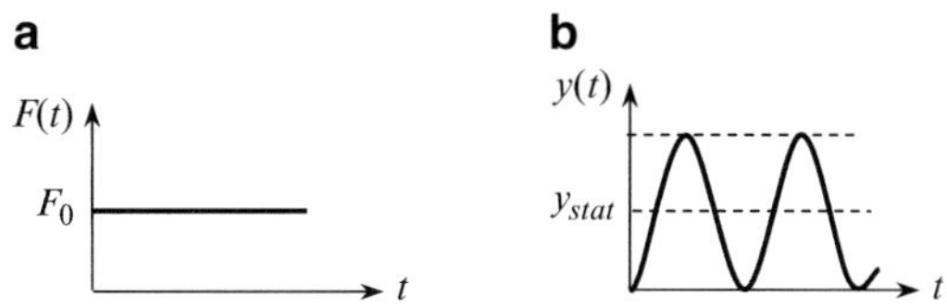

Fig. 7.1 Constant force excitation and the corresponding forced response

Integral (7.3) becomes

$$y^*(t) = \frac{F_0}{\omega m} \int_0^t \sin \omega(t-\tau)\, \mathrm{d}\tau = \frac{F_0}{\omega^2 m}(1-\cos \omega t) = y_{\text{stat}}(1-\cos \omega t), \quad (7.8)$$

where y_{stat} is the displacement of the mass caused by static load F_0, i.e.,

$$y_{\text{stat}} = \frac{F_0}{k} \quad \text{and} \quad \omega = \sqrt{\frac{k}{m}}.$$

According to (7.8), the displacement of the mass occurs by harmonic law with frequency ω of the free vibration around position $y = y_{\text{stat}}$. Dynamical effect of such application of load, as shown in Fig. 7.1b, is twice as much as the static one, i.e.,

$$y_{\max} = 2y_{\text{stat}}(F_0).$$

For bar structure, expression (7.7) becomes

$$y(x,t) = \omega\, X(x) \int_0^t \varphi(\tau) \sin \omega(t-\tau) \mathrm{d}\tau = X(x)(1-\cos \omega t) = y_{\text{stat}}(1-\cos \omega t),$$

$$v(x,t) = \frac{\mathrm{d}y(x,t)}{\mathrm{d}t} = \omega X(x) \sin \omega t. \quad (7.9)$$

The bar executes harmonic vibration around the static elastic curve $X(x)$. For any point on the bar, the maximum displacement is twice more than the static displacement at the same point.

Pulse of Duration τ

The load $q(x)$ is suddenly appeared at time $t = 0$. This load remains constant until the time $t = \tau$ and then suddenly disappeared (rectangular pulse). Assume that the duration of load actions τ and the period of free vibration T satisfy the condition $\tau \leq 0.25\,T$. In this case, a maximum displacement occurs after the load is disappeared.

Table 7.1 Dynamical coefficient vs. parameter τ/T [Rab54a]

τ/T	0.01	0.02	0.05	0.10	0.167	0.20	0.30	0.40	0.50	>0.50
μ_{dyn}	0.0628	0.126	0.313	0.618	1.000	1.175	1.617	1.902	2.000	2.000

Within the interval $t < \tau$, the displacement is described by (7.9). For $t > \tau$, it may be considered that the structure is subjected to actions of two loads $q(x)$ and $-q(x)$. This again allows us to apply solution (7.9)

$$\begin{aligned} y(x,t) &= X(x)(1-\cos\omega t) - X(x)[1-\cos\omega(t-\tau)] \\ &= X(x)\left[\cos\omega(t-\tau) - \cos\omega t\right]. \end{aligned}$$

This solution may be rewritten in the following form

$$y(x,t) = 2X(x)\,\sin\frac{\omega\tau}{2}\,\sin\omega\left(t-\frac{\tau}{2}\right).$$

Thus, the structure executes vibration around unloaded state with frequency of the free vibration ω and amplitude $2X(x)\,\sin(\omega\tau/2)$. Therefore,

$$y(x,t)_{\text{max,min}} = \pm 2X(x)\,\sin\left(\frac{\omega\tau}{2}\right) \tag{7.10}$$

and dynamical coefficient is

$$\mu_{\text{dyn}} = \frac{y_{\max}}{y_{\text{stat}}} = \frac{y_{\max}}{X} = 2\sin\frac{\pi\tau}{T}, \tag{7.11}$$

where T is the period of free vibration. Thus, the effect of the short-term load depends on the parameter τ/T. Maximum of the dynamical coefficient is equal to 2. For $\tau < T/6$, dynamical coefficient is $\mu_{\text{dyn}} < 1$ (Table 7.1).

Impulse Excitation

A bar structure is subjected to a distributed load $q(x)$. This load acts within a very small time interval τ. Impulse of the elementary load $q(x)\mathrm{d}x$ is equal to

$$\tau\, q(x)\mathrm{d}x = s(x)\mathrm{d}x, \text{ thus } q(x) = s(x)/\tau.$$

According to (7.10), the maximum displacement becomes

$$y_0 = y(x,t)_{\text{max,min}} = \pm 2X(x)\,\sin\left(\frac{\omega\tau}{2}\right) = \pm X(x)\,\omega\tau\frac{\sin(\omega\tau/2)}{\omega\tau/2}, \text{ and}$$

$$y = y_0 \sin\omega t.$$

Since $X(x)$ is an elastic curve, which corresponds to a static load $q(x)$, then $X(x)\,\omega\tau$ is the elastic curve due to the static load $\omega\,q(x)\,\tau = \omega\,s(x)$. Thus, the action of any distributed impulse $s(x)$ is equivalent to a static distributed load $q_{\mathrm{eq}}(x)$ with the intensity at any point of a structure being

$$q_{\mathrm{eq}}(x) = \pm\,\omega\,s(x)\,\frac{\sin(\omega\,\tau/2)}{\omega\tau/2}. \tag{7.12}$$

The unit of distributed impulse $s(x)$ is (kNs/m). If impulse S(kNs) is applied at a single point, then equivalent static force F_{eq} at the same point is

$$F_{\mathrm{eq}} = \omega\,S\,\frac{\sin(\omega\,\tau/2)}{\omega\tau/2}. \tag{7.12a}$$

Since $[(\sin\alpha)/\alpha) < 1$ and $\lim_{\alpha\to 0}[(\sin\alpha)/\alpha] = 1$, then for two impulses with equal S but different τ, the impulse with smaller τ is more dangerous. Therefore, the instantaneous impulse is most dangerous. Equivalent load for this case is

$$\begin{aligned} q_{\mathrm{eq}}(x) &= \pm\,\omega\,s(x),\\ F_{\mathrm{eq}} &= \pm\,\omega\,S. \end{aligned}$$

With the decrease of time τ, effect of the constant force F trends to zero, but the effect of constant impulse S trends to a maximum. In the first case, the impulse of a force $S = F\tau$, with decreasing of τ, trends to zero, while in the second case the impulse S for any τ retains its magnitude.

Equivalent static load depends not only on the values of the impulse, but also on the properties of the structure. The greater the stiffness and the lighter the structure, the greater is its frequency ω, and thus the greater is the equivalent load.

Example 7.1. Determine the effect of instantaneous impulse s which acts on the uniform simply supported beam. This impulse is uniformly distributed within the entire span l. Parameters of the beam are: flexural rigidity EI, mass per unit length μ.

Solution. The frequency of the free vibration is $\omega = (\pi^2/l^2)\sqrt{EI/\mu}$. Equivalent static load on the beam is

$$q_{\mathrm{eq}} = \omega\,s = \frac{\pi^2 s}{l^2}\sqrt{\frac{EI}{\mu}}, \tag{7.13}$$

so the reactions of the beam are

$$R_{\max,\min} = \frac{q_{\mathrm{eq}}l}{2} = \omega\,s = \pm\frac{\pi^2 s}{2l}\sqrt{\frac{EI}{\mu}}.$$

Bending moment and shear force are

$$M_{\max,\min} = \pm\frac{q_{\mathrm{eq}}l^2}{8} = \pm\frac{\pi^2 s}{8}\sqrt{\frac{EI}{\mu}},$$

$$Q_{\max,\min} = \pm\frac{q_{\mathrm{eq}}l}{2} = \pm\frac{\pi^2 s}{2l}\sqrt{\frac{EI}{\mu}}.$$

Displacement of the beam is

$$y_{\max} = \frac{5}{384}\frac{q_{\mathrm{eq}}l^4}{EI} = \frac{5\pi^2}{384}\frac{sl^2}{\sqrt{EI\mu}}.$$

Maximum normal stresses occur at the extreme fibers of the beam and are

$$\sigma_n = \pm\frac{M}{I}e = \pm\frac{M}{Ar^2}e = \pm\frac{\pi^2 s}{8}\frac{e}{r}\sqrt{\frac{E}{A\mu}},$$

where e is the distance from the neutral line to an extreme fiber of the beam, r is the radius of inertia, and A is the area of the cross section of the beam.

Maximum shear stress occurs at the neutral line and is

$$\sigma_\tau = \pm\frac{QU}{bI} = \pm\frac{\pi^2 s}{2l}\frac{U}{br}\times\sqrt{\frac{E}{A\mu}},$$

where b is the width of the cross section at the neutral line and U is the first moment of the cross-sectional area above (or below) of the neutral axis with respect to the neutral axis.

Discussion [Rab54a]

Comparison of the static loading and impulsive excitation has fundamental differences:

1. Even if the entire beam presents a statically determinate structure, in the case of impulsive excitation, the reactions and internal forces depend on the flexural rigidity EI of the beam, while in case of static loading these parameters do not depend on the rigidity EI.
2. Increasing of flexural rigidity EI of a beam by n times leads to decreasing of displacement by n times for any static loading and by $\sqrt{n}$ times at the impulsive excitation.

3. In the case of impulsive excitation, the expressions for the reactions, internal forces, stresses, and displacements have a factor $1/\sqrt{\mu}$ while in the case of static loading, these expressions have a factor μ. Indeed, the uniformly distributed load q may be presented in term of the mass per unit length as $q = \mu\, g$, where g is the acceleration due to gravity.
4. The bending moment due to impulsive loading does not depend on the length of the beam, while the shear force is inversely proportional to the length of the beam. In the case of a simply supported beam of length l, subjected to static, uniformly distributed load, the bending moment and shear are proportional to l^2 and l, respectively.

Note: on the basis of (7.1), detailed response of the structure subjected to numerous types of different impact loads is presented in [Har88].

Harmonic Excitation

Such excitation of the bar system may be realized, particularly, by a harmonic distributed load $q(x,t) = q(x)\cos\theta\, t$ or a concentrated force $F(t) = F_0 \cos\theta\, t$, where θ is a frequency of an external excitation. Duhamel's integral (7.7) leads to the following result

$$\begin{aligned} y(x,t) &= \omega X(x) \int_0^t \cos\theta\tau \times \sin\omega(t-\tau)\mathrm{d}\tau \\ &= \frac{X(x)}{1-(\theta/\omega)^2}(\cos\theta\, t - \cos\omega\, t). \end{aligned} \tag{7.14}$$

where $X(x)$ is the elastic curve which satisfies the boundary conditions. Vibrations (7.14) are combination of forced vibration with frequency θ of disturbing force (the fist term) and natural vibration with frequency ω of free vibration (the second term). Both vibrations have amplitude

$$A = \frac{X(x)}{1 \quad (\theta/\omega)^2}.$$

In practice, due to the inevitable presence of various damping forces, the natural vibrations with frequency ω rapidly disappear. The steady-state forced vibration

$$y(x,t) = A\cos\theta\, t$$

is a sustained periodic motion with an amplitude A and a frequency θ of the frequency of the disturbing force. The factor

$$\mu_{\mathrm{dyn}} = \frac{1}{1-(\theta/\omega)^2}$$

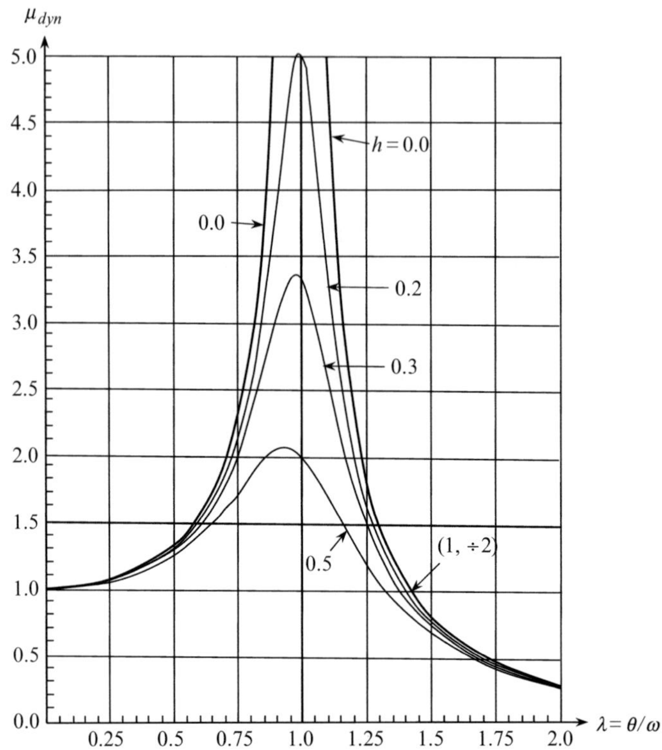

Fig 7.2 Dynamic coefficient μ_{dyn} vs. $\lambda = \theta/\omega$; parameter $h = 2n/\omega$

is called the dynamical coefficient. This coefficient is the ratio of maximum displacement of any point of the structure and displacement due to the amplitude of the static load $q(x)$. The factor μ_{dyn} vs. $\lambda = \theta/\omega$ is presented in Fig. 7.2; for undamped vibration the parameter $2n/\omega$ is 0.0.

1. If $\lambda = \theta/\omega$ is almost unity, the amplitude of forced vibrations becomes very large. This phenomenon is called resonance. Precise analysis shows that in this case

$$y(x,t) = X(x)\frac{\omega t}{2}\sin\omega t = X(x)\frac{\omega t}{2}\cos\left(\omega t - \frac{\pi}{2}\right). \tag{7.14a}$$

   The first term of (7.14) and formula (7.14a) shows the following: at resonance, the displacement $y(x,t)$ and disturbing force $q(x,t) = q(x)\cos\theta t$ (or $F(t) = F_0\cos\theta t$) have a phase shift $\pi/2$. It means that the displacements become extreme at the moments when disturbing force is zero.
2. If the frequency ratio $\lambda = \theta/\omega$ is very large, the dynamical coefficient becomes very small. This case is of special interest for the problem of the suppression of the forced vibrations in structures.

Damped forced vibration. Now let us consider the dynamical system with one degree of freedom and take the resisting force into account. Assume that this force is proportional to the first degree of the velocity, i.e., $R = -\beta\dot{y}$ (the minus sign indicates that force R and velocity $\dot{y}$ are oppositely directed). Let c is a stiffness of a system. The mass m is subjected to a disturbed harmonic force $F(t) = F_0 \sin\theta\, t$. Forced vibration is described by the equation

$$\ddot{y} + 2n\dot{y} + \omega^2 y = P_0 \sin\theta\, t, \quad 2n = \frac{\beta}{m}, \quad \omega^2 = \frac{c}{m}, \quad P_0 = \frac{F_0}{m}.$$

Duhamel's integral should be presented in the modified form, which takes into account the resisting force [Clo75].

Partial solution of this equation (forced vibration) is

$$y^*(t) = A\sin(\theta\, t - \gamma).$$

This is a steady-state forced vibration, a sustained periodic motion with an amplitude A, and a frequency θ which is equal to the frequency of the disturbing force. The quantity γ characterizes the phase shift of forced vibration with respect to the disturbing force. It is easy to show that

$$A = \frac{\delta_0}{\sqrt{\left(1-\lambda^2\right)^2 + h^2\lambda^2}}, \quad \tan\gamma = \frac{\lambda h}{1-\lambda^2},$$

where $\delta_0 = F_0/c$ is a static displacement of mass m caused by the load F_0; the frequency ratio and a quantity characterizing the damping effect are

$$\lambda = \frac{\theta}{\omega}, \quad h = 2\frac{n}{\omega}.$$

Thus, the amplitude and phase shift depend on two dimensionless parameters λ and h. As before, the quantity A/δ_0 presents the dynamic coefficient

$$\mu_{\text{dyn}} = \frac{A}{\delta_0} = \frac{1}{\sqrt{\left(1-\lambda^2\right)^2 + h^2\lambda^2}}.$$

Graphs of dynamic coefficient and for phase shift for certain values of λ are given in Figs. 7.2 and 7.3 [Clo75].

1. If $\lambda = \theta/\omega < 1$, then the phase shift is $\gamma < 90°$, and amplitude of forced vibration is $A > \delta_0$. However, if the frequency ratio λ is very small ($\theta \ll \omega$), then the amplitude of forced vibration is approximately equal to the static displacement of the mass m, i.e., $A \cong \delta_0$ and the phase shift is $\gamma = 0$. Starting approximately from $\theta/\omega = 0.25$ the dynamical coefficients and the phase shift rapidly increase. Note that an increase of the damping parameter h leads to a decrease of dynamic coefficient and an increase of the phase shift.

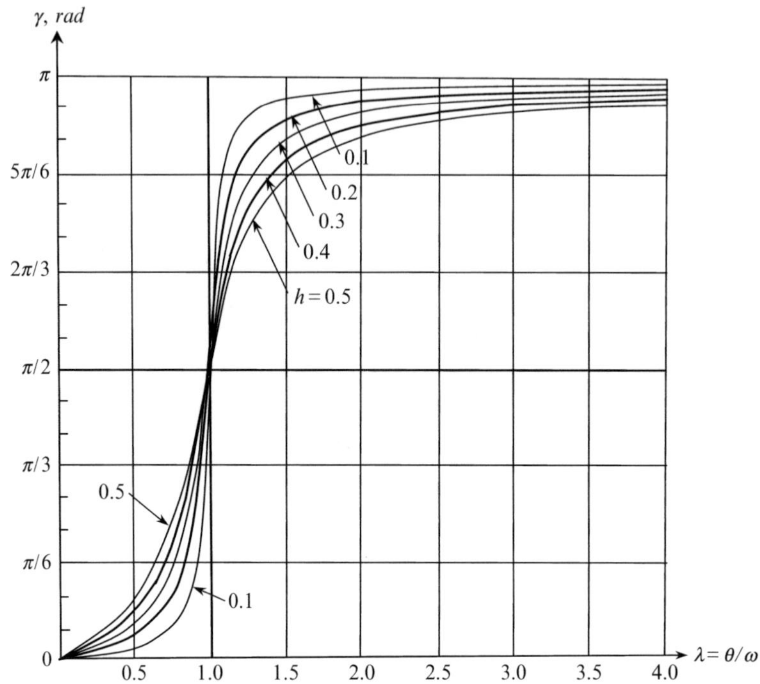

Fig. 7.3 Phase shift γ, rad, vs. $\lambda = \theta/\omega$; parameter $h = 2n/\omega$

2. At resonance ($\lambda = 1$), the amplitude of forced vibration and phase shift are $A_{\text{res}} = \delta_0/2h$ and $\gamma_{\text{res}} = \pi/2$, respectively. In the case of damped forced vibration, the displacement lags behind the disturbing force by 90°. As this place, the maximum values of dynamic coefficients are shifted left from the vertical line $\lambda = 1$. The damping forces significantly decrease the amplitudes at the resonance.
3. If the frequency ratio is very large ($\theta \gg \omega$), the dynamic coefficient is $\mu_{\text{dyn}} \ll 1$, i.e., the amplitude A of forced vibration is very small. For example, if $\lambda = \theta/\omega = 2.5$, and $h = 0$, then $A = 0.2\delta_0$.
4. The smaller coefficient of resistance n (or dimensionless parameter h), the larger the amplitude A of forced vibration.
5. The phase angle smoothly changes from zero until 180°, with the change occurring very rapidly in the resonance zone.

7.3 The Steady-State Vibrations of the Structure with a Finite Number of Degrees of Freedom

When a structure is subjected to any type of an exciting force, the vibration of the structure presents a combination of free and forced vibrations. Vibrations, which are caused by the disturbing forces only, are called the steady-state vibrations.

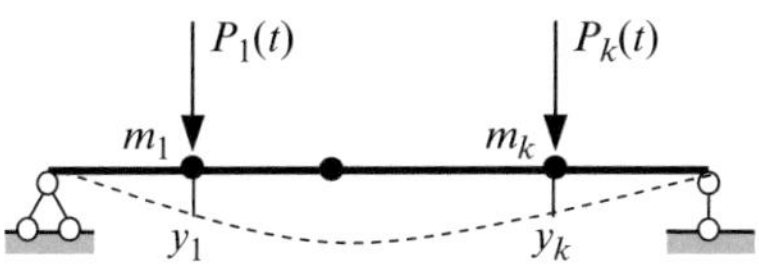

Fig. 7.4 Harmonic excitation of a structure with a finite number of degrees of freedom

The vibrations that occur in the structure, before it fully achieves a steady state, are called transient vibrations. This paragraph considers only steady-state vibrations of the arch, while the following paragraph considers transient vibrations.

If a structure with a finite number of degrees of freedom is subjected to a harmonic load, then the reactions, internal forces, and displacements will also change by a harmonic law. With this, their amplitudes will depend on the relationship between the frequency of free vibration and frequency of the harmonic force. If a structure is subjected to several harmonic loads, and all of them have the same frequency and act in the same phase, then inertial forces and displacements approach their respective extreme values simultaneously. Dynamic analysis involves computation of the amplitudes for internal forces and displacements, as well as testing the structure for possible resonance. The Force method in canonical form may be applied for these purposes.

7.3.1 Application of the Force Method

Elastic structure with lumped masses $m_k(k = 1, \ldots, n)$ is subjected to harmonic forces $P_k(t) = P_k \sin \theta t$ (Fig. 7.4).

Displacement of any mass m_i at time t is given by the formula

$$y_i = \delta_{i1}X_1 + \delta_{i2}X_2 + \cdots + \delta_{ii}X_i + \cdots + \delta_{in}X_n + \Delta_{i\text{P}}, \tag{7.15}$$

where X_i are inertial forces of the corresponding masses, $\delta_{i1}, \delta_{i2}, \ldots, \delta_{in}$ are displacements in the directions of the force X_i caused by unit forces $X_1, X_2, \ldots, X_n$, $\Delta_{i\text{P}}$ is displacement in the direction of X_i caused by amplitude values of harmonic loads.

Displacement of mass m_i and its acceleration for forced harmonic vibrations occurs with the frequency of exciting force θ according to the expressions

$$y_i = A_i \sin \theta t, \; \ddot{y}_i = -A_i\theta^2 \sin \theta t = -y_i\theta^2.$$

Since the inertial force for mass m_i is $X_i = -m_i\ddot{y}_i = m_i y_i \theta^2$, then $y_i = X_i / \left(m_i\theta^2\right)$. Now (7.15) may be rewritten in the form

$$\delta_{i1}X_1 + \delta_{i2}X_2 + \cdots + \delta_{ii}^{*}X_i + \cdots + \delta_{in}X_n + \Delta_{i\text{P}} = 0, \tag{7.16}$$

where $\delta_{ii}^{*} = \delta_{11} - 1/\left(m_i\theta^2\right)$.

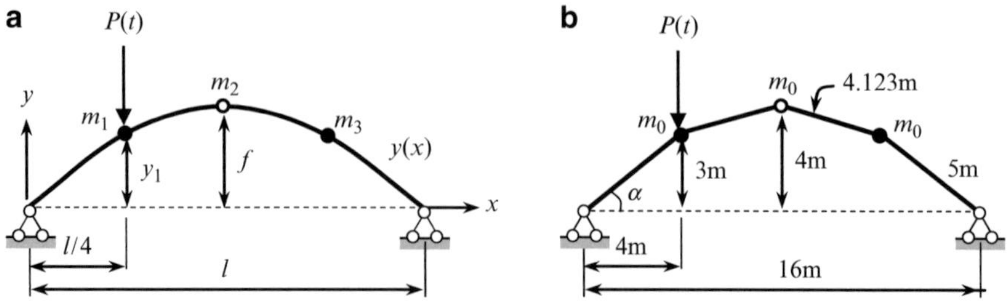

Fig. 7.5 (**a**) Design diagram of three-hinged arch with lumped masses and (**b**) modified design diagram (all dimensions are in meters)

As a result, we get a set of equations to calculate the amplitudes of the inertial forces X_i.

$$\begin{aligned}
&\delta_{11}^{*}X_1 + \delta_{12}X_2 + \cdots + \delta_{1n}X_n + \Delta_{1P} = 0,\\
&\delta_{21}X_1 + \delta_{22}^{*}X_2 + \cdots + \delta_{2n}X_n + \Delta_{2P} = 0,\\
&\cdots\cdots\cdots\cdots\cdots\cdots\cdots\cdots\cdots\\
&\delta_{n1}X_1 + \delta_{n2}X_2 + \cdots + \delta_{nn}^{*}X_n + \Delta_{nP} = 0.
\end{aligned} \tag{7.17}$$

Once the amplitudes of the inertial forces X_i are calculated from (7.17), dynamic bending moments are

$$M = \bar{M}_1X_1 + \bar{M}_2X_2 + \cdots + \bar{M}_nX_n + M_P. \tag{7.18}$$

7.3.2 The Steady-State Vibrations of the Arch

Parabolic three-hinged uniform symmetric arch with three equal lumped masses $m_i = m_0$ is subjected to harmonic excitation $P(t) = P_0 \sin\theta t,\ P_0 = 2$ kN as shown in Fig. 7.5. The span and rise of the arch are $l = 16$ m, $f = 0.25l = 4$ m [Pro48]. Let us show an application of (7.17)–(7.18) for the computation of dynamic bending moments for steady-state vibrations of the arch.

Free vibration of this structure has been considered in Sect. 6.2.3, Example 6.3. Group unknowns X_1 and X_3 and single unknowns X_2 and X_4 and corresponding unit bending moment diagram are shown in Fig. 6.7c. Unit displacements and frequencies of the antisymmetrical and symmetrical free vibrations are

$$\delta_{11} = \frac{9.50}{EI},\quad \delta_{22} = \frac{6.08}{EI},\quad \delta_{33} = \frac{38.01}{EI},\quad \delta_{44} = \frac{1.52}{EI},\quad \delta_{12} = \delta_{21} = -\frac{7.60}{EI},$$
$$\delta_{13} = \delta_{31} = 0,\quad \delta_{14} = \delta_{41} = 0,\quad \delta_{23} = \delta_{32} = 0,\quad \delta_{24} = \delta_{42} = 0.$$

$$\omega_{\text{antisym}} = 0.221\sqrt{\frac{EI}{m_0}},\ \omega_{\text{sym}} = 0.304\sqrt{\frac{EI}{m_0}}.$$

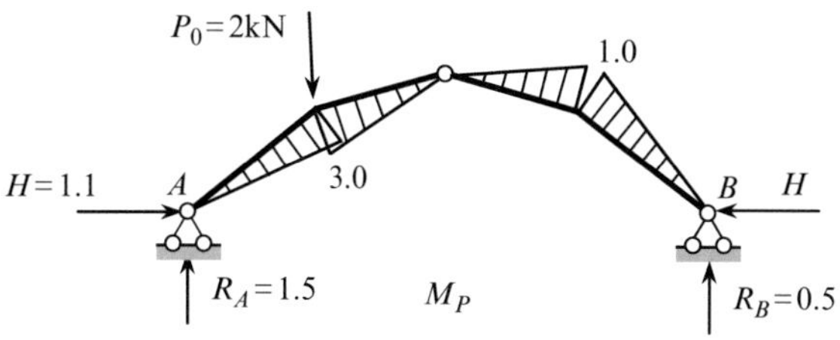

Fig. 7.6 Bending moment diagram caused by the amplitude of exciting force

Let the frequency of the harmonic excitation be $\theta = k\sqrt{\mathrm{EI}/m_0}$, so $k^2\mathrm{EI} = \theta^2 m_0$, where k is any number.

Therefore, main displacements become

$$\delta_{11}^* = \delta_{11} - \frac{2}{m_0\theta^2} = \delta_{11} - \frac{2}{k^2EI} = \frac{1}{EI}\left(9.5 - \frac{2}{k^2}\right), \quad \delta_{22}^* = \delta_{22} - \frac{1}{m_0\theta^2} = \frac{1}{EI}\left(6.08 - \frac{1}{k^2}\right),$$
$$\delta_{33}^* = \delta_{33} - \frac{2}{m_0\theta^2} = \frac{1}{EI}\left(38.0 - \frac{2}{k^2}\right), \quad \delta_{44}^* = \delta_{44} - \frac{1}{m_0\theta^2} = \frac{1}{EI}\left(1.52 - \frac{1}{k^2}\right).$$

Bending moment diagram caused by the amplitude force $P_0 = 2$ kN is shown in Fig. 7.6.

Loading terms are

$$\begin{aligned} EI\Delta_{1\mathrm{P}} =& \bar{M}_1 \times M_\mathrm{P} = \frac{1}{2} \times 5 \times 1.25 \times \frac{2}{3} \times 3 + \frac{1}{2} \times 4.123 \times 1.25 \times \frac{2}{3} \times 3 \\ & - \frac{1}{2} \times 5 \times 1.25 \times \frac{2}{3} \times 1 - \frac{1}{2} \times 4.123 \times 1.25 \times \frac{2}{3} \times 1 = 7.60. \end{aligned}$$

$$\Delta_{2\mathrm{P}} = \frac{\bar{M}_2 \times M_\mathrm{P}}{EI} = -\frac{6.08}{EI}; \quad \Delta_{3\mathrm{P}} = \frac{30.41}{EI}; \quad \Delta_{4\mathrm{P}} = \frac{6.08}{EI}.$$

Equation (7.17) serve for the calculation of amplitudes of the inertial forces X_i. In our case, these equations fall down into two separate subsystem for antisymmetrical and symmetrical vibrations.

Antisymmetrical vibrations

$$\begin{aligned} &\left(38.0 - \frac{2}{k^2}\right)X_3 + 7.6X_4 + 30.41 = 0, \\ &7.60X_3 + \left(1.52 - \frac{1}{k^2}\right)X_4 + 6.08 = 0. \end{aligned} \tag{7.19}$$

Symmetrical vibrations

$$\begin{aligned} &\left(9.5 - \frac{2}{k^2}\right)X_1 - 7.6X_2 + 7.6 = 0, \\ &-7.6X_1 + \left(6.08 - \frac{1}{k^2}\right)X_2 - 6.08 = 0. \end{aligned} \tag{7.20}$$

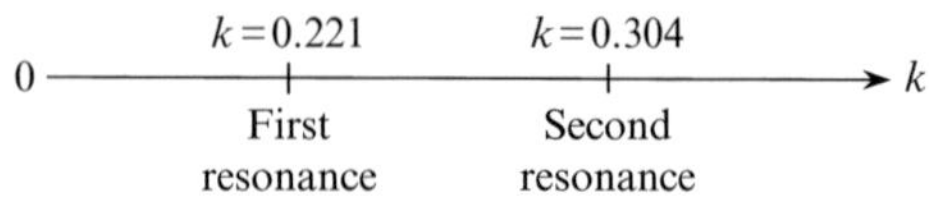

Fig. 7.7 Three frequency intervals

Table 7.2 Amplitude values of inertial forces X_i for different parameters k, $\theta = k\sqrt{EI/m_0}$

Type of unknowns	Unknowns	0.10	0.20	0.221	0.25	0.28	0.304	0.32	1.0	Equation
Symmetric	X_1	0.04	0.22	*	0.73	1.97	∞	4.10	4.35	(7.20)
	X_2	−0.07	−0.18	*	−1.17	−3.15	∞	6.55	5.32	
Antisymmetric	X_3	0.19	1.97	∞	−3.56	−1.95	*	−2.33	−0.81	(7.19)
	X_4	0.08	0.90	∞	−1.34	−0.74	*	1.58	−0.16	

For resonance parameters k, the values of inertial forces (shown by *asterisks*) are not calculated because two other inertial forces equal infinity

Two points, $k = 0.221$ and 0.304, which correspond to the resonance frequencies, are shown on a frequency axis. These frequencies partition the frequency axis into three intervals (Fig. 7.7).

At $k = 0.221$, the determinant of system (7.19), for antisymmetrical vibration, becomes

$$\begin{vmatrix} 38.0 - \dfrac{2}{k^2} & 7.6 \\ 7.6 & 1.52 - \dfrac{1}{k^2} \end{vmatrix} = \begin{vmatrix} -2.949 & 7.6 \\ 7.6 & -18.95 \end{vmatrix} = 55.88 - 57.76 \cong 0.$$

At $k = 0.304$, the determinant of system (7.20), for symmetrical vibration, becomes

$$\begin{vmatrix} 9.5 - \dfrac{2}{k^2} & -7.6 \\ -7.6 & 6.08 - \dfrac{1}{k^2} \end{vmatrix} = \begin{vmatrix} -12.14 & -7.6 \\ -7.6 & -4.74 \end{vmatrix} = 57.55 - 57.75 \cong 0.$$

Therefore, at $k = 0.221$, the forces $X_3 = X_4 = \infty$ and at $k = 0.304$, the forces $X_1 = X_2 = \infty$. Amplitude values of inertial forces for different parameter k are presented in Table 7.2.

Dynamical bending moment diagram for each parameter k may be constructed by the formula

$$M_{\text{dyn}} = M_P + \sum \bar{M}_i X_i.$$

Bending moment diagram for $k = 0.20$ and 1.0 are shown in Fig. 7.8; the index of nodal points is shown by bold numbers.

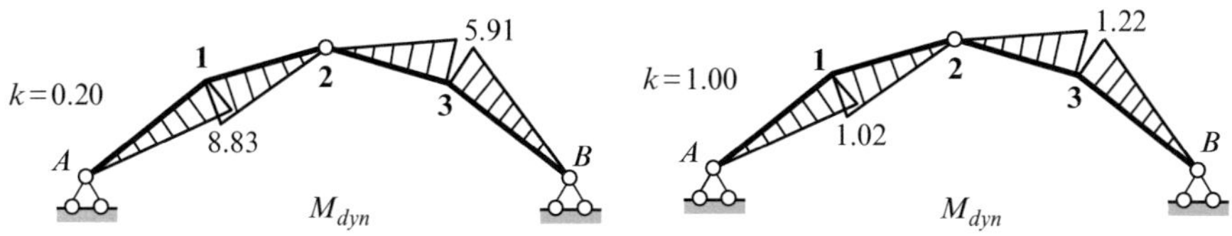

Fig. 7.8 Dynamic bending moment diagram

Dynamic coefficient at points 1 and 3 for $k = 0.20$ are $8.83/3.0 = 2.94$ and $5.91/1.0 = 5.91$, respectively.

Note: Nonlinear vibration of a shallow arch as a jump-system (Fig. 4.3) has been considered in detail by Kazakevich [Kaz89]. In particularly, he described a period-doubling for such system. Further details of this unique property is described in [Stro94].

7.4 Transient Vibration of the Arch

Rabinovich's method allows us to investigate the transient vibration of the arch and for any type of exciting force to construct the dynamical bending moment diagram without calculating displacements of the arch. The method is based on resolving the bending moment diagram into mode shapes of vibration.

7.4.1 Procedure of Analysis

Let the arch be subjected to an arbitrary distributed and/or to concentrated time-dependent loading.

To construct the dynamical bending moment diagram, the following procedure is used [Rab58], [Sni66].

1. For the given arch, construct a parabolic polygon and determine its parameters (parameters of polygon for six portions are presented in Table A.36).
2. Arbitrary dynamic loads on the arch should be presented as a set of concentrated time-dependent forces, which are applied at the nodal points of the Rabinovich's model. These forces may be written in the form $P_B f(t)$, $P_C f(t)$, ... where $f(t)$ is a certain time-dependent function, and P_B, P_C, ... may be treated as static loads at the nodal points B, C,. . .
3. Resolve the arbitrary nodal loads into symmetrical and antisymmetrical components. The following steps should be performed for each component of load separately; for now let us consider symmetrical loading.
4. We initially consider the nodal forces as static $(f(t) = 1)$. Construct the bending moment diagram for symmetrical loading; this diagram is denoted M^{stat}.

5. Resolve bending moment diagram M^{stat} for symmetrical loading into two components by the first and second mode of symmetrical vibrations. Ordinates of these diagrams for three-hinged arch are presented in Tables A.37 and A.38; for two-hinged arch in Tables A.41 and A.42.
6. Form the following equations for two nodal points B and C of the arch

$$\begin{aligned} k_1 M_{B1} + k_2 M_{B2} &= M_B^{\text{stat}}, \\ k_1 M_{C1} + k_2 M_{C2} &= M_C^{\text{stat}}. \end{aligned} \tag{7.21}$$

   and calculate coefficients k_1 and k_2. For a two-hinged arch, either of the two equations (7.21) may be replaced by the similar equation for point D at the crown, namely, $k_1 M_{D1} + k_2 M_{D2} = M_D^{\text{stat}}$.
7. For any function $f(t)$, the dynamic bending moment for each point may be presented as a sum of two diagrams

$$M(t) = \sum k_i M_i \omega_i \int_0^t f(\tau) \sin \omega_i (t - \tau) \mathrm{d}\tau, \quad i = 1, 2. \tag{7.22}$$

where the first and second terms of (7.22) correspond to the first and second forms of symmetrical vibrations.

Duhamel integrals in (7.22), for some loadings, are presented in Sect. 7.2.3.

The steps 1–7 and (7.22) allow us to construct the bending moment diagram as a sum of two diagrams corresponding to the symmetrical modes of vibrations of the arch. It is very important that in the case of any excitation *f*(*t*), we *do not* need to determine the displacements of the arch.

7.4.2 Impulsive Load

Parabolic symmetrical three-hinged uniform arch A–A' of span l and rise $f = 0.2l$ is subjected to the impulsive excitation as shown in Fig. 7.9a; maximum ordinate at crown D equals s(Ns/m). Let us calculate the dynamic bending moment at specified points.

The span of the arch is divided into six equal portions. Ordinates at the specific points are $y_B = l/9$ and $y_C = 8l/45$ (Fig. 7.9b) [Rab58], [Sni66].

Step 1. First of all we will consider the given load as static. Distributed load is replaced by a set of the nodal forces. For this purpose, we will consider three simply supported beams; the span of each beam is $\lambda = l/6$. General design diagram of each beam of span λ and corresponding reactions R_1 and R_2 are shown in Fig. 7.9b. Specific forces for each nodal point of the arch are shown in Fig. 7.9c, with factor $k = sl/108$.

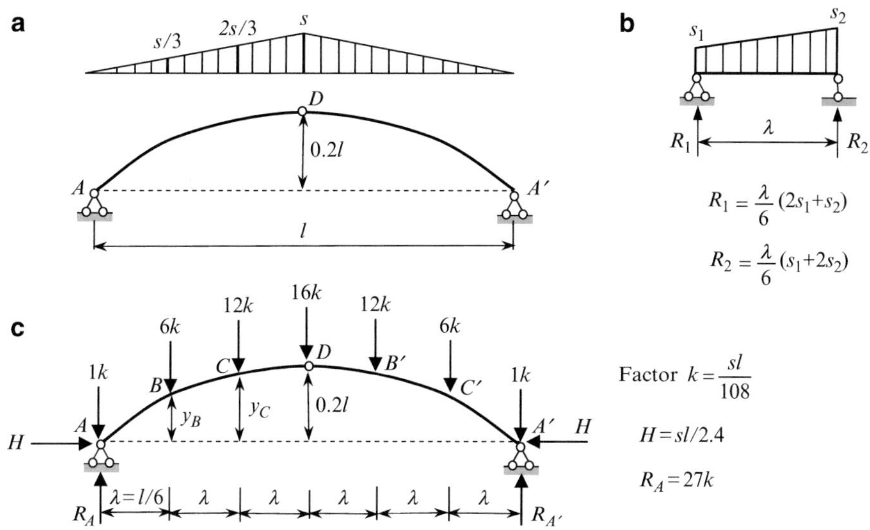

Fig. 7.9 (**a**) Design diagram of the arch subjected to impulsive excitation; (**b**) loading and reactions for any portion of span λ; and (**c**) loading of the arch by loads at the nodal points

For example, the force at node B is

$$F_B = \underbrace{\frac{\lambda}{6}(s_1 + 2s_2)}_{R_2 \text{ portion } AB} + \underbrace{\frac{\lambda}{6}(2s_1 + s_2)}_{R_1 \text{ portion } BC} = \frac{\lambda}{6}\left(0 + 2\frac{s}{3}\right) + \frac{\lambda}{6}\left(2\frac{s}{3} + \frac{2s}{3}\right) = \frac{s\,l}{18} = 6k.$$

The factor k becomes $k = s\,l/108$. Vertical reactions of the arch are $R_A = R_{A'} = 27k$. Bending moment at D for substitute beam is $M_D^0 = sl^2/12$, so the thrust of the arch becomes $H = M_D^0/f = (sl/2.4)(\text{kN})$.

Bending moments at joints B and C are

$$M_B^{\text{stat}} = (R_A - 1)\,k \times \lambda - H \times y_B = -0.0062sl^2,$$
$$M_C^{\text{stat}} = (R_A - 1)\,k \times 2\lambda - 6k \times \lambda - H \times y_C - -0.0031sl^2.$$

The bending moments at the nodal points B and C for the first and second mode of vibrations, according to Tables A.37 and A.38, are

$$M_{B1} = 78.12EI/l^2, \quad M_{B2} = 253.96EI/l^2,$$
$$M_{C1} = 51.51EI/l^2, \quad M_{C2} = -348.62EI/l^2.$$

Equations (7.21) for nodal points B and C become

$$(78.12k_1 + 253.96k_2)EI/l^2 = -0.0062sl^2,$$
$$(51.51k_1 - 348.62k_2)EI/l^2 = -0.0031sl^2.$$

Solution of these equations is

$$k_1 = -7.314 \times 10^{-5} \frac{sl^4}{EI}, \quad k_2 = -0.1914 \times 10^{-5} \frac{sl^4}{EI}.$$

Step 2. Dynamic bending moment at joint B is

$$M_B^{\text{dyn}}(t) = k_1 M_{B1} \omega_1 \int_0^t f(\tau) \sin \omega_1 (t - \tau) \mathrm{d}\tau + k_2 M_{B2} \omega_2 \int_0^t f(\tau) \sin \omega_2 (t - \tau) \mathrm{d}\tau.$$

Duhamel integral for different types of loads is presented in [Har88]. Taking into account Duhamel integral for a given impulsive loading, the last expression becomes

$$M_B^{\text{dyn}}(t) = k_1 M_{B1} \omega_1 \sin \omega_1 t + k_2 M_{B2} \omega_2 \sin \omega_2 t.$$

According to Tables A.37 and A.38, the first and second frequency of symmetrical vibration are

$$\omega_1 = \frac{43.54}{l^2} \sqrt{\frac{EI}{\mu}}, \quad \omega_2 = \frac{138.73}{l^2} \sqrt{\frac{EI}{\mu}}.$$

Finally, for bending moment at point B, we need to use the maximum ordinate of impulse s, thus, we get

$$M_B^{\text{dyn}}(t) = -s(0.259 \sin \omega_1 t + 0.0674 \sin \omega_2 t) \sqrt{\frac{EI}{\mu}} (\text{Nm}).$$

Note that this expression contains the span l in the frequencies ω_i only.

Similarly we can calculate the dynamic bending moment at joint C. Finding the extremes of these functions poses no difficulty. In the case of antisymmetrical vibration of three-hinged arch, we need to use Tables A.39 and A.40. This procedure is the same for other types of arches.

Let us consider another excitation. Assume that triangle impulse in Fig. 7.9a acts within a short time τ and after that vanishes. The maximum displacement occurs after the disappearance of the load [Tho81], [Wea90]. As this takes place

$$M(t) = 2 \sum k_i \frac{\omega_i \tau}{2} \sin \omega_i \left(t - \frac{\tau}{2} \right), \quad i = 1, 2.$$

Similar expressions may be formed for any joint of the Rabinovich model in the case of an antisymmetric vibration. Procedure for vibration analysis of the other types of arches is the same.

Part IV
Special Arch Problems

Chapter 8
Special Statics Topics

This chapter contains two topics, which significantly extend the classic static analysis of arches. They are the plastic analysis of arches and analysis of arched structures with one-sided constraints.

These problems are nonlinear. The reason for their nonlinearity for both topics is the same – some constraints are eliminated from the work of a structure. However, the *reasons* for this elimination for both cases are different. In the case of plastic analysis, this occurs because the bearing capability of these constrains become exhausted. In the case of structures with one-sided constraints, we deal with constraints which are unable to perceive the internal force of a certain sign.

In both cases, the design diagram of a structure, in the process of loading, is undergoing change, i.e., we have a system with a variable structure. In such problems, application of the superposition principle as well as general theorems of elastic structures requires rigorous justification.

Analysis of such problems falls in the category of the most difficult problems in structural mechanics.

8.1 Plastic Analysis of the Arches

In the previous chapters, we considered structures taking into account only elastic properties of materials for all members of a structure. Analysis of a structure based on elastic properties of material is called the elastic (or linear) analysis. Elastic analysis does not allow us to find out the *reserve* of strength of the structure beyond its elastic limit. Also this analysis cannot answer the question: what would happen with the structure, if the stresses in its members will be larger than the proportional limit? Therefore, a problem concerning to the actual strength of a structure cannot be solved using elastic analysis. Plastic analysis allows us to use the reserves of strength of material, which remains unused considering material of structure as elastic. Therefore, plastic analysis allows us to define the limit load on the structure and to design a more economical structure.

I.A. Karnovsky, *Theory of Arched Structures: Strength, Stability, Vibration*,
DOI 10.1007/978-1-4614-0469-9_8, © Springer Science+Business Media, LLC 2012

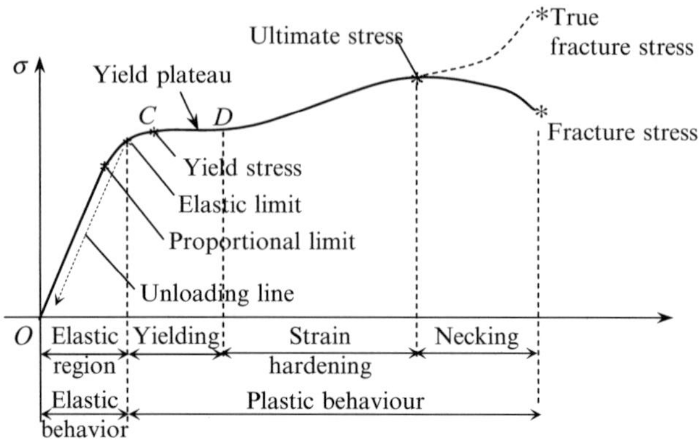

Fig. 8.1 Typical stress–strain diagram for structural steel

Fundamental idea of the plastic analysis is discussed using the direct method. Kinematical and statical methods of calculation of the limit loads are considered. Detailed plastic analysis of the arches is presented.

8.1.1 Idealized Stress–Strain Diagrams

The typical stress–strain diagram for the specimen of structural steel is presented in Fig. 8.1. Elastic analysis corresponds to the initial straight portion of the σ–ε diagram. If a specimen is loaded into the proportional limit (or below) and then released, then material will unload along the loading path back to the origin. So, there are no residual strains. This property of unloaded specimen to return to its original dimensions is called *elasticity*, and material in this region is called the *linearly elastic*. Within the elastic region, a relationship between stress and strain obeys to Hooke's law $\sigma = E\varepsilon$.

Let a specimen is loaded into the elastic limit. The stress at this point slightly exceeds the proportional limit. From this point, the material unloads along the line that is parallel to straight portion of the diagram and thus, the material has the very small *residual* strain (Fig. 8.1).

Plastic behavior starts at the elastic limit. The region *CD* is referred as the *perfect plastic* zone. In this region, the specimen continues to elongate without any increase in stress. Above the yield plateau, starting from point *D*, the behavior of the specimen is described by nonlinear relationships σ–ε. If the specimen will be unloaded at point *A* (Fig. 8.2), then unloading line will be parallel to the load straight line, so the specimen returns only partially to its original length. Total strain of the specimen is ON, while the strain MN has been recovered elastically and the strain OM remains as residual one [Cra00].

If the material remains within the elastic region, it can be loaded, unloaded, and loaded again without significantly changing the behavior. However, when the load is reapplied in a plastic region, the internal structure of material is altered,

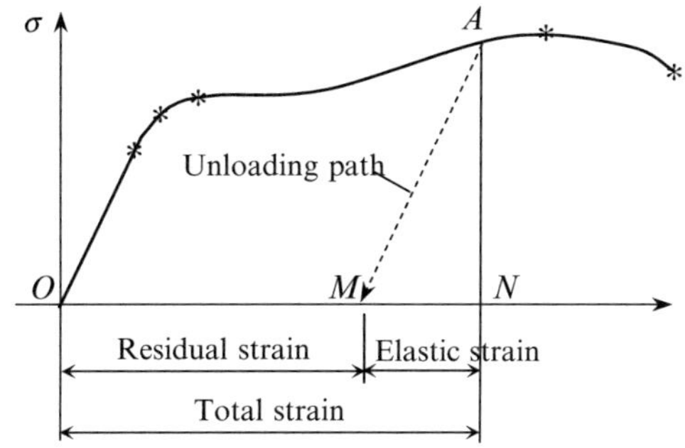

Fig. 8.2 Loading–unloading diagram

Table 8.1 Idealized relationship σ–ε for axially loaded members

Material	Elasto-plastic	Rigid-plastic	Elasto-plastic with linear hardening	Rigid-plastic with linear hardening
Diagram	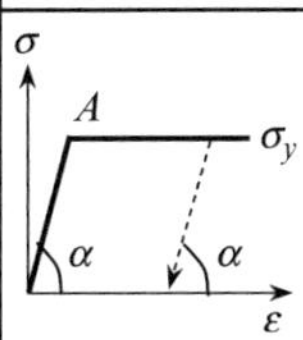	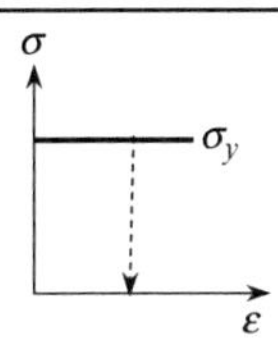	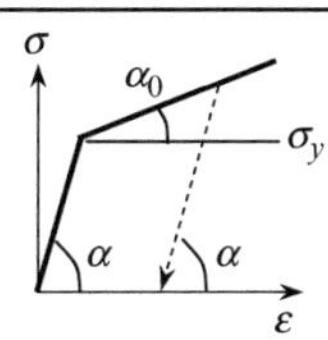	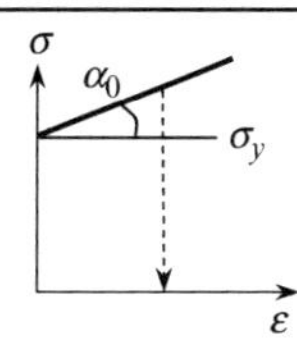

its properties change, and the material obeys to Hook's law within the straight line *MA*; it means that the proportional limit of the material has been increased. This process is referred to as the *strain-hardening*.

For plastic analysis, we change the typical diagram by its idealized diagram. Different idealized diagrams are considered in engineering practice [Bir68], [Cra00]. Some of idealized models are presented in Table 8.1.

For further analysis, we will consider idealized elasto-plastic material and rigid-plastic material. We starts from elasto-plastic material; corresponding diagram is called Prandtl diagram. This diagram has two portions – linear "stress–strain" part and the yield plateau. Elastic properties of material are holds up to yield point stress σ_y. The yield plateau shows that displacement of material can become indeterminate large under the same stress. Idealized elasto-plastic material does not have effect of hardening. This diagram may be applicable for a structural steel and for reinforced concrete. Structural analysis on the basis of idealized diagram is referred as the *plastic analysis*. The quantitative results of plastic analysis are much closer to the actual behavior of a structure than the results obtained on the basis of elastic properties of material.

In case of statically determinate structure, yielding of any members leads to the failure of the structure as a whole. Other situation occurs in case of statically indeterminate structure. Assume that for all members of the structure, the Prandtl diagram is applicable. On the first stage, when loads are small, behavior of all members follows the first portion of the Prandtl diagram. Proportional increase of all loads leads to the yielding in the most loaded member. It means that the degree

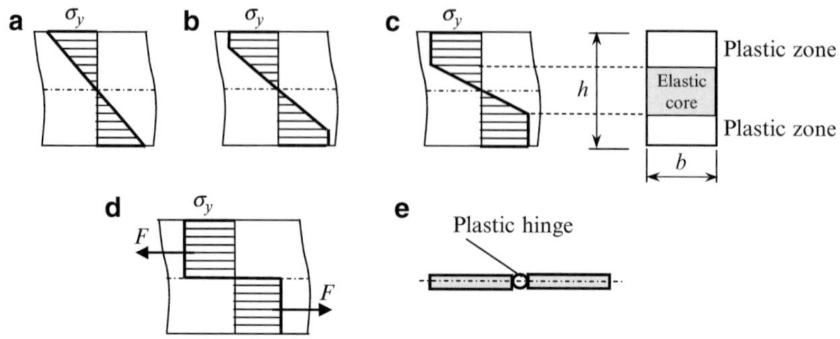

Fig. 8.3 Distribution of normal stresses within the height of a beam

of statical indeterminacy is decreased by one. The following proportional increase of all loads leads to the following effect: the internal force in the *yielding* member remains the same, while the forces in the *other* members will be increased. This effect will be continues until the next member starts to yield. Finally, the structure becomes statically determinate and yielding of any member of this structure immediately leads to the failure of the structure, since the structure is transformed into a mechanism. In general, if the structure has n redundant constrains, then its failure occurs when the number of yielding member becomes $n + 1$. It means that capability of a structure to carry out the increasing load has been exhausted. This condition is called *limit equilibrium condition*. In this condition, the limit loads and internal forces satisfy to equilibrium condition. The following increase of a load is impossible. In this condition, the displacement of the structure becomes undefined. While the linear portion of typical stress–strain diagram leads to linear problems of structural analysis (elastic problems), the Prandtl diagram leads to nonlinear problems of plastic behavior of structures. Indeed, the design diagram of a structure is changed upon different levels of loads. Transition from one design diagram to another happens abruptly.

Let us consider the plane bending of a beam with a rectangular ($b \times h$) cross section. In the elastic region of the stress–strain diagram, the normal stresses are distributed within the height of a cross section of the beam linearly. The maximum tensile and compressed stresses are located at the extreme fibers of the beam. The stress σ_y corresponds to the yield plateau (Fig. 8.3a). Increasing of the load leads to appearance and developing of the yield zone and decreasing of the "elastic core" of the section of the beam. Figure 8.3b, c corresponds to partially plastic bending of a beam, which means that the middle part of the cross section is in elastic condition, while the bottom and top parts of the beam are in plastic condition [Cra00].

Further increasing of load leads to complete plastic state (Fig. 8.3d), which corresponds to the limit equilibrium, i.e., we are talking about the appearance of so-called *plastic hinge* (Fig. 8.3d, e). It is obvious that all sections of the beam are in different states. Defining the location of the plastic hinge is an additional problem of plastic analysis. This problem will be considered below.

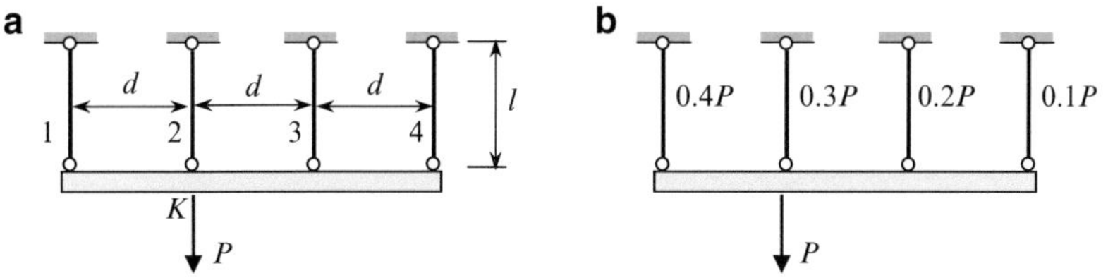

Fig. 8.4 (**a**, **b**) Design diagram and distribution of internal forces according to elastic analysis

What is the difference between plastic and ideal hinge? First, the plastic hinge disappears if the structure is unloaded, so the plastic hinge may be considered as fully recoverable or one-sided hinge. Second, in the ideal hinge, the bending moment equals to zero, while the plastic hinge is characterized by the appearance of bending moment, which is equal to the limit (or plastic) moment of internal forces $F = \sigma_y(bh/2)$ (Fig. 8.3d). A bearing capability of a structure is characterized by the plastic moment $M_P = F(h/2) = \sigma_y(bh^2/4)$.

Plastic analysis involves determination of limit load, which structure can resist before full failure due to yielding of some elements. The limiting load does not depend on settlements of supports, errors of fabrication, prestressed tension, and temperature changes; this is the fundamental difference between plastic and elastic analysis. In the following sections, we consider different methods of determining plastic loads.

8.1.2 *Direct Method of Plastic Analysis*

The fundamental concept of plastic analysis of a structure may be clearly presented using the direct method [Kar10]. Let us consider the structure shown in Fig. 8.4a subjected to load P at point K [Rzh82]. The horizontal rod is absolutely rigid. All hangers have constant stiffness EA. The plastic analysis must be preceded by the elastic analysis.

Elastic Analysis

This analysis should be performed on the basis of any appropriate method of analysis of statically indeterminate structures. Omitting this analysis, which is familiar for reader and presents no difficulties, the distribution of internal forces in members 1–4 of the structure is as follows (Fig. 8.4b):

$$N_1 = 0.4P; \quad N_2 = 0.3P; \quad N_3 = 0.2P; \quad \text{and} \quad N_4 = 0.1P.$$

Plastic Analysis

Step 1. Increasing of load leads to the appearance of yield stresses. They are reached in the most highly stressed member. In our case, this member is element 1.

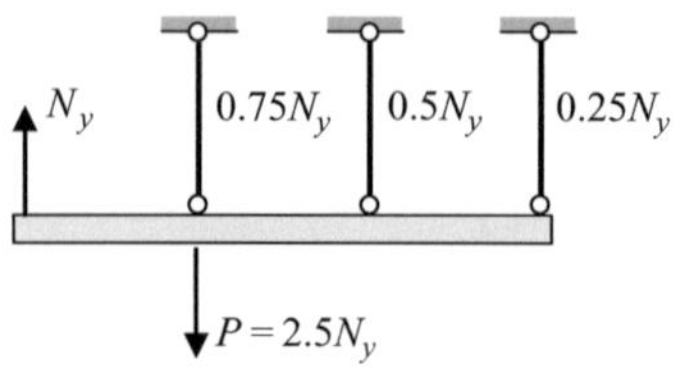

Fig. 8.5 Step 1 – plastic state in the member 1 and internal forces in the rest members

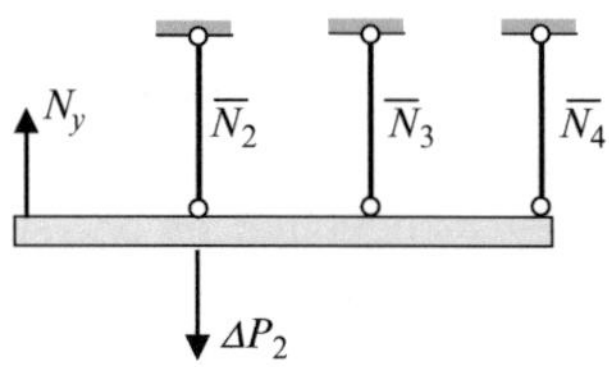

Fig. 8.6 Step 2 – internal forces in the members 2–4 due to load ΔP_2

Let N_1 become equal to limit load, i.e., $N_1 = N_y$. Since $N_1 = 0.4P$, then it occurs if external load would be equal to $P = (N_y/0.4) = 2.5N_y$. For this load P, the limit tension will be reached in the first hanger. Internal forces in other members are (Fig. 8.5).

$$N_2 = 0.3P = 0.3 \times 2.5N_y = 0.75N_y; \quad N_3 = 0.5N_y; \quad N_4 = 0.25N_y.$$

Step 2. If load P will be increased by the value ΔP_2, then $N_1 = N_y$ remains without changes. It means that additional load will be distributed between the three members 2–4, i.e., the design diagram had been changed (Fig. 8.6). This structure is statically indeterminate to the first degree. Elastic analysis of this structure due to load ΔP_2 leads to the following internal forces

$$\bar{N}_2 = 0.833\Delta P_2; \quad \bar{N}_3 = 0.333\Delta P_2; \quad \bar{N}_4 = -0.167\Delta P_2.$$

As always, the most highly stressed member will reach the yield stress first. Since first hanger is already in yield condition (and cannot resist any additional load), the most highly stressed member due to load ΔP_2 is the second hanger. The total limit load in this element equals

$$N_2 = 0.75N_y + 0.833\Delta P_2.$$

In this formula, the first term corresponds to initially applied load $P = 2.5N_y$ (Fig. 8.5), while the second term corresponds to additional load ΔP_2. The limit load for the second hanger is $N_2 = N_y$. Thus, the equation

$$N_2 = 0.75N_y + 0.833\Delta P_2 = N_y$$

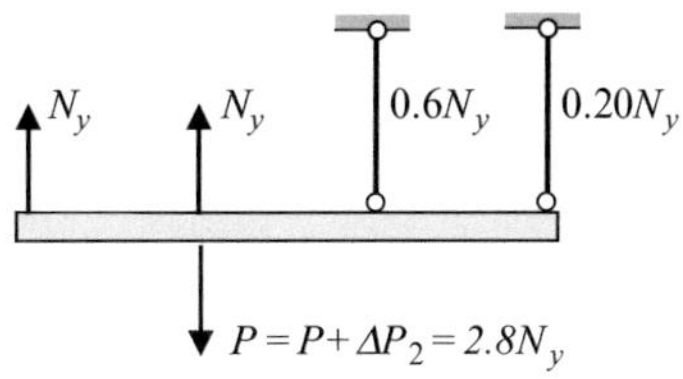

Fig. 8.7 Step 2 – plastic state in the members 1 and 2

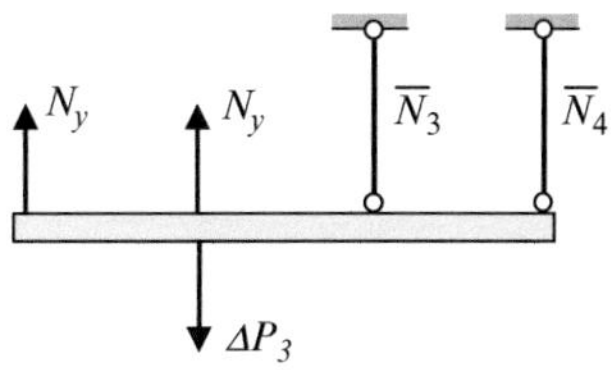

Fig. 8.8 Step 3 – internal forces in the members 3–4 due to load ΔP_3

leads to the following value for the increment of load $\Delta P_2 = (0.25N_y)/0.833 = 0.3N_y$. Thus, the value $\Delta P_2 = 0.3N_y$ represents additional load, which is required so that the second hanger reaches its yielding state. Therefore, if load

$$P = 2.5N_y + 0.3N_y = 2.8N_y,$$

then both members 1 and 2 reach their limit state. As this takes place (Fig. 8.7), the internal forces in hangers 3 and 4 are the following

$$N_3 = 0.5N_y + 0.333 \times 0.3N_y = 0.6N_y;$$
$$N_4 = 0.25N_y - 0.167 \times 0.3N_y = 0.20N_y.$$

Step 3 Since internal forces in hangers 1 and 2 reached the limit values, then the following increase of the load by the value ΔP_3 (Fig. 8.8) affects the members 3 and 4 only. Elastic analysis of this statically determinate structure leads to the following internal forces in members 3 and 4: $\bar N_3 = 2\Delta P_3$ and $\bar N_4 = -\Delta P_3$.

Step 4. Similarly as above, the limit state for this case occurs if internal force in hanger 3 reaches its limit value

$$N_3 = 0.6N_y + 2\Delta P_3 = N_y.$$

This equation leads to the following value for the increment of load

$$\Delta P_3 = \frac{0.4N_y}{2} = 0.2N_y.$$

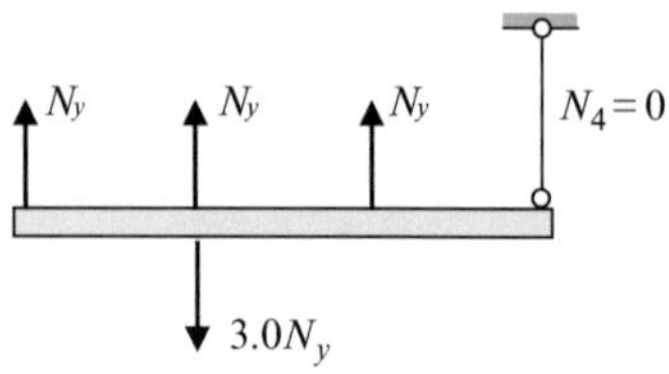

Fig. 8.9 Step 4 – plastic state in the members 1–3

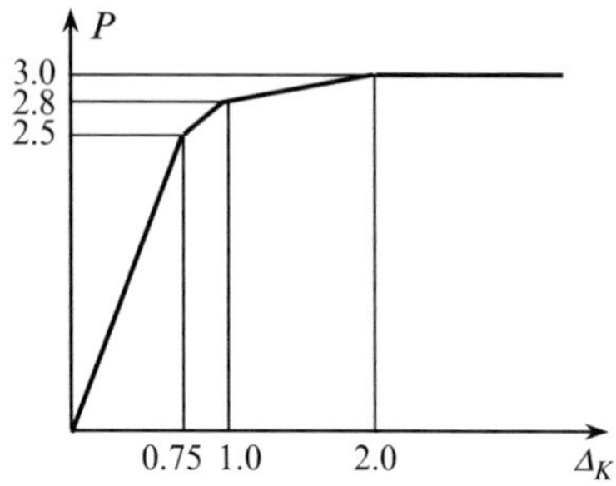

Fig. 8.10 $P - \Delta_K$ diagram in plastic analysis

The total value of external force (Fig. 8.9) is given as follows

$$P = 2.5N_y + \Delta P_2 + \Delta P_3 = 2.5N_y + 0.3N_y + 0.2N_y = 3.0N_y.$$

The first term in this formula corresponds to limit load in the first member; increment of the force by $0.3N_y$ leads to the limit state in the second member. The following increment of the force by $0.2N_y$ leads to the limit state in the third member. After that the load-carrying capacity of the structure is exhausted. From the equilibrium equation for the entire structure, we can see that on this stage $N_4 = 0$ (Fig. 8.9).

All forces satisfy to equilibrium condition. Plastic behavior analysis leads to the increment of the limit load by $[(3 - 2.5)/2.5]\,100\% = 20\%$.

Plastic displacements. If internal force in some of the elements reached its limiting value and the load continues to increase, then we cannot determine displacements of the system using only elastic analysis. However, plastic analysis allows calculating displacements of a structure on each stage of loading. Let us show the graph of displacement of the point application of force P (point K).

If load $P = 2.5N_y$, then internal force in second element equals $0.75N_y$ (Fig. 8.5) and vertical displacement of point K is $\Delta_K = 0.75N_yl/EA$.

If load $P = 2.8N_y$, then internal force in second element equals N_y (Fig. 8.7) and vertical displacement of point K is $\Delta_K = N_yl/EA$.

If load $P = 3.0N_y$, then internal force in third element equals N_y (Fig. 8.9) and deflection of this element equals N_yl/EA. Since internal force in fourth element equals zero, its deflection is zero and required displacement $\Delta_K = 2N_yl/EA$.

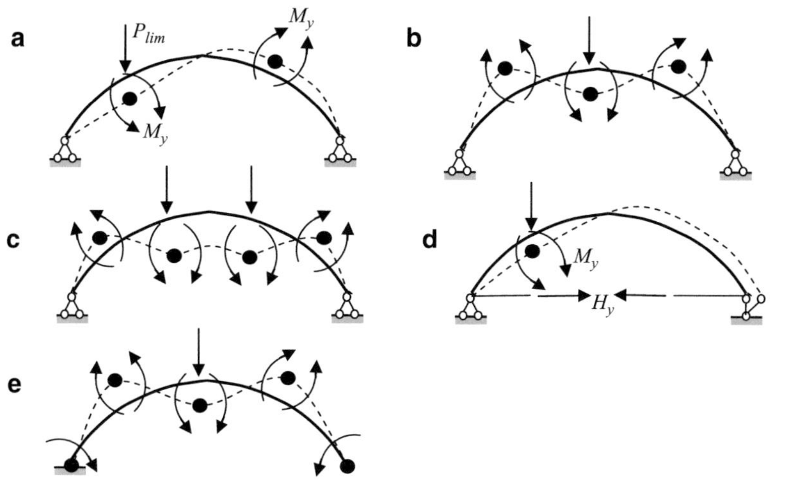

Fig. 8.11 Arch mechanisms of failure

Corresponding $P - \Delta_K$ diagram is shown in Fig. 8.10; the factors l/EA and N_y for horizontal and vertical axis, respectively. This diagram shows that $P - \Delta_K$ relationship is nonlinear. This is typical for plastic analysis.

8.1.3 Mechanisms of Failure in Arches

We will consider in-plane failure of the arches. A fundamental difficulty of the plastic analysis of the arches is determining the real form of their failure, i.e., determining location of the plastic hinges.

For a two-hinged arch, the different mechanisms of failure are shown in Fig. 8.11a–c. In case of main form of failure, the arch exhibits the mechanism of sideway displacement (Fig. 8.11a). Symmetric forms of failure are also possible (Fig. 8.11b, c). However, such mechanisms may be realized if additional constraints, which prevent sideway failure, are introduced into the structure.

For two-hinged arch with tie mechanism of failure can be characterized by the appearance of the yield in the tie and one plastic hinge in the arch (Fig. 8.11d).

If a hingeless arch is subjected to load P at the crown (Fig. 8.11e), then the mechanism of failure is characterized by the appearance of five plastic hinges [Kle80]. Their locations (except at the supports) are initially unknown.

8.1.4 Limiting Plastic Analysis of Parabolic Arches

Two-hinged uniform parabolic arch is subjected to force P as shown in Fig. 8.12a. Determine the limit load P and find the location of the plastic hinges, if $l = 12$ m,

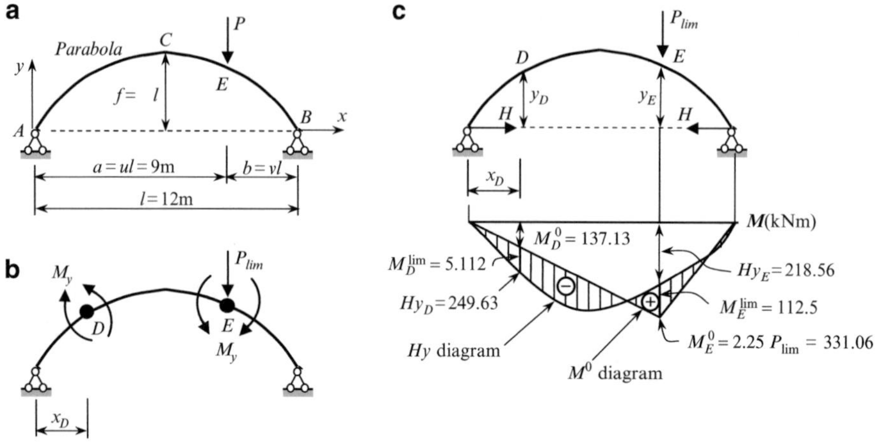

Fig. 8.12 Design diagram of the arch and mechanism of failure

$f = 3$ m, and bearing capacity of all cross sections within the arch is $M_y = 112.5$ kN m

Solution. First, let us consider the behavior of structure subjected to given load and the failure mechanism.

First stage. From the elastic analysis of the arch we know, that the maximum moment occurs at the point of application of load P, i.e., at point E. Therefore, just here will be located the first plastic hinge and the entire two-hinged arch is transformed into three-hinged nonsymmetrical arch. The maximum possible bending moment at the point E will be equal to the limit bending moment M_y (Fig. 8.12b).

Second stage. Again, we will increase the load P. The second plastic hinge occurs at point D with the maximum bending moment. However, the location of this point is unknown yet. We note that on the arch, in addition to the two support hinges A and B, two plastic hinges E and D are also added. The nature of these hinges is different. Hinge at the support points are ideal ones, while two other hinges are plastic ones and they are result of exhausted bearing capability of the beam.

The first plastic hinge with opening below occurs at point E. The bending moment at this point for the reference beam is

$$M_E^0 = P_{\lim}\frac{ab}{l} = P_{\lim}uvl = P_{\lim}u(1-u)l = P_{\lim} \times 0.75 \times 0.25 \times 12 = 2.25P_{\lim}.$$

The bending moment at any section x caused by thrust H is

$$\begin{aligned} M_x(H) &= -Hy = H\frac{4f}{l^2}x(l-x) = H\frac{4\xi}{l}x(l-x) = H\frac{4 \times 0.25}{12}x(12-x) \\ &= Hx\left(1 - \frac{x}{12}\right). \end{aligned} \tag{8.1}$$

At point E, we get

$$M_E(H) = H \times 9\left(1 - \frac{9}{12}\right) = 2.25H.$$

Finally, the bending moment at section E is

$$M_E = M_E^0 - M_E(H) = 2.25P_{\lim} - 2.25H. \tag{8.2}$$

For computation of the thrust H, the elastic analysis methods cannot be applied; therefore, for now the thrust H remains unknown.

The second plastic hinge with opening above occurs at point D. The bending moment at point D for the reference beam is

$$M_D^0 = M_E^0 \frac{x_D}{ul} = P_{\lim} u(1-u)\, l \frac{x_D}{ul} = 0.25P_{\lim} x_D.$$

The bending moment at point D caused by thrust H according to (8.1) is

$$M_D(H) = Hy_D = H\left(x_D - \frac{x_D^2}{12}\right).$$

The bending moment at section D becomes

$$M_D = M_D^0 - M_D(H) = 0.25P_{\lim} x_D - H\left(x_D - \frac{x_D^2}{12}\right). \tag{8.3}$$

The maximum bending moment occurs when $\mathrm{d}M_D/\mathrm{d}x_D = 0.25P_{\lim} - H(1-(x_D/6)) = 0$. Solution of this equation is

$$x_D = 6 - \frac{1.5P_{\lim}}{H}. \tag{8.4}$$

Substituting (8.4) into (8.3) yields

$$M_D = 1.5P_{\lim} - 3H - \frac{0.1875P_{\lim}^2}{H}. \tag{8.5}$$

Taking into account the limit plastic moment M_y we get

$$M_E^{\lim} = 2.25P_{\lim} - 2.25H = M_y = 112.5.$$

Thus, we obtain the following relationships between thrust and limit load

$$H = P_{\lim} - 50. \tag{8.6}$$

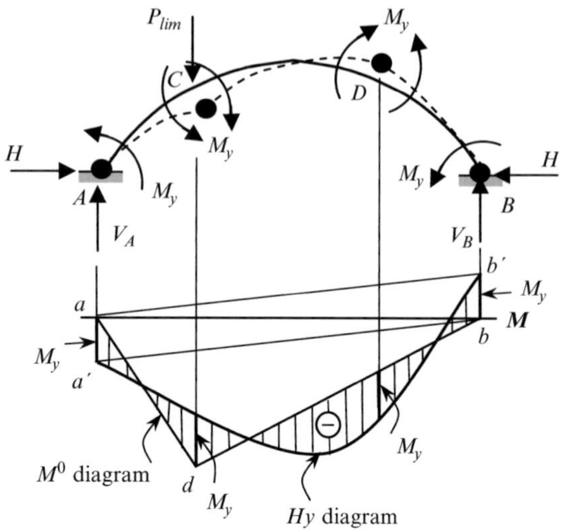

Fig. 8.13 Design diagram of the hingeless arch, mechanism of failure, and bending moment diagram in plastic state

Substitution of (8.6) into (8.5) and taking into account the limit plastic moment $M_D^{\lim} = -112.5$, we get a following algebraic equation with respect to $P_{\lim}$

$$1.6875P_{\lim}^2 - 337.5P_{\lim} + 13,125 = 0.$$

This equation leads to the limit load

$$P_{\lim\ 1} = 147.14\,\text{kN and } P_{\lim\ 2} = 52.86\,\text{kN}.$$

The first root leads to the following results for thrust and location of the plastic hinge

$$H_1 = P_{\lim\ 1} - 50 = 97.14\ (\text{kN}); x_{D\,1} = 6 - \frac{1.5P_{\lim\ 1}}{H_1} = 3.728\ (\text{m}).$$

The second root $P_{\lim\ 2}$ leads to the $x_{D\,2} = -21.7(\text{m})$ and should be discarded.

General expression for bending moment in plastic condition $M = M^0 - Hy$ is realized in Fig. 8.12c. Here M^0 is a bending moment diagram for reference beam, the term $Hy = Hx[1 - (x/12)]$ for points E and D becomes 218.56 and 249.63, respectively. In fact, this diagram presents the static method of plastic analysis [Kar10].

In case of nonsymmetrical loading of the hingeless arch, four plastic hinges appear in the system (Fig. 8.13).

A bending moment diagram in plastic state is combined from four diagrams. We construct the bending moment diagram *a–d–b* for substitute beam (M^0 diagram) from the base line *a–b*. Bending moment diagrams *a–a′–b* and *a–b–b′* are diagrams caused by plastic moments M_y at the supports. Bending moment diagram due to thrust is denoted as *Hy* diagram. Limit moments M_y at the sections *A–D* have alternating signs.

8.2 Arched Structures with One-Sided Constraints

So far it has been assumed that all constraints of the structure are two-sided. It means that if a constraint prevents displacements in some direction, then this constraint also prevents displacements in the opposite direction. However, the one-sided constraints often appear in design practice. The word "constraints" means not only support constraints, but any member of a structure, for example, the diagonal members of a truss.

A classic example of a structure with one-sided constraints is a beam on an elastic foundation. The foundation acts back on the beam within the portion where the beam touches the ground, and ceases to act on the beam in the zone where the beam is separated from the ground. The zone of contact of the beam with the foundation (or separation of beam from the ground) depends on the load value and its location.

Analysis of structures with one-sided constraints falls into a category of the most difficult problems in structural mechanics.

8.2.1 General Properties of Structures with One-Sided Constraints

One-sided constraint, which perceives the internal force, is called an active one-sided constraint. One-sided constraint which cannot resist the load is known as an inactive one-sided constraint. Active one-sided constraint works, while inactive one-sided constraint does not work. It is obvious that a one-sided constraint may be active for some special location and value of a load and for other location and value of a load the same one-sided constraint becomes inactive.

Assume that one-sided constraints are absent among absolutely required constraints and is present only among the redundant constraints. It means that the structure is geometrically unchangeable [Kar10].

Some properties of structures with one-sided constraints:

1. A structure with one-sided (even ideal-elastic) constraints is a nonlinear structure.
2. As the load that acts on the structure changes, some constraints become inactive (excluded from work) while others, previously inactive, become active (included in work); therefore the principle of superposition is not applicable.
3. As a one-sided active constraint becomes inactive or as a one-sided inactive constraint becomes active, the reciprocal works and reciprocal displacements theorems can become invalid.

If a structure contains n one-sided constraints, then considering different combinations of the members as active and inactive members, it is possible to generate 2^n different systems. For example, if $n = 10$, then the number of possible structures is $2^n = 1,024$ [Rab54a]. The main problem of analysis of structures with one-sided constraints is to determine the constraints which are active (works) for a given load [Pro48], [Rab50].

8.2.2 Criteria of the Working System

Assume that a system has n one-sided constraints, and under the given load the one-sided redundant constraints $C_1, C_2, ..., C_m$ are active, while constraints $C_{m+1}, C_{m+2}, ..., C_n$ are inactive. If we assume that active constraints $C_1, C_2, ..., C_m$ are two-sided, we can determine internal forces (reactions) in these constraints by any method for analysis of redundant structures and after that determine displacements along the inactive constraints $C_{m+1}, C_{m+2}, ..., C_n$. Assume that internal forces and deflections of the one-sided constraints in the operating state are positive.

If forces in active (working) constraints $C_1, C_2, ..., C_m$ are positive, while displacements along the inactive constraints $C_{m+1}, C_{m+2}, ..., C_n$ are negative, then the working system is chosen correctly. It means that forces and deflections of the one-sided constraints correspond to the entire working system.

The first paper devoted to analysis of structures with one-sided constraints was published in 1852. The author, D.I. Jourawsky, considered a truss with crossed diagonals. He showed that wood diagonals can resist only compressive internal forces, while the ends of the extended diagonals being dislodged from their original positions, and cease to work in the entire structure as an element which can resist internal forces [Tim53], [Ber57]. Some fundamental theorems related to analysis of structures with one-sided constraints are presented in ref. [Rab50].

8.2.3 Analysis of Structures with One-Sided Constraints

Analysis of structures with one-sided constraints contains minimum of two steps.

In the first step, we assume that all constraints are two-sided. Then we analyze the structure and mark those constraints, which are not be able to resist the given load. For example, the hangers of the cable bridge are constraints with compressed internal forces.

In the second step, we change the design diagram of the structure. For this, we exclude the inactive constraints and perform new analysis of the structure, considering all remaining constraints as two-sided. After that we verify the criteria of the working system. If this criteria is satisfied, then the analysis is finished. Otherwise it is necessary to adopt a new working system and repeat the entire procedure.

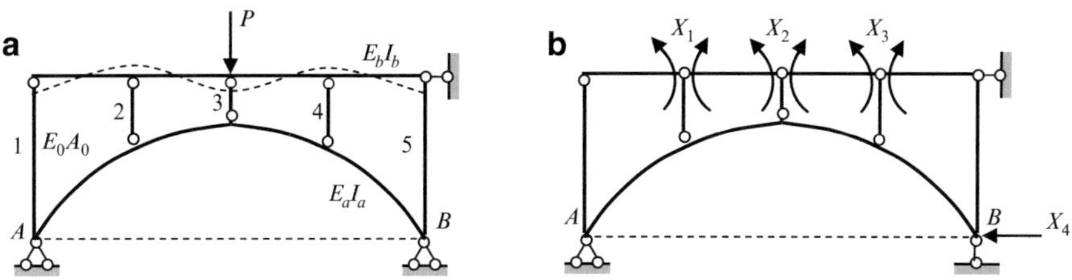

Fig. 8.14 Design diagram of arched structure and primary system

Herein lies the difficulty in analyzing systems with one-sided constraints: to this day, it is impossible to chose the correct system that satisfies the criteria of the working system without first performing the analysis described in Step 1.

The procedure for construction of influence lines for a system with one-sided constraints becomes especially difficult.

Let us consider the arched structure which consists of the arch itself, a beam, and the vertical poles. Each pole and beam are connected by means of simple hinges (Fig. 8.14a). Therefore, the beam may be treated as a continuous beam on elastic supports, and the structure in whole becomes statically indeterminate. Assume that the poles can resist tensioned and compressed forces. Analysis of this structure may be performed using the Force method.

The equation of the neutral line of the arch is $y = y(x)$. Let l, f, and h_k be the span of the arch, rise, and length of the kth pole, respectively. The flexural stiffness of the arch is E_aI_a, of beam it is E_bI_b and axial stiffness of poles is E_0A_0. One version of the primary system is shown in Fig. 8.14b.

Next we need to construct the internal force diagrams for primary system due to the unit primary unknowns, as well as the diagram due to the external load. Unit and loaded displacements are calculated using graph multiplication method. In doing so we take into account the bending moments in the beam and arch as well as the axial forces in the poles. For multiplication of diagrams along the arch, the length of the arch is approximated by a set of the chords with length $s = \sqrt{(\Delta x)^2 + (\Delta y)^2}$. Canonical equations are

$$\delta_{11}X_1 + \cdots + \delta_{14}X_4 + \Delta_{1P} = 0$$

$$. \quad . \quad . \quad . \quad . \quad .$$

$$\delta_{41}X_1 + \cdots + \delta_{44}X_4 + \Delta_{4P} = 0.$$

Solution of these equations are X_i. Bending moment diagram for beam is constructed immediately since the primary unknowns are bending moments at the elastic supports. Bending moments for arch may be calculated by the formula $M = \sum \bar{M}_i X_i + M_P^0$. Axial forces in the poles may be constructed by the formula

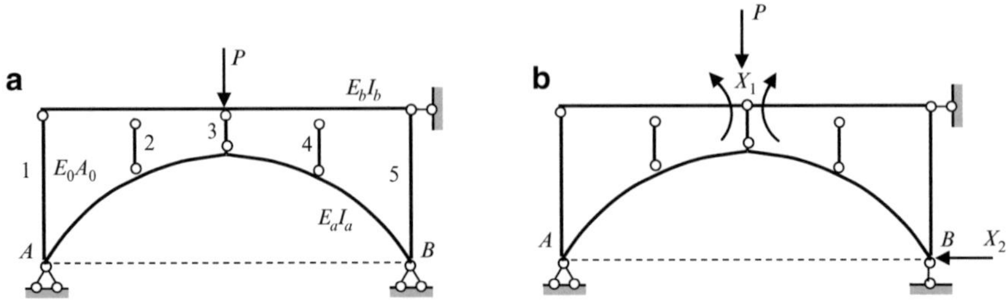

Fig. 8.15 Design diagram of arched structure without inactive constraints 2 and 4 and the primary system

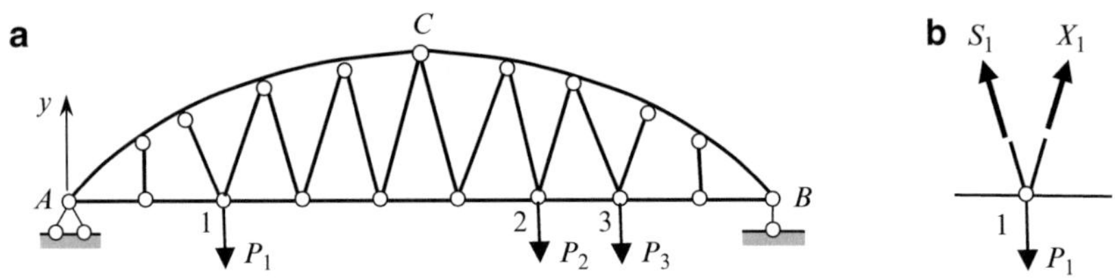

Fig. 8.16 Design diagram of arched structure with inclined hangers

$N = \sum \bar{N}_i X_i + N_P^0$, where $\bar{N}_i$ is axial forces in the ith unit state. This classic procedure describes analysis of a structure with two-sided constraints.

Let for a given location of the force P, an elastic line of the beam is shown by a dotted line (Fig. 8.14a). If we assume that the poles can resist only compressed forces then the design diagram in Fig. 8.14a does not reflect a real behavior of the structure. Therefore, the extended poles should be omitted from the following analysis. Let the external force P be located over pole 3. In this case, it is obvious that poles 2 and 4 has the positive internal forces. Design diagram which corresponds to omitted extended constraints is shown in Fig. 8.15a. The primary system is shown in Fig. 8.15b.

Now we need to perform analysis of structure in Fig. 8.15b, determine internal forces in members 1, 3, and 5 and displacements along omitted constraints 2 and 4, and finally, apply the criteria of the working system.

Another example of the arched structure with one-sided constraints is shown in Fig. 8.16. The hangers are connected with the arch by simple hinges (except crown C) and by multiple hinges with the horizontal part AB. Inclined hangers distribute the forces in the arch better than the vertical hangers. The degree of redundancy of such a structure equals to the number of joints on the horizontal part AB, which contains two hangers (in our case, degree of redundancy equals to six).

It is easy to show that internal forces in the hangers which are concurrent in the unloaded joint of the part AB are equal to zero. Indeed, equilibrium equation $\sum Y = 0$ for each unloaded joint shows that one of the two hangers is extended while other is compressed. A compressed hanger may be discarded, considering it as a one-sided constraint. Then internal force in the second hanger turns out to equals to zero, because the joint becomes a three-member unloaded joint.

Now let us consider the *loaded* joints of contour *AB* with two *paired* hangers; such joints in Fig. 8.16a are denoted as 1, 2, 3. One force for each such joint is considered as the primary unknown X_1, X_2, ... (Fig. 8.16b). Assuming that all hungers are two-sided, form the canonical equations of the Force method and determine all primary unknowns X_1, X_2, ... Given this, if all primary unknowns X_i and all internal forces S_i turn out to be positive, then this means that the assumed scheme is operational. If one or several internal forces turns out to be negative then the corresponding one-sided constrains should be omitted from consideration. As result, we obtain a new design diagram with a new number of primary unknowns; eventually a set of calculations allow us to find the working system for the entire design diagram.

Chapter 9
Special Stability and Dynamic Topics

This chapter contains two topics which extend the stability and vibration analysis of the arches. They are the dynamical stability of arch and dynamic action of the moving load on the arched structure.

The feature of these problems lies in the fact that dynamic loading of the arch in both cases leads to an effect which can be treated as a loss of stability.

9.1 Dynamical Stability of Arches

Analysis of the elastic structures from the point of view of its dynamical stability is considered to be an important problem of the theory of structures. This analysis allows us to determine the parameters of a "structure-load" system which lead to the intense vibration of the structure; such vibration can lead to its collapse. The task of the engineer is to prevent such vibrations.

The problem of dynamical stability of an elastic structure is as follows: structure is loaded in such a way that its stiffness changes. Therefore, such loading (and the problem as a whole) often is called parametric excitation [New89].

The first analysis of dynamical stability of simply supported column was performed by Beljaev (1924) [Smi47]. Later outstanding scientists Krylov and Bogoliubov (1935), Chelomey (1939), Smirnov (1947), and many others devoted theirs works to this problem. This problem has been investigated in great detail by Bolotin [Bol64]. Dynamical stability of arches was investigated by Morgaevsky [Mor61]. A detailed literature review on this issue is presented by Kazakevich [Kaz09].

I.A. Karnovsky, *Theory of Arched Structures: Strength, Stability, Vibration*,

DOI 10.1007/978-1-4614-0469-9_9,

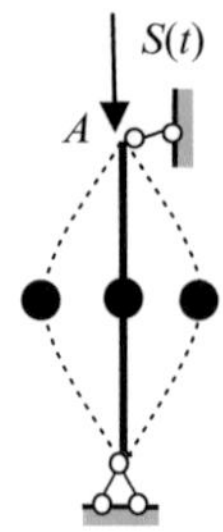

Fig. 9.1 Design diagram of the column subjected to axial harmonic force $S(t)$

9.1.1 Dynamical Stability of a Simply Supported Column

Let us consider a column which is subjected to the axial harmonic force $S(t)$; the form of transversal vibration is shown in Fig. 9.1. This vibration of a structure is called parametric vibration. The meaning of this term will become clear later. At the certain frequency of excitation, the force in the system causes significant increase of transversal vibrations. Such phenomena is called the dynamical loss of stability.

The scheme in Fig. 9.1 allows us to show a fundamental difference of the parametric vibration from pure forced vibration: if the vibrations of hinge A in the axial direction occur with frequency θ of the disturbing force $S(t)$, then the vibrations of mass m in the transversal direction occur with frequency $\varphi_0 = \theta/2$ [Smi47].

From Bernoulli–Euler theory, the transversal vibration of the beam with distributed mass is described by the equation $EI[(\partial^4 y)/\partial x^4] + \mu[(\partial^2 y)/\partial t^2] = 0$, where μ is the mass per unit length of the beam. If we take into account the axial force, the behavior of the beam is described by the equation

$$EI\frac{\partial^4 y}{\partial x^4} + S(t)\frac{\partial^2 y}{\partial x^2} + \mu\frac{\partial^2 y}{\partial t^2} = 0. \tag{9.1}$$

If $S = S(t) = S_0 \cos\theta t$ [Now61], [New89], then (9.1) becomes

$$EI\frac{\partial^4 y}{\partial x^4} + S_0 \cos\theta t\frac{\partial^2 y}{\partial x^2} + \mu\frac{\partial^2 y}{\partial t^2} = 0. \tag{9.2}$$

This equation has a fundamental difference from the equation of forced vibrations. Indeed, the disturbing force appears in the left-hand side of the equation as the coefficient of the second derivative term. In this manner, the disturbing force influences the *parameters* of the system, and therefore such excitation is called parametric excitation.

The transverse displacement of a beam depends on the axial coordinate x and time t, i.e., $y = y(x, t)$. A solution of differential equation (9.2) may be presented in the form

$$y(x,t) = T(t)\sin\frac{\pi x}{l}, \tag{9.3}$$

where $T(t)$ is some time-dependent function and $\sin(\pi x/l)$ is a function which satisfies the boundary condition of a simply supported beam.

Substitution of this expression into (9.2) leads to the equation [Fil70]

$$\mu\frac{\mathrm{d}^2T}{\mathrm{d}t^2} + EI\frac{\pi^4}{l^4}\left[1 - \frac{l^2}{EI\pi^2}S(t)\right]T(t) = 0$$

or

$$\frac{\mathrm{d}^2T}{\mathrm{d}t^2} + \omega^2\left[1 - \frac{S_0}{P_{\mathrm{cr}}}\cos\theta t\right]T(t) = 0, \tag{9.4}$$

where $\omega = (\pi^2/l^2)\sqrt{EI/\rho A}$ and $P_{\mathrm{cr}} = (\pi^2 EI)/l^2$ are the frequency of transversal vibration (assuming the axial force vanishes) and the Euler critical force for simply supported beam, respectively. Equation (9.4) is an ordinary differential equation of the second order. This equation describes the response of a linear dynamical structure with time-varying stiffness. Indeed, this equation contains time-dependent stiffness which depends on the parameters of structure itself (frequency ω of transversal vibration and the Euler critical force) as well as the parameters of disturbing force (amplitude value S_0 and frequency θ of axial force).

Equation (9.4) takes its name from the first paper by Mathieu [Mat68]. Solution of this equation have been studied extensively by Haupt [Hau19], Goldshtein [Gol27], Strutt [Str32], McLachlan [Mcl47], and Stoker [Sto50].

9.1.2 Ince–Strutt Diagram

The main feature of Mathieu equation lies in the fact that depending on the relationship between the parameters of the system and the disturbing force, the solutions of the equation can be bounded or grow infinitely in time.

Let $z = \theta t$. Substituting this change of variable into (9.4), we get

$$\frac{\mathrm{d}^2T}{\mathrm{d}z^2} + \frac{\omega^2}{\theta^2}\left[1 - \frac{S_0}{P_E}\cos z\right]T = 0$$

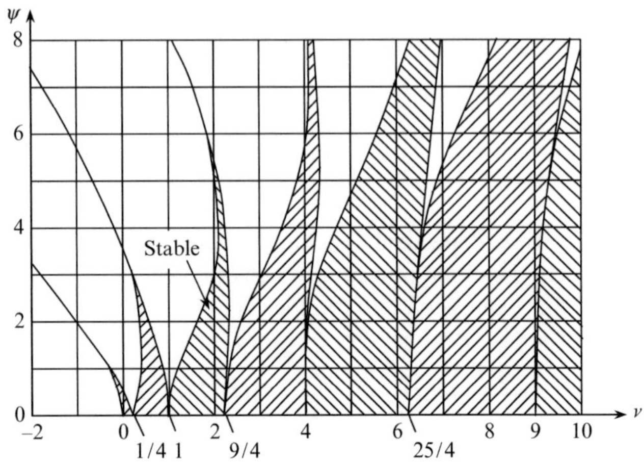

Fig. 9.2 Ince–Strutt diagram

or

$$\frac{d^2T}{dz^2} + (\nu - \psi \cos z)T = 0, \quad \nu = \frac{\omega^2}{\theta^2}, \quad \psi = \frac{\omega^2}{\theta^2}\frac{S_0}{P_E}. \tag{9.5}$$

Solutions of this equation are a periodic functions with period π or 2π. These solutions are presented in $\nu - \psi$ Ince–Strutt diagram [Inc25-27], [Str32] by solid lines (Fig. 9.2). These lines divide the $\nu - \psi$ plane into regions. The diagram in Fig. 9.2 presents the stable and unstable regions of (9.5) as a function of parameters ν and ψ; this diagram is symmetric about the horizontal axis. Equation of solid lines may be presented in terms of ν and ψ in algebraic form [Bol78]; this book also contains the useful technical application of parametric excitations.

Stability region of solution to Mathieu equation is indicated by cross-hatched area. Boundaries are unstable. If parameters of the system ν and ψ turn out to be in the unshaded region, the system exhibits vibrations which grow in time (parametric resonance). If damping is present in the system, the stability region will be larger than if no damping was present in the system.

9.1.3 Dynamical Stability of Circular Arch

Let us consider a circular two-hinged uniform arch which is subjected to a radial load $q(t) = q_0 \cos \theta t$ (Fig. 9.3). The problem of dynamical stability of the arch we will consider in a simplest formulation.

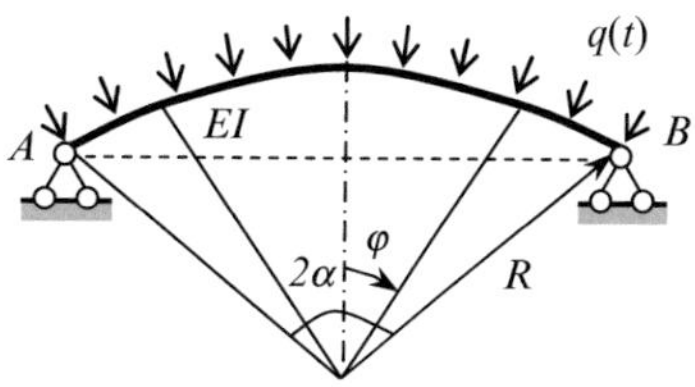

Fig. 9.3 Design diagram of the arch

Governing equation of vibration with respect to tangential displacement u of the points on an axis of the arch (equation of dynamic stability) becomes [Chu52], [Bir68], [Bol64]

$$L = \frac{\partial^6 u}{\partial \varphi^6} + 2\frac{\partial^4 u}{\partial \varphi^4} + \frac{\partial^2 u}{\partial \varphi^2} + \frac{qR^3}{EI}\left(\frac{\partial^4 u}{\partial \varphi^4} + \frac{\partial^2 u}{\partial \varphi^2}\right) + \frac{mR^4}{EI}\frac{\partial^2}{\partial t^2}\left(\frac{\partial^2 u}{\partial \varphi^2} - u\right) = 0, \quad (9.6)$$

where

u is a tangential displacement of a point on the axis of the arch and m is mass per unit length of the arch.

φ $(-\alpha \le \varphi \le \alpha)$ is the angle that determines location of the point on the nondeformable axis of the arch. The boundary conditions for hinged ends are $u = 0, \quad \partial u/\partial \varphi = 0, \quad \partial^3 u/\partial \varphi^3 = 0.$

Two-Hinged Arch

Assume the approximate solution for two-hinged arch in the form

$$u(t) = -f(t)\frac{\alpha}{\pi}\left(1 + \cos\frac{\pi\varphi}{\alpha}\right),$$

where $f(t)$ depends on time. This expression satisfies all the boundary conditions. Apply Bubnov–Galerkin procedure to (9.6)

$$\int_{-\alpha}^{\alpha} L\left(1 + \cos\frac{\pi\varphi}{\alpha}\right)\mathrm{d}\varphi = 0, \quad (9.7)$$

where L is an operator in the left part of expression (9.6). As a result, we obtain

$$(\xi^2 + 1)\frac{mR^4}{EI}\ddot{f}(t) + \xi^2(\xi^2 - 1)^2 f(t) - \frac{qR^3}{EI}(\xi^4 - \xi^2)f(t) = 0.$$

This equation may be rewritten as

$$\ddot{f}(t) + \omega^2\left(1 - \frac{q}{mR\omega^2}\gamma\right)f(t) = 0 \quad \text{or} \quad \ddot{f}(t) + \omega^2\left(1 - \frac{q}{q_{\mathrm{cr}}}\right)f(t) = 0, \quad (9.7a)$$

where

$$\omega = \frac{\xi(\xi^2 - 1)}{\sqrt{\xi^2 + 1}} \frac{1}{R^2} \sqrt{\frac{EI}{m}}, \quad \gamma = \frac{\xi^4 - \xi^2}{\xi^2 + 1},$$

$$q_{\mathrm{cr}} = \frac{mR\omega^2}{\gamma} = \frac{\left(\xi^3 - \xi\right)^2}{\xi^4 - \xi^2} \frac{EI}{R^3}, \quad \xi = \frac{\pi}{\alpha}.$$

Formulas for critical radial load and frequency of free vibration have been obtained earlier; see (4.11) and (6.16a), respectively.

Since $q = q_0 \cos \theta t$, then (9.7) becomes

$$\ddot{f}(t) + \omega^2 \left(1 - \frac{q_0}{q_{\mathrm{cr}}} \cos \theta t\right) f(t) = 0.$$

Dimensionless parameters of Ince–Strutt diagram are $v = \omega^2/\theta^2$ and $\psi = (\omega^2/\theta^2)(q_0/q_{\mathrm{cr}})$.

If disturbing harmonic load has a constant term q_0, i.e., $q = q_0 + q_1 \cos \theta t$, then we get the following equation of parametric vibration

$$\frac{\mathrm{d}^2 f}{\mathrm{d}t^2} + \omega^2 \left(1 - \frac{q_0}{mR\omega^2}\gamma - \frac{q_1}{mR\omega^2}\gamma \cos \theta t\right) f(t) = 0$$

or

$$\frac{\mathrm{d}^2 f}{\mathrm{d}t^2} + \omega^2 \left(1 - \frac{q_0}{q_{\mathrm{cr}}} - \frac{q_1}{q_{\mathrm{cr}}} \cos \theta t\right) f(t) = 0.$$

This equation may also be easily presented in standard forms (9.4) and (9.5).

Arch with Fixed Ends

For hingeless arch $u(\varphi, t) = \sum_j f(t)\, u_j(\varphi)$,

$$u_j(\varphi) = 1 - a_j \cos \theta + b_j \cos k_j \theta,$$

$$a_j = \frac{k_j^2}{k_j^2 - 1} \sec \alpha; \quad b_j = \frac{1}{k_j^2 - 1} \sec k_j \alpha,$$

where k_j are roots of the transcendental equation $k_j \tan \alpha = \tan k_j \alpha$.

Detailed solution of this problem, using Bubnov–Galerkin procedure, is presented in [Guz61].

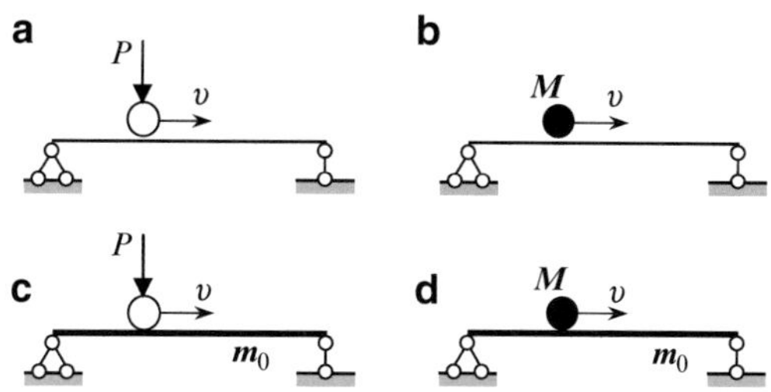

Fig. 9.4 Design diagrams for beam subjected to moving load

9.2 Arched Structure Subjected to Moving Loads

This section is devoted to the analysis of structures subjected to moving loads. So far the moving load was associated only with the *position* of a load and any dynamical effects have not been taken into account; this fact is a fundamental assumption of influence line concepts.

In Sect. 9.2.1, we consider different *dynamic* states of problems regarding moving loads and discuss the important case of loading of the beam (quasi-static state). In Sect. 9.2.2, following the Morgaevsky procedure, we consider quasi-static loading of the arch.

9.2.1 Beam with a Traveling Load

Figure 9.4 presents the four possible design diagrams for dynamical analysis of the beam subjected to a moving concentrated load. Parameters that are taken into account (mass of moving load M and distributed mass of a beam m_0) are shown in bold and thick solid lines. The scheme (a) does not take into account neither the mass of the beam nor the mass of the load; therefore, the inertial forces are absent. This case corresponds to static loading, and speed υ only means that force P may be located at any point; this case of loading is assumed for the construction of influence lines. Case (b) takes into account only the mass of the moving load. Case (c) corresponds to the motion of a massless force along the beam with distributed mass per unit length. Case (d) takes into account the mass of the load and mass of the beam. The difficulty in solving these dynamical problems increases from case (b) to case (d).

Chester bridge (England) was designed by Robert Stephenson. The bridge collapsed when a train was passing over it (May 24, 1847). This catastrophe led scientists to focus on the problem of dynamic stresses arising in beams subjected to moving loads. Solution of the classical problem in Fig. 9.4a was independently solved by Winkler and Mohr in 1868. The problem in Fig. 9.4b was formulated by Willis in 1849 [Tim53]; Stokes solved this problem (1849) and derived the expression for dynamical coefficient $\mu = 1 + Ml\upsilon^2/3\,EI$. The problem in Fig. 9.4c

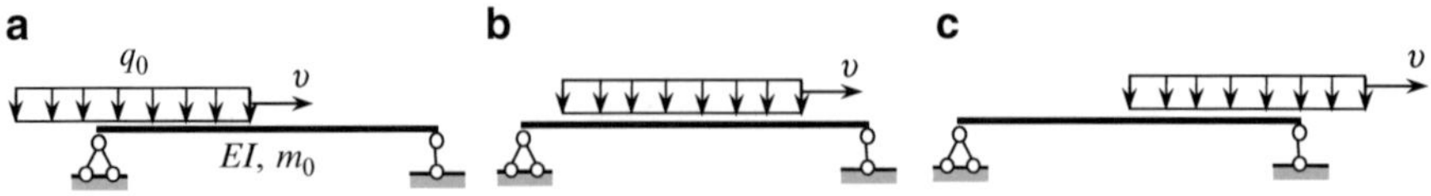

Fig. 9.5 Nonstationary loading of a beam by a moving distributed load

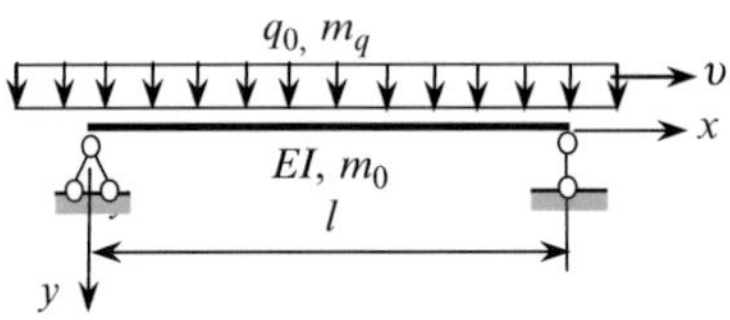

Fig. 9.6 Stationary loading of a beam by a moving load

was thoroughly examined by Kryloff [Kru05]. Problem in Fig. 9.4d was explored by Schallenkamp [Sch37], Inglis [Ing34], and Bolotin [Bol64].

It is possible to have situations where a *distributed* load is moving. A beam is subjected to a uniformly distributed load, with intensity q_0, which is moving along the beam with a constant speed v; the mass of this load per unit length is m_q. Assume that the load covers only *part* of the beam, so the following cases of loading are possible: load oncoming onto the beam (Fig. 9.5a), moving load within a beam (Fig. 9.5b), or outgoing load from the beam (Fig. 9.5c). In these cases, the vibration of the beam takes on a nonstationary character; these vibrations are described by partial differential equations with variable coefficients.

Let us consider the quasi-static loading of the beam by a traveling inertial load. Such loading arises if a beam is *fully* covered by the traveling load (Fig. 9.6) [Kar70].

The vibrations of the above beam are governed by the differential equation

$$EI\frac{\partial^4 y}{\partial x^4} = q. \tag{9.8}$$

Intensity of the external load q is the sum of the inertial forces of the elements of the beam and inertial forces of the moving load

$$q = -m_0\frac{\partial^2 y}{\partial t^2} + \left[q_0 - m_q\frac{\mathrm{d}^2 y}{\mathrm{d}t^2}\right]. \tag{9.9}$$

Determine the inertial forces of the moving load. For this we need to calculate the total derivative of the displacement with respect to time. It should be kept in mind that an element of the load executes a complex motion. Indeed, we have two frames of reference. One of them is a moving frame (vibrating beam) and another we assumed is fixed (nonvibrating beam). The element of load performs the relative motion with respect to the moving coordinate system; the transport motion is performed by the moving frame together with the load with respect to fixed system, and absolute motion with respect to fixed coordinate system. Coordinate x of the element of the load depends on time, i.e., $x = \upsilon t$. Therefore, projection of velocity of the moving load onto the y-axis equals

$$\frac{\mathrm{d}y}{\mathrm{d}t} = \frac{\partial y}{\partial t} + \frac{\partial y}{\partial x}\frac{\partial x}{\partial t} = \frac{\partial y}{\partial t} + \frac{\partial y}{\partial x}\upsilon.$$

Similarly, the vertical acceleration of the element of the moving load is

$$\frac{\mathrm{d}^2 y}{\mathrm{d}t^2} = \frac{\partial}{\partial t}\left(\frac{\partial y}{\partial t}\right) + \frac{\partial}{\partial t}\left(\frac{\partial y}{\partial x}\upsilon\right).$$

After rearrangement of each term we get

$$\frac{\mathrm{d}}{\mathrm{d}t}\left(\frac{\partial y}{\partial t}\right) = \frac{\partial^2 y}{\partial t^2} + \frac{\partial^2 y}{\partial t \partial x}\frac{\partial x}{\partial t} = \frac{\partial^2 y}{\partial t^2} + \frac{\partial^2 y}{\partial t \partial x}\upsilon,$$

$$\frac{\mathrm{d}}{\mathrm{d}t}\left(\frac{\partial y}{\partial x}\upsilon\right) = \frac{\partial^2 y}{\partial t \partial x}\upsilon + \frac{\partial^2 y}{\partial x^2}\frac{\partial x}{\partial t}\upsilon = \frac{\partial^2 y}{\partial x \partial t}\upsilon + \frac{\partial^2 y}{\partial x^2}\upsilon^2.$$

Thus, the vertical acceleration of the element of the moving load becomes

$$\frac{\mathrm{d}^2 y}{\mathrm{d}t^2} = \frac{\partial^2 y}{\partial t^2} + 2\upsilon\frac{\partial^2 y}{\partial t \partial x} + \upsilon^2\frac{\partial^2 y}{\partial x^2}. \tag{9.10}$$

Here, the first term represents the transport acceleration, the second term represents the supplementary or Coriolis acceleration, and third term is a normal (relative) acceleration. It can be seen that intensity of the load q depends on the speed of the load. If we assume that Coriolis acceleration may be neglected [Kar70], then substitution of (9.9) and (9.10) into (9.8) leads to the differential equation

$$EI\frac{\partial^4 y}{\partial x^4} + m_0\frac{\partial^2 y}{\partial t^2} + m_q\left(\frac{\partial^2 y}{\partial t^2} + \upsilon^2\frac{\partial^2 y}{\partial x^2}\right) + q_0 = 0. \tag{9.11}$$

Assume that the solution of this equation may be presented in the form

$$y(x,t) = f(t)\sin\frac{\pi x}{l},$$

where $f(t)$ is an unknown time-depending function. This expression satisfies the following boundary conditions $y(0) = y(l) = 0$ and $y''(0) = y''(l) = 0$. Corresponding derivatives are

$$\frac{\partial^2 y}{\partial x^2} = -f(t)\left(\frac{\pi}{l}\right)^2 \sin\left(\frac{\pi x}{l}\right), \quad \frac{\partial^4 y}{\partial x^4} = f(t)\left(\frac{\pi}{l}\right)^4 \sin\left(\frac{\pi x}{l}\right), \quad \frac{\partial^2 y}{\partial t^2} = \ddot{f}(t)\sin\left(\frac{\pi x}{l}\right).$$

Bubnov–Galerkin procedure

$$\int_0^l \left[EI\frac{\partial^4 y}{\partial x^4} + m_0\frac{\partial^2 y}{\partial t^2} + m_q\left(\frac{\partial^2 y}{\partial t^2} + \upsilon^2\frac{\partial^2 y}{\partial x^2}\right) + q_0\right]\sin\frac{\pi x}{l}\,\mathrm{d}x = 0$$

leads to the following ordinary differential equation with respect to the function $f(t)$

$$\ddot{f}(t)\,(m_0 + m_q) + f(t)\left[EI\left(\frac{\pi}{l}\right)^2 - m_q\upsilon^2\right]\left(\frac{\pi}{l}\right)^2 = q_0\frac{2l}{\pi}.$$

The frequency of vibration squared is

$$\omega^2 = \frac{1}{(m_0 + m_q)}\left[EI\left(\frac{\pi}{l}\right)^2 - m_q\upsilon^2\right]\left(\frac{\pi}{l}\right)^2.$$

If $EI(\pi/l)^2 - m_q\upsilon^2 = 0$, then the beam losses its stability. The critical speed is

$$\upsilon_{\mathrm{cr}} = \frac{\pi}{l}\sqrt{\frac{EI}{m_q}}.$$

Notice that the loss of stability occurs even if the beam is not subjected to a compressed load.

Formulation of the problem about dynamical influence of a moving load can be complicated by taking into account various factors. Among them the axial compressed force [New89], elastic foundation [Fil70], variable speed of the moving load, one-sided constraints, etc.

9.2.2 *Arch Subjected to Inertial Traveling Load: Morgaevsky Solution*

Design diagram of the arch with an over-arch construction is shown in Fig. 9.7. Loaded contour may be within the rise of an arch, on the same level of the crown or above the crown (as shown in Fig. 9.7). The uniform circular arch of radius R has a central angle of 2α. The flexural stiffness of the arch and mass per unit length of the arch are EI, q_0 and m_0, respectively. The over-arch structure is loaded by the uniformly distributed, moving load. The intensity of the moving load, its mass per unit length, and speed are q_1, m_q, and υ, respectively. We consider a stationary vibration of the arch, which occurs if the arch is *totally* covered by the moving load.

A feature of this problem: the mass of the moving load is taken into account. It means that a moving load is *inertial*.

The total load on the arch consists of two parts. They are the constant static load $q^{\mathrm{st}}(\varphi)$ (the weight of the arch itself, weight of the over-arch structure, and *weight* of the moving load) and disturbing load $q^t(\varphi, t)$, i.e., the total load is $q = q^{\mathrm{st}}(\varphi) + q^t(\varphi, t)$.

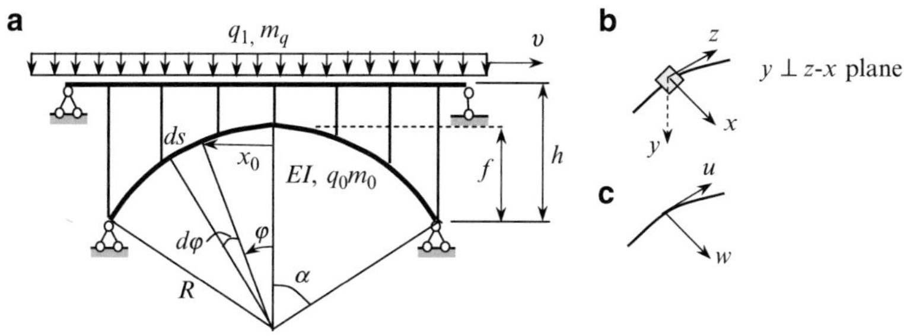

Fig. 9.7 (**a**) Stationary uploading of the arch by the moving load; (**b**) moving axis; and (**c**) displacement components

Assumptions

1. The moving load *totally* covers the entire span of the arch.
2. Only linear (geometrical and physical) vibrations of arch are considered.
3. The bend of the over-arch-structure is neglected, therefore, displacements of the element of the moving load are equal to the displacements of u and w of the element ds of an arc of the arch.

Notations

The moving axis x and z are directed along a radius of the arch and along a tangent to the axis of the arch, respectively; the moving axis y is perpendicular to the x–z plane of the arch (Fig. 9.7b). The tangential and radial displacements of any point of the arch are denoted by u and w, respectively (Fig. 9.7c).

M_y^{st}, Q_x^{st}, N_z^{st} are bending moment in the plane of the arch, shear, and axial forces due to constant static load $q^{\mathrm{st}}(\varphi)$.

M_y^t, Q_x^t, N_z^t are the same internal forces due to the load $q^t(\varphi, t)$.

m_y is the intensity of the distributed moment.

Morgaevsky equation. Stationary vibration of a circular arch subjected to moving vertical inertial load is described by the Morgaevsky equation [Mor59], [Mor60].

$$L(\varphi,t) = \frac{\partial^3 M_y^t}{\partial \varphi^3} + \frac{\partial M_y^t}{\partial \varphi} - \frac{R^2}{EI}\frac{\partial}{\partial \varphi}\left(N_z^{\mathrm{st}} M_y^t\right) - \frac{R^2}{EI}\frac{\partial}{\partial \varphi}\left(Q_x^{\mathrm{st}} M_y^t\right) + m_0 R^2 \frac{\partial^2}{\partial t^2}\left(\frac{\partial w}{\partial \varphi} - u\right)$$
$$-R^2\left(\frac{\partial q_x^t}{\partial \varphi} - q_z^t\right) + R\left(\frac{\partial^2 m_y}{\partial \varphi^2} + m_y\right) = 0.$$

This equation corresponds to the tracking load (Fig. 4.1a).

Internal forces are

$$N_z^{\mathrm{st}} = -q^{\mathrm{st}}R(\eta\cos\varphi + \sin^2\varphi),$$
$$Q_x^{\mathrm{st}} = q^{\mathrm{st}}R(\eta\sin\varphi - 0.5\sin 2\varphi),$$

where parameter $\eta = H/(q^{\mathrm{st}}R)$ is a dimensionless trust of the arch.

Assume that disturbing forces q_x^t and q_z^t are equal to the inertial forces of the moving load on the displacements u and w; these distributed forces in projection onto the moving axis x and z, taking into account the expressions for total derivatives, are

$$q_x^t = -m_q\left[\frac{\partial^2}{\partial t^2}\left(\frac{\partial u}{\partial\varphi}\right)\cos\varphi + 2\frac{\upsilon}{R}\frac{\partial}{\partial t}\left(\frac{\partial^2 u}{\partial\varphi^2} + u\right) + \frac{\upsilon^2}{R^2}\left(\frac{\partial^2 u}{\partial\varphi^2} + \frac{\partial u}{\partial\varphi}\right)\sec\varphi\right.$$
$$\left. + \tan\varphi\sec\varphi\left(\frac{\partial^2 u}{\partial\varphi^2} + u\right)\right],\quad q_z^t = -m_q\left[\frac{\partial^2 u}{\partial t^2}\cos\varphi - \frac{\upsilon^2}{R^2}\left(\frac{\partial^2 u}{\partial\varphi^2} + u\right)\sec\varphi\right].$$

These forces are applied on the level of the loaded contour. They are transferred into the center of gravity of the cross section of the arch. If inertia of rotation is neglected, then for distributed moment we get the following expression

$$m_y = R\left[q_z^t\cos\varphi(1-\cos\varphi) - q_x^t\sin\varphi(1-\cos\varphi)\right].$$

In the case of uniform motion, the velocity and acceleration of the element of the moving load in direction u become

$$\frac{\mathrm{d}u}{\mathrm{d}t} = \frac{\partial u}{\partial t} + \upsilon\frac{\partial u}{\partial x_0},\quad \frac{\mathrm{d}^2 u}{\mathrm{d}t^2} = \frac{\partial^2 u}{\partial t^2} + 2\upsilon\frac{\partial^2 u}{\partial t\,\partial x_0} + \upsilon^2\frac{\partial^2 u}{\partial x_0^2}.$$

Similarly, we can determine the first and second total derivatives of displacement w. As in the case of the beam, these formulas present the total velocity and acceleration of a point (absolute velocity and acceleration) in the two systems of coordinates, one of them is moving and second is conventionally stationary.

Tangential displacement which satisfy the boundary condition for antisymmetrical vibration of the arch is

$$u(x,t) = f(t)\frac{\alpha}{\pi}\left(1 + \cos\frac{\pi}{\alpha}\varphi\right).$$

Bubnov–Galerkin procedure is

$$\int_{-\alpha}^{+\alpha} L(\varphi,t)\,(1+\cos n\varphi)\,\mathrm{d}\varphi = 0.$$

Table 9.1 Parameters μ_1, K, and C_1 for arch with central angle 2α

2α	f/l	μ_1	K	C_1
45°	0.094	0.9781	63.10	61.161
90°	0.207	0.9242	16.22	13.930
120°	0.289	0.8770	8.293	7.724

If location of a loaded contour satisfies the condition $f/3 \le h \le 2f/3$, then the distributed moments may be neglected, so $R\left[\left(\left(\partial^2 m_y\right)/\partial\varphi^2\right) + m_y\right] = 0$. In this case, Bubnov–Galerkin procedure leads to the following ordinary differential equation of the second order, with respect to the function $f(t)$

$$\left(1+\frac{m_q}{m_0}\eta_1\right)\ddot{f}(t)+\omega_0^2\left(1-\frac{K_q}{K}-\frac{\upsilon^2}{R^2}\frac{m_q}{m_0}\frac{1}{\omega_0^2}C_1\right)f(t)=0,$$

$$\omega_0^2=\frac{n^2\left(n^2-1\right)^2}{\left(n^2+3\right)}\frac{EI}{R^4 m_0},\qquad n=\frac{\pi}{\alpha},\qquad K_q=\frac{(q_0+q_1)R^3}{EI}.$$

Coefficient K presents the stability coefficient for a uniform circular two-hinged arch, subjected to a uniform vertical load within the entire span of the arch, $q_{\mathrm{cr}} = KEI/R^3$. This coefficient, as well as parameter C_1, is presented in Table 9.1.

Expression for ω_0 coincides with Bolotin's formula (6.33) for first frequency of the antisymmetric vibration of the uniform arch without external loading.

The frequency of antisymmetric vibration of the arch, taking into account the mass and speed of the moving load, is

$$\omega_\upsilon=\frac{\omega_0}{\sqrt{1+(m_q/m_0)\mu_1}}\sqrt{1-\frac{K_q}{K}-\frac{\upsilon^2}{R^2}\frac{m_q}{m_0}\frac{1}{\omega_0^2}C_1}.$$

With the additional fixed ($\upsilon = 0$) distributed load q_1 at the loaded contour, the frequency of antisymmetric vibration of the arch is

$$\omega_{\upsilon=0}=\frac{\omega_0}{\sqrt{1+(m_q/m_0)\mu_1}}\sqrt{1-\frac{K_q}{K}}.$$

Condition $\omega_\upsilon = 0$ leads to the formula for critical speed of the moving load

$$\upsilon_{\mathrm{cr}}=\frac{R\omega_0}{\sqrt{C_1}}\sqrt{\left(1-\frac{K_q}{K}\right)\frac{m_0}{m_q}}.$$

In this case, the in-plane vibration of the arch becomes unstable.

Parameters μ_1, K, and C_1 for arch with central angle 2α are presented in Table 9.1 [Mor59], [Mor60].

If a loaded contour is located on the level of the crown, then the distributed moments m_y cannot be neglected. Corresponding results are presented in [Mor60].

Dynamical action of moving loads on different structures, such as continuous beams, arches, combined arched structures, plates, and shells are presented in [Mor61], [Kol77], [Kar70], [Fil70], etc. Optimal parametric suppression of vibration of shells subjected to inertial moving loads is presented in [Kar89].

Chapter 10
Conclusion

To this day, when it comes down to analyzing arches for their strength, stability, and vibration, the theory of arches is well developed and mature. On the other hand, in some cases, real-world engineering requirements are unable to be satisfied with the current theory of arches. These requirements are related to arch-like structures becoming more and more complex, being subjected to different loading conditions, and generally being exploited to a greater extent, and as a result exhibiting a completely new pattern of behavior. We list several problems related to the theory of arches that require extra careful attention from engineers and researchers.

1. As opposed to static problems, reference data for problems related to stability of arches is scarcely available. Analytical solutions are only available for the classical arches and for arches with a very simplistic loading scheme. In the framework of classical analysis of arches that is presented in this book (Dinnik's, Smirnov's and displacement methods), the peculiarities of arches (asymmetry, skewness, complex ties, overarched structures, etc.) and their methods of loading suffer from some serious, sometimes intractable drawbacks. It is these scenarios where the finite element method becomes the only possible solution to tackle these problems. There is a vast amount of literature available on the finite element method, with some of the fundamental textbooks being Weaver and Johnson [Wea84], Bathe [Bat96], to name a few.
2. To this day, analysis of the arches subjected to combined in-plane and out-of-plane bending still awaits a solution. Such complex loading of arches leads to warping torsion with inevitable appearances of bi-moments, and as a result, additional stresses. It is very interesting to expand the warping torsion problems of arches to the cases of stability and vibrations. These problems become very important in the cases when external loads are only partial determined. The general theory of warping torsion of straight bars was originally developed by Vlasov [Vla40], [Vla59] and presented in detail by Feodosiev [Feo70], Urban [Urb55], Tatur [Tat66], to name a few.
3. To this day, there are very few problems related to the synthesis of arches. These problems can be formulated as optimization problems that must satisfy some

I.A. Karnovsky, *Theory of Arched Structures: Strength, Stability, Vibration*,
DOI 10.1007/978-1-4614-0469-9_10,

predefined boundary conditions. In the most simplistic cases, these can be formulated as purely static problems. However, in more complicated cases, one must incorporate the dynamical characteristics, stability, and reliability parameters into the model. One must pay special attention to the formulation of certain optimization criteria and restrictions. Problems including arches that must simultaneously satisfy several criteria are of great interest. The relevant fields of optimization theory are well developed and quite mature, but their applications to the theory of arches are scarcely present in the literature. Some serious works related to mathematical optimization of deformable objects can be found in Haug and Arora [Hau79]; Armand [Arm72]; Gill et al. [Gil81]; Rozvany [Roz76]; Troitskiy and Petukhov [Tro82]; Grinev and Filippov [Gri79], to name a few. Several simplistic static problems related to optimizing the shape of the arch may be found in Ju Gol'dshtein and Solomeshch [Gol80].

4. Structural failure can lead to significant monetary losses and even to the loss of life. The designer of a structure must take into account the expected lifespan of a structure and make the structure reliable enough so that it can withstand it's required loading tasks. Statistical analysis of reliability, stability, and vibrations of arches becomes a very important problem. Analyzing structures by statistical methods becomes especially important when ones must design structures that are subjected to a type of loading that caries a probabilistic component (wind gusts, snow, earthquakes, etc.). Considerable effort and progress has been made on this subject by Clough and Penzien [Clo75]; Rzhanitsun [Rzh82]; Bolotin [Bol84] to name a few.
5. It should be noted that among the most important problems of modern engineering is the problem of vibration protection, or more specifically, controlling vibrations of deformable structures. A successful solution to this problem will allow the engineer to lower the overall vibrations to a specified tolerance and thus ensuring normal functionality of the structure. The results of passive vibration protection can be significantly improved if one utilizes the theory of optimal control of vibrations. There exist many scenarios where one must use a probabilistic approach to solve the problem of vibration protection of arches. Some of the fundamental works utilizing this approach are Athans and Falb [Ath66]; Komkov [Kom72]; Pontryagin et. al. [Pon62]; Luzin and Kuznetsov [Luz51], to name a few. A lot of these problems are very complex in nature and require a combination of deterministic/probabilistic approaches, utilizing a wide range of mathematical tools.

By all means, this short list is not an exhaustive description of all real-life problems and approaches.

Appendix

Arches of Different Shapes. Equation of Neutral Line Tables A.1 and A.2

Table A.1 Equation of neutral line in terms of span l and arch rise f [Lee89]

Type of arch	Notation	Equation of neutral line
Parabolic	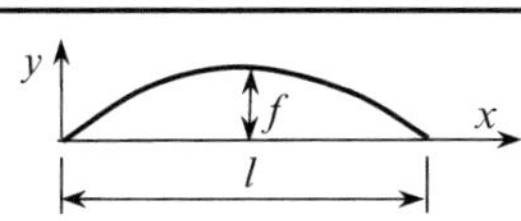	$y = \frac{4f}{l^2}(l - x)x$
Circle	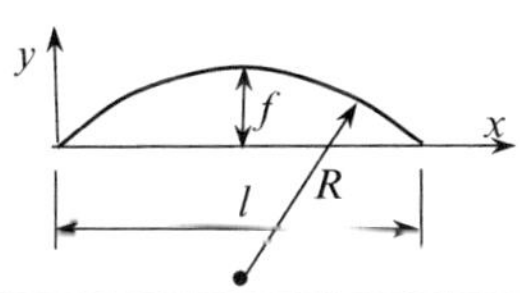	$y = \sqrt{R^2 - \left(\frac{l}{2} - x\right)^2} - R + f$

I.A. Karnovsky, *Theory of Arched Structures: Strength, Stability, Vibration*,
DOI 10.1007/978-1-4614-0469-9, © Springer Science+Business Media, LLC 2012

Table A.2 Geometric relationships of the arches with different equations of the neutral line. The radius of curvature is given by the functional relation $R(\alpha) = R_0 \cos^n \alpha$ where R_0 is the radius of curvature at $\alpha = 0$ and n is an integer specified for a typical line [Rom72]

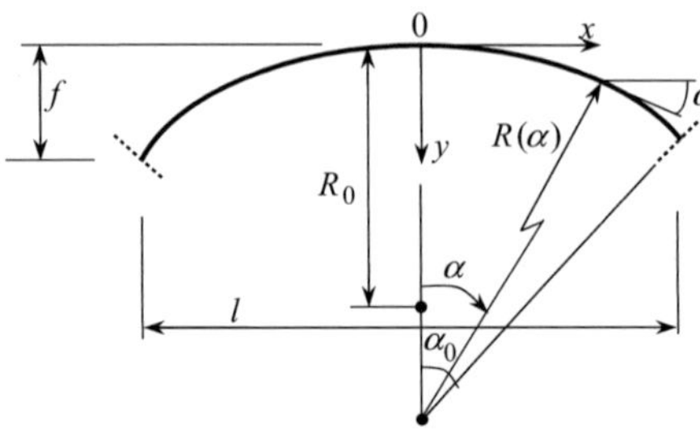

Curve	Parameter n	Equation of the neutral line	$\frac{l}{R_0}$	$\frac{f}{R_0}$
Parabola	−3	$y = \frac{x^2}{2R_0}$	$2 \tan \alpha_0$	$\frac{1}{2}\tan^2 \alpha_0$
Catenary	−2	$y = R_0(\cosh x/R_0 - 1)$	$2 \sinh^{-1}(\tan \alpha_0)$	$1/\cos \alpha_0 - 1$
Spiral	−1	$y = -R_0 \ln \cos x/R_0$	$2\alpha_0$	$-\ln \cos \alpha_0$
Circle	0	$y = R_0 - \sqrt{R_0^2 - x^2}$	$2 \sin \alpha_0$	$1 - \cos \alpha_0$
Cycloid	1	$x = \frac{R_0}{4} \arccos\left(\frac{4y}{R_0} - 1\right) \pm \sqrt{y\left(\frac{R_0}{2} - y\right)} - \frac{R_0 \pi}{4}$	$\alpha_0 + \frac{1}{2}\sin 2\alpha_0$	$1 - \cos 2\alpha_0$

For an elliptic arch, the equations of neutral line are in standard form and the radius of curvature are as follows:

$$\frac{x^2}{a^2} + \frac{y^2}{b^2} = 1, \qquad R = a^2 b^2 \left(\frac{1 + \tan^2 \alpha}{a^2 \tan^2 \alpha + b^2}\right)^{3/2}.$$

The instantaneous radius of curvature R, of an axis of any type of arch is expressed in terms of polar coordinates ρ and α as follows:

$$R = \frac{\left[\rho^2 + (\rho')^2\right]^{3/2}}{\rho^2 - \rho\rho'' + 2(\rho')^2}, \quad \rho' = \frac{d\rho}{d\alpha}, \quad \rho'' = \frac{d^2\rho}{d\alpha^2}.$$

Geometrical Parameters of Arches (Tables A.3–A.5)

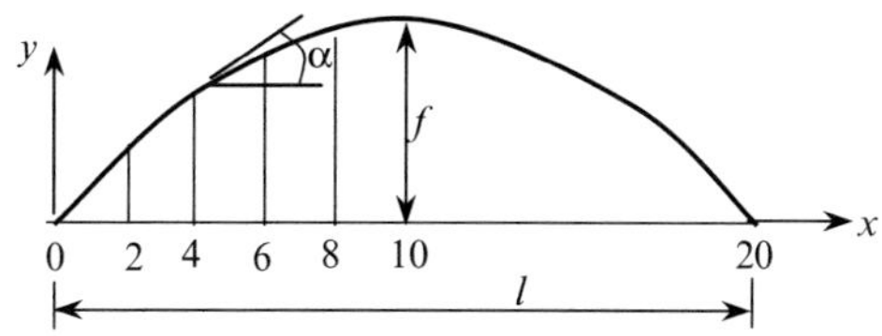

Table A.3 Geometrical parameters of a parabolic arch; $y = (4f/l^2)x(l - x)$, $\tan\alpha = (4f/l^2)\,(l - 2x)$

# of section[a]	x	y	$\tan\alpha$	Section x	x	y	$\tan\alpha$
0	0.00	0.00	4.00	0	0.000	0.00	4.00
1	0.05	0.19	3.60	1/8	0.125	0.438	3.00
2	0.10	0.36	3.20	1/6	0.167	0.556	2.667
3	0.15	0.51	2.80	1/4	0.250	0.750	2.00
4	0.20	0.64	2.40	3/8	0.375	0.938	1.00
5	0.25	0.75	2.00	1/2	0.500	1.000	0.00
6	0.30	0.84	1.60	5/8	0.625	0.938	−1.00
7	0.35	0.91	1.20	3/4	0.750	0.750	−2.00
8	0.40	0.96	0.80	5/6	0.833	0.556	−2.667
9	0.45	0.99	0.40	7/8	0.875	0438	−3.00
10	0.50	1.00	0.00	1.0	1.000	0.000	−4.00
Factor	l	l	$\frac{f}{l}$	l	l	f	$\frac{f}{l}$

[a] The total span is divided into 20 equal portions

Table A.4 Trigonometric functions of angle α for parabolic arch

$\frac{f}{l}$		$\frac{x}{l}$: 0.0	0.05	0.10	0.15	0.20	025	030	0.35	0.40	0.45	0.50
1/2	sin α	0.894	0.874	0.839	0.814	0.768	0.707	0.625	0.515	0.371	0.196	0.0
	cos α	0.447	0.485	0.545	0.581	0.640	0.707	0.781	0.857	0.928	0.981	1.00
1/3	sin α	0.800	0.768	0.730	0.682	0.625	0.555	0.470	0.371	0.258	0.132	0.0
	cos α	0.600	0.640	0.684	0.731	0.781	0.832	0.882	0.928	0.966	0.991	1.00
1/4	sin α	0.707	0.669	0.625	0.574	0.515	0.447	0.371	0.287	0.196	0.100	0.0
	cos α	0.707	0.743	0.781	0.819	0.857	0.894	0.928	0.958	0.981	0.995	1.00
1/5	sin α	0.625	0.584	0.539	0.489	0.433	0.371	0.305	0.233	0.158	0.080	0.0
	cos α	0.781	0.812	0.833	0.872	0.901	0.928	0.952	0.972	0.987	0.996	1.00
1/6	sin α	0.555	0.515	0.470	0.423	0.371	0.316	0.258	0.196	0.132	0.067	0.0
	cos α	0.832	0.857	0.882	0.906	0.928	0.949	0.966	0.981	0.991	0.997	1.00
1/7	sin α	0.496	0.457	0.416	0.371	0.324	0.275	0.223	0.168	0.113	0.057	0.0
	cos α	0.868	0.889	0.909	0.928	0.946	0.961	0.975	0.986	0.994	0.998	1.00
1/8	sin α	0.447	0.410	0.371	0.330	0.287	0.242	0.196	0.148	0.100	0.050	0.0
	cos α	0.894	0.912	0.928	0.944	0.958	0.970	0.981	0.989	0.995	0.999	1.00
1/10	sin α	0.371	0.339	0.305	0.270	0.233	0.196	0.158	0.119	0.080	0.040	0.0
	cos α	0.928	0.941	0.952	0.963	0.972	0.981	0.987	0.983	0.997	0.999	1.00
1/12	sin α	0.316	0.287	0.258	0.227	0.196	0.165	0.132	0.100	0.067	0.015	0.0
	cos α	0.949	0.958	0.966	0.974	0.981	0.986	0.991	0.995	0.998	0.999	1.00

Table A.5 Geometrical parameters of a circular arch

$\frac{f}{l}$	α_0		$\frac{x}{l}$ 0.0	0.10	0.20	0.30	0.40	0.50	$s{:}l$	$R{:}l$	$e{:}l$
1/2	90°	$y{:}f$	0.0	0.600	0.800	0.916	0.980	1.000	1.5708	0.5000	0.000
		sin α	1.0	0.800	0.600	0.400	0.200	0.0			
		cos α	0.0	0.600	0.800	0.916	0.980	1.000			
1/2.383	80°	$y{:}f$	0.0	0.535	0.765	0.902	0.976	1.000	1.4179	0.5077	0.0881
		sin α	0.985	0.788	0.591	0.394	0.197	0.0			
		cos α	0.174	0.616	0.807	0.919	0.980	1.000			
1/2.856	70°	$y{:}f$	0.0	0.482	0.735	0.889	0.973	1.000	1.3001	0.5321	0.1820
		sin α	0.940	0.752	0.564	0.376	0.188	0.0			
		cos α	0.342	0.659	0.826	0.927	0.982	1.000			
1/3	72°22′48″	$y{:}f$	0.0	0.471	0.728	0.885	0.972	1.000	1.2740	0.5417	0.2083
		sin α	0.923	0.738	0.554	0.367	0.185	0.0			
		cos α	0.385	0.674	0.832	0.929	0.983	1.000			
1/3.464	60°	$y{:}f$	0.0	0.442	0.709	0.876	0.970	1.000	1.2092	0.5774	0.2887
		sin α	0.866	0.693	0.520	0.346	0.173	0.0			
		cos α	0.500	0.721	0.854	0.938	0.985	1.000			
1/4	53°7′48″	$y{:}f$	0.0	0.421	0.693	0.868	0.968	1.000	1.1591	0.6250	0.3750
		sin α	0.800	0.640	0.480	0.320	0.160	0.0			
		cos α	0.600	0.768	0.877	0.947	0.987	1.000			
1/5	43°36′10″	$y{:}f$	0.0	0.398	0.675	0.859	0.965	1.000	1.1033	0.7250	0.5250
		sin α	0.690	0.552	0.414	0.276	0.138	0.0			
		cos α	0.724	0.834	0.910	0.961	0.990	1.000			
1/6	36°52′10″	$y{:}f$	0.0	0.386	0.665	0.854	0.964	1.000	1.0731	0.8333	0.6667
		sin α	0.600	0.480	0.360	0.240	0.120	0.0			
		cos α	0.800	0.877	0.933	0.971	0.993	1.000			
1/7	32°53′27″	$y{:}f$	0.0	0.379	0.658	0.850	0.963	1.000	1.0536	0.9464	0.8036
		sin α	0.528	0.423	0.317	0.211	0.106	0.0			
		cos α	0.849	0.906	0.948	0.977	0.994	1.000			
1/8	28°04′20″	$y{:}f$	0.0	0.375	0.654	0.848	0.962	1.000	1.0411	1.0625	0.9375
		sin α	0.471	0.378	0.282	0.188	0.094	0.0			
		cos α	0.882	0.926	0.959	0.982	0.996	1.000			
1/10	22°37′5″	$y{:}f$	0.0	0.369	0.649	0.845	0.961	1.000	1.0265	1.3000	1.200
		sin α	0.385	0.308	0.231	0.154	0.077	0.0			
		cos α	0.923	0.951	0.973	0.988	0.997	1.000			

Nomenclature: span l, rise f, radius R, central angle 2α, length of the arc s. Origin on the left support; $e = R - f$

Internal Forces and Displacements of Uniform In-Plane Loading Circular Bar (Tables A.6 and A.7)

Table A.6 Internal forces and displacements of uniform in-plane loading circular bar subjected to distributed loads ($a_y = R/EI$) [Uma72]

Internal forces and displacements			Design diagram of a circular bar		
			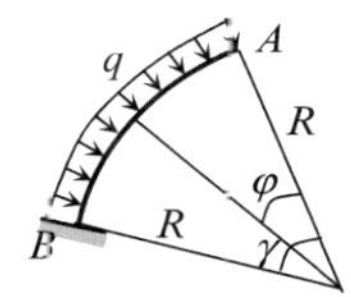	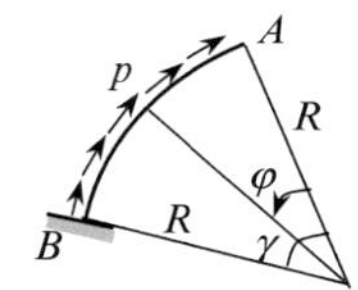	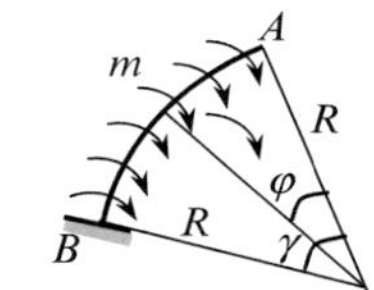
Forces at a section φ	Radial force	1	$qR \sin \varphi$	$pR(1 - \cos \varphi)$	0
	Axial force	2	$-qR(1 - \cos \varphi)$	$pR \sin \varphi$	0
	Bending moment	3	$qR^2(1 - \cos \varphi)$	$pR^2(\varphi - \sin \varphi)$	$mR\varphi$
Displacements of a free end	Radial	4	$qR^3 a_y \frac{(1 - \cos \gamma)^2}{2}$	$pR^3 a_y \left(\sin \gamma + \frac{\sin 2\gamma}{4} - \gamma \cos \gamma - \frac{\gamma}{4} \right)$	$mR^2 a_y (\sin \gamma - \gamma \cos \gamma)$
	Axial	5	$qR^3 a_y \left(\frac{3\gamma}{2} - 2 \sin \gamma + \frac{\sin 2\gamma}{4} \right)$	$pR^3 a_y \left(\frac{\gamma^2}{2} + \frac{\sin^2 \gamma}{2} - \gamma \sin \gamma \right)$	$mR^2 a_y \left(1 - \frac{\gamma^2}{2} - \cos \gamma - \gamma \sin \gamma \right)$
	Angle of rotation	6	$qR^2 a_y (\gamma - \sin \gamma)$	$pR^2 a_y \left(\frac{\gamma^2}{2} + \cos \gamma - 1 \right)$	$mRa_y \frac{\gamma^2}{2}$

Table A.7 Internal forces and displacements of uniform in-plane loading circular bar subjected to concentrated loads ($a_y = R/\mathrm{EI}$)

Internal forces and displacements			Design diagram of a circular bar		
			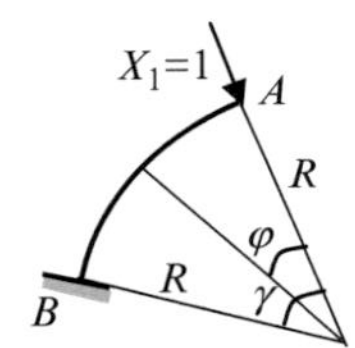	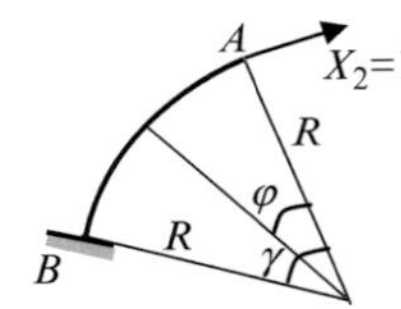	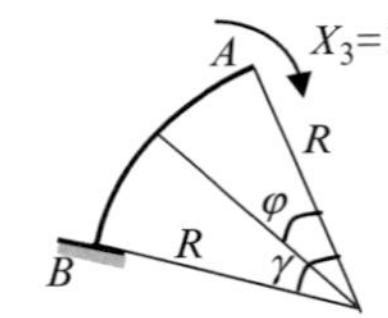
Unit forces at a section φ	Radial force	1	$\cos\varphi$	$\sin\varphi$	0
	Axial force	2	$-\sin\varphi$	$\cos\varphi$	0
	Bending moment	3	$R\sin\varphi$	$R(1-\cos\varphi)$	1
Unit displacements of a free end	Radial	4	$\delta_{11} = R^2 a_y\left(\frac{\gamma}{2} - \frac{\sin 2\gamma}{4}\right)$	$\delta_{12} = R^2 a_y \frac{(1-\cos\gamma)^2}{2}$	$\delta_{13} = Ra_y(1-\cos\gamma)$
	Axial	5	$\delta_{21} = R^2 a_y \frac{(1-\cos\gamma)^2}{2}$	$\delta_{22} = R^2 a_y\left(\frac{3\gamma}{2} + \frac{\sin 2\gamma}{4} - 2\sin\gamma\right)$	$\delta_{23} = Ra_y(\gamma - \sin\gamma)$
	Angle of rotation	6	$\delta_{31} = Ra_y(1-\cos\gamma)$	$\delta_{32} = Ra_y(\gamma-\sin\gamma)$	$\delta_{33} = a_y\gamma$

Note: Internal forces and displacements of uniform out-of-plane loading fixed-free circular bar are presented in [Roa75], [You89], [Uma72]

Three-Hinged Parabolic Arch. Bending Moments, Reactions, and Thrust Due to Force *P* (Tables A.8 and A.9)

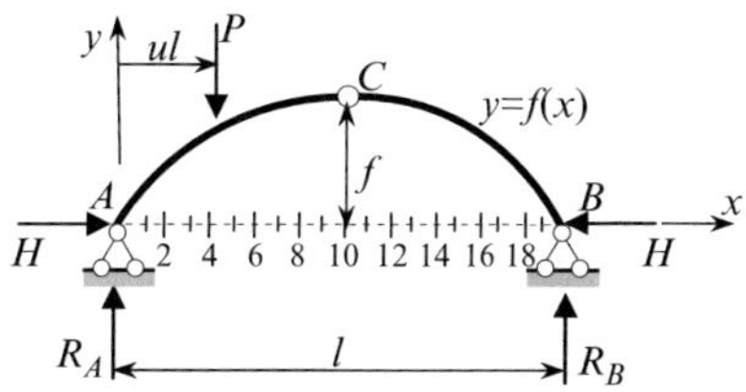

$$y = 4fx(l - x)\frac{1}{l^2}$$

Table A.8 Bending moments (factor *Pl*) [Uma72]

Section	$u = 0.10$	$u = 0.15$	$u = 0.20$	$u = 0.25$	$u = 0.30$	$u = 0.35$	$u = 0.40$	$u = 0.45$	$u = 0.50$
1	0.036	0.029	0.021	0.014	0.006	0.000	−0.008	−0.015	−0.023
2	0.072	0.058	0.044	0.030	0.016	0.002	−0.012	−0.026	−0.040
3	0.060	0.090	0.069	0.049	0.028	0.009	−0.012	−0.032	−0.053
4	0.048	0.072	0.096	0.070	0.044	0.018	−0.008	−0.034	−0.060
5	0.043	0.057	0.075	0.094	0.062	0.032	0.000	−0.031	−0.063
6	0.028	0.042	0.056	0.070	0.084	0.048	0.012	−0.024	−0.060
7	0.020	0.030	0.039	0.047	0.058	0.069	0.028	−0.012	−0.053
8	0.012	0.018	0.024	0.030	0.036	0.042	0.048	0.004	−0.040
9	0.006	0.009	0.011	0.014	0.016	0.020	0.022	0.025	−0.023
10(*C*)	0.000	0.000	0.000	0.000	0.000	0.000	0.000	0.000	0.000
11	−0.005	−0.006	0.009	0.011	−0.014	−0.015	−0.018	−0.023	−0.023
12	−0.008	−0.012	−0.016	−0.020	−0.021	−0.028	−0.032	−0.036	−0.040
13	−0.011	−0.015	−0.021	−0.026	−0.032	−0.036	−0.042	−0.047	−0.053
14	−0.012	−0.018	−0.024	−0.030	−0.036	−0.042	−0.048	−0.054	−0.060
15	−0.013	−0.018	−0.025	−0.031	−0.038	−0.043	−0.050	−0.056	−0.063
16	−0.012	−0.018	−0.024	−0.030	−0.036	−0.042	−0.048	−0.054	−0.060
17	−0.011	−0.015	−0.021	−0.026	−0.032	−0.036	−0.042	−0.017	−0.053
18	−0.008	−0.012	−0.016	−0.020	−0.024	−0.028	−0.032	−0.036	−0.040
19	−0.005	−0.007	−0.009	−0.011	−0.014	−0.015	−0.018	−0.020	−0.023

This table may be used for the construction of influence lines for bending moment. Two position of load $P = 1$ should be considered.

1. If load *P* is located on the left part of the arch ($u < 0.5$), then ordinates of IL are taken from the table immediately.
2. If load *P* is located on the right part of the arch ($u > 0.5$), then we need to take the ordinates for symmetrical section: for section 5 we use ordinates of section 15.

Table A.9 Reactions (factor *P*) and thrust (factor *Pl/f*)

Reactions	$u = 0.10$	$u = 0.15$	$u = 0.20$	$u = 0.25$	$u = 0.30$	$u = 0.35$	$u = 0.40$	$u = 0.45$	$u = 0.50$
R_A	0.900	0.850	0.800	0.750	0.700	0.650	0.600	0.550	0.500
R_B	0.100	0.150	0.200	0.250	0.300	0.350	0.400	0.450	0.500
H	0.050	0.075	0.100	0.125	0.150	0.175	0.200	0.225	0.250

Symmetrical Three-Hinged Arch of any Shape. Reactions, Thrust, and Bending Moments Due to Different Vertical Loads (Table A.10)

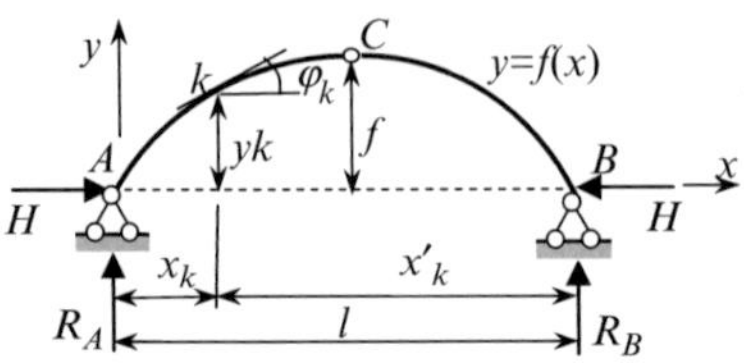

$$M_k = M_k^0 - Hy_k,$$
$$Q_k = Q_k^0 \cos\varphi_k - H \sin\varphi_k,$$
$$N_k = -Q_k^0 \sin\varphi_k - H \cos\varphi_k,$$
$$\xi = \frac{x_k}{l}, \quad \xi' = \frac{x'_k}{l} = 1 - \xi,$$
$$\eta = \frac{y_k}{f}.$$

Table A.10 Special cases of loading [Uma72]

	Loading	R_A	R_B	H	M_k
1	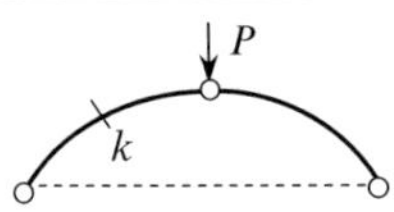	$\frac{P}{2}$	$\frac{P}{2}$	$\frac{Pl}{4f}$	$\frac{Pl}{4}(2\xi - \eta)$
2	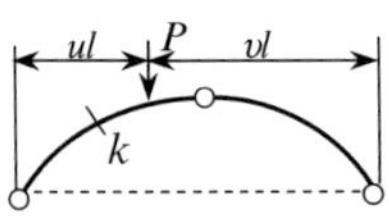	$P\upsilon$	Pu	$\frac{Pul}{2f}$	$\frac{Pul}{2}\left(2\frac{\upsilon}{u}\xi - \eta\right)$
3	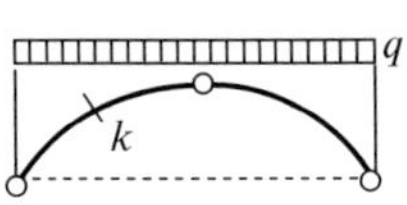	$\frac{ql}{2}$	$\frac{ql}{2}$	$\frac{ql^2}{8f}$	$\frac{ql^2}{8}\left[4(\xi - \xi^2) - \eta\right]$ $M_k = 0$ if arch shape is a quadratic parabola
4	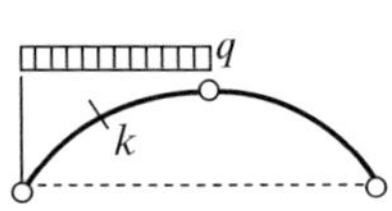	$\frac{3ql}{8}$	$\frac{ql}{8}$	$\frac{ql^2}{16f}$	$\frac{ql^2}{16}\left[8(\xi - \xi^2) - 2\xi - \eta\right]$
5	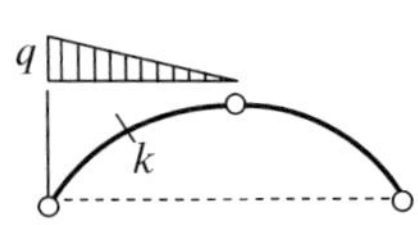	$\frac{5ql}{24}$	$\frac{ql}{24}$	$\frac{ql^2}{48f}$	$\frac{ql^2}{48}\left[2\xi + 8\left(\xi' - \xi'^3 - \xi + \xi^3\right) - \eta\right]$
6	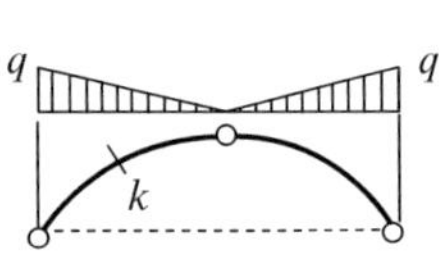	$\frac{ql}{4}$	$\frac{ql}{4}$	$\frac{ql^2}{24f}$	$\frac{ql^2}{24}\left[2\xi + 4\left(\xi' - \xi'^3 - \xi + \xi^3\right) - \eta\right]$
7	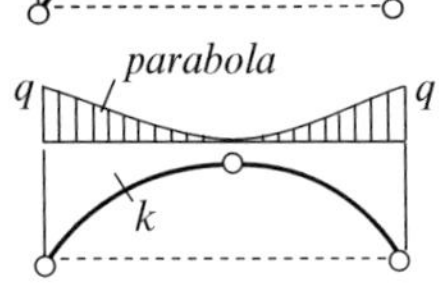	$\frac{ql}{6}$	$\frac{ql}{6}$	$\frac{ql^2}{48f}$	$\frac{ql^2}{48}\left[8(\xi - \xi^2)(1 - 2\xi + 2\xi^2) - \eta\right]$

M_k^0 and Q_k^0 represent the bending moment and shear force at the section k for the reference beam

Three-Hinged Parabolic Arch. Bending Moments, Reactions, and Thrust Due to A One-side Distributed Load q Within Some Portion of the Arch (Tables A.11 and A.12)

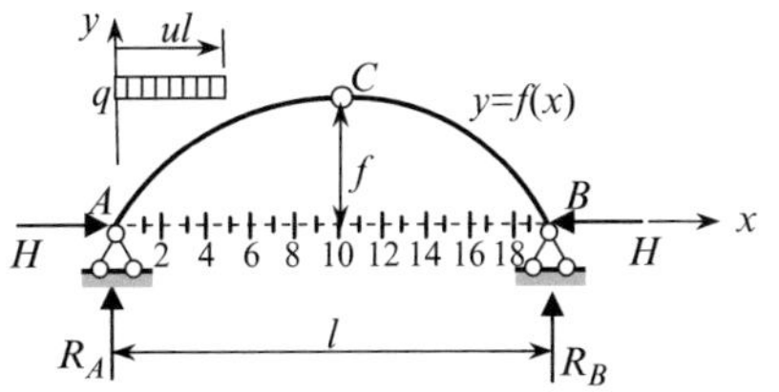

$$y = 4fx(l - x)\frac{1}{l^2}$$

Table A.11 Bending moments (factor ql^2) [Uma72]

Section	$u = 0.10$	$u = 0.15$	$u = 0.20$	$u = 0.25$	$u = 0.30$	$u = 0.35$	$u = 0.40$	$u = 0.45$	$u = 0.50$	$u = 1.00$
1	0.0030	0.0046	0.0069	0.0170	0.0071	0.0073	0.0072	0.0065	0.0056	0.00
2	0.0036	0.0069	0.0094	0.0112	0.0122	0.0127	0.0126	0.0115	0.0100	0.00
3	0.0030	0.0065	0.0105	0.0136	0.0153	0.0163	0.0164	0.0151	0.0131	0.00
4	0.0026	0.0052	0.0094	0.0138	0.0163	0.0180	0.0184	0.0172	0.0150	0.00
5	0.0019	0.0041	0.0075	0.0115	0.0152	0.0177	0.0188	0.0177	0.0156	0.00
6	0.0014	0.0030	0.0056	0.0085	0.0122	0.0157	0.0174	0.0169	0.0150	0.00
7	0.0010	0.0021	0.0039	0.0059	0.0084	0.0115	0.0144	0.0145	0.0131	0.00
8	0.0006	0.0012	0.0024	0.0035	0.0049	0.0074	0.0096	0.0106	0.0100	0.00
9	0.0003	0.0006	0.0011	0.0016	0.0025	0.0033	0.0044	0.0051	0.0056	0.00
10(C)	0.0000	0.0000	0.0000	0.0000	0.0000	0.0000	0.0000	0.0000	0.0000	0.00
11	−0.0002	−0.0005	−0.0009	−0.0015	−0.0022	−0.0028	−0.0036	−0.0050	−0.0056	0.00
12	−0.0004	−0.0010	−0.0016	−0.0027	−0.0041	−0.0054	−0.0064	−0.0086	−0.0100	0.00
13	−0.0005	−0.0012	−0.0021	−0.0034	−0.0051	0.0068	−0.0084	−0.0110	−0.0131	0.00
14	−0.0006	−0.0014	−0.0024	−0.0039	−0.0058	−0.0077	−0.0096	−0.0125	−0.0150	0.00
15	−0.0006	−0.0018	−0.0025	−0.0040	−0.0060	−0.0080	−0.0100	−0.0130	−0.0156	0.00
16	−0.0006	−0.0014	−0.0024	−0.0038	−0.0057	−0.0076	−0.0096	−0.0124	−0.0150	0.00
17	−0.0005	−0.0012	−0.0021	−0.0033	−0.0049	−0.0066	−0.0084	0.0108	−0.0131	0.00
18	−0.0004	−0.0009	−0.0016	0.0026	−0.0038	−0.0051	−0.0064	−0.0083	−0.0100	0.00
19	−0.0002	−0.0005	−0.0009	−0.0014	−0.0021	−0.0028	−0.0036	−0.0046	−0.0056	0.00

Table A.12 Vertical reactions R (factor ql) and thrust H (factor ql^2/f)

	$u = 0.05$	$u = 0.10$	$u = 0.15$	$u = 0.20$	$u = 0.25$	$u = 0.30$	$u = 0.35$	$u = 0.40$	$u = 0.45$	$u = 0.50$	$u = 1.00$
R_A	0.049	0.095	0.139	0.180	0.219	0.255	0.289	0.320	0.349	0.375	0.500
R_B	0.001	0.005	0.011	0.020	0.031	0.045	0.061	0.080	0.101	0.125	0.500
H	0.0006	0.0025	0.0056	0.0100	0.0156	0.0230	0.0310	0.0400	0.0510	0.0625	0.1250

Example 1. The parabolic arch is subjected to distributed loads q_1 and q_2. Calculate the reactions and bending moment at section 3.

The principle of superposition and symmetry property of an arch are used.

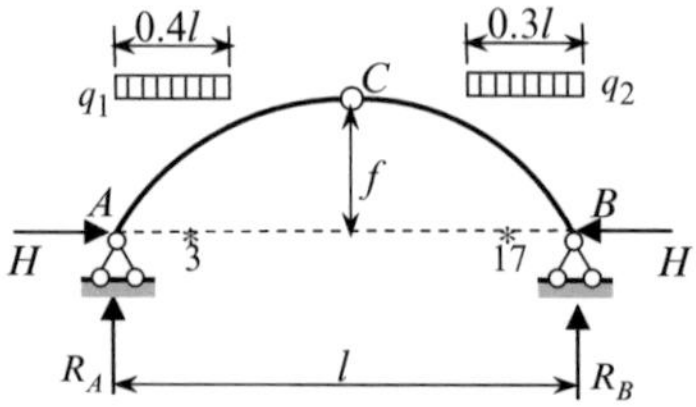

Reaction $R_A = 0.320q_1l + 0.045q_2l$

Thrust $H = (0.0400q_1 + 0.023q_2)\left(\frac{l^2}{f}\right)$

Bending moment $M_3 = 0.0164q_1l^2 - 0.0049q_2l^2$

Example 2. The parabolic arch is subjected to distributed loads q. Calculate the reactions and bending moment at section 3.

The given load may be considered as resulting from two loads: distributed load q with parameter $u = 0.5$ (load within the entire left part of the arch) and distributed load $q_1 = q$ with parameter $u = 0.2$ and directed upward. Principle of superposition leads to the following results.

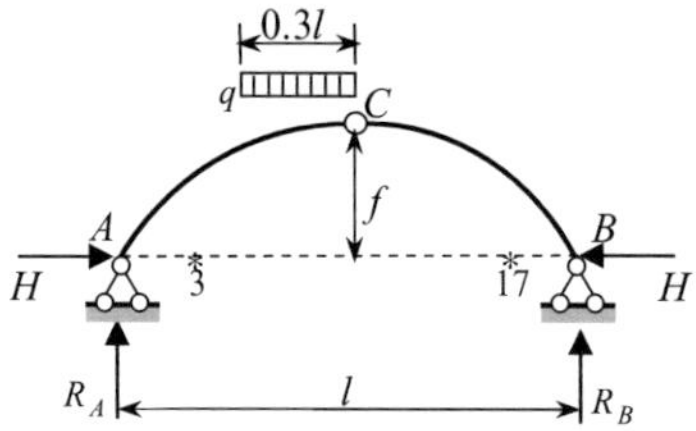

Reaction $R_A = 0.375ql - 0.180q_1l = 0.195ql$

Thrust $H = (0.0625q - 0.010q_1)(l^2/f) = 0.0525(ql^2/f)$

Bending moment $M_3 = 0.0131ql^2 - 0.0105q_1l^2 = 0.0026ql^2$

Three-Hinged Parabolic Arch. Bending Moments, Reactions, and Thrust Due to Symmetrical Distributed Load q Within Some Portion of the Arch $y = 4fx(l - x)(1/l^2)$ (Tables A.13–A.15)

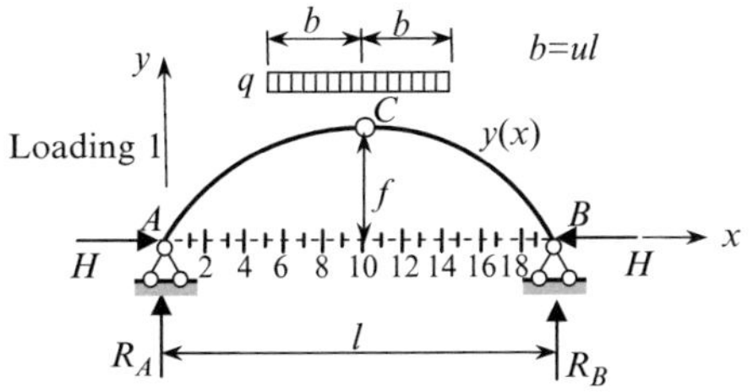

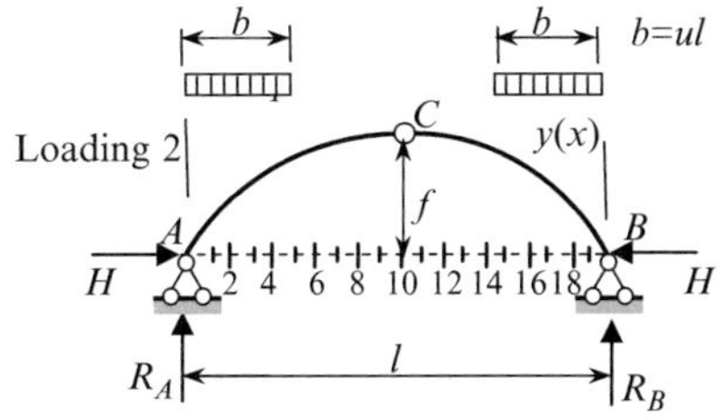

Table A.13 Bending moments (factor ql^2); for loading 2 use parameter u in brackets and change signs for moments [Uma72]

Sections	$u = 0.45$ (0.05)	$u = 0.40$ (0.10)	$u = 0.35$ (0.15)	$u = 0.30$ (0.20)	$u = 0.25$ (0.25)	$u = 0.20$ (0.30)	$u = 0.15$ (0.35)	$u = 0.10$ (0.40)	$u = 0.05$ (0.45)
1	−0.0009	−0.0028	−0.0041	−0.0060	−0.0056	−0.0050	−0.0045	−0.0036	−0.0019
2	−0.0008	−0.0032	−0.0061	−0.0078	−0.0086	−0.0084	−0.0076	−0.0062	−0.0032
3	−0.0006	−0.0025	−0.0053	−0.0084	−0.0103	−0.0104	−0.0097	−0.0080	−0.0043
4	−0.0004	−0.0020	−0.0038	−0.0070	−0.0088	−0.0106	−0.0104	−0.0088	−0.0048
5	−0.0002	−0.0013	−0.0023	−0.0050	−0.0075	−0.0092	−0.0097	−0.0088	−0.0047
6	−0.0002	−0.0008	−0.0016	−0.0032	−0.0046	−0.0064	−0.0087	−0.0078	−0.0044
7	−0.0002	−0.0005	−0.0009	−0.0018	−0.0025	−0.0033	−0.0047	−0.0060	−0.0035
8	0.0000	−0.0002	−0.0002	−0.0008	−0.0008	−0.0008	−0.0020	−0.0032	−0.0020
9	0.0000	−0.0001	−0.0001	−0.0002	−0.0001	−0.0003	−0.0005	−0.0008	−0.0001

Table A.14 Loading 1: Reactions R (factor ql) and thrust H (factor ql^2/f)

	$u = 0.45$	$u = 0.40$	$u = 0.35$	$u = 0.30$	$u = 0.25$	$u = 0.20$	$u = 0.15$	$u = 0.10$	$u = 0.05$
$R_A = R_B$	0.45	0.40	0.35	0.30	0.25	0.20	0.15	0.10	0.05
H	0.1283	0.1200	0.1138	0.1050	0.0938	0.0790	0.0630	0.0450	0.0230

Table A.15 Loading 2: Reactions R (factor ql) and thrust H (factor ql^2/f)

	$u = 0.05$	$u = 0.10$	$u = 0.15$	$u = 0.20$	$u = 0.25$	$u = 0.30$	$u = 0.35$	$u = 0.40$	$u = 0.45$
$R_A = R_B$	0.05	0.10	0.15	0.20	0.25	0.30	0.35	0.40	0.45
H	0.0012	0.0050	0.0112	0.0200	0.0312	0.0460	0.0620	0.0800	0.1020

Two-Hinged Parabolic Nonuniform Symmetric Arch. Bending Moments, Reactions, and Thrust Due to Different Loads (Tables A.16–A.18)

Notation

I_C, A_C – the moment of inertia and cross-sectional area of the arch at the crown C
E, E_t – the modulus of elasticity of the material of the arch and tie
A_t – cross-sectional area of the tie

Coefficient k takes into account the axial force in the arch (Tables A.16 and A.17).

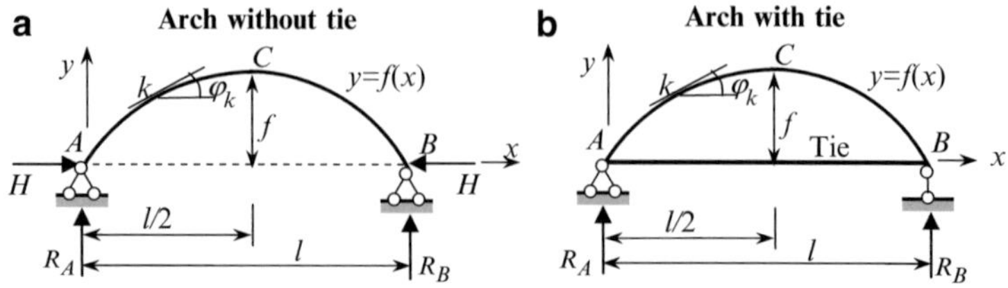

$$y = 4fx(l - x)\frac{1}{l^2},\ I_k \cos\varphi_k = I_C,\ \beta = \frac{EI_C}{E_t A_t}$$

The formulae in Tables A.16 and A.17 may also be used for uniform arches [Uma72].

Table A.16 Reactions, thrust, and bending moments [Uma72]

Loading	Reactions and moments	Loading	Reactions and moments
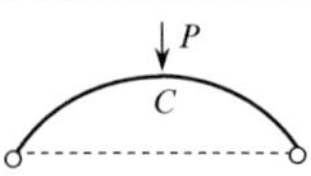	$H = \frac{25}{128}P\frac{l}{f}k$ $M_C = \left(0.25 - \frac{25}{128}k\right)Pl$	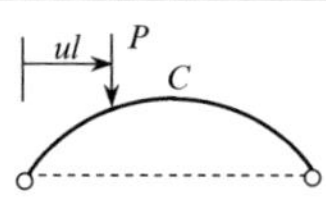	$H = \frac{5}{8}P\frac{l}{f}k\left(u - 2u^3 + u^4\right)$ $M_C = \frac{Pl}{8}\left[4u - 5k\left(u - 2u^3 + u^4\right)\right]$
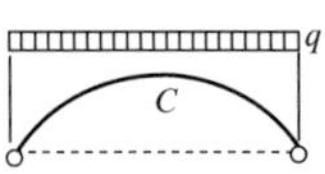	$H = \frac{ql^2}{8f}k$ $M_C = \frac{ql^2}{8}(1 - k)$	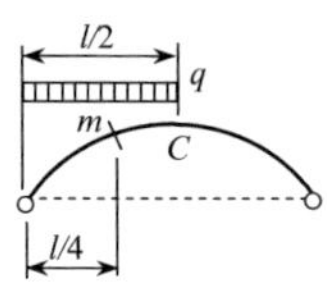	$H = \frac{ql^2}{16f}k$ $M_C = \frac{ql^2}{16}(1 - k)$ $M_m = \left(\frac{1}{16} - \frac{3}{64}k\right)ql^2$
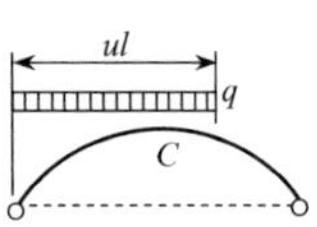	$H = \frac{ql^2}{16f}u^2k\left(5 - 5u^2 + 2u^3\right)$ $R_A = \frac{ql}{2}u(2 - u)$	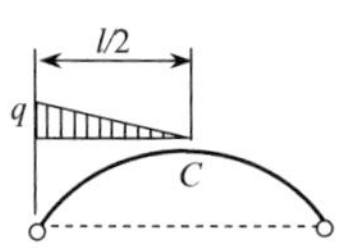	$R_A = \frac{5}{24}ql;\quad R_B = \frac{1}{24}ql$ $H = 0.0228\frac{ql^2}{f}k$
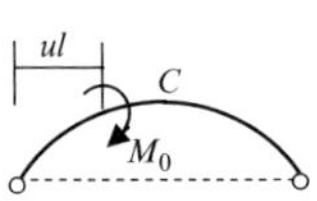	$R_A = -\frac{M_0}{l};\quad R_B = \frac{M_0}{l}$ $H = \frac{5}{8}\frac{M_0}{f}k\left(1 - 6u^2 + 4u^3\right)$	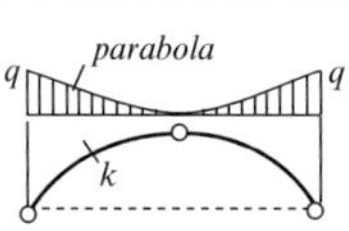	$R_A = R_B = \frac{ql}{6}$ $H = 0.024\frac{ql^2}{f}k$
Relative horizontal displacement of supports 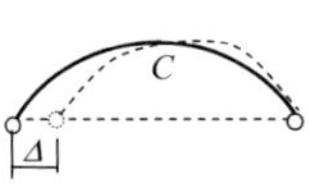	$R_A = R_B = 0$ $H = \frac{15}{8}\frac{EI_C}{f^2 l}k\Delta$ $M_C = -Hf$	Uniform change of temperature by $t°$ 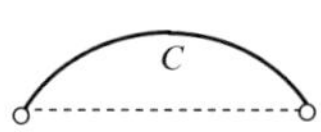 α is thermal coefficient	$R_A = R_B = 0$ $H = \frac{15}{8}\frac{EI_C}{f^2}k\alpha t$ $M_C = -Hf$

Table A.17 Coefficient k

Type of the arch	Axial force influence on the arch and tie	Axial force influence only on the tie
Arch without tie	$k = \dfrac{1}{1+v}$; $v = \dfrac{15}{8}\dfrac{I_C n}{A_C f^2}$	$k = 1$; $v = 0$
Arch with tie	$k = \dfrac{1}{1+v_1}$; $v_1 = \dfrac{15}{8}\dfrac{I_C n}{A_C f^2}\left(\dfrac{EA_C}{E_t A_t n} + 1\right)$	$k = \dfrac{1}{1+v_2}$; $v_2 = \dfrac{15}{8}\dfrac{\beta}{f^2}$

Coefficients n depend on the ratio f/l (Table A.18).

Table A.18 Coefficient n

f/l	1/3	1/4	1/5	1/6	1/7	1/8	1/9	1/10	1/15	1/20
n	0.6960	0.7852	0.8434	0.8812	0.9110	0.9306	0.9424	0.9521	0.9706	0.9888

Two-Hinged Symmetric Parabolic Nonuniform Arch. Bending Moments, Reactions, and Thrust Due to A Single Force *P* (Tables A.19–A.21)

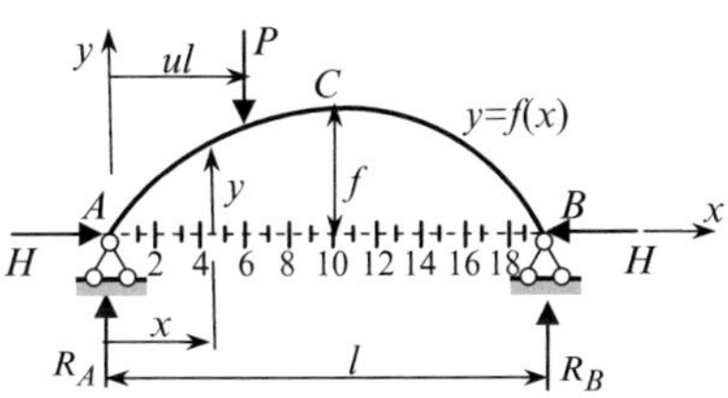

$$y = 4fx(l - x)\frac{1}{l^2}$$

$$I_k \cos \varphi_k = I_C$$

Bending moment at any section of the arch

$$M_x = M_x^0 - Hyk$$

Coefficient k should be taken from Table A.17.

Table A.19 Bending moments M_x^0 for simply supported beam (factor Pl) [Uma72]

Section	$u = 0.10$	$u = 0.15$	$u = 0.20$	$u = 0.25$	$u = 0.30$	$u = 0.35$	$u = 0.40$	$u = 0.45$	$u = 0.50$
1	0.0450	0.0425	0.0400	0.0375	0.0350	0.0325	0.0300	0.0275	0.0250
2	0.0900	0.0850	0.0800	0.0750	0.0700	0.0650	0.0600	0.0550	0.0500
3	0.0850	0.1275	0.1200	0.1125	0.1050	0.0975	0.0900	0.0825	0.0750
4	0.0800	0.1200	0.1600	0.1500	0.1400	0.1300	0.1200	0.1100	0.1000
5	0.0750	0.1125	0.1500	0.1875	0.1750	0.1625	0.1500	0.1375	0.1250
6	0.0700	0.1050	0.1400	0.1750	0.2100	0.1950	0.1800	0.1650	0.1500
7	0.0650	0.0975	0.1300	0.1625	0.1950	0.2275	0.2100	0.1925	0.1750
8	0.0600	0.0900	0.1200	0.1500	0.1800	0.2100	0.2400	0.2200	0.2000
9	0.0550	0.0825	0.1100	0.1375	0.1650	0.1925	0.2200	0.2475	0.2250
10(*C*)	0.0500	0.0750	0.1000	0.1250	0.1500	0.1750	0.2000	0.2250	0.2500
11	0.0450	0.0675	0.0900	0.1125	0.1350	0.1575	0.1800	0.2025	0.2250
12	0.0400	0.0600	0.0800	0.1000	0.1200	0.1400	0.1600	0.1800	0.2000
13	0.0350	0.0525	0.0700	0.0875	0.1050	0.1225	0.1400	0.1575	0.1750
14	0.0300	0.0450	0.0600	0.0750	0.0900	0.1050	0.1200	0.1350	0.1500
15	0.0250	0.0375	0.0500	0.0625	0.0750	0.0875	0.1000	0.1125	0.1250
16	0.0200	0.0300	0.0400	0.0500	0.0600	0.0700	0.0800	0.0900	0.1000
17	0.0150	0.0225	0.0300	0.0375	0.0450	0.0525	0.0600	0.0675	0.0750
18	0.0100	0.0150	0.0200	0.0250	0.0300	0.0350	0.0400	0.0450	0.0500
19	0.0050	0.0075	0.0100	0.0125	0.0150	0.0175	0.0200	0.0225	0.0250

Table A.20 Complex *Hy* (factor *Pl*)

Section	$u = 0.10$	$u = 0.15$	$u = 0.20$	$u = 0.25$	$u = 0.30$	$u = 0.35$	$u = 0.40$	$u = 0.45$	$u = 0.50$
1	0.0117	0.0171	0.0220	0.0264	0.0302	0.0332	0.0353	0.0367	0.0371
2	0.0221	0.0324	0.0418	0.0501	0.0572	0.0628	0.0670	0.0695	0.0703
3	0.0313	0.0458	0.0592	0.0710	0.0810	0.0890	0.0949	0.0984	0.0996
4	0.0392	0.0575	0.0742	0.0891	0.1016	0.1117	0.1190	0.1235	0.1250
5	0.0460	0.0674	0.0870	0.1044	0.1191	0.1309	0.1395	0.1447	0.1465
6	0.0516	0.0755	0.0974	0.1169	0.1334	0.1466	0.1562	0.1621	0.1641
7	0.0558	0.0818	0.1056	0.1266	0.1445	0.1589	0.1693	0.1756	0.1777
8	0.0589	0.0863	0.1114	0.1336	0.1525	0.1676	0.1786	0.1853	0.1875
9	0.0607	0.0890	0.1148	0.1378	0.1572	0.1728	0.1842	0.1911	0.1934
10(*C*)	0.0613	0.0899	0.1160	0.1392	0.1588	0.1746	0.1860	0.1930	0.1953

Table A.21 Reactions (factor *P*) and thrust (factor *Pl*/*f*)

Reactions	$u = 0.10$	$u = 0.15$	$u = 0.20$	$u = 0.25$	$u = 0.30$	$u = 0.35$	$u = 0.40$	$u = 0.45$	$u = 0.50$
R_A	0.90	0.85	0.80	0.75	0.70	0.65	0.60	0.55	0.50
R_B	0.10	0.15	0.20	0.25	0.30	0.35	0.40	0.45	0.50
H	0.0613	0.0898	0.1160	0.1392	0.1588	0.1146	0.1860	0.1930	0.1953

Example. Two-hinged symmetrical parabolic nonuniform arch with tie is subjected to load P at $u = 0.30$. Parameters of the arch are the following: $I_k \cos\varphi_k = I_C$, the length l and rise f, $f/l = 1/6$. Parameters of the tie are E_t and A_t. Calculate the bending moments at sections 4 and 16, taking into account the influence of the axial force on the arch and tie. Required bending moments are

$$M_4 = 0.140Pl - 0.1016Plk$$
$$M_{16} = 0.060Pl - 0.1016Plk,$$

where $k = 1/(1 + .v_1)$, $v_1 = (15/8)(I_C n/A_C f^2)(EA_C/E_t A_t n + 1)$, (Table A.17), and $n = 0.8812$ (Table A.18).

Two-Hinged Parabolic Symmetric Nonuniform Arch. Bending Moments, Reactions, and Thrust Due to Partial Distributed Load *q* (Tables A.22–A.24)

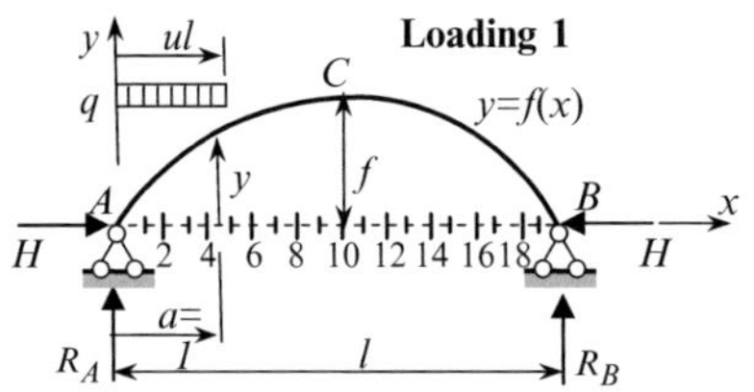

$$y = 4fx(l - x)\frac{1}{l^2}$$

$$I_k \cos\varphi_k = I_C$$

Bending moment at any section of the arch

$$M_x = M_x^0 - Hyk$$

Coefficient k should be taken from Table A.17.

Table A.22 Bending moments M_x^0 for simply supported beam (factor $ql^2 \times 10^{-2}$) [Uma72]

Section	$u = 0.10$	$u = 0.15$	$u = 0.20$	$u = 0.25$	$u = 0.30$	$u = 0.35$	$u = 0.40$	$u = 0.45$	$u = 0.50$
1	0.3500	0.5688	0.7750	0.9688	1.1500	1.3188	1.4750	1.6188	1.7500
2	0.4500	0.8875	1.300	1.6875	2.0500	2.3875	2.7000	2.9875	3.2500
3	0.4250	0.9563	1.5750	2.1563	2.7000	3.2063	3.6750	4.0063	4.5000
4	0.4000	0.9000	1.6000	2.3750	3.1000	3.7750	4.4000	4.9750	5.5000
5	0.3750	0.8438	1.5000	2.3238	3.2500	4.0938	4.8750	5.5938	6.2500
6	0.3500	0.7875	1.4000	2.1875	3.1500	4.1625	5.1000	5.9625	6.7500
7	0.3250	0.7313	1.3000	2.0313	2.9250	3.9813	5.0750	6.0813	7.0000
8	0.3000	0.6750	1.2000	1.8750	2.7000	3.6750	4.8000	5.9500	7.0000
9	0.2750	0.6188	1.1000	1.7188	2.4750	3.3688	4.4000	5.5688	6.7500
10(*C*)	0.2500	0.5625	1.0000	1.5625	2.2500	3.0625	4.0000	5.0625	6.2500
11	0.2250	0.5063	0.9000	1.4063	2.0250	2.7563	3.6000	4.5563	5.6250
12	0.2000	0.4500	0.8000	1.2500	1.8000	2.4500	3.2000	4.0500	5.0000
13	0.1750	0.3938	0.7000	1.0938	1.5750	2.1438	2.8000	3.5438	4.3750
14	0.1500	0.3375	0.6000	0.9375	1.3500	1.8375	2.4000	3.0375	3.7500
15	0.1250	0.2813	0.5000	0.7813	1.1250	1.5313	2.0000	2.5313	3.1250
16	0.1000	0.2250	0.4000	0.6250	0.9000	1.2250	1.6000	2.0250	2.5000
17	0.0750	0.1688	0.3000	0.4688	0.6750	0.9188	1.2000	1.5188	1.8750
18	0.0500	0.1125	0.2000	0.3125	0.4500	0.6125	0.8000	1.0125	1.2500
19	0.0250	0.0563	0.1000	0.1563	0.2250	0.3063	0.4000	0.5063	0.6250

Table A.23 Complex *Hy* (factor $ql^2 \times 10^{-2}$)

Section	$u = 0.10$	$u = 0.15$	$u = 0.20$	$u = 0.25$	$u = 0.30$	$u = 0.35$	$u = 0.40$	$u = 0.45$	$u = 0.50$
1	0.0588	0.1309	0.2288	0.3510	0.4921	0.6507	0.8216	1.0027	1.1875
2	0.1114	0.2478	0.4334	0.6650	0.9323	1.2329	1.5566	1.8998	2.2500
3	0.1579	0.3510	0.6140	0.9421	1.3208	1.7467	2.2052	2.6914	3.1875
4	0.1981	0.4405	0.7706	1.1822	1.6574	2.1919	2.7674	3.3775	4.0000
5	0.2321	0.5162	0.9030	1.3854	1.9423	2.5686	3.2430	3.9580	4.6875
6	0.2600	0.5781	1.0114	1.5516	2.1754	2.8769	3.6322	4.4330	5.2500
7	0.2817	0.6263	1.0956	1.6809	2.3567	3.1166	3.9348	4.8024	5.6875
8	0.2971	0.6607	1.1558	1.7733	2.4862	3.2878	4.1510	5.0663	6.0000
9	0.3064	0.6813	1.1920	1.8287	2.5639	3.3906	4.2808	5.2246	6.1875
10(*C*)	0.3095	0.6882	1.2040	1.8472	2.5898	3.4248	4.3240	5.2773	6.2500

Table A.24 Reactions (factor *ql*) and thrust (factor $(ql^2/f) \times 10^{-2}$)

Reactions	$u = 0.10$	$u = 0.15$	$u = 0.20$	$u = 0.25$	$u = 0.30$	$u = 0.35$	$u = 0.40$	$u = 0.45$	$u = 0.50$
R_A	0.095	0.139	0.180	0.219	0.255	0.289	0.320	0.349	0.375
R_B	0.005	0.011	0.020	0.031	0.045	0.061	0.080	0.101	0.125
H	0.3095	0.6882	1.2040	1.8472	2.5898	3.4248	4.3240	5.2273	6.250

Two-Hinged Parabolic Symmetric Nonuniform arch. Bending Moments, Reactions, and Thrust Due to Partial Distributed Load q (Tables A.25–A.27)

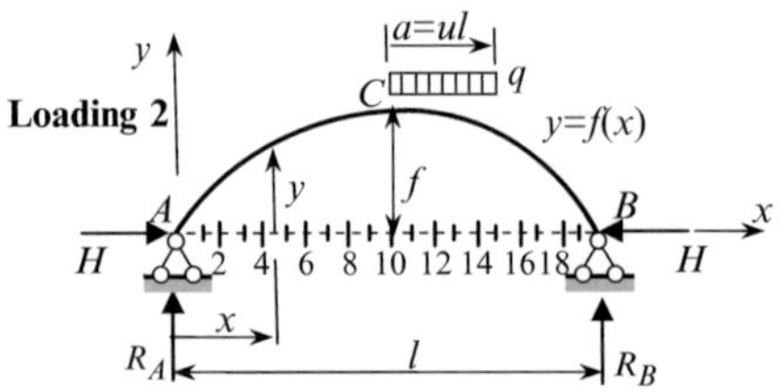

$$y = 4fx(l-x)\frac{1}{l^2}$$

$$I_k \cos\varphi_k = I_C$$

Bending moment at any section of the arch

$$M_x = M_x^0 - Hyk$$

Coefficient k should be taken from Table A.17.

Table A.25 Bending moments M_x^0 for simply supported beam (factor $ql^2 \times 10^{-2}$) [Uma72]

Section	$u = 0.10$	$u = 0.15$	$u = 0.20$	$u = 0.25$	$u = 0.30$	$u = 0.35$	$u = 0.40$	$u = 0.45$	$u = 0.50$
1	0.2750	0.4313	0.6000	0.7813	0.9750	1.1813	1.4000	1.6313	1.7500
2	0.5500	0.8625	1.2000	1.5625	1.9500	2.3625	2.8000	3.1388	3.2500
3	0.8250	1.2938	1.8000	2.3438	2.9250	3.5438	4.1750	4.3938	4.5000
4	1.1000	1.7250	2.4000	3.1250	3.9000	4.6000	5.1000	5.4000	5.5000
5	1.3750	2.1568	3.0000	3.9263	4.7500	5.4063	5.8750	6.1564	6.2500
6	1.6500	2.5875	3.6000	4.6625	5.3500	5.9625	6.4000	6.6625	6.7500
7	1.9250	3.0188	4.0750	4.9688	5.7000	6.2688	6.6750	6.9188	7.0000
8	2.2000	3.3250	4.3000	5.1250	5.8000	6.3250	6.7000	6.9250	7.0000
9	2.3500	3.4438	4.2750	5.0313	5.6500	6.1313	6.4750	6.6813	6.7500
10(C)	2.2500	3.1875	4.0000	4.6875	5.2500	5.6875	6.0000	6.1875	6.2500
11	2.0250	2.8688	3.6000	4.2188	4.7000	5.1188	5.4000	5.5688	5.6250
12	1.8000	2.5500	3.2000	3.7500	4.2000	4.5500	4.8000	4.9500	5.0000
13	1.5750	2.2313	2.8000	3.2813	3.6750	3.9813	4.2000	4.3313	4.3750
14	1.3500	1.9125	2.4000	2.8125	3.1500	3.4125	3.6000	3.7125	3.7500
15	1.1250	1.5938	2.0000	2.3438	2.6250	2.8438	3.0000	3.0938	3.1250
16	0.9000	1.2750	1.6000	1.8750	2.1000	2.2750	2.4000	2.4750	2.5000
17	0.6750	0.9563	1.2000	1.4063	1.5750	1.7063	1.8000	1.8563	1.8750
18	0.4500	0.6375	0.8000	0.9375	1.0500	1.1375	1.2000	1.2375	1.2500
19	0.2250	0.3188	0.4000	0.4688	0.5250	0.5688	0.6000	0.6188	0.6250

Table A.26 Complex Hy (factor $ql^2 \times 10^{-2}$)

Section	$u = 0.10$	$u = 0.15$	$u = 0.20$	$u = 0.25$	$u = 0.30$	$u = 0.35$	$u = 0.40$	$u = 0.45$	$u = 0.50$
1	0.3659	0.5368	0.6955	0.8365	0.9587	1.0567	1.1287	1.1727	1.1875
2	0.6934	1.0175	1.3177	1.5850	1.8166	2.0022	2.1386	2.2220	2.2500
3	0.9823	1.4408	1.8667	2.2454	2.5735	2.8365	3.0297	3.1478	3.1875
4	1.2326	1.8081	2.3426	2.8178	3.2294	3.5595	3.8019	3.9501	4.0000
5	1.4445	2.1189	2.7452	3.3021	3.7845	4.1713	4.4554	4.6291	4.6875
6	1.6178	2.3731	3.0746	3.6984	4.2386	4.6719	4.9900	5.1845	5.2500
7	1.7527	2.5709	3.3308	4.0066	4.5919	5.0612	5.4059	5.6166	6.6875
8	1.8490	2.7121	3.5138	4.2267	4.8442	5.3393	5.7029	5.9252	6.0000
9	1.9067	2.7969	3.6237	4.3588	4.9955	5.5062	5.8811	6.1104	6.1875
10(C)	1.9260	2.8252	3.6603	4.4028	5.0460	5.5618	5.9405	6.1721	6.2500

Table A.27 Reactions (factor ql) and thrust (factor $(ql^2/f) \times 10^{-2}$)

Reactions	$u = 0.10$	$u = 0.15$	$u = 0.20$	$u = 0.25$	$u = 0.30$	$u = 0.35$	$u = 0.40$	$u = 0.45$	$u = 0.50$
R_A	0.055	0.086	0.120	0.156	0.195	0.236	0.280	0.326	0.375
R_B	0.045	0.064	0.080	0.094	0.105	0.114	0.120	0.124	0.125
H	1.9260	2.8252	3.6603	4.4028	5.0460	5.5618	5.9405	6.1721	6.250

Two-Hinged Circle Uniform Arch. Bending Moments and Thrust Due to Single Force *P* (*EI* = const) (Table A.28)

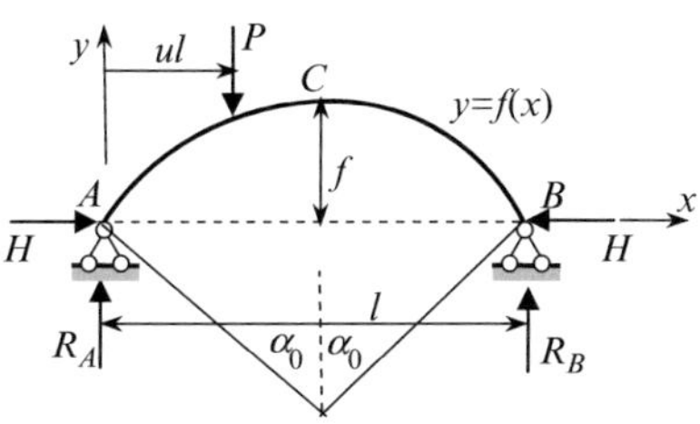

$$y = \sqrt{R^2 - \left(\frac{l}{2} - x\right)^2} - R + f; \quad R = \frac{f}{2} + \frac{l^2}{8f}$$

Bending moment, shear, and axial force
$M_x = M_x^0 - Hy$, M_x^0 see Table A.19.

$$Q_k = Q_k^0 \cos \varphi_k - H \sin \varphi_k$$

$$N_k = -Q_k^0 \sin \varphi_k - H \cos \varphi_k$$

Table A.28 Thrust (factor Pk), coefficient k should be taken from Table A.17 [Uma72]

f/l or α_0	$u = 0.10$	$u = 0.15$	$u = 0.20$	$u = 0.25$	$u = 0.30$	$u = 0.35$	$u = 0.40$	$u = 0.45$	$u = 0.50$
1/2	0.1146	0.1623	0.2037	0.2387	0.2673	0.2896	0.3055	0.3151	0.3182
80°	0.1428	0.2023	0.2539	0.2976	0.3333	0.3610	0.3809	0.3928	0.3967
70°	0.1768	0.2504	0.3142	0.3682	0.4124	0.4468	0.4714	0.4861	0.4910
1/3	0.1861	0.2636	0.3308	0.3877	0.4341	0.4692	0.4962	0.5117	0.5169
60°	0.2195	0.3110	0.3902	0.4573	0.5122	0.5549	0.5854	0.6037	0.6097
1/4	0.2541	0.3600	0.4517	0.5924	0.5929	0.6407	0.6776	0.6988	0.7059
1/5	0.3256	0.4612	0.5787	0.6783	0.7592	0.8209	0.8681	0.8952	0.9043
1/6	0.3911	0.5541	0.6953	0.8149	0.9125	0.9862	1.0429	1.0755	1.0865
1/7	0.4595	0.6510	0.8169	0.9574	1.0721	1.1587	1.2254	1.2636	1.2765
1/8	0.5272	0.7468	0.9371	1.0983	1.2299	1.3292	1.4057	1.4497	1.4644
1/10	0.6682	0.9466	1.1878	1.3921	1.5588	1.6847	1.7817	1.8374	1.8561
1/12	0.7950	1.1261	1.4131	1.6562	1.8546	2.0044	2.1197	2.2871	2.2082

Formulas for internal forces for hingeless circular uniform arches loaded normal to the plane of curvature are presented in [Uma72-73], [You89], [Roa75].

Two-Hinged Circular Symmetric Uniform Arch. Bending Moments, Reactions, and Thrust Due to Partial Distributed Load *q* (Tables A.29–A.30)

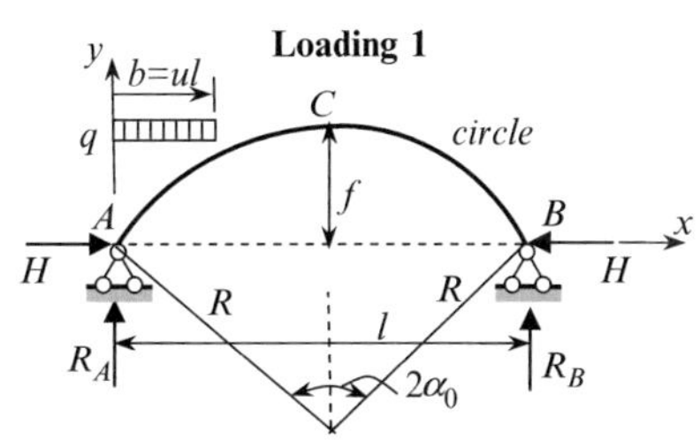

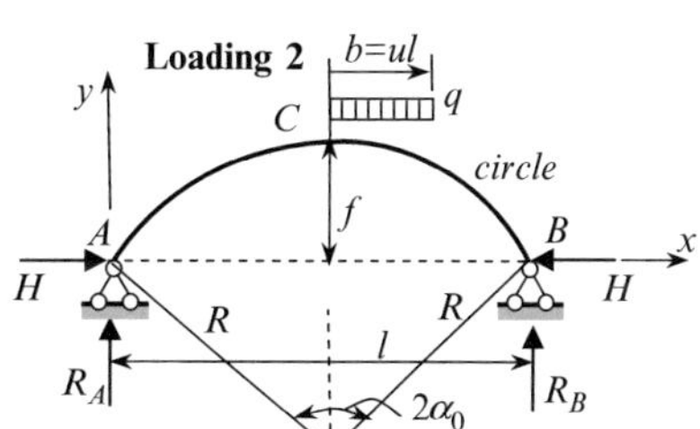

The shape equation is $y = \sqrt{R^2 - \left(\frac{l}{2} - x\right)^2} - R + f; \quad R = \frac{f}{2} + \frac{l^2}{8f}.$

The bending moment, shear, and axial forces at any section of the arch are

$$M_x = M_x^0 - Hy,$$
$$Q_k = Q_k^0 \cos \varphi_k - H \sin \varphi_k$$
$$N_k = -Q_k^0 \sin \varphi_k - H \cos \varphi_k.$$

Tables A.29 and A.30 contains thrust H for two cases of loading. If arch ratio $f < l/6$ then for a thrust H, the factor k should be included (Table A.17).

Table A.29 Loading 1: Thrust (factor ql) [Uma72]

f/l or α_0	$u = 0.10$	$u = 0.15$	$u = 0.20$	$u = 0.25$	$u = 0.30$	$u = 0.35$	$u = 0.40$	$u = 0.45$	$u = 0.50$
1/2	0.0059	0.0128	0.0220	0.0330	0.0457	0.0596	0.0744	0.0899	0.1061
80°	0.0074	0.0160	0.0274	0.0412	0.0571	0.0734	0.0920	0.1120	0.1322
70°	0.0091	0.0198	0.0339	0.0510	0.0705	0.0919	0.1149	0.1388	0.1636
1/3	0.0096	0.0208	0.0356	0.0536	0.0741	0.0957	0.1208	0.1460	0.1718
60°	0.0113	0.0246	00422	0.0634	0.0877	0.1144	0.1429	0.1727	0.2032
1/4	0.0131	0.0284	0.0487	0.0732	0.1012	0.1321	0.1640	0.1984	0.2346
1/5	0.0167	0.0364	0.0624	0.0928	0.1297	0.1692	0.2114	0.2555	0.3006
1/6	0.0201	0.0437	0.0749	0.1127	0.1558	0.2033	0.2540	0.3069	0.3611
1/7	0.0236	0.0514	0.0880	0.1324	0.1831	0.2389	0.2985	0.3607	0.4243
1/8	0.0271	0.0589	0.1010	0.1519	0.2100	0.2740	0.3423	0.4137	0.4867
1/10	0.0343	0.0747	0.1280	0.1925	0.2662	0.3473	0.4339	0.5244	0.6169
1/12	0.0409	0.0889	0.1523	0.2290	0.3168	0.4132	0.5163	0.6239	0.7339

Table A.30 Loading 2: Thrust (factor ql)

f/l or α_0	$u = 0.10$	$u = 0.15$	$u = 0.20$	$u = 0.25$	$u = 0.30$	$u = 0.35$	$u = 0.40$	$u = 0.45$	$u = 0.50$
1/2	0.0317	0.0465	0.0604	0.0731	0.0841	0.0933	0.1002	0.1046	0.1061
80°	0.0402	0.0588	0.0751	0.0910	0.1048	0.1162	0.1248	0.1303	0.1322
70°	0.0487	0.0717	0.0931	0.1126	0.1297	0.1438	0.1545	0.1613	0.1636
1/3	00510	0.0761	0.0977	0.1182	0.1362	0.1510	0.1622	0.1693	0.1718
60°	0.0603	0.0888	0.1155	0.1398	0.1610	0.1786	0.1919	0.2003	0.2032
1/4	0.0706	0.1025	0.1334	0.1614	0.1859	0.2062	0.2215	0.2313	0.2346
1/5	0.0892	0.1314	0.1709	0.2078	0.2382	0.2642	0.2839	0.2963	0.3006
1/6	0.1071	0.1578	0.2053	0.2484	0.2862	0.3174	0.3410	0.3559	0.3611
1/7	0.1258	0.1854	0.2412	0.2919	0.3363	0.3729	0.4007	0.4182	0.4243
1/8	0.1444	0.2127	0.2767	0.3348	0.3857	0.4278	0.4596	0.4798	0.4867
1/10	0.1830	0.2696	0.3507	0.4244	0.4889	0.5422	0.5826	0.6081	0.6169
1/12	0.2176	0.3207	0.4171	0.5049	0.5816	0.6450	0.6930	0.7234	0.7339

Example. The arch is loaded by a uniformly distributed load q across the entire span l. The central angle of the arch is 120°. The thrust of the arch equals $H = 2 \times 0.2032ql = 0.4064ql$. Bending moment at C is $M_C = ql^2/8 - 0.4064qlf$.

Hingeless Parabolic Nonuniform Arch (Table A.31)

Table A.31 Reactions and bending moments due to different types of loads [Uma72][a]
$I_x = (I_C/\cos\alpha)$, $\nu = (45/4)(I_C/A_C f^2)$, $k = 1/(1-\nu)$;
I_C, A_C are moment of inertia and area of cross section at the crown C. If axial forces are neglected, then $\nu = 0$, $k = 1$

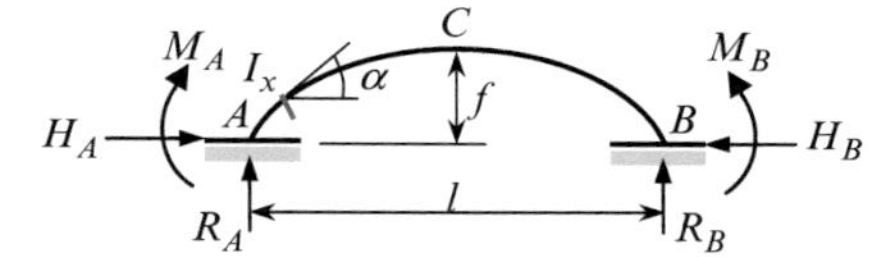

	Loading	R_A	H	M_A	M_B	M_c
1	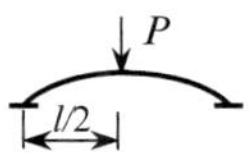	$\frac{P}{2}$	$\frac{15}{64}\frac{Pl}{f}k$	$\frac{Pl}{8}\left(\frac{5}{4}k-1\right)$	$\frac{Pl}{8}\left(\frac{5}{4}k-1\right)$	$\frac{Pl}{8}\left(1-\frac{5}{8}k\right)$
2[b]	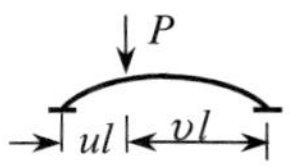	$P\upsilon^2(1+2u)$ $\upsilon = 1-u$	$\frac{15}{4}\frac{Pl}{f}u^2\upsilon^2 k$	$Plu\upsilon^2\left(\frac{5k}{2}u-1\right)$	$Plu^2\upsilon\left(\frac{5k}{2}\upsilon-1\right)$	$u \leqslant 0.5$ $\frac{Pl}{2}u^2\left(1-\frac{5k}{2}\upsilon^2\right)$
3	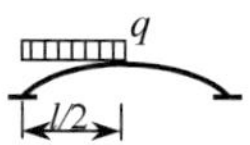	$\frac{13}{32}ql$	$\frac{ql^2}{16f}k$	$-\frac{ql^2}{192}(11-8k)$	$\frac{ql^2}{192}(8k-5)$	$\frac{ql^2}{48}(1-k)$
4		$-\frac{6EI_C}{l^2}\varphi$	$\frac{15}{2}\frac{EI_C}{lf}\varphi$	$\frac{9EI_C}{l}\varphi$	$\frac{3EI_C}{l}\varphi$	$-\frac{3}{2}\frac{EI_C}{l}\varphi$
5		0	$\frac{45}{4}\frac{EI_C}{lf^2}\Delta$	$\frac{15}{2}\frac{EI_C}{lf}\Delta$	$\frac{15}{2}\frac{EI_C}{lf}\Delta$	$-\frac{15}{4}\frac{EI_C}{lf}\Delta$
6[c]	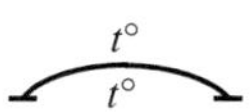 Uniform increase of temperature	0	$\frac{45}{4}\frac{EI_C}{f^2}\alpha_t tk$	$\frac{15}{2}\frac{EI_C}{f}\alpha_t tk$	$\frac{15}{2}\frac{EI_C}{f}\alpha_t tk$	$-\frac{15}{4}\frac{EI_C}{f}\alpha_t tk$

[a] This table may be used for uniform parabolic arches or arches with cross section slightly changing throughout the span, if $f \leq 0.25l$
[b] Row 2 may be used for the construction of influence lines of the vertical reactions, thrust, and moments at supports and crown
[c] α_t is the thermal expansion coefficient

Hingeless Parabolic Nonuniform Arch. Influence Lines for Reactions, Thrust, and Bending Moments at Support and Crown; $I_x = I_C/\cos\varphi_x$ [Uma72], [Kar10] **(Table A.32)**

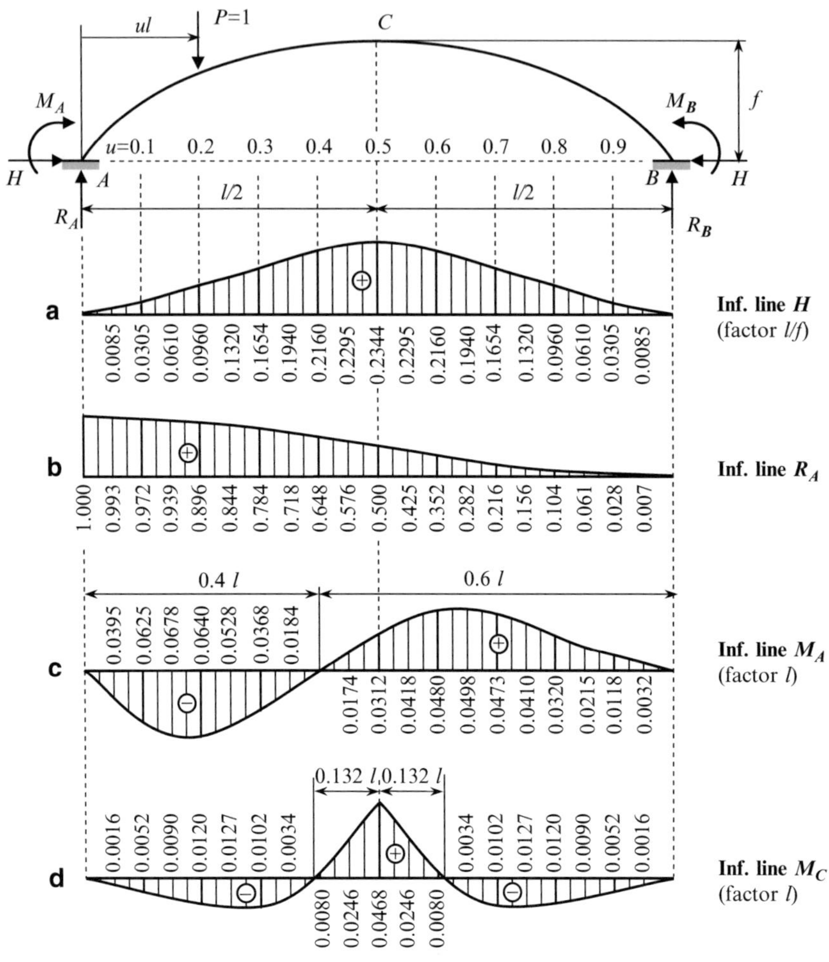

Table A.33 Reactions of uniform beams subjected to compressed load and unit settlement of support, $\upsilon = l\left(\sqrt{P/EI}\right)$, $i = EI/l$, EC – elastic curve [Bol64], [Smi84], [Kar10]

	Pinned-clamped beam	Clamped-clamped beam
1	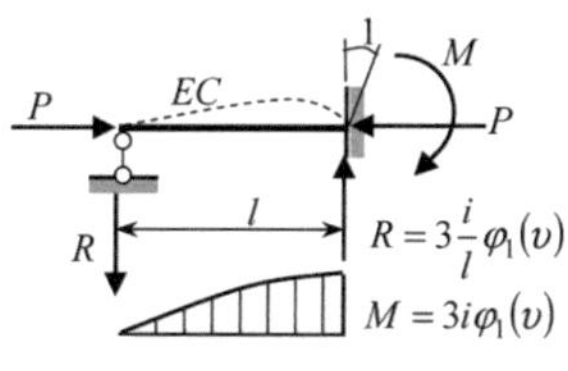	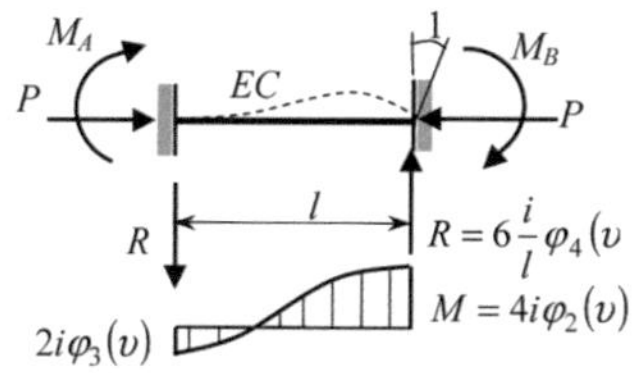
2	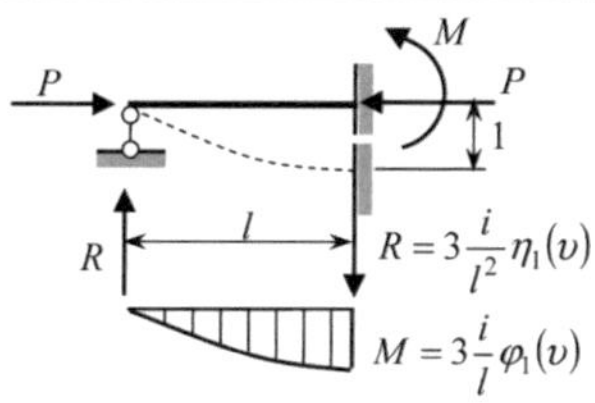	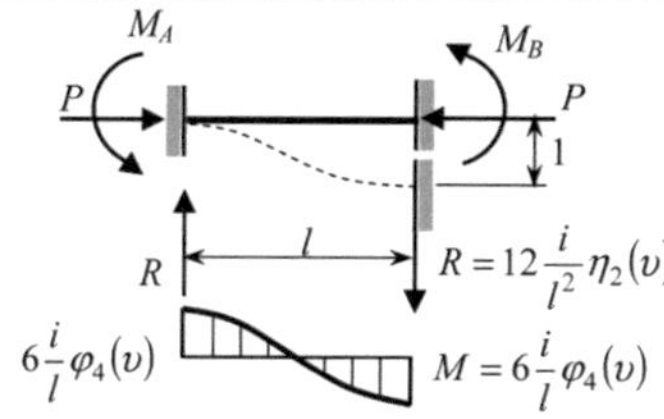
3	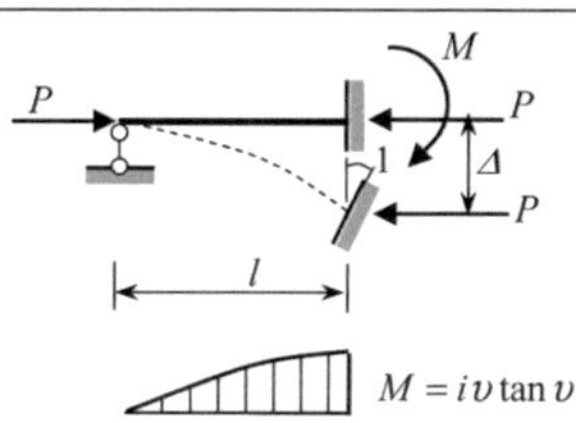	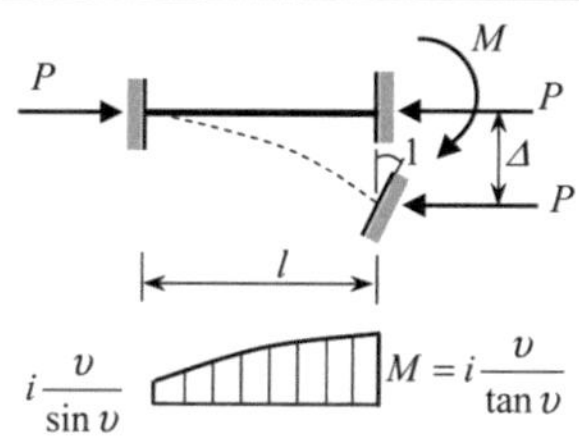
4	**Pinned-pinned beam** 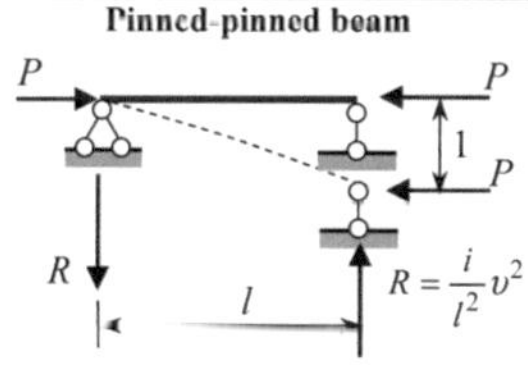	**Clamped-free beam** 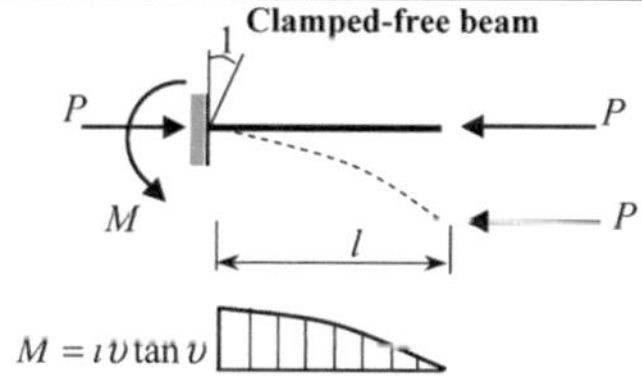

Table A.33a Special functions for stability analysis [Smi47], [Bol64]

Functions	Form 1	Form 2	Maclaurin series
$\varphi_1(\upsilon)$	$\dfrac{\upsilon^2 \tan \upsilon}{3(\tan \upsilon - \upsilon)}$	$\dfrac{1}{3}\dfrac{\upsilon^2 \sin \upsilon}{\sin \upsilon - \upsilon \cos \upsilon}$	$1 - \dfrac{\upsilon^2}{15} - \dfrac{\upsilon^4}{525} + \cdots$
$\varphi_2(\upsilon)$	$\dfrac{\upsilon(\tan \upsilon - \upsilon)}{8 \tan \upsilon\left(\tan\dfrac{\upsilon}{2} - \dfrac{\upsilon}{2}\right)}$	$\dfrac{1}{4}\dfrac{\upsilon \sin \upsilon - \upsilon^2 \cos \upsilon}{2 - 2 \cos \upsilon - \upsilon \sin \upsilon}$	$1 - \dfrac{\upsilon^2}{30} - \dfrac{11\upsilon^4}{25200} + \cdots$
$\varphi_3(\upsilon)$	$\dfrac{\upsilon(\upsilon - \sin \upsilon)}{4 \sin \upsilon\left(\tan\dfrac{\upsilon}{2} - \dfrac{\upsilon}{2}\right)}$	$\dfrac{1}{2}\dfrac{\upsilon(\upsilon - \sin \upsilon)}{2 - 2 \cos \upsilon - \upsilon \sin \upsilon}$	$1 + \dfrac{\upsilon^2}{60} + \dfrac{13\upsilon^4}{25200} + \cdots$
$\varphi_4(\upsilon)$	$\varphi_1\left(\dfrac{\upsilon}{2}\right)$	$\dfrac{1}{6}\dfrac{\upsilon^2 \sin \upsilon}{2 \sin \upsilon - \upsilon - \upsilon \cos \upsilon}$	$1 - \dfrac{\upsilon^2}{60} - \dfrac{\upsilon^4}{84000} + \cdots$
$\eta_1(\upsilon)$	$\dfrac{\upsilon^3}{3(\tan \upsilon - \upsilon)}$	$\dfrac{1}{3}\dfrac{\upsilon^3 \cos \upsilon}{\sin \upsilon - \upsilon \cos \upsilon}$	$1 - \dfrac{2\upsilon^2}{5} - \dfrac{\upsilon^4}{525} + \cdots$
$\eta_2(\upsilon)$	$\eta_1\left(\dfrac{\upsilon}{2}\right)$	$\dfrac{1}{12}\dfrac{\upsilon^3(1 + \cos \upsilon)}{2 \sin \upsilon - \upsilon - \upsilon \cos \upsilon}$	$1 - \dfrac{\upsilon^2}{10} - \dfrac{\upsilon^4}{8400} + \cdots$
$\dfrac{\upsilon}{\sin \upsilon}$	$\dfrac{\upsilon}{\sin \upsilon}$	$\dfrac{\upsilon}{\sin \upsilon}$	$1 + \dfrac{\upsilon^2}{6} + \dfrac{7\upsilon^4}{360} + \cdots$
$\dfrac{\upsilon}{\tan \upsilon}$	$\dfrac{\upsilon}{\tan \upsilon}$	$\dfrac{\upsilon \cos}{\sin \upsilon}$	$1 - \dfrac{\upsilon^2}{3} - \dfrac{\upsilon^4}{45} + \cdots$
$\upsilon \tan \upsilon$	$\upsilon \tan \upsilon$	$\dfrac{\upsilon \sin \upsilon}{\cos \upsilon}$	$0 + \upsilon^2 + \dfrac{\upsilon^4}{3} + \cdots$

Numerical values of these functions in terms of dimensionless parameter υ, $0<\upsilon<2\pi$, may be found in [Smi47], [Kar10]

Critical Loads for Circular and Parabolic Arches with Different Boundary Conditions (Tables A.34–A.35)

Table A.34 Critical parameter K for circular arch subjected to uniform radial load q. Critical load $q_{cr} = K(EI/R^3)$, $EI = \text{const}$, 2α is a central angle [Sni66]

$\alpha°$ (half of a central angle)	Coefficient K					
		Two-hinged arch			Hingeless arch	
	Three-hinged arch	First form	Second form	One-hinged arch	First form	Second form
15	107	143	320	162	294	484
30	27.1	35	79.2	40.2	73.3	120
45	12.0	15	34.7	17.9	32.4	53.2
60	6.75	8	19.1	10.2	18.1	29.7
75	4.32	4.76	11.9	–	11.6	18.8
90	3.0	3.0	8.0	4.61	8.0	12.9

Table A.35 Critical parameters K, K_1 for parabolic symmetrical arch subjected to vertical uniform load q within the entire span, EI = const, span l, rise f [Sni66]

Critical load (first form) $q_{cr} = K\frac{EI}{l^3}$					Critical thrust $H_{cr} = K_1\frac{EI}{l^2}$				
	Coefficient K					Coefficient K_1			
f/l	Three-hinged	Two-hinged	One-hinged	Hingeless	f/l	Three-hinged	Two-hinged	One-hinged	Hingeless
0.1	22.5	28.5	33.8	60.7	0.1	28.1	35.6	42.2	75.8
0.2	39.6	45.4	59.0	101.0	0.2	24.8	28.4	36.8	63.0
0.3	47.3	46.5	84.0	115.0	0.3	19.7	19.4	33.0	48.0
0.4	49.2	43.9	96.0	111.0	0.4	15.4	13.5	30.0	34.7
0.5	43.0	38.4	87.0	97.4	0.5	11.5	9.6	23.0	24.4
0.6	38.0	30.5	80.0	83.8	0.6	7.9	6.4	16.8	17.5
0.8	28.8	20.0	63.0	59.1	0.8	4.5	3.1	9.7	9.2
1.0	22.1	14.1	48.0	43.7	1.0	2.8	1.8	6.0	5.5

Table A.36 Parameters of parabolic polygon ($k = 6$) [Rab58]

	Ratio $\frac{f}{l}$					
Parameters	0.1	0.2	0.3	0.4	0.5	Factor
y_B	0.0556	0.1111	0.1667	0.2222	0.2778	l
y_C	0.0889	0.1778	0.2667	0.3556	0.4444	l
y_D	0.1000	0.2000	0.3000	0.4000	0.5000	l
$\tan\beta_1$	0.3333	0.6667	1.0000	1.3333	1.6667	
$\tan\beta_2$	0.2000	0.4000	0.6000	0.8000	1.0000	
$\tan\beta_3$	0.0667	0.1333	0.2000	0.2667	0.3333	
s_1	0.1757	0.2003	0.2357	0.2778	0.3240	l
s_2	0.1700	0.1795	0.1944	0.2134	0.2360	l
s_3	0.1670	0.1681	0.1700	0.1725	0.1760	l
m_B	0.1728	0.1899	0.2150	0.2456	0.2800	μl
m_C	0.1685	0.1738	0.1822	0.1930	0.2060	μl
m_D	0.0835	0.0840	0.0850	0.0863	0.0880	μl

Nomenclature: Span l, rise f, m_B, m_C, m_D are lumped masses at joints B, C, D (Figs. 6.15 and 6.16), EI = const, μ is mass per unit length

Three-Hinged Parabolic Uniform Arch. Frequencies, Displacements, Bending Moments (Tables A.37 and A.38)

Table A.37 The first mode of symmetrical vibration [Rab58] (Span l, rise f, $EI = \text{const}$)

		Ratio f/l					
Parameters		0.1	0.2	0.3	0.4	0.5	Factor
Frequency ω_1		48.58	43.54	36.82	29.98	23.98	$\frac{1}{l^2}\sqrt{\frac{EI}{\mu}}$
Displacements	w'_B	1.0	1.0	1.0	1.0	1.0	
	w'_C	0.131	0.096	0.057	0.023	−0.003	
	w'_D	−2.262	−2.192	−2.114	−2.046	−1.994	
	u'_B	0.333	0.667	1.000	1.333	1.667	
	u'_C	0.160	0.305	0.434	0.552	0.664	
	u'_D	0.0	0.0	0.0	0.0	0.0	
Bending moments	M_B	82.28	78.12	71.94	65.30	58.80	EI/l^2
	M_C	60.63	51.51	41.47	32.88	26.46	EI/l^2

Table A.38 The second mode of symmetrical vibration [Rab58]

		Ratio f/l					
Parameters		0.1	0.2	0.3	0.4	0.5	Factor
Frequency ω_2		154.16	138.73	121.25	105.25	91.18	$\frac{1}{l^2}\sqrt{\frac{EI}{\mu}}$
Displacements	w'_B	1.0	1.0	1.0	1.0	1.0	
	w'_C	−1.417	−1.616	−1.967	−2.768	−3.041	
	w'_D	0.832	1.231	1.935	2.935	4.088	
	u'_B	0.333	0.667	1.000	1.333	1.667	
	u'_C	−0.150	−0.379	−0.780	−1.441	−2.377	
	u'_D	0.0	0.0	0.0	0.0	0.0	
Bending moments	M_B	254.80	253.96	258.26	294.46	287.80	EI/l^2
	M_C	−313.60	−348.62	−408.14	−551.22	−566.50	EI/l^2

Three-Hinged and Two-Hinged Parabolic Uniform Arches. Frequencies, Displacements, Bending Moments (Tables A.39 and A.40)

Table A.39 The first mode of antisymmetrical vibration [Rab58] (Span l, rise f, EI = const, μ is mass per unit length)

		Ratio f/l					
Parameters		0.1	0.2	0.3	0.4	0.5	Factor
Frequency ω_1		36.72	28.60	22.51	17.52	13.13	$\frac{1}{l^2}\sqrt{\frac{EI}{\mu}}$
Displacements	w'_B	1.0	1.0	1.0	1.0	1.0	
	w'_C	0.923	0.963	0.922	0.677	0.874	
	w'_D	0.0	0.0	0.0	0.0	0.0	
	u'_B	0.333	0.667	1.000	1.333	1.667	
	u'_C	0.318	0.652	0.953	1.075	1.541	
	u'_D	0.256	0.524	0.769	0.894	1.249	
Bending moments	M_B	42.976	40.307	37.974	34.90	31.173	EI/l^2
	M_C	40.539	37.512	31.585	26.48	23.774	EI/l^2

Table A.40 The second mode of antisymmetrical vibration [Rab58]

		Ratio f/l					
Parameters		0.1	0.2	0.3	0.4	0.5	Factor
Frequency ω_2		147.42	131.67	108.22	92.22	74.56	$\frac{1}{l^2}\sqrt{\frac{EI}{\mu}}$
Displacements	w'_B	1.0	1.0	1.0	1.0	1.0	
	w'_C	−1.294	−1.554	−1.762	−2.112	−2.351	
	w'_D	0.0	0.0	0.0	0.0	0.0	
	u'_B	0.333	0.667	1.000	1.333	1.667	
	u'_C	−0.0925	−0.355	−0.657	−1.156	−1.684	
	u'_D	−0.0172	−0.148	−0.305	−0.593	−0.901	
Bending moments	M_B	224.91	234.87	221.40	224.62	204.65	EI/l^2
	M_C	−237.62	−270.89	−281.80	−317.22	−307.64	EI/l^2

Two-Hinged Parabolic Uniform Arches. Frequencies, Displacements, Bending Moments (Tables A.41 and A.42)

Table A.41 The first mode of symmetrical vibration [Rab58] (span l, rise f, EI = const, μ is mass per unit length)

Parameters		Ratio f/l 0.1	0.2	0.3	0.4	0.5	Factor
Frequency ω_1		82.84	72.26	59.25	47.14	37.23	$\frac{1}{l^2}\sqrt{\frac{EI}{\mu}}$
Displacements	w'_B	1.0	1.0	1.0	1.0	1.0	
	w'_C	−0.223	−0.255	−0.289	−0.316	−0.335	
	w'_D	−1.554	−1.490	−1.422	−1.368	−1.329	
	u'_B	0.333	0.667	1.000	1.333	1.667	
	u'_C	0.0887	0.165	0.267	0.281	0.331	
	u'_D	0.0	0.0	0.0	0.0	0.0	
Bending moments	M_B	128.12	118.36	107.08	95.54	85.32	EI/l^2
	M_C	−4.232	−12.44	−20.070	−25.00	−27.53	EI/l^2
	M_D	−133.94	−119.30	−104.19	−92.19	−83.30	EI/l^2

Table A.42 The second mode of symmetrical vibration [Rab58]

Parameters		Ratio f/l 0.1	0.2	0.3	0.4	0.5	Factor
Frequency ω_2		196.14	183.83	168.65	152.94	136.83	$\frac{1}{l^2}\sqrt{\frac{EI}{\mu}}$
Displacements	w'_B	1.0	1.0	1.0	1.0	1.0	
	w'_C	−2.379	−2.857	−3.799	−5.166	−6.848	
	w'_D	2.658	3.714	5.598	8.332	11.696	
	u'_B	0.333	0.667	1.000	1.333	1.667	
	u'_C	−0.332	−0.876	−1.879	−3.599	−6.181	
	u'_D	0.0	0.0	0.0	0.0	0.0	
Bending moments	M_B	321.85	341.31	380.22	436.62	498.57	EI/l^2
	M_C	−660.37	−794.14	−1024.03	−1329.35	−1659.89	EI/l^2
	M_D	822.55	1043.65	1428.34	1964.85	2575.90	EI/l^2

Circular Uniform Arches with Different Boundary Conditions. Frequencies of Free Vibration (Tables A.43–A.44)

$\omega_i = (k_i/R^2\alpha_0^2)\sqrt{EI/m}$, α_0 is a central angle, $\beta = \alpha_0/\pi$, EI is a const, m is mass per unit length.

Table A.43 Parameter k_i for two-hinged arch [Uma72,73]

Mode of vibration	Shape of vibration	Parameter k_i
First anti-symmetrical		$\frac{4\pi^2 - \alpha_0^2}{\sqrt{1+0.75\beta^2}}$
First symmetrical		$\frac{9\pi^2 - \alpha_0^2}{\sqrt{1+0.1652\beta^2}}$
Second anti-symmetrical		$\frac{16\pi^2 - \alpha_0^2}{\sqrt{1+0.1875\beta^2}}$

Table A.44 Parameter k_i for hingeless arch

Mode of vibration	Shape of vibration	Parameter k_i
First anti-symmetrical		$\sqrt{\frac{3,803.2 - 92.101\alpha_0^2 + \alpha_0^4}{1+0.06054\alpha_0^2}}$
First symmetrical		$\sqrt{\frac{14,620 - 197.84\alpha_0^2 + \alpha_0^4}{1+0.01227\alpha_0^2}}$
Second anti-symmetrical		$\sqrt{\frac{39,942 - 343.16\alpha_0^2 + \alpha_0^4}{1+0.02148\alpha_0^2}}$

Parabolic Nonuniform Hingeless Arch

Frequency of symmetrical vibration $\omega_i = (4k_i^2/l^2)\sqrt{EI_C/m}(\mathrm{s}^{-1})$; EI_C is the flexural rigidity at crown, $I_x = I_{cr}\sec\varphi$, span l, rise f, r_{cr} is the radius of inertia of the cross-section at the crown, m is the mass per unit length [Bon52b], [Uma72-73].

Table A.45 Paramters k for first and second frequencies of symmetrical vibration

f/r_{cr}	k_1	k_2	f/r_{cr}	k_1	k_2
2	2.55	5.51	12	4.44	5.67
4	2.93	5.52	14	4.69	5.75
6	3.36	5.54	16	4.88	5.89
8	3.76	5.57	18	4.99	6.26
10	4.12	5.61	20	5.05	9.00

Pressure Curve

Curve of pressure (funicular polygon) allows us to obtain comprehensive information about functionality of the three-hinged arch. Construction of the curve of pressure starts from construction of the force polygon.

Force Polygon

Three-hinged arch ACB is subjected to forces P_i ($i = 1, 2, 3$) as shown in Fig. A.1a (supports are not shown). Polygon P_1–P_2–P_3–R_B–R_A in Fig. A.1b presents force polygon. Construction of this polygon is based on the principle of superposition. Initially, we consider forces P_1–P_2 on the left-hand side of the arch. Their resultant is R; corresponding reaction B_1 passes through a hinge C. Intersection point of R and B_1 (point K_1) defines the direction of reaction A_1. Considering the polygon P_1–P_2–B_1–A_1 (or R–B_1–A_1) we obtain reactions B_1 and A_1.

Force polygon P_3–A_2–B_2 defines the vectors A_2 and B_2. The vector A_2 is translated to the initial point of the vector A_1; this vector is shown by dotted line. Initial point of vector A_2 defines the point K (Fig. A.1b). Both vectors A_2–A_1 give us the total reaction R_A. Similarly, the vector B_1 is translated to the end point of the vector B_2. Thus, we get the total reaction $\boldsymbol{R_B}$. Intersection point of the vectors R_A and R_B is point $\boldsymbol{K}$. The forces R_A–P_1–P_2–P_3–R_B define a *closed* polygon.

Next we connect the point K with the vertices of the force polygon. These lines are denoted as $\boldsymbol{R_A}$, 1–2 (between forces P_1 and P_2), 2–3, and $\boldsymbol{R_B}$. It is obvious that resultant of the forces $\boldsymbol{R_A}$ and P_1 is presented by the vector 1–2, and resultant of the forces $\boldsymbol{R_A}$, P_1, and P_2 is presented by the vector 2–3.

Line of Pressure

The first ray a (Fig. A.1a) passes through support hinge A parallel to reaction $\boldsymbol{R_A}$, the next ray 1–2 between forces P_1 and P_2 is parallel to the corresponding vector 1–2 of the force polygon. The last ray b is parallel to the reaction R_B. A ray between the forces P_2 and P_3 pass through the hinge C. This polygon is called the polygon of pressure; in the case of distributed load the pressure polygon becomes the curve of pressure.

> Any line of the polygon of pressure represents the direction of the resultant of all forces which act to the left (or to the right) of the section under consideration. The magnitude of this resultant is determined from the force polygon in the scale of forces P_i.

Pressure polygon provides a very clear picture of the functionality of the arch. Figure A.1a shows that the pressure polygon (or the resultant of all the one-sided forces) for the left part of the arch is placed *above* of the center line of the arch. It means that the curvature within the left-hand part of the arch will decrease.

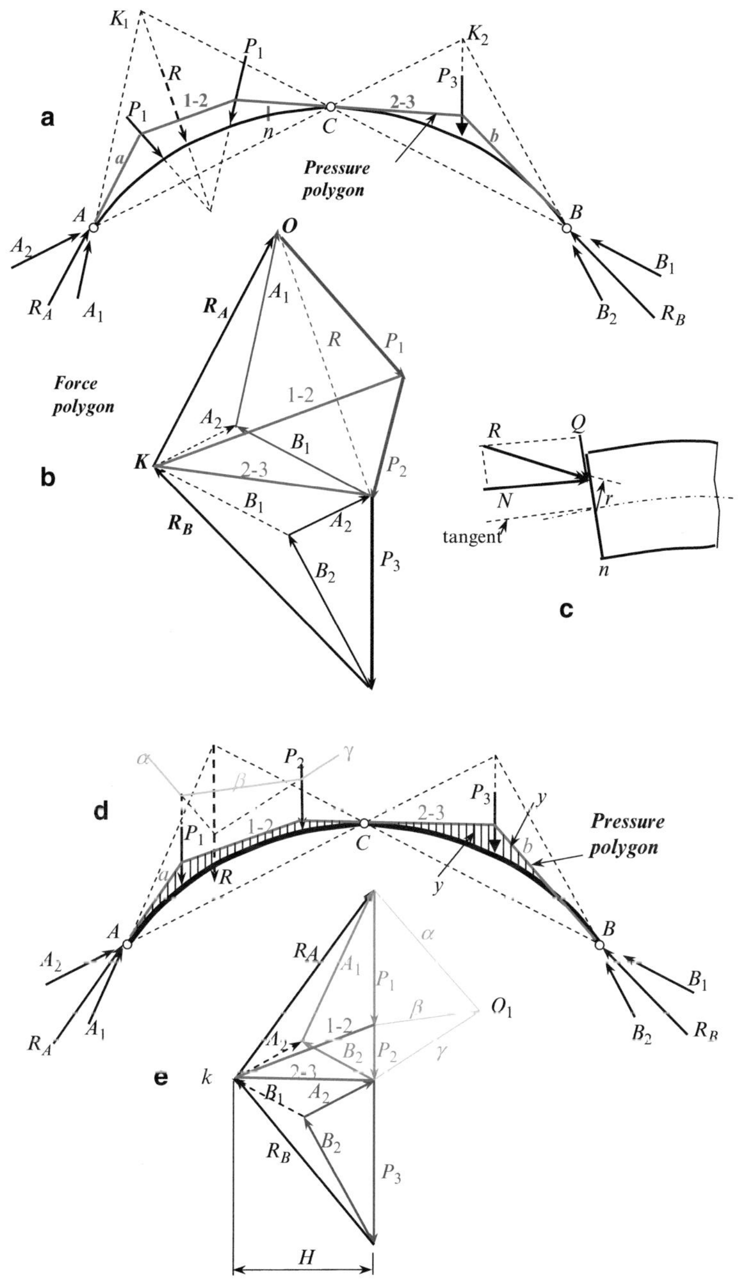

Fig. A.1 (**a**) Pressure polygon; (**b**) force polygon; (**c**) internal forces at the section *n* of the arch; (**d**, **e**) pressure polygon in the case of vertical forces

The force polygon and pressure polygon allow us to determine all the internal forces in the any section of the arch. The bending moment at any section n equals Rr, where R is a resultant and r is an arm of the resultant about the center of gravity of the cross section (Fig. A.1c). Given this, resultant of the one-sided forces is taken from the force polygon, while the arm is taken from the pressure polygon. In order to determine the axial and shear forces at any section n, we need to resolve the resultant, which acts in this section into two components; one of them is parallel to the tangent at section n and other is perpendicular to the tangent.

Construction of the pressure polygon in the case of an arch which is subjected to vertical forces only is shown in Fig. A.1d. The following procedure may be recommended [Rab60], [Dar89].

1. Determine the location of the resultant R_1 of all external forces (without reactions) which act on the left part of the arch. A funicular polygon α–β–γ with arbitrary center O_1 may be used for this purpose.
2. The resultant R_2 of all external forces (without reactions) which act on the right part of the arch $R_2 = P_{3.}$
3. Determine reactions R_A and R_B caused by resultants R_1 and R_2.
4. Construct the force polygon and pressure polygon.

Since all external forces lie on one vertical (Fig. A.1e), then the total reactions R_A and R_B have the same horizontal projection H, which is a thrust of the arch.

It can be shown that in the case of the parallel forces all the vertical segments y between the central line of the arch and pressure polygon (Fig. A.1d) in the constant scale for all arch comprise the bending moment diagram; this diagram is constructed on the compressed fibers. The closer the pressure polygon is to the center line of the arch, the smaller the eccentricity of the compressed force about neutral axis of the arch. In the case of an optimal shape of the arch, the axial line coincides with a pressure curve. In this case the bending moments are zero.

Bibliography

[Arm72] Armand. J.-L.P.: Application of the theory of optimal control of distributed –parameter systems to structural optimization. NASA CR-2044 (1972)

[Ath66] Athans, M., Falb, P.I.: Optimal Control: An Introduction to the Theory and its Applications. McGraw-Hill, New York (1966)

[Bat96] Bathe, K.J.: Numerical Methods in Finite Element Analysis, 3rd edn. Prentice-Hall, Englewood Cliffs, New Jersey (1996)

[Ber57] Bernshtein, S.A.: (Бернштейн СА. Очерки по истории строительной механики. М.: Госстройиздат, 1957)

[Bir68] Birger, I.A., Panovko, J.G.: (Биргер ИА, Пановко ЯГ, Ред. Прочность, устойчивость, колебания. Справочник в трех томах. М.: Машиностроение, 1968)

[Ble79] Blevins, R.D.: Formulas for Natural Frequency and Mode Shape. Van Nostrand Reinhold, New York (1979)

[Bol64] Bolotin, V.V.: The Dynamic Stability of Elastic Systems. Holden-Day, San Francisco (1964)

[Bol78] Bolotin, V.V.: (Болотин ВВ, Ред. Вибрации в технике. Справочник в шести томах; Том1, Колебания линейных систем. М.: Машиностроение, 1978)

[Bol84] Bolotin, V.V.: Random Vibrations of Elastic Systems. Martinus Nijhoff, The Hague (1984)

[Bon50] Bondar, N.G.: (Бондарь НГ. О частоте свободных пространственных колебаний бесшарнирных арок и сводов массивных мостов. Инженерный сборник , т. 8. Изд-во АН СССР, 1950)

[Bon52a] Bondar, N.G.: (Бондарь НГ. О частоте плоских свободных незатухающих колебаний бесшарнирных параболических и цепных арок переменного сечения. Инженерный сборник , т. X1. Изд-во АН СССР, 1952)

[Bon52b] Bondar, N.G.: (Бондарь НГ. Динамическая устойчивость и колебания бесшарнирных параболических арок. Инженерный сборник. т. XIII. Изд-во АН СССР, 1952)

[Bou83] Boussinesq, J.V.: Rèsistance d'un anneau à la flexion, quand sa surface extèrieure supporte une pression normale, constante par unite de longueur de sa fibre. Comptes rendus des séances de l'Acadèmie des sciences **97**(15) (1883)

[Bre54] Bresse, J.A.C.: Recherches analytiques sur la fléxion et la résistance des piéces courbes, Paris (1854)

[Bre59] Bresse, J.A.C.: Cours de Mécanique Appliquée, t. I-III, 2nd edn (1866), 3rd edn (1880), Paris (1859)

[Bro06] Brockenbrough, R.L., Merritt, F.S. (eds.): Structural Steel Engineers' Handbook, 4th edn. McGraw-Hill, New York (2006)

[Bry88] Bryan, G.H.: On the stability of elastic systems. Proceedings of the Cambridge Philosophical Society (mathematical and physical sciences), vol. 6 (1888), Cambridge

[Bur01] Burden, R.L., Faires, D.J.: Numerical Analysis 7th edn. Brooks/Cole (2001)

[Cha69] Chang, T.C., Volterra, E.: Upper and lower bounds or frequencies of elastic arcs. J. Acoust. Soc. Am. **46**(5) (Part 2) (1969)

[Chu52] Chudnovsky, V.G.: (Чудновский ВГ. Методы расчета колебаний и устойчивости стержневых систем. Киев: Изд-во АН УССР, 1952)

[Clo75] Clough, R.W., Penzien, J.: Dynamics of Structures. McGraw-Hill, New York (1975)

[Cra00] Craig, R.R.: Mechanics of Materials. Wiley, New York (2000)

[Dar89] Darkov, A.V. (ed.): Structural Mechanics, 6th edn. Mir Publishers, Moscow (1989)

[Dem49] Demidovich, B.P.: (Демидович БП. Колебания стержня, изогнутого по дуге круга. Инженерный сборник, Том V, вып 2. М-Л.: Изд-во АН СССР, ОТН, 1949)

[DeR91] De Rosa, M.A.: The influence of the support flexibilities on the vibration frequencies of arches. J. Sound Vib. **146**(1) (1991)

[Den28] Den Hartog, J.P.: The lowest natural frequencies of circular arcs. Philosophical Magazine Series, **75** (1928)

[Din46] Dinnik, A.N.: (Динник АН. Устойчивость арок. М-Л.: ОГИЗ ГОСТЕХИЗДАТ, 1946) (1946)

[Duh43] Duhamel, J.M.C.: J. école polytech, cahier 25, Paris (1843)

[Ing34] Inglis, C.E.: Mathematical Treatise on Vibration in Railway Bridges. The University Press, Cambridge and Macmillan, New York (1934)

[Fed34] Federhofer, K.: Ueber die Eigenschwingungen und Knicklasten des Bogens. Wien (1934)

[Feo70] Feodosiev, V.I.: (Феодосьев ВИ. Сопротивление материалов, изд.5. М.: Наука, 1970)

[Fil70] Filippov, A.P.: (Филиппов АП. Колебания деформируемых систем. М.: Машиностроение, 1970)

[Fil88] Filipich, C.P., Laura, P.A.A.: First and second natural frequencies of hinged and clamped circular arcs: a discussion of a classical paper. J. Sound Vib. **125** (1988)

[Fil90] Filipich, C.P., Rosales, M.B.: In-plane vibration of symmetrically supported circumferential Rings. J. Sound Vib. **136**(2) (1990)

[Gab34] Gaber, E.: Knicksicherheit vollwandiger Bogen. Bautechnik (1934)

[Gil81] Gill, P.E., Murray, W., Wright, M.H.: Practical Optimization. Academic, London (1981)

[God04] Godoy, L.A.: Arches: a neglected topic in structural analysis courses. ASEE Southeast Section Conference (2004), Auburn, Alabama

[God00] Godoy, L.A.: Theory of Elastic Stability: Analysis and Sensitivity. Taylor and Francis, Philadelphia (2000)

[Gol27] Goldshtein, S.: Mathieu functions. Trans. Cambridge Phil. Soc. **23**, 303–336 (1927)

[Gol80] Goldshtein Ju, B., Solomeshch, M.A.: (Гольдштейн ЮБ, Соломещ МА. Вариационные задачи статики оптимальных стержневых систем. Л.: Изд-во ЛГУ, 1980)

[Gri79] Grinev, V.B., Filippov, A.P.: (Гринев ВБ, Филиппов АП. Оптимизация стержней по спектру собственных значений. Киев: Наукова Думка, 1979)

[Guz61] Guzov, L.A.: (Гузов ЛА. Динамическая устойчивость бесшарнирных круговых арок. Днепропетровск: Труды ДМЕТИ, вып 42, 1961)

[Gut89] Gutierrez, R.H., Laura, P.A.A., Rossi, R.E., Bertero, R., Villaggi, A.: In-plane vibrations of non-circular arcs of non-uniform cross-section. J. Sound Vib. **129**(2) (1989)

[Ham35] Hamilton, W.: On a general method in dynamics. Phil. Trans. R. Soc., London (1835) (Collected Papers, v.II, Cambridge, 1940, pp.103–211)

[Hau79] Haug, E.J., Arora, J.S.: Applied Optimal Design. Mechanical and Structural Systems. Wiley, New York (1979)

[Hau19] Haupt, O.: Über Lineare Homogene Differentialgleichungen Zweiter Ordung Mit Periodischen Koeffizienten. Math. Ann. D. **79**, 278 (1919)

[Har61] Harris, C.M., Crede, C.E.: Shock and Vibration Handbook. McGraw-Hill, New York (1961, 1976)

[Har88] Harris, C.M.: Shock and Vibration Handbook. McGraw-Hill, New York (1988)

[Hen81] Henrich, J.: The Dynamics of Arches and Frames. Elsevier, Amsterdam (1981)

[Inc25-27] Ince, E.L.: Researches into the characteristic numbers of the Mathieu equation. Proc. R. Soc. Edinburgh, **46** (1925–1926), **47**(1926–1927)

[Iri82] Irie, T., Yamada, G., Tanaka, K.: Free out-of-plane vibration of arcs. ASME J. Appl. Mec. **49** (1982)

[Kar10] Karnovsky, I.A., Lebed, O.: Advanced Methods of Structural Analysis. Springer, New York (2010)

[Kar04] Karnovsky, I.A., Lebed, O.: Non-Classical Vibrations of Arches and Beams. Eigenvalues and Eigenfunctions. McGraw-Hill Engineering Reference, New York (2004)

[Kar01] Karnovsky, I.A., Lebed, O.: Formulas for Structural Dynamics. Tables, Graphs and Solutions. McGraw-Hill, New York (2001)

[Kar70] Karnovsky, I.A.: (Карновский ИА. Колебания пластин и оболочек несущих подвижную нагрузку. Автореф. дисс. на соиск.уч. степ. канд.техн. наук, Дн-ск: Инж-строит. институт, 1970)

[Kar89] Karnovsky, I.A.: (Карновский ИА. Оптимальная виброзащита деформируемых элементов конструкций, машин и приборов. Автореф. дисс. на соиск.уч. степ. докт. техн. наук, Тбилиси: Груз. политехн. институт, 1989)

[Kaz09] Kazakevich, M.I.: (Казакевич МИ. Избранное. Днепропетровск) (2009)

[Kir76] Kirchhoff, G.R.: Vorlesungen über mathematische Physic. Mechanik, Bd.1, Leipzig (1876)

[Kis60] Kiselev, V.A.: (Киселев ВА. Строительная механика. М.: Стройиздат, 1960)

[Kis80] Kiselev, V.A.: (Киселев ВА. Специальный курс: Динамика и устойчивость сооружений, изд. 3. М.: Стройиздат, 1980)

[Kle80] Klein, G.K.: (Клейн ГК , Ред. Руководство к практическим занятиям по курсу строительной механики: Статика стержневых систем, изд. 4. М.: Высшая школа, 1980)

[Kle72] Klein, G.K., Rekach, V.G., Rozenblat, G.I.: (Клейн ГК, Рекач ВГ, Розенблат ГИ. Руководство к практическим занятиям по курсу строительной механики. Специальный курс. Основы теории устойчивости, динамики сооружений и расчета пространственных систем, изд. 2. М.: Высшая школа, 1972)

[Kol77] Kolesnik, I.A.: (Колесник ИА. Колебания комбинированных арочных систем под действием подвижных нагрузок. Киев: Вища школа, 1977)

[Kom72] Komkov, V.: Optimal control theory for the damping of vibrations of simple elastic systems. Lecture Notes in Mathematics, vol. 253, Springer, Berlin and New York (1972)

[Kru05] Kruloff, A.N.: Über die erzwungenen Schwingungen von gleichformigen Staben. Mathematische Annalen **61** (1905)

[Lau87a] Laura, P.A.A., Maurizi, M.J.: Recent research on vibrations of arch-type structures. Shock Vib. Digest **19**(1) (1987)

[Lau87b] Laura, P.A.A., Filipich, C.P., Cortinez, V.H.: In-plane vibrations of an elastically cantilevered circular arc with a tip mass. J. Sound Vib. **115**(3) (1987)

[Lau88a] Laura, P.A.A., Verniere De Irassar, P.L., Carnicer, R., Bertero, R.: A note on vibrations of a circumferential arch with thickness varying in a discontinuous fashion. J. Sound Vib. **120**(1) (1988a)

[Lau88b] Laura, P.A.A., Verniere De Irassar, P.L.: A note on in-plane vibrations of a arch-type structures of non-uniform cross-section: the case of linearly varying thickness. J. Sound Vib. **124**, 1–12 (1988)

[Lau88c] Laura, P.A.A., Bambill, E., Filipich, C.P., Rossi, R.E.: A note on free flexural vibrations of a non-uniform elliptical ring in its plane. J. Sound Vib. **126**(2) (1988c)

[Lee89] Lee, B.K., Wilson, J.F.: Free vibrations of arches with variable curvature. J. Sound Vib. **136**(1), 75–89 (1989)

[Lev84] Lèvy, M. Mèmoire sur un nouveau cas integrable du probleme de l'elastique et l'une de ses applications. J. de mathèmatiques pures et appliqués (Liouville Journal) 10(3) (1884)

[Lok34] Lokshin, A.S.: (Локшин АС, Об устойчивости стержня с криволинейной осью. Прикладная математика и механика, том 2(1), 1934)

[Lov20] Love, A.E.H.: A Treatise on the Mathematical Theory of Elasticity, 3rd edn. The University Press, Cambridge (4th edn. 1927) (1920)

[Luz51] Luzin, N.N., Kuznetsov, P.I.: (Лузин НН, Кузнецов ПИ. К абсолютной инвариантности и инвариантности до ε в теории дифференциальных уравнений, ДАН СССР, т.80 (3), 1951)

[May13] Mayer, R.: Űber Elastizität und Stabilität des geschlossenen und offenen Kreisbogens. Zeitschrift für Mathematic und Physic, Bd 1, H.3 (1913)

[Mcl47] McLachlan, N.W.: Theory and Application of Mathieu Functions. Clarendon, Oxford (1947)

[Mat68] Mathieu, E.: Mémoire sur le Mouvement Vibratoire d'une Membrane de Forme Elliptique. J. de Math. Pures et Appliqués (J.de Liouville) 13(137) (1868)

[Mau90] Maurizi, M.J., Rossi, R.E., Belles, P.M.: Lowest natural frequency of clamped circular arcs of linearly tapered width. J. Sound Vib. **144**(2) (1990)

[Mel31] Melan, J., Gesteschi, T.: Bogenbrücken, Handbuch für Eisenbetonbau, **11** (1931)

[Mor40] Morgaevsky, A.B.: (Моргаевский АБ. Колебания параболических арок. Днепропетровск: Труды ДМЕТИ, вып.IV, 1940)

[Mor59] Morgaevsky, A.B.: (Моргаевский АБ. Критические скорости для круговых арок. Дн-ск: Труды ДМЕТИ, вып. 41, 1959)

[Mor61] Morgaevsky, A.B.: (Моргаевский АБ. Колебания, статическая и динамическая устойчивость арок. Дисс. на соиск.уч. степ. докт. техн. наук. М.: МИСИ, 1961)

[Mor60] Morgaevsky, A.B.: (Моргаевский АБ. О колебаниях арок под действием подвижной нагрузки. Строительная механика и расчет сооружений, вып. 2. Научно-техн. журнал Акад. стр-ва и архит., 1960)

[Mor39] Morgaevsky, A.B.: (Моргаевский АБ. Устойчивость арок при неплоской деформации. Прикладная математика и механика, т.II (3), 1939)

[Mor38] Morgaevsky, A.B.: (Моргаевский АБ. Об устойчивости двухшарнирных и трехшарнирных арок с учетом поведеия нагрузки. М.: Сб. Вестник инженеров, вып.1, 1938)

[Mor73] Morgaevsky, A.B.: (Моргаевский АБ. Устойчивость линейно упругих колец и арок. В книге: Справочник проектировика, ред. А.А. Уманский, раздел 17.12. М.: том 2, изд-во лит.по строительству, 1973)

[New89] Newland, D.E.: Mechanical Vibration Analysis and Computation. Longman Scientific & Technical, NY (1989)

[Nik18] Nikolaee, E.L.: (Николаи ЕЛ. Об устойчивости кругового кольца и арки. Petrograd: Изв. Политехн. Института, т. XXVII, 1918)

[Now61] Nowacki, W.: Dynamika Budowli. Arcady, Warszawa (1961)

[Olh77] Olhoff, N.: Maximizing higher order eigenfrequencies of beams with constraints on the design geometry. J. Struct. Mech. **5**(2) (1977)

[Pav66] Pavlenko, G.L., Vol'per, D.B.: Testing the stability of parabolic ferroconcrete arches. J. Int. Appl. Mech. **2**(8), Springer, New York (1966)

[Pav51] Pavlenko, G.L., and Morgaevsky, A.B.: (Павленко ГЛ, Моргаевский АБ. Устойчивость упругой параболической арки с затяжкой. Сб. Исследования по теории сооружений, том 5. М.: Госстройиздат, 1951)

[Pfl50] Pflüger, A.: Stabilitätsprobleme der elastostatic. Spinger, Berlin (1950, 2-te Auflage 1964)

[Pi02] Pi, Y.L., Bradford, M.A., Uy, B.: In-plane stability of arches. Int. J. Solids Struct. **39**(1) (2002)

[Pon62] Pontryagin, L.S., Boltyanskii, V.G., Gamkrelidze, R.V., Mishchenko, E.F.: The Mathematical Theory of Optimal Processes. Wiley, New York (1962)

[Pra52] Pratusevich, J.A.: (Пратусевич ЯА. О колебаниях упругих арок. Труды МИИТ, вып.76. М.: Трансжелдориздат, 1952)

[Pro48] Prokofiev, I.P., Smirnov, A.F.: (Прокофьев ИП, Смирнов АФ. Теория сооружений, ч.3. М.: Трансжелдориздат, 1948)

[Rab51] Rabinovich, I.M.: (Рабинович ИМ. Приближенный способ определения частот и форм собственных колебаний параболических и других арок. Исследования по теории сооружений, вып. 5. М.: Госстройиздат, 1951)

[Rab54a] Rabinovich, I.M.: (Рабинович ИМ. Курс строительной механики стержневых систем, Часть 2, 2 изд. М.: Госстройиздат, 1954)

[Rab54b] Rabinovich, I.M.: (Рабинович ИМ. Частоты и формы обратносимметричных собственных колебаний двухшарнирных параболических арок. Исследования по теории сооружений, вып. 6, М.: Госстройиздат, 1954)

[Rab58] Rabinovich, I.M., Sinitcun, A.P., Terenin, B.M.: (Рабинович ИМ, Синицын АП, Теренин БМ. Расчет сооружений на действие кратковременных и мгновенных сил., М.: Военно-инженерная Академия им. Куйбышева, Часть 1–1956, Ч.2-1958)

[Rab50] Rabinovich, I.M.: (Рабинович ИМ. Некоторые вопросы теории сооружений, содержащих односторонние связи, М.: Инженерный сборник, вып. VI, 1950) (1950)

[Rab60] Rabinovich, I.M.: (Рабинович ИМ. Основы строительной механики стержневых систем, 3 изд. М.: Госстройиздат, 1960)

[Ray77] Rayleigh, J.W.S.: The theory of sound. Macmillan, London; 2nd edn. (1945) Dover, NewYork (1877)

[Rek73] Rekach, V.G.: (Рекач ВГ. Руководство к решению задач прикладной теории упругости, М.: Высшая школа, 1973)

[Ric99] Richardson, G.S.: Arch bridges. In: Brockenbrough, R.L., Merritt, F.S. (eds.) Structural Steel Designers' Handbook, 3rd edn, chap. 14. McGraw-Hill, New York (1999)

[Roa75] Roark, R.J.: Formulas for Stress and Strain, 5th edn. McGraw-Hill, New York (1975)

[Roz76] Rozvany, G.I.N.: Optimal Design of Flexural Systems: Beams, Grillages, Slabs, Plates and Shells. Pergamon, Oxford (1976)

[Rom72] Romanelli, E., Laura, P.A.: Fundamental frequencies of non-circular, elastic hinged arch. J. Sound Vib. **24**(1) (1972)

[Ros89] Rossi, R.E., Laura, P.A.A., Verniere De Irassar, P.L.: In-plane vibrations of cantilevered non- circular arcs of non-uniform cross-section with a tip mass. J. Sound Vib. **129**(2) (1989)

[Rzh55] Rzhanitsun, A.R.: (Ржаницын АР. Устойчивость равновесия упругих систем. М.: ГОСИЗДАТ, 1955)

[Rzh82] Rzhanitsun, A.R.: (Ржаницын АР. Строительная механика, М.: Высшая школа, 1982)

[Sak85] Sakiyama, T.: Free vibrations of arches with variable cross section and non-symmetrical axis. J. Sound Vib. **102**, 448–452 (1985)

[Sch37] Schallenkamp, A.: Schwingungen von Trägern bei bewegten lasten. Ingenier-Archiv, #8, 187 (1937)

[Sch80] Schodek, D.L.: Structures, 5th edn. Pearson Education, New Jersey (1980)

[Sht35] Shtaerman, I.J.: (Штаерман ИЯ. Загальна теорія стійкості арок. Сб. Стійкость арок, Кіев, 1935)

[Smi47] Smirnov, A.F.: (Смирнов АФ. Статическая и динамическая устойчивость сооружений. М.: Трансжелдориздат, 1947)

[Smi84] Smirnov, A.F., Aleksandrov, A.V., Lashchennikov, B.J., Shaposhnikov, N.N.: (Смирнов АФ и др. Строительная механика. Динамика и устойчивость сооружений. М.: Стройиздат, 1984)

[Sni57] Snitko, N.K.: (Снитко НК. Метод перемещений в теории колебаний любых незакрепленных рам при использовнии кинематичской цепи. Исследования по теории сооружений, т.6, М.: Госстройиздат, 1957)

[Sni66] Snitko, N.K.: (Снитко НК. Строительная механика, М.: Высшая школа, 1966)

[Sou13] Southwell, R.H.: On the general theory of elastic stability. Phil. Trans. R. Soc. Lond. A **213**, 501 (1913)

[Sto50] Stoker, J.J.: Nonlinear Vibrations in Mechanical and Electrical Systems. Interscience, New York (1950)

[Str27] Strassner, A.: Der Bogen und das Brückengewölbe, 3rd edn, vol. 2. Berlin (1927)

[Stro94] Strogatz, S.H.: Nonlinear Dynamics and Chaos. Westview Press, (1994)

[Str32] Strutt, M.O.: Lamésche, Mathieusche, und verwandte Funktionen in Physik und Technik. Ergeb Math. Drenzg, **1**–**3**, Berlin (1932)

[Suz78] Suzuki, K., Aida, H., Takahashi, S.: Vibrations of curved bars perpendicular to their planes. Bull. JSME **21**(162) (1978)

[Tat66] Tatur, G.K.: (Татур ГК. Курс сопротивления материалов, ч.3. Минск: Высшая школа, 1966)

[Ter54] Terenin, B.M.: (Теренин БМ. Колебания круговых арок. М.: Военно-инженерная Академия им. Куйбышева, 1954)

[Tho81] Thomson, W.T.: Theory of Vibration with Applications, 2nd edn. Prentice-Hall, New Jersey (1981)

[Tim61] Timoshenko, S.P., Gere, J.M.: Theory of Elastic Stability. McGraw-Hill, New York (1961)

[Tim72] Timoshenko, S.P.: (Тимошенко СП. Курс теории упругости. Киев: Наукова думка, 1972)

[Tim53] Timoshenko, S.P.: History of Strength of Materials. McGraw-Hill, New York (1953)

[Tod60] Todhunter, I., Pearson, K.: A History of the Theory of Elasticity and of the Strength of Materials, vol. I –II. Dover, New York (originally published by the Cambridge University Press in 1886 and 1893) (1960)

[Tro82] Troitskiy, V.A., Petukhov, L.V.: (Троицкий ВА и Петухов ЛВ. Оптимизация формы упругих тел. М.: Наука, 1982)

[Tuf98] Tüfekci, E., Arpaci, A.: Exact solution of in-plane vibrations of circular arches with account taken of axial extension, transverse shear and rotary inertia effects. J. Sound Vib. **209**(5) (1998)

[Uma72-73] Umansky, A.A.: (Уманский АА. Ред. Справочник проектировщика, изд.2. М.: Стройиздат, т. 1–1972, т.2-1973)

[Urb55] Urban, I.V.: (Урбан ИВ. Теория расчета стержневых тонкостенных конструкций. М.: Трансжелдориздат, 1955)

[Vla40] Vlasov, V.Z.: (Власов ВЗ. Тонкостенные упругие стержни, М.: Физматлит, 1940)

[Vol67] Vol'mir, A.S.: (Вольмир АС. Устойчивость деформирумых систем, изд. 2. М.: Наука, 1967) (1967)

[Wea90] Weaver, W., Timoshenko, S.P., Young, D.H.: Vibration Problems in Engineering, 5th edn. Wiley, New York (1990)

[Wea84] Weaver Jr., W., Johnson, P.R.: Finite Elements for Structural Analysis. Prentice-Hall, Englewood Cliffs, New Jersey (1984)

[Vol60] Volterra, E., Morell, J.D.: A note on the lowest natural frequencies of elastic arcs. Am. Soc. Mech. Eng. J. Appl. Mech. **27**, 744–746 (1960)

[Vol61a] Volterra, E., Morell, J.D.: Lowest natural frequencies of elastic hinged arcs. J. Acoust. Soc. Am. **33**(12) (1961)

[Vol61b] Volterra, E., Morell, J.D.: Lowest natural frequencies of elastic arcs outside the plane of initial curvature. ASME J. Appl. Mech. **28**(4), 624–627 (1961)

[Wal34] Waltking, F.W.: Schwingungszahlen und Schwingungsformen von Kreisbogenträgern. Ingenieur Archiv, Bd.**V**(6) (1934)

[Wan72] Wang, T.M.: Lowest natural frequencies of clamped parabolic arcs. Proc. Am. Soc. Civil Eng. **98**(1), 407–411 (1972)

[Wan75] Wang, T.M.: Effect of variable curvature on fundamental frequency of clamped parabolic arcs. J. Sound Vib. **41**, 247–251 (1975)

[Wan73] Wang, T.M., Moore, J.A.: Lowest natural extensional frequency of clamped elliptic arcs. J. Sound Vib. **30**, 1–7 (1973)

[Was78] Wasserman, Y.: Spatial symmetrical vibrations and stability of circular arches with flexibly supported ends. J. Sound Vib. **59**, 181–194 (1978)

[Wol71] Wolf, J.A.: Natural frequencies of circular arches. Proc. Am. Soc. Civil Eng. **97**(9), 2337–2350 (1971)

[You89] Young, W.C.: Roark's Formulas for Stress and Strain, 6th edn. McGraw-Hill, New York (1989)

Index

MIX
Papier aus verantwortungsvollen Quellen
Paper from responsible sources
FSC® C105338

FSC
www.fsc.org

If you have any concerns about our products,
you can contact us on
ProductSafety@springernature.com

In case Publisher is established outside the EU,
the EU authorized representative is:
Springer Nature Customer Service Center GmbH
Europaplatz 3, 69115 Heidelberg, Germany

Printed by Libri Plureos GmbH
in Hamburg, Germany